Engineering Mathematics with Examples and Applications

Engineering Mathematics with Examples and Applications

Xin-She Yang
Middlesex University
School of Science and Technology
London, United Kingdom

Academic Press is an imprint of Elsevier
125 London Wall, London EC2Y 5AS, United Kingdom
525 B Street, Suite 1800, San Diego, CA 92101-4495, United States
50 Hampshire Street, 5th Floor, Cambridge, MA 02139, United States
The Boulevard, Langford Lane, Kidlington, Oxford OX5 1GB, United Kingdom

Library of Congress Cataloging-in-Publication Data
A catalog record for this book is available from the Library of Congress

British Library Cataloguing-in-Publication Data
A catalogue record for this book is available from the British Library

ISBN: 978-0-12-809730-4

For information on all Academic Press publications
visit our website at https://www.elsevier.com

Publisher: Nikki Levy
Acquisition Editor: Graham Nisbet
Editorial Project Manager: Susan Ikeda
Production Project Manager: Mohanapriyan Rajendran
Designer: Matthew Limbert

Typeset by VTeX

Contents

Part III
Vectors and Matrices

8. Vectors and Vector Algebra

9. Matrices

Part IV
Calculus

10. Differentiation

11. Integration

12. Ordinary Differential Equations

13. Partial Differentiation

14. Multiple Integrals and Special Integrals

15. Complex Integrals

Part V
Fourier and Laplace Transforms

16. Fourier Series and Transform

17. Laplace Transforms

Part VI
Statistics and Curve Fitting

18. Probability and Statistics

About the Author

Xin-She Yang obtained his DPhil in Applied Mathematics from the University of Oxford. He then worked at Cambridge University and National Physical Laboratory (UK) as a Senior Research Scientist. Now he is Reader at Middlesex University London, Adjunct Professor at Reykjavik University (Iceland) and Distinguished Guest Professor at Xi'an Polytechnic University (China).

He is an elected Bye-Fellow and College Lecturer at Downing College, Cambridge University. He is also the IEEE Computer Intelligence Society (CIS) Chair for the Task Force on Business Intelligence and Knowledge Management and Director of the International Consortium for Optimization and Modelling in Science and Industry (iCOMSI).

With more than 20 years' research and teaching experience, he has authored 10 books and edited more than 15 books. He published more than 200 research papers in international peer-reviewed journals and conference proceedings with more than 17,000 citations. According to Thomson Reuters' *Web of Science* in 2016, he is a highly cited researcher. He serves on the Editorial Boards of many international journals including *International Journal of Bio-Inspired Computation* and Elsevier's *Journal of Computational Science (JoCS)*. He is also the Editor-in-Chief of the *International Journal of Mathematical Modelling and Numerical Optimisation*.

Preface

Engineering mathematics is an essential part of core training in engineering. Any quantitative analysis in engineering requires mathematical analysis to a certain degree. Many mathematical methods and techniques are essential to the understanding and calculations in engineering applications.

Engineering mathematics is a compulsory course for all undergraduates studying in all disciplines of engineering, and students have to develop all necessary mathematical skills and problem-solving practice needed for all other engineering courses such as control theory, signal processing, structural analysis, circuit theory, solid mechanics, numerical methods and many other advanced courses.

Though there are currently over a dozen good books on engineering mathematics, some are very lengthy (over 1000 pages), and some have very specific audience. Most students find a thick book of over a thousand pages intimidating, especially for the first-year undergraduates. Though all the topics may be important in such big books, no one can be expected to learn all the topics in a short time. In addition, topics have to be selected well to be truly relevant to the courses in engineering and thus there is no need to provide the same level of details for all topics. On the other hand, specialized books on a selection of special topics may be useful to some students, especially for senior undergraduates, but these books can be frustrating to most readers when certain topics may be unfamiliar or with overwhelming in-depth mathematical analysis and terminologies. Therefore, this is a strong need for a textbook with a good balance of topic coverage and the right level of details so that most students can use it with confidence and be able to use engineering mathematics for problem-solving in its familiar context of engineering and real-world applications.

This book tries to fill in the gap with a compact and concise approach by starting with the basic level of secondary mathematics and gradually develop to the advanced level of mathematics that is all necessary for engineering disciplines. The main objective of this book is to cover all the major topics in engineering mathematics and thus it is more suitable for students to build these key skills in a more confident way. This book will also be useful for graduates in engineering, computer science, natural sciences and management sciences to refresh their basic mathematical background and skills.

The main approach and style of this book can be considered as informal, theorem-free and practical. By using an informal and theorem-free approach, all fundamental mathematics topics required for engineering are covered, and students can gain such basic knowledge of all important topics without worrying about rigorous (often boring) proofs. Some proofs and derivations will be presented in an informal way by direct, straightforward mathematical operations and calculations, and students will gain the same level of fundamental knowledge without going through too many tedious steps of mathematical proofs and analyses.

In addition, with a practical approach, this book provides over 250 worked examples with step-by-step explanation so that students can see how each step of the mathematical problems can be derived without any gap or jump in steps. Thus, students can build their understanding and mathematical confidence gradually and in a step-by-step manner. Furthermore, such an informal and practical style can also enable students to learn engineering mathematics by self-study and at their own pace.

Therefore, this book aims for undergraduates to rapidly develop all the fundamental knowledge of engineering mathematics. It can also be used by graduates to review and refresh their mathematical skills. The step-by-step worked examples will help the students to gain more insight and build sufficient confidence in engineering mathematics for problem-solving in real-world applications.

Some of the contents and examples have been used in my teaching in the last 15 years at Cambridge University and Middlesex University. The valuable comments and feedback by my students have greatly influenced the ways how I prepare these examples and organize the overall contents. I am really grateful to my students and colleagues for their helpful comments over these years. Obviously, I will be more than happy to receive any comments from readers and I will strive to incorporate such comments in a future edition of the book.

Xin-She Yang
December 2016

Acknowledgment

I would like to thank all my students and colleagues who have given valuable feedback and comments on some of the examples and contents of this book. I also would like to thank my editors, Graham Nisbet and Susan Ikeda, and the staff at Elsevier for their professionalism. Last but not least, I thank my family for all the help and support.

Xin-She Yang
December 2016

Part I

Fundamentals

Chapter 1

Equations and Functions

Chapter Points

- Introduce basic concepts of real numbers and sets.
- Cover functions, specially univariate functions, including linear functions, modulus function and power functions.
- Introduce the basic equations, quadratic equation and simultaneous equations.

1.1 NUMBERS AND REAL NUMBERS

From the basic mathematics we learned in school, we know that whole numbers such as -10, 0, 2 and 77 are integers, while the positive integers such as 1, 2, 3, ... are often called natural numbers. Numbers such as $\frac{1}{2}$ and $\frac{2}{5}$ are called fractions.

In mathematics, we often use some symbolic notation to denote a number so as to ease our discussions. For example, we can use letter n to denote any number in $1, 2, 3, \cdots$. That is, $n = 1, 2, 3, \cdots$.

Among all the natural numbers, there is a special class of numbers, called prime numbers. A prime number is a positive integer that can only be divisible by 1 and itself. In other words, a prime number p can have only two factors: 1 and itself p. It is easy to check that 2, 3, 5, 7, 11, 13, 17 and 19 are all prime numbers. The only even prime number is 2, and all the rest prime numbers are odd numbers.

The ratio of any two non-zero integers m and n in general is a fraction m/n where the denominator n should not be zero (that is, $n \neq 0$). For example, $\frac{1}{2}$, $-\frac{4}{3}$ and $\frac{22}{7}$ are fractions. If $m < n$, the fraction is called a proper fraction. All the integers and fractions make up the rational numbers.

Numbers such as $\sqrt{2}$ and π are called irrational numbers because they cannot be expressed in terms of a fraction. For example, we cannot express

$$\pi = 3.14159265358979323846264338327950288...$$

as a ratio, and the common used expression $22/7 \approx 3.142857142857...$ is just an approximation; it is not exactly π. In fact, $\pi \neq 22/7$. The decimal digits of an irrational number do not repeat periodically. If any part of the decimal repeats, such as 0.3333333..., it always possible to be written as a fraction such as $0.3333... = 1/3$. Here, $22/7 = 3.142858142857...$ has a digit sequence 142857 that repeats.

1.1.1 Notes on Notations and Conventions

As we have seen above, we have used some notations such as $\approx$ and $\cdots$ without much explanation. Before we proceed, it is better to explain briefly the conventions and notations we often use in mathematics and certainly in this book.

We know that $=$ means equality such as $1 + 2 = 3$ and $\neq$ means 'not equal' such as $1 + 2 \neq 4$. However, we often use $\approx$ to mean 'is approximately equal to'. For example, we can write $4.99998 \approx 5$.

In order to avoid writing out all the things out, we often use $n = 1, 2, ..., 10$ to mean exactly $n = 1, 2, 3, 4, 5, 6, 7, 8, 9, 10$. Similarly, when we write $n = 1, 2, \cdots$, we mean all the numbers from 1 until any large numbers. Basically, we use $\cdots$ or ... to mean 'etc.'. In addition, we occasionally wish to include the infinity (∞) in our discussion, and there is a positive infinity $+\infty$ and a negative infinity $-\infty$. Any number n is less than $+\infty$, that is, $n < +\infty$. Any number is also greater than $-\infty$ (i.e., $-\infty < n$). Therefore, all numbers are between $-\infty$ and $+\infty$.

For simplicity to denote multiplication, we use ab to means $a \times b$ and also $a \cdot b = ab = a \times b$ where a and b are any two numbers.

Sometimes, to write more compactly, we use $\pm$ to mean either $+$ or $-$. Thus, 2 ± 3 means $2 + 3$ or $2 - 3$.

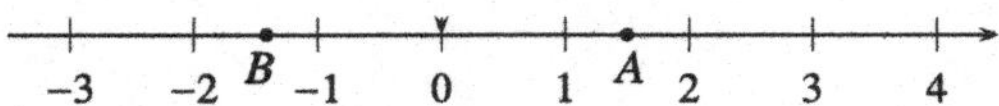

FIGURE 1.1 Real numbers and their representations (as points) on the number line.

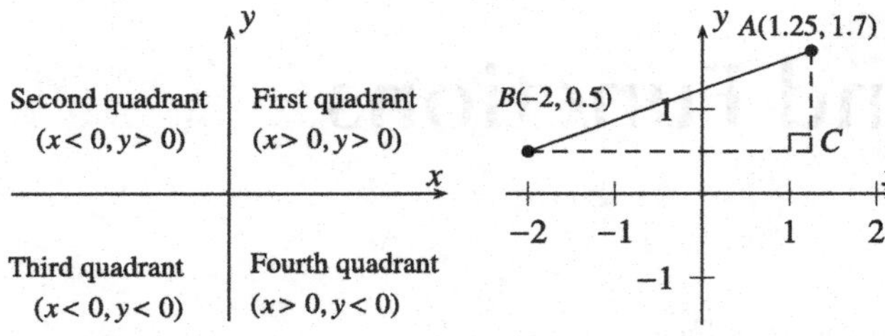

FIGURE 1.2 Cartesian coordinates and the four quadrants.

Now let us come back to the discussion of numbers. The proof that $\sqrt{2}$ is irrational is among the first mathematical proofs, as given in Euclid's *Elements* about 2300 years ago. Interested readers can refer to any book on number theory or prime numbers.

Example 1.1

Since $\sqrt{5}$ is an irrational number, the famous golden ratio

$$\phi = \frac{1+\sqrt{5}}{2} = 1.6180339887...,$$

is also irrational.

Another well-know irrational number is Euler's number

$$e = 2.71828182845904523536028 7...,$$

which has an infinite number of decimal places.

Rational numbers and irrational numbers together make up the real numbers. In mathematics, all the real numbers are often denoted by $\mathbb{R}$ or $\Re$, and a real number corresponds to a unique point or location in the number line (see Fig. 1.1). For example, $3/2$ corresponds to point A and $-\sqrt{2}$ corresponds to point B.

On a plane, any point can be located by a pair of two real numbers (x, y), which are called Cartesian coordinates (see Fig. 1.2). For ease of reference to locations or different parts on the plane, we conventionally divide the coordinate system into four quadrants.

From the first quadrant with $(x > 0, y > 0)$ in an anti-clockwise manner, we consecutively call them the second, third and fourth quadrants as shown in Fig. 1.2. Such representation makes it straightforward to calculate the distance d, often called Cartesian distance, between any two points $A(x_i, y_i)$ and $B(x_j, y_j)$. The distance is the line segment AB, which can be obtained by the Pythagoras's theorem for the right-angled triangle ABC. We have

$$d = \sqrt{(x_i - x_j)^2 + (y_i - y_j)^2}, \tag{1.1}$$

which is always non-negative. That is to say, $d \geq 0$.

Example 1.2

For example, the point $A(1.25, 1.7)$ lies in the first quadrant, while point $B(-2, 0.5)$ lies in the second quadrant. The distance between A and B is the length along the straight line connecting A and B, and we have

$$d = \sqrt{[1.25 - (-2)]^2 + (1.7 - 0.5)^2} = \sqrt{12.0025} \approx 3.4645,$$

which should have the same unit as the coordinates themselves. Here the sign '$\approx$' means 'is approximately equal to'.

1.1.2 Rounding Numbers and Significant Digits

In calculations, we often have to round up numbers. For example, it is not possible to write out all the digits of $\sqrt{2}$ because it is an irrational number. However, for first 12 digits of $\sqrt{2}$, we have

$$\sqrt{2} = 1.41421356237..., \tag{1.2}$$

which has 11 decimal places. That is

$$\sqrt{2} = 1 \underbrace{.}_{\text{decimal point}} \overbrace{4}^{\text{1st decimal figure}} \underbrace{1}_{\text{2nd decimal figure}} 42135623 \underbrace{7}_{\text{11th decimal figure}} ..., \tag{1.3}$$

where the decimal figures are also called decimal places or decimal digits. Thus, the first decimal figure is also called the first decimal place or decimal digit.

Obviously, we do not need so many decimal places, and thus we need to carry out the so-called rounding of numbers. For a given number of decimal places, the basic rule for rounding numbers is to look at the number at next decimal place and decide by the following rule:

- If it is less than 5, round down.
- If it is 5 or more, then round up.

Example 1.3

For example, for $\sqrt{2} = 1.41421356237...$, if we want to round it to 3 decimal places, the current number at the 3rd decimal place is 4. Now we look at the number at the next decimal place, which is 2, so we round it down and we have

$$1.41\underline{4}21356237 \approx 1.414.$$

If we wish to keep 6 decimal places, the current number at the 6th decimal place is 3, and now we look at the next decimal place, which is 5. Thus, we round it up and we have

$$1.41421\underline{3}56237 \approx 1.414214.$$

Sometimes, we are not sure how many digits we should keep. In science and engineering, we often have to measure a quantity such as temperature, then the accuracy determines the number of decimal places. Thus, we have to deal with significant digits. If the temperature in a room is 22.15 °C, it has four significant digits. When we say the purity of gold is 99.999%, it has five significant digits or five significant figures. However, care should be taken when dealing with digit 0. For example, 100 has one significant digit, while 100.0 has four significant digits as '.0' signifies something important. Similarly, if we write '100.', it means that it has three significant digits. In addition, 10005 has 5 significant digits, while 0.00005 has only one significant digit.

Thus, all nonzero digits are significant, so are the zeros between digits. The zeros to the right of the decimal point are also significant, but zeros to the left the nonzero digits are not.

Example 1.4

1234500 has five significant digits, while 123450.0 has seven significant digits. 0.01234 has four significant digits, while 0.01230 has four significant digits because the final zero in the decimal place is significant.

When adding up numbers with different significant digits, we have to round them to the appropriate number of significant digits. For example,

$$3.1415 + 3.7 = 6.8415, \tag{1.4}$$

is not appropriate because 3.1415 has five significant digits and is accurate to the 4th decimal place, while 3.7 has two significant digits and is accurate to the first decimal place. Thus, the final sum should also be accurate to the first decimal place. We should write

$$3.1415 + 3.7 \approx 6.8. \tag{1.5}$$

For multiplication, for example, we have

$$3.1415 \times 0.37 \times 9 = 10.461195, \tag{1.6}$$

but numbers of significant digits of 3.1415, 0.37 and 9 are 5, 2 and 1, respectively. Thus, we have to round the final answer to one significant digit, which means that $10.461195 = 10$. Thus, we have

$$3.1415 \times 0.37 \times 9 \approx 10, \tag{1.7}$$

where we have used $\approx$ to show such rounding and approximation.

It is worth pointing out that the rule of addition and subtraction focuses on the decimal place, while the multiplication and division focus on the number of significant digits. Let us look at another example.

Example 1.5

If we naively do the calculations, we have

$$\begin{aligned}(2.71828 - 1.23) \times 3.1415 \div 11 &= (1.48828 \times 3.1415)/11 \\ &= 4.67543162/11 = 0.425039238182.\end{aligned}$$

This is not the answer with the appropriate number of significant digits.

Since 2.71828 has six significant digits and is accurate to the 5th decimal place, while 1.23 is only accurate to the 2nd decimal place, we should have $2.717828 - 1.23 = 1.49$. In addition, we have $1.49 \times 3.1415 = 4.68$ (three significant digits) because 3.1415 has five significant digits and 1.49 has three significant digits. Therefore, the answer

$$4.68/11 = 0.4254545...$$

should have only two significant digits because 11 has two significant digits. Finally, we have

$$4.68/11 = 0.43.$$

That is

$$(2.71828 - 1.23) \times 3.1415 \div 11 \approx 0.43.$$

1.1.3 Concept of Sets

Let us first introduce some basic concepts in set theory. A set is any well-defined collection of objects or elements, and the elements are the members or objects in a set. We conventionally use the upper-case letters to denote sets and lower-case letters for elements, and the listed elements are enclosed in the curly brace {}.

Example 1.6

For example, the following numbers: $1, 1.2, 2, 3, -1, 4.9, 77$ can be considered as a set $\mathcal{A} = \{1, 1.2, 2, 3, -1, 4.9, 77\}$, and the number 1.2 is a member of the set, and we can write this membership as

$$1.2 \in \mathcal{A} = \{1, 1.2, 2, 3, -1, 4.9, 77\}. \tag{1.8}$$

In general, the membership in a set is denoted using $\in$, thus

$$x \in \mathcal{A}, \tag{1.9}$$

means that 'x is a member of the set $\mathcal{A}$', while

$$\mathbf{x} \notin \mathcal{A}, \tag{1.10}$$

means that 'x is not a member of the set $\mathcal{A}$'.

A special set is the empty set or null set which has no element and is denoted by

$$\emptyset = \{\}, \tag{1.11}$$

which should not be confused with a non-empty set which consists of a single element {0}. Also, note the special notation $\emptyset$.

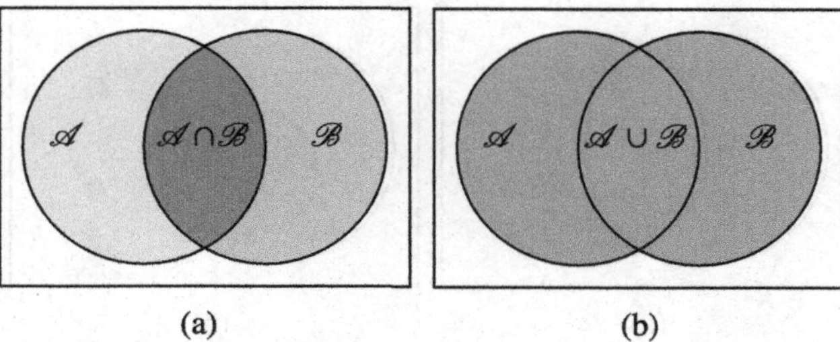

FIGURE 1.3 Venn diagrams: (a) $\mathscr{A}\cap\mathscr{B}$, (b) $\mathscr{A}\cup\mathscr{B}$.

We say that $\mathscr{A}$ is a subset of $\mathscr{B}$ if $a \in \mathscr{A}$ implies that $a \in \mathscr{B}$. That is to say that all the members of $\mathscr{A}$ are also members of $\mathscr{B}$. We denote this relationship as

$$\mathscr{A} \subseteq \mathscr{B}. \tag{1.12}$$

If all the members of $\mathscr{A}$ are also members of $\mathscr{B}$, but there exists at least one element b such that $b \in \mathscr{B}$ while $b \notin \mathscr{A}$, we say $\mathscr{A}$ is a *proper* subset of $\mathscr{B}$, and denote this relationship as

$$\mathscr{A} \subset \mathscr{B}. \tag{1.13}$$

When combining sets, we say that $\mathscr{A}$ union $\mathscr{B}$, denoted by

$$\mathscr{A} \cup \mathscr{B}$$

forms a set of all elements that are in $\mathscr{A}$, *or* $\mathscr{B}$, *or both*. On the other hand, $\mathscr{A}$ intersect $\mathscr{B}$, written as

$$\mathscr{A} \cap \mathscr{B},$$

is the set of all elements that are in both $\mathscr{A}$ *and* $\mathscr{B}$.

A universal set Ω is the set that consists of all the elements under consideration. The complement set of $\mathscr{A}$ or *not* $\mathscr{A}$, denoted by $\bar{\mathscr{A}}$, is the set of all the elements that are not in $\mathscr{A}$. The set $\mathscr{A} - \mathscr{B}$ or $\mathscr{A}$ minus $\mathscr{B}$ is the set of elements that are in $\mathscr{A}$ and not in $\mathscr{B}$, this is equivalent to removing or subtracting from $\mathscr{A}$ all the elements that are in $\mathscr{B}$. This leads to

$$\mathscr{A} - \mathscr{B} = \mathscr{A} \cap \bar{\mathscr{B}}, \tag{1.14}$$

and

$$\bar{\mathscr{A}} = \Omega - \mathscr{A}. \tag{1.15}$$

Example 1.7

For two sets

$$\mathscr{A} = \{2, 3, 5, 7\}, \quad \mathscr{B} = \{2, 4, 6, 8, 10\},$$

and a universal set

$$\Omega = \{1, 2, 3, 4, 5, 6, 7, 8, 9, 10\},$$

it is straightforward to check that

$$\mathscr{A} \subset \Omega, \quad \mathscr{B} \subset \Omega, \quad \mathscr{A} - \mathscr{B} = \{3, 5, 7\},$$
$$\mathscr{A} \cup \mathscr{B} = \{2, 3, 4, 5, 6, 7, 8, 10\}, \quad \mathscr{A} \cap \mathscr{B} = \{2\},$$
$$\bar{\mathscr{A}} = \Omega - \mathscr{A} = \{1, 4, 6, 8, 9, 10\},$$

and finally $\bar{\mathscr{A}} \cap \mathscr{A} = \emptyset$.

The set operations seem too abstract, there is a better way to represent such mathematical operations between sets, that is to use the Venn diagrams as the topological representation. Fig. 1.3 represents the intersect $\mathscr{A} \cap \mathscr{B}$ and union $\mathscr{A} \cup \mathscr{B}$, while Fig. 1.4 represents $\mathscr{B} - \mathscr{A}$, $\bar{\mathscr{A}} = \Omega - \mathscr{A}$ and $\mathscr{B} \subset \mathscr{A}$.

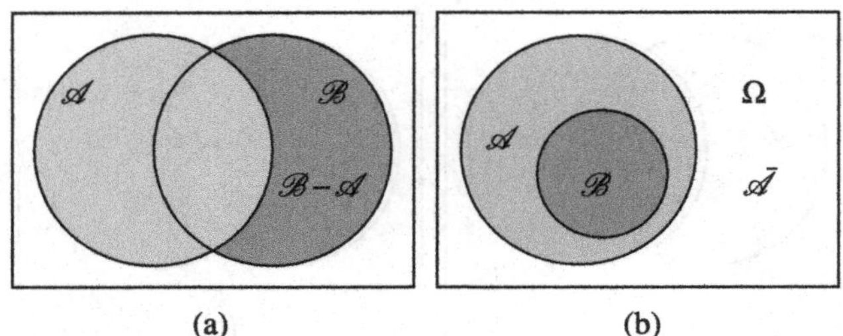

FIGURE 1.4 Venn diagrams: (a) $\mathcal{B}-\mathcal{A}$, (b) $\bar{\mathcal{A}}=\Omega-\mathcal{A}$ and $\mathcal{B}\subset\mathcal{A}$.

1.1.4 Special Sets

Some common sets in mathematics are used so often that they deserve special names or notations. These include:

- $\mathbb{N}=\{1,2,3,...\}$ or $\mathbb{N}=\{0,1,2,...\}$ denotes the set of all natural numbers;
- $\mathbb{Z}=\{...,-2,-1,0,1,2,...\}$ is the set of all integers;
- $\mathbb{P}=\{2,3,5,7,11,...\}$ is the set of all primes;
- $\mathbb{Q}=\{\frac{m}{n}: m,n\in\mathbb{Z}, n\neq 0\}$=$\{\frac{-5}{11},2,\frac{7}{19},...\}$ is the set of all rational numbers;
- $\mathbb{R}$ is the set of all real numbers consisting of all the rational numbers and all the irrational numbers such as $\sqrt{2},\pi,e$.
- Sometimes, we may have to deal with division by zero and/or infinity ∞ in mathematics. By including the positive infinity $(+\infty)$ and negative infinity $(-\infty)$, the real number system becomes the so-called affinely extended real number system, denoted by $\bar{\mathbb{R}}$. That is $\bar{\mathbb{R}}=\mathbb{R}\cup\{-\infty,+\infty\}$.

All these sets have an infinite number of elements. From the definitions, it is straightforward to check that

$$\mathbb{P}\subset\mathbb{N}\subset\mathbb{Z}\subset\mathbb{Q}\subset\mathbb{R}\subset\bar{\mathbb{R}}. \tag{1.16}$$

1.2 EQUATIONS

In general, mathematical expressions such as $x+2$ often contain unknown quantity such as x, and this unknown quantity is called a variable, which can be any value in the real numbers. To determine the actual value of the variable, a condition is required and such a condition is often written as an equation.

An equation is a mathematical statement that is written in terms of symbols and an equal sign '='. All the terms on the left of '=' are collectively called the left-hand side (LHS), while those on the right of '=' are called the right-hand side (RHS). For example, $x+2=7$ is a simple equation with its left-hand side being $x+2$ and right hand-side being 7.

In general, an equation has the following properties:

- Any quantity can be added to, subtracted from or multiplied by both sides of the equation;
- A non-zero quantity can divide both sides;
- Any function such as power (such as x^2) and surds (such as $\sqrt{2}$) can be applied to both sides equally.

After such manipulations, an equation may be converted into a completely different equation, though we can obtain the original equation if we can carefully reverse the above procedure, though special care is needed in some cases.

The idea of these manipulations is to transform the equation to a much simpler form whose solution can be found easily.

Example 1.8

For example, from equation

$$x+2=7, \tag{1.17}$$

we can subtract 2 from both sides, which leads to

$$x+2-2=7-2, \tag{1.18}$$

which gives

$$x=5. \tag{1.19}$$

For another equation

$$3x + 5 = 17 - x,$$

we can first add $+x$ and we have

$$\underbrace{3x}\ \underbrace{+5}\ \underbrace{+x} = 17 \underbrace{-x + x},$$

which becomes

$$4x + 5 = 17.$$

Subtracting 5 from both sides, we have

$$4x + 5 - 5 = 17 - 5, \quad \text{or} \quad 4x = 12.$$

Dividing both sides by 4, we have

$$\frac{4x}{4} = \frac{12}{4},$$

which gives the final solution

$$x = 3.$$

Similarly, from equation

$$x\sqrt{2} = \sqrt{14} + \sqrt{2}, \tag{1.20}$$

we can divide both sides by $\sqrt{2}$ (or multiply by $1/\sqrt{2}$) and we have

$$\frac{x\cancel{\sqrt{2}}}{\cancel{\sqrt{2}}} = \frac{\sqrt{14}}{\sqrt{2}} + \frac{\cancel{\sqrt{2}}}{\cancel{\sqrt{2}}}, \tag{1.21}$$

which gives

$$x = \frac{\sqrt{14}}{\sqrt{2}} + 1 = \frac{\sqrt{2 \times 7}}{\sqrt{2}} + 1 = \frac{\cancel{\sqrt{2}} \times \sqrt{7}}{\cancel{\sqrt{2}}} + 1 = \sqrt{7} + 1.$$

Equations can have many applications and many physical laws and rules are stated using equations. For example, an excellent example is Newton's second law of motion $F = ma$ states that the external force F is equal to the mass m of the object multiplied by its acceleration a.

Example 1.9

If a body has a mass of $m = 70$ kg, what is its weight on the Earth's surface? We know the acceleration due to gravity g is $g = 9.80665\ \text{m/s}^2$, the force acting on the mass is

$$F = mg = 70 \times 9.80665 \approx 686.465\ \text{N},$$

that is about 686 newtons.

In many applications, when we try to determine some unknown quantities, we have to re-arrange the equation slightly so as to solve the unknown variable or quantity from known quantities. Let us see a simple example.

Example 1.10

Little's law of a queuing system states that the average number L of items in the system is equal to the average arrival rate λ multiplied by the average waiting time W that an item spends in the queuing system. That is

$$L = \lambda W.$$

For example, a coffee shop on the campus has an arrival rate of 2 customers per minute, and the average queue length is 8 people. What the average waiting time for a customer?

We know that $\lambda = 2$ and $L = 8$, so

$$W = \frac{L}{\lambda} = \frac{8}{2} = 4,$$

which means that a customer usually waits 4 minutes before being served.

1.2.1 Modular Arithmetic

Sometime, we may need to calculate the reminder of the division of two integers, which requires the modular arithmetic operation in general. For example, 6 hours after 9 o'clock is 3 o'clock, which can be obtained by finding the reminder after dividing $9 + 6 = 15$ by 12. Here, 12 is the modulus. In general, two numbers p and q are said to be congruent modulo n can be written as

$$p \equiv q \pmod{n}, \tag{1.22}$$

which means that the reminder of p divided by n is q. Alternatively, we can say that $p - q$ is a multiple of n.

The above example about time can be written as

$$15 \equiv 3 \pmod{12}. \tag{1.23}$$

That is the reminder of 15 divided by 12 is 3. In other words, we have

$$15 = 1 \times 12 + 3. \tag{1.24}$$

There are some interesting properties about modular arithmetic manipulations. For

$$p_1 \equiv q_1 \pmod{n}, \quad p_2 \equiv q_2, \pmod{n}, \tag{1.25}$$

we have

$$p_1 \pm p_2 \equiv q_1 \pm q_2 \pmod{n}, \tag{1.26}$$

where $\pm$ means either $+$ or $-$. That is, the above equation is equivalent to two equations:

$$p_1 + p_2 \equiv q_1 + q_2 \pmod{n}, \tag{1.27}$$

or

$$p_1 - p_2 \equiv q_1 - q_2 \pmod{n}. \tag{1.28}$$

In addition, we can also have

$$p_1 \cdot p_1 \equiv q_1 \cdot q_2 \pmod{n}. \tag{1.29}$$

Let us look at some simple examples.

Example 1.11

For

$$23 \equiv 2 \pmod{7},$$

and

$$6 \equiv -1 \pmod{7},$$

we have

$$23 + 6 \equiv 2 + (-1) \pmod{7},$$

which is simply

$$29 \equiv 1 \pmod{7}.$$

We also have

$$23 - 6 \equiv 2 - (-1) \pmod{7},$$

FIGURE 1.5 A sphere with a radius of r has a volume $4\pi r^3/3$.

or simply

$$17 \equiv 3 \pmod 7.$$

In addition, we have

$$23 \times 6 \equiv 2 \times (-1) \pmod 7,$$

or

$$138 \equiv -2, \pmod 7,$$

which can also be written as

$$138 \equiv 5 \pmod 7.$$

Another related mathematical operation is the round-up to the nearest integer, which is very useful in many numerical analysis. For example, 2.34 can be round down to the nearest integer 2, which can be written as

$$\lfloor 2.34 \rfloor = 2. \tag{1.30}$$

Mathematically speaking, a floor operation $\lfloor x \rfloor$ is the largest integer less than or equal to x. What about 2.89, should we round it up to 3? After all, 3 is the nearest integer to 2.89. In fact, there is a ceiling function $\lceil x \rceil$ which is the smallest integer greater than or equal to x. So we can write

$$\lceil 2.89 \rceil = 3, \quad \lceil 2.34 \rceil = 3. \tag{1.31}$$

Similarly, we have

$$\lfloor 2.89 \rfloor = 2. \tag{1.32}$$

It is worth pointing out that such floor and ceil operations are a special kind of functions because they take into a real number and output an integer. In general, a function takes a real number and churns out another number.

1.3 FUNCTIONS

A function is a quantity (say y) which varies with another independent quantity x in a deterministic way. The volume y of a sphere with a radius r (see Fig. 1.5) is simply $4\pi/3r^3$, and the surface area of a sphere is $4\pi r^2$.

To use the most generic notation for an independent variable x, we now use x to represent the radius, so we have

$$y = \frac{4\pi}{3}x^3, \tag{1.33}$$

which is an example of a cubic function. Similarly, the surface area of the sphere can be written as $4\pi x^2$. For any given value of x, there is a unique corresponding value of y. By varying x smoothly, we can vary y in such a manner that the point (x, y) will trace out a curve on the $x - y$ plane (see Fig. 1.6). Thus, x is called the independent variable, and y is called the dependent variable or function. Sometimes, in order to emphasize the relationship, we use $f(x)$ to express a general function, showing that it is a function of x. This can also be written as $y = f(x)$.

For any real number x, there is a corresponding unique value $y = f(x) = 4\pi x^3/3$, and although the negative volume is meaningless physically, it is valid mathematically. This relationship is a one-to-one mapping from x to y. That is to say, for a given value of x such as $x = 1$, there is a unique value of y that corresponds to the case of $x = 1$.

In mathematics, when we say x is a real number, we often write $x \in \Re$. Here $\Re$ means the set of all real numbers, and $\in$ stands for 'is a member of' or 'belongs to'. Thus, $x \in \Re$ is equivalent to saying that x is a real number or a member of the set of real numbers.

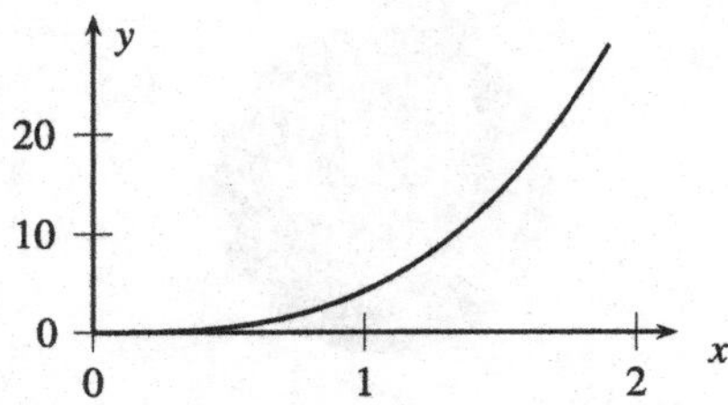

FIGURE 1.6 The graph of the function $y = f(x) = 4\pi x^3/3$.

It is worth pointing out that the one-to-one mapping from x to y usually does not guarantee a one-to-one mapping from y to x. For example, $y = x^2$ is a one-to-one mapping, but there are two values of x (say, ± 2) correspond to the same value of y (say, 4). We will discuss this in more detail in its proper context later in this book.

Strictly speaking, functions such as x^2 and x^3 are power functions. We will discuss in more detail when index notations and polynomials are introduced in the next chapter.

1.3.1 Domain and Range

The domain of a function is the set of numbers x for which the function $f(x)$ is defined validly. If a function is defined over a range $a \le x \le b$, we say its domain is $[a, b]$ which is called a closed interval. If a and b are not included, we have $a < x < b$ which is denoted as (a, b), and we call this interval an open interval. If b is included, while a is not, we have $a < x \le b$, and we often write this half-open and half-closed interval as $(a, b]$. Thus, the domain of function $f(x) = x^2$ is the set of all real numbers $\Re$ which is also written

$$f(x) = x^2, \qquad -\infty < x < +\infty. \tag{1.34}$$

Here the notation ∞ means infinity. The values that a function can take for a given domain is called the range of the function. Thus, the range of $y = f(x) = x^2$ is $0 \le y < \infty$ or $[0, \infty)$. Here, the closed bracket '[' means that the value (here 0) is included as part of the range, while the open round bracket ')' means that the value (here ∞) is not part of the range.

Sometimes, it is more convenient to write a function in a concise mathematical language, and we write

$$f : x \mapsto \frac{1}{2}x^2 - 15, \qquad x \in \Re, \tag{1.35}$$

which is called a mapping. This means that $f(x)$ is a function of x and this function will turn any real number x (input) in the domain into another number (output) $\frac{1}{2}x^2 - 15$.

1.3.2 Linear Function

The simplest general function is probably the linear function

$$y = kx + c, \tag{1.36}$$

where k and c are real constants. For example, when $k = 1$ and $c = 0$, we have $y = x$, which corresponds a simple straight line through the origin $(0, 0)$ with a gradient $k = 1$.

In general, for any given values of k and c, this will usually lead to a straight line if we plot values y versus x on a Cartesian coordinate system. In this case, k is the gradient of the line and c is the intercept (see Fig. 1.7).

Example 1.12

The relationship between the Fahrenheit (°F) and degree Celsius (°C) for temperature can be written as a linear function

$$F = \frac{9}{5}C + 32 = 1.8C + 32,$$

which gives the intercept of 32 and the gradient of 1.8. The above equation can be written as

$$C = \frac{F - 32}{1.8} = \frac{5(F - 32)}{9},$$

to convert from Fahrenheit to degree Celsius.

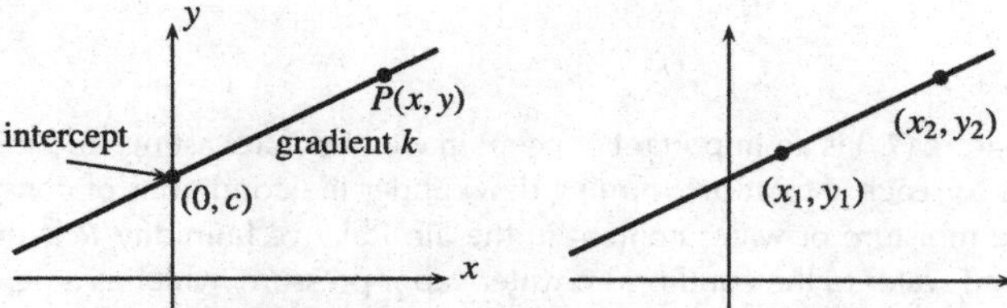

FIGURE 1.7 The equation $y = kx + c$ and a line through two points.

The normal human body temperature is about $C = 36.8\,°\text{C}$, what is the temperature in Fahrenheit? From the above equation, we have

$$F = 1.8C + 32 = 1.8 \times 36.8 + 32 = 98.24,$$

which is about 98 °F. Obviously, the temperature of a human body will vary slightly, depending on where the measurement is taken.

Now let us ask if there is a temperature at which both the Fahrenheit and Celsius readings will be the same?

From $F = 1.8C + 32$, if the readings are the same (i.e., $F = C$), we have

$$F = 1.8F + 32,$$

which gives

$$(1 - 1.8)F = 32,$$

or

$$-0.8F = 32,$$

which leads to

$$F = \frac{32}{-0.8} = -40.$$

That is, the reading at $-40\,°\text{C}$ is the same as $-40\,°\text{F}$.

For any two different points $A(x_1, y_1)$ and $B(x_2, y_2)$, we can always draw a straight line. Its gradient can be calculated by

$$k = \frac{y_2 - y_1}{x_2 - x_1}, \qquad (x_1 \neq x_2). \tag{1.37}$$

For any point $P(x, y)$ on the straight line, the gradient must be the same as k. Thus, the equation for the straight line can be written as

$$\frac{y - y_1}{x - x_1} = k = \frac{y_2 - y_1}{x_2 - x_1}. \tag{1.38}$$

This is

$$y = y_1 + \frac{(y_2 - y_1)}{(x_2 - x_1)}(x - x_1) = \frac{(y_2 - y_1)}{(x_2 - x_1)}x + \Big[y_1 - \frac{(y_2 - y_1)}{(x_2 - x_1)}x_1\Big], \tag{1.39}$$

which gives the intercept

$$c = y_1 - \frac{(y_2 - y_1)}{(x_2 - x_1)}x_1. \tag{1.40}$$

In a special case $y_1 = y_2$ while $x_1 \neq x_2$, we have $k = 0$. The line becomes horizontal, and the equation simply becomes $y = y_1$ which does not depend on x as x does not appear in the equation.

In another special case, $x_1 = x_2$, k is very large or approaching infinity, which we write as $k \to \infty$. It becomes a vertical line. Any small change in x will lead to an infinite change in y. For example, if we have $x_1 = x_2 = 2$, we have a vertical line through $x = 2$, and we simply write the equation as $x = 2$, and y does not appear in the equation in this case.

Example 1.13

The dewpoint or dewpoint temperature (T_d) is an important concept in weather forecasting, and it is a special temperature to which the moist air must be cooled so as to reach saturation (forming dew) under the conditions of constant pressure and constant water content. It is closely related to the moisture or water content in the air. Relative humidity R is the ratio of the partial pressure of water vapor in the mixture of air and water to the equilibrium water vapor pressure, which is a percentage between 0 to 100%.

When the relative humidity is above 50%, there is an approximate linear function to estimate the dewpoint temperature for given dry air temperature T and R. We have

$$T_d \approx T - \frac{(100 - R)}{5}.$$

As the fraction (the second term) on the right-hand side of the above equation is always positive for $R < 100$, this means that the dewpoint temperature is always less than the air temperature. When $R = 100\%$ (that is the air is saturated with water vapor), we have $T_d = T$ for $R = 100$.

For example, for $T = 30\,°\text{C}$ and $R = 60\%$, we have

$$T_d \approx 30 - \frac{(100 - 60)}{5} = 30 - \frac{40}{5} = 30 - 8 = 22,$$

which is 22 °C. When the dewpoint is above 20 °C, it will be uncomfortable for most people. The most comfortable dewpoints for most people range from 10 °C to 16 °C.

1.3.3 Modulus Function

A special function is the modulus function $|x|$ which is defined by

$$|x| = \begin{cases} x & \text{if } x \geq 0, \\ -x & \text{if } x < 0. \end{cases} \tag{1.41}$$

That is to say, $|x|$ is always non-negative. For example, $|5| = 5$ and $|-5| = 5$.

The modulus function has the following properties

$$|a \times b| = |a| \times |b|, \qquad a, b \in \Re, \tag{1.42}$$

which means that $|a^2| = |a|^2 = a^2 \geq 0$ for any $a \in \Re$. Similarly, for any real numbers a and $b \neq 0$, we have

$$\left|\frac{a}{b}\right| = \frac{|a|}{|b|}. \tag{1.43}$$

Example 1.14

For example, when $a = -2$ and $b = 3$, we have

$$|a| = |-2| = 2, \quad |b| = |3| = 3,$$

so we have

$$|a \times b| = |-2 \times 3| = |-6| = 6,$$

and

$$|a| \times |b| = 2 \times 3 = 6,$$

which indeed means that

$$|a \times b| = |a| \times |b|.$$

Similarly, we have

$$\left|\frac{a}{b}\right| = \left|\frac{-2}{3}\right| = \frac{2}{3},$$

which is the same as $\frac{|a|}{|b|} = \frac{2}{3}$.

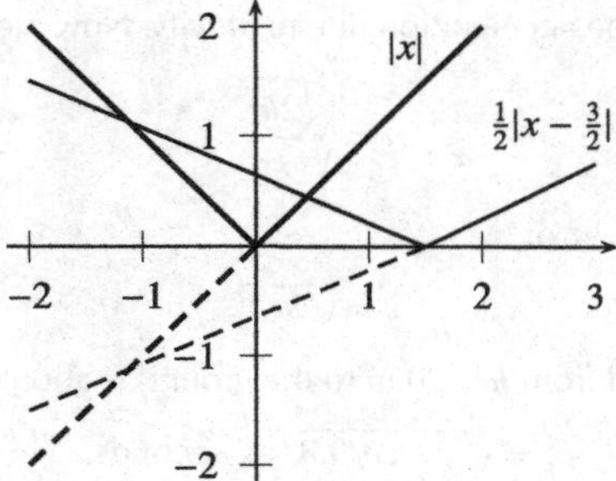

FIGURE 1.8 Graphs of modulus functions.

But for addition in general, we have $|a+b| \neq |a|+|b|$. Here, the sign $\neq$ means that the left-hand side is not equal to the right-hand side.

In fact, there is an inequality in this case: $|a+b| \leq |a|+|b|$. The sign $\leq$ means less than or equal to. Obviously, the equality holds in the above case when either $a=0$ or $b=0$. We will discuss this in detail in the vector analysis in later chapters.

Example 1.15

From the previous example, we know that

$$|a+b| = |-2+3| = |1| = 1,$$

while

$$|a|+|b| = |-2|+|3| = 2+3 = 5,$$

which gives that

$$|a|+|b| > |a+b|.$$

Two examples of modulus functions $f(x)=|x|$ and $f(x)=\frac{1}{2}|x-\frac{3}{2}|$ are shown as solid lines in Fig. 1.8. The dashed lines correspond to x and $\frac{1}{2}(x-\frac{3}{2})$ (without the modulus operator), respectively.

1.3.4 Power Function

When we discussed the volume of a sphere, we mentioned that x^2 and x^3 are power functions and their exponents (2 and 3 here) are integers. In general, a power function can be defined as

$$f(x) = x^p, \tag{1.44}$$

where p is a constant, call the exponent of the power function. We usually read x^p as x to the pth power or simply x to the p. In the case when p is a positive integer, it becomes the familiar powers x, x^2, x^3 as we learned in school. In this case, x can be any real numbers, thus the domain is $x \in \Re$ if p is a positive integer.

In general, p can be any real number. For example, when $p=1/2$, we have the square root function

$$f(x) = x^{1/2} = \sqrt{x}. \tag{1.45}$$

In this case, its domain is $x \geq 0$ because the negative does not have a root in the real domain.

Example 1.16

As an example, we now try to calculate the time taken for an object to fall freely from a height of h to the ground. If we assume the object is released with a zero velocity at h, we know from the basic physics that the distance h it travels can be determined by

$$h = \frac{1}{2}gt^2, \tag{1.46}$$

where t is the time taken, and $g = 9.8$ m/s^2 is the acceleration due to gravity. Now we have

$$t = \sqrt{\frac{2h}{g}}. \tag{1.47}$$

The velocity just before it hits the ground is $v = gt$ or

$$v = \sqrt{2gh}. \tag{1.48}$$

For example, the time taken for an object to fall from $h = 20$ m to the ground is about

$$t = \sqrt{2 \times 20/9.8} \approx 2 \text{ seconds}.$$

The velocity it approaches the ground is

$$v \approx \sqrt{2 \times 9.8 \times 20} \approx 19.8 \text{ m/s}.$$

Obviously, as h increases, t will also increase, so does the velocity.

1.4 QUADRATIC EQUATIONS

When an equation involves one unknown variable x, it may not be always linear. Some terms may involve x^2, in this case, we have to deal with quadratic functions or equations.

Example 1.17

Imagine that we are trying to find a special number x such that its reciprocal (i.e., 1 divided by this number or $1/x$) plus 1 is the special number x itself. Mathematically, it requires that

$$x = \frac{1}{x} + 1.$$

So what is the value of x?

Obviously, if x is positive (i.e., $x > 0$), then $1/x$ is positive, which means $x > 1$ so as to satisfy the above equation. Multiplying both sides by x, we have

$$x^2 = 1 + x,$$

which means that x must be less than 2. Otherwise, the left-hand side x^2 is much larger than the right-hand side $1 + x$. Therefore, we have

$$1 < x < 2.$$

Re-arranging the above equation by moving all the terms on the right-hand side to the left, we have

$$x^2 - x - 1 = 0,$$

which is a well-known quadratic equation, leading to the famous golden ratio

$$\phi = \frac{1 + \sqrt{5}}{2} \approx 1.618.$$

We will solve this equation in detail later in this section.

Quadratic functions are widely used in many applications. In general, a quadratic function $y = f(x)$ can be written as

$$f(x) = ax^2 + bx + c, \tag{1.49}$$

where the coefficients a, b and c are real numbers. Here we can assume $a \neq 0$. If $a = 0$, it reduces to the case of linear functions that we have discussed earlier. Two examples of quadratic functions are shown in Fig. 1.9 where both $a > 0$ and $a < 0$ are shown, respectively.

Depending on the combination of the values of a, b, and c, the curve may cross the x-axis twice, once (just touch), and not at all. The points at which the curve crosses the x-axis are the roots of

$$f(x) = ax^2 + bx + c = 0. \tag{1.50}$$

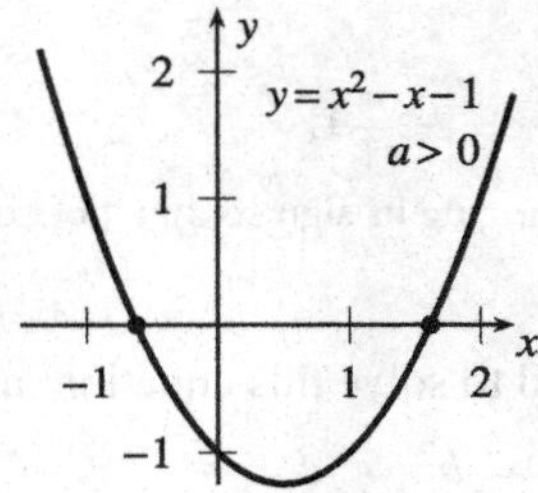

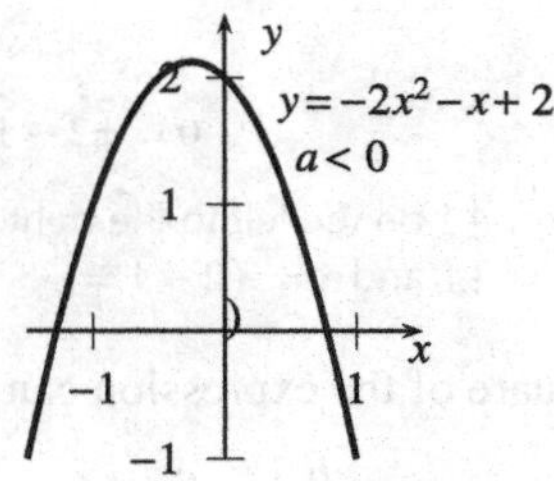

FIGURE 1.9 Graphs of quadratic functions.

As we can assume $a \neq 0$, we have

$$x^2 + \frac{b}{a}x + \frac{c}{a} = 0. \tag{1.51}$$

Of course, we can use factorization to find the solution in many cases. For example, we can factorize $x^2 - 5x + 6 = 0$ as

$$x^2 - 5x + 6 = (x - 2)(x - 3) = 0, \tag{1.52}$$

and the solution is thus either

$$(x - 2) = 0,$$

or

$$(x - 3) = 0.$$

That is $x = 2$ or $x = 3$. However, sometimes it is difficult (or even impossible) to factorize even seemingly simple expressions such as $x^2 - x - 1$. In this case, it might be better to address the quadratic equation in a generic manner using the so-called complete-square method.

Example 1.18

For example, from equation

$$x^2 + 2x - 3 = 0, \tag{1.53}$$

we can factorize the above equation as

$$(x - 1)(x + 3) = 0. \tag{1.54}$$

In order for the left-hand side to be equal to the right-hand side zero, either factor must be zero, which means either $x - 1 = 0$ or $x + 3 = 0$. This gives

$$x = 1, \quad \text{or} \quad x = -3. \tag{1.55}$$

Another way to solve this equation is to use the idea of completing the square. First, we can add the same term 4 to both sides, and we have

$$x^2 + 2x - 3 + 4 = 0 + 4, \quad \text{or} \quad x^2 + 2x + 1 = 4. \tag{1.56}$$

Since $(x + 1)^2 = x^2 + 2x + 1$, we get

$$(x + 1)^2 = 4. \tag{1.57}$$

We can then take the square root of both sides, we have

$$\sqrt{(x + 1)^2} = x + 1 = \pm\sqrt{4} = \pm 2. \tag{1.58}$$

Here we have to keep $\pm$ as the square root of 4 can be either $+2$ or -2 (otherwise, we just find one root). Now we subtract $+1$ from (or add -1 to) both sides, we have

$$x \underbrace{+1 - 1}_{\text{sum}=0} = \pm 2 - 1, \tag{1.59}$$

which leads to

$$x+0=\pm 2-1, \quad \text{or} \quad x=\pm 2-1. \tag{1.60}$$

This is equivalent to moving the term $+1$ on the left to the right and changing its sign so that it becomes -1. Now we have obtained the same two solutions $x=+2-1=+1$ and $x=-2-1=-3$.

In general, completing the square of the expression can be used to solve this equation, and we have

$$x^2+\frac{b}{a}x+\frac{c}{a}=\left(x+\frac{b}{2a}\right)^2-\frac{b^2}{4a^2}+\frac{c}{a}$$
$$=\left(x+\frac{b}{2a}\right)^2-\frac{(b^2-4ac)}{4a^2}=0, \tag{1.61}$$

which leads to

$$\left(x+\frac{b}{2a}\right)^2=\frac{(b^2-4ac)}{4a^2}. \tag{1.62}$$

Taking the square root of the above equation, we have either

$$x+\frac{b}{2a}=+\sqrt{\frac{b^2-4ac}{4a^2}}, \quad \text{or} \quad x+\frac{b}{2a}=-\sqrt{\frac{b^2-4ac}{4a^2}}. \tag{1.63}$$

Using $\sqrt{4a^2}=+2a$ or $-2a$, we have

$$x=-\frac{b}{2a}\pm\frac{\sqrt{b^2-4ac}}{2a}=\frac{-b\pm\sqrt{b^2-4ac}}{2a}. \tag{1.64}$$

In some books, the expression b^2-4ac is often denoted by Δ. That is $\Delta=b^2-4ac$.

- If $\Delta=b^2-4ac<0$, there is no solution which corresponds to the case where the curve does not cross the x-axis at all.
- In the special case of $b^2-4ac=0$, we have a single solution $x=-b/(2a)$, which corresponds to the case where the curve just touches the x-axis.
- If $b^2-4ac>0$, we have two distinct real roots.

Let us look at an example.

Example 1.19

Now we revisit an earlier example about the quadratic equation leading to the golden ratio. We have $f(x)=x^2-x-1=0$, which corresponds to $a=1$, $b=-1$ and $c=-1$. Since

$$\Delta=b^2-4ac=(-1)^2-4\times 1\times(-1)=5>0,$$

it has two different real roots. We have

$$x_{1,2}=\frac{-(-1)\pm\sqrt{\Delta}}{2}=\frac{1\pm\sqrt{5}}{2}.$$

So the two roots are $(1-\sqrt{5})/2$ and $(1+\sqrt{5})/2$ which are marked as solid circles on the graph in Fig. 1.9.

The positive root is the famous golden ratio

$$\phi=\frac{1+\sqrt{5}}{2}=1.6180339887...,$$

and this ratio appears in many natural phenomena, from flower petals and DNA structures to star formation and spiral galaxies.

The other root $\Phi=(1-\sqrt{5})/2=-0.618...$ is the so-called golden ratio conjugate, whose absolute value is the reciprocal of the golden ratio, that is

$$|\Phi|=\frac{\sqrt{5}-1}{2}=0.618...,$$

which can also be expressed as

$$|\Phi|=\phi-1=\frac{(1+\sqrt{5})}{2}-1=\frac{(1+\sqrt{5})-2}{2}=\frac{\sqrt{5}-1}{2}$$

and

$$\frac{1}{\phi} = \frac{2}{\sqrt{5}+1} = \frac{2}{\sqrt{5}+1} \times \frac{\sqrt{5}-1}{\sqrt{5}-1} = \frac{2(\sqrt{5}-1)}{(\sqrt{5})^2 - 1^2}$$
$$= \frac{2(\sqrt{5}-1)}{4} = \frac{\sqrt{5}-1}{2} = |\Phi| = 0.618...,$$

where we have used $(\sqrt{5}+1)(\sqrt{5}-1) = (\sqrt{5})^2 - 1^2 = 5 - 1 = 4$.

Obviously, equations can have more than one unknown. In this case, we have to deal with simultaneous equations.

1.5 SIMULTANEOUS EQUATIONS

The fundamental ideas of mathematical operations can also be applied to a system of equations. In order to solve two simultaneous equations for the unknowns x and y

$$\begin{cases} 5x & +y & = -7, \\ -4x & -2y & = 2, \end{cases} \tag{1.65}$$

we first multiply both sides of the first equation by 2, we have

$$2 \times (5x + y) = 2 \times (-7), \tag{1.66}$$

and we get

$$10x + 2y = -14. \tag{1.67}$$

Now we add this equation (thinking of the same quantity on both sides) to the second equation of the system. We have

$$\begin{array}{lll} 10x & +2y & = -14 \\ -4x & -2y & = 2 \\ \hline 6x & +0 & = -12. \end{array} \tag{1.68}$$

This means that

$$6x = -12, \quad \text{or} \quad x = \frac{-12}{6} = -2. \tag{1.69}$$

Now substitute this solution $x = -2$ into one of the equations (say, the first); we have

$$5 \times (-2) + y = -7, \tag{1.70}$$

or

$$-10 + y = -7. \tag{1.71}$$

This simply leads to

$$y = -7 + 10 = 3. \tag{1.72}$$

Thus, we finally get $x = -2$ and $y = 3$. As each equation corresponds to a line in the two-dimensional (2D) Cartesian plane, the cross point $(-2, 3)$ of the two lines is the solution we just obtained.

Now let us look again at the example about the temperature conversion studied earlier.

Example 1.20

The conversion from Fahrenheit F to Celsius C is given by

$$C = \frac{5}{9}(F - 32). \tag{1.73}$$

Conversely, from Celsius to Fahrenheit, we have

$$F = \frac{9C}{5} + 32. \tag{1.74}$$

So the melting point of ice $C = 0\,°\text{C}$ is equivalent to $F = \frac{9\times 0}{5} + 32 = 32\,°\text{F}$, while the boiling point of water $C = 100\,°\text{C}$ becomes $9 \times 100/5 + 32 = 212\,°\text{F}$.

So the room temperature 20 °C is equivalent to

$$F = \frac{9 \times 20}{5} + 32 = 68\,°\text{F}.$$

We can see that these two temperature scales are very different. However, there is a single temperature at which these two coincide. Let us now try to determine this temperature. We have $C = F$, that is

$$\frac{5}{9}(F - 32) = \frac{9C}{5} + 32 = \frac{9F}{5} + 32,$$

which becomes

$$(\frac{5}{9} - \frac{9}{5})F = 32 + \frac{160}{9},$$

or

$$-\frac{56}{45}F = \frac{448}{9}.$$

The solution is simply $F = C = -40$, which is exactly the same solution we obtained earlier. This means that −40 °F and −40 °C represent the same temperature on both scales.

It is worth pointing out that the two equations (1.73) and (1.74) in this example look like two simultaneous equations, but they are not. In fact, one equation is the re-arrangement of the other. In other words, the two equations are not independent. The independency of equations is important. Let us look at the following two equations:

$$x + 2y = 5, \tag{1.75}$$

$$2x + 4y = 10. \tag{1.76}$$

If we try to solve this set of simultaneous equations, we will not get any unique solutions if we follow the procedure discussed above. In fact, the second equation is simply the first equation multiplied by 2 on both sides, which means they are not independent and essentially become the same equation. We have two unknowns and one equation. Therefore, there is not enough information to determine two unknowns. In general, in order to determine two (or n) unknowns, we need two (or n) independent linear equations to determine the unknowns uniquely. We will discuss this issue further in the linear system after we have learned the fundamentals of matrix algebra in later chapters.

In mathematics, expressions can be complicated, even for expressions with one unknown or variable x. For example, $x^3 + 2x^2 + 3x + 4$ is a cubic expression, while $x^5 + x^4 + 2x^3 - 4x^2 + 1$ is of the fifth-order. In general, we are dealing with polynomials which use the notation of indices and manipulations of equations extensively, and we will discuss them in the next chapter.

EXERCISES

1.1 Which of the following numbers is irrational?

$$\sqrt{3},\ \frac{\sqrt{8}}{\sqrt{2}},\ 2\pi,\ \sqrt{100},\ 3.1415,\ 0.123456789.$$

1.2 Solve $x^2 - 101x + 100 = 0$.

Hint. Factorize to get $x = 1, 100$.

1.3 Show that $x^2 + x + k = 0$ has no solution for any $x \in \Re$ if $k > 1/4$.

Hint. Use $\Delta = b^2 - 4ac < 0$.

1.4 Show that the two equations $x + 5y = 11$ and $7x - 2y = 3$ have a solution $x = 1$ and $y = 2$.

1.5 Solve $|x - 2| - 3 = 0$.

Hint. Discuss $x > 2$ and $x < 2$, and then explain $x = 5$ and $x = -1$.

Chapter 2

Polynomials and Roots

Chapter Points

- Index notations will be introduced, which makes writing of many things simpler.
- Floating-point numbers are introduced due to their wide use in mathematics and scientific computing.
- Polynomials and multiple roots will be discussed with a detailed presentation on how to solve quadratic and cubic equations in general.

2.1 INDEX NOTATION

Before we proceed to study polynomials, let us first introduce the notations of indices. For the discussion of the product of $2 \times 2 \times 2 \times 2 \times 2 \times 2 \times 2$ (2 multiplied by itself 7 times), it is more compact to write it as

$$2^7 \equiv 2 \times 2 \times 2 \times 2 \times 2 \times 2 \times 2. \tag{2.1}$$

Here, we use '$\equiv$' to mean the definition as the left-hand side and the right-hand side are always equal in this case. The compact index notation is not only just to save time and space, but it also looks better. Imagine that writing 2^{100} using the multiplication symbol '$\times$' 99 times.

For higher-order products such as $a \times a \times a \times a \times a$ or $aaaaa$, it might be more economical to write it as a simpler form a^5. In general, we use the index form to express the product

$$a^n = \overbrace{a \times a \times \ldots \times a}^{n}, \tag{2.2}$$

where a is called the base, and n is called the index or exponent. Conventionally, we write $a \times a = aa = a^2$, $a \times a \times a = aaa = a^3$ and so on and so forth. For example, if $n = 100$, it is obviously advantageous to use the index form. A very good feature of index notation is the multiplication rule

$$a^n \times a^m = \overbrace{a \times a \times \ldots \times a}^{n} \times \overbrace{a \times \ldots \times a}^{m} = \overbrace{a \times a \times \ldots \times a}^{n+m} = a^{n+m}.$$

Thus, the product of factors with the same base can easily be carried out by adding their indices.

If we interpret $a^{-n} = 1/a^n$, we then have the division rule

$$a^n \div a^m = \frac{a^n}{a^m} = a^{n-m}. \tag{2.3}$$

In the special case when $n = m$, it requires that $a^0 = 1$ because any non-zero number divided by itself should be equal to 1.

Example 2.1

In order to calculate $2^5 \times 2^7 \times 2^9$, we have

$$2^5 \times 2^7 \times 2^9 = 2^{5+7+9} = 2^{21} = 2097152.$$

Similarly, we can calculate

$$2^{-5} \times 2^9 = 2^{-5+9} = 2^4 = 16,$$

and

$$2^{-10} \times 2^2 \times 2^5 = 2^{-10+2+5} = 2^{-3} = \frac{1}{2^3} = \frac{1}{8} = 0.125.$$

If we replace a by $a \times b$ and use $a \times b = b \times a$, we can easily arrive at the factor rule

$$(a \times b)^n = a^n \times b^n = a^n b^n. \tag{2.4}$$

Similarly, using a^m to replace a in the expression (2.2), it is straightforward to verify the following power-on-power rule

$$(a^m)^n = a^{n \times m} = a^{nm}. \tag{2.5}$$

Here, the exponents m and n are integers and they can be positive or negative or zero. Fractional exponents can be treated by certain convention. For example, the fraction exponent $1/n$ means to take the nth root of a. That is,

$$a^{\frac{1}{n}} = a^{1/n} = \sqrt[n]{a}. \tag{2.6}$$

Thus, we have

$$a^{\frac{m}{n}} = a^{m/n} = \sqrt[n]{a^m} = (\sqrt[n]{a})^m, \tag{2.7}$$

which gives the following examples:

$$2^{\frac{4}{3}} = \sqrt[3]{2^4} = \sqrt[3]{16}, \quad 4^{-2/3} = \frac{1}{4^{2/3}} = \frac{1}{\sqrt[3]{4^2}} = \frac{1}{\sqrt[3]{16}}. \tag{2.8}$$

However, the exponents can also be generalized to be any real values such as 0.5 or $-1/2$. For example, $a^{0.5} = \sqrt{a}$ and $a^{-1/2} = 1/\sqrt{a}$. Therefore, in the rest of this book, we will treat the exponents in the most general sense; that is, the exponents can be real numbers.

Now we can revisit the power function

$$f(x) = x^p. \tag{2.9}$$

There are four cases:

- $p > 1$: The power x^p increases faster as x becomes large.
- $p = 1$: It becomes a linear function or a straight line x.
- $0 < p < 1$: The power increases slower compared to the increase of x.
- $p < 0$: It decreases to zero as x becomes large.

Care should be taken when p is a fraction, especially negative fraction. For example, if $p = -2/3$, we have

$$f(x) = x^{-2/3} = \frac{1}{x^{2/3}} = \frac{1}{\sqrt[3]{x^2}}, \tag{2.10}$$

whose domain can be any real number, except for $x = 0$. This means that this function is not defined at $x = 0$. Similarly, when $p = 1/4$, we have

$$f(x) = x^{1/4} = \sqrt[4]{x}, \tag{2.11}$$

whose domain is $x \geq 0$ because we cannot find an even root for a negative number.

Example 2.2

The surface-area-to-volume ratio R_{sv} is a very important factor in engineering and chemistry. For a sphere of radius a, its surface area is $S = 4\pi a^2$ and its volume is $V = 4\pi a^3/3$, so the surface-to-volume ratio is

$$R_{sv} = \frac{S}{V} = \frac{4\pi a^2}{\frac{4\pi a^3}{3}} = \frac{3}{a}.$$

Similarly, for a cube of side length a, its volume is $V = a^3$ and surface area is $S = 6a^2$, so the ratio is simply

$$R_{sv} = \frac{6a^2}{a^3} = \frac{6}{a}.$$

It leaves an exercise to check that most object shapes will have a surface-to-volume ratio that is inverse proportional to its size.

We may wonder what happens when $p = 0$. For consistency in mathematics, it requires that $x^0 = 1$ for any value of $x \neq 0$. In addition, in a very special case when $x = 0$, we have 0^0. In this case, 0^0 is undefined (that is, its value is not defined with a single fixed value).

One way to look at 0^0 is that for any non-zero a, $a^0 = 1$ as mentioned earlier. Another way to look at it is that for $0^2 = 0$, $0^1 = 0$ or $0^a = 0$ for any $a > 0$. Thus, it seems that either $0^0 = 0$ or $0^0 = 1$ is possible. Some mathematicians suggest that $0^0 = 1$ can allow certain formulas to be expressed more compactly. Though a mathematical curiosity, 0^0 does not make any difference in practice because it rarely occurs in any real-world applications.

Many phenomena in nature obey the so-called power laws. Both gravitation and electromagnetism obey the inverse-square law. For example, Newton's law of gravitation can be written as

$$F = \frac{G m_1 m_2}{r^2}, \tag{2.12}$$

where m_1 and m_2 are the masses of two objects and r is the distance between the centers of the masses. G is the universal gravitational constant. The force F between two point masses will decrease as an inverse-square law as their distance increases.

The inverse-square law has an exponent -2. In reality, many phenomena can have various exponents.

Example 2.3

For example, the well-know Zipf's law in linguistics takes the following form

$$f = k\,r^{-\alpha}, \tag{2.13}$$

where f is the frequency of the rth popular word in the language. α is the exponent and k is a scaling constant. For example, in the English language, the top ten most used words are: 'the', 'of', 'be' (including its variations such as 'are' and 'is'), 'to', 'and', 'a', 'in', 'that', 'have' and 'I', which consist of about 25% of the recorded language according to the Oxford English Corpus project. In this case, $\alpha \approx 1$. For example, 'the' appears about 7% of all word occurrences and 'of' appears about 3.5% of all words. That is, the most used word 'the' appears almost as twice as the second most frequent.

Zipf's law $\alpha = 1.07$ for cities, relating the ranks of the world's largest cities to their population sizes (Zipf GK, *Human Behaviour and the Principle of Least Effort*, Addison-Wesley, 1949; also Scarrott GS, Will Zipf join Gauss? *New Scientist*, pp. 402–404, 16 May 1975). In fact, Zipf-type power laws have been observed in the numbers of hits on web sites, numbers of telephone calls received by companies, magnitudes of earthquakes, diameters of moon craters with an exponent of about $\alpha = 3.1$ and many others.

Example 2.4

How fast can a raindrop fall? The velocity of a falling raindrop can be estimated using the following formula:

$$v = \sqrt{1.1 \times 10^4 d / C},$$

where d is the diameter of the raindrop in meters and C is the drag coefficient. Usually, $C = 0.5$ for raindrops in the air. Typically, the largest raindrops can reach to $d = 5$ mm or $d = 0.005$ m, which gives

$$v = \sqrt{1.1 \times 10^4 \times 0.005/0.5} = \sqrt{55/0.5} \approx 10.5 \text{ m/s},$$

which is about $10.5 \times 3600/1000 \approx 38$ km per hour or about 24 mph.

2.2 FLOATING POINT NUMBERS

In science and engineering, we often have to deal with very large or very small numbers, and thus floating point notations become a very useful tool. For example, the mass of the Earth is about 5.972×10^{24} kg, while the mass of a proton is about $m_p \approx 1.67262 \times 10^{-27}$ kg.

In general, a floating-point number has three parts: significand, base and exponent. Here, the mass of the Earth

$$\text{significand} \times \text{base}^{\text{exponent}} = 5.972 \times 10^{24}, \tag{2.14}$$

where 5.972 is the significand or significant digits, 10 is the base and 24 is the exponent. Such notations are also called the scientific notation, standard form or standard index form.

Floating-point numbers provide a convenient way to deal with any real numbers (large or small), and thus are widely used in scientific computing and engineering applications.

It is worth pointing out that the base 10 is usually written as E in the scientific computing and numeric outputs in many programming languages. For example, 5.972×10^{24} is often written as $5.972E+24$ or $5.972E24$.

In addition, depending on the precision or accuracy of the calculations, there is a tradeoff using such notations. For example, we know the basic charge of an electron is

$$q = (1.602176487 \pm 0.000000040) \times 10^{-19} \tag{2.15}$$

which has 10 significant digits. However, if we are concerned with the basic estimates, we can approximate the above simply as

$$q = 1.602 \times 10^{-19}, \quad \text{or even} \quad 1.6 \times 10^{-19}, \tag{2.16}$$

depending on the precision we may want.

Example 2.5

The black-body radiation of an object at an absolute temperature T emits the energy E per unit area per unit time, which obeys the Stefan-Boltzmann law

$$E_b = \sigma T^4,$$

where $\sigma = 5.67 \times 10^{-8}\ \mathrm{W K^{-4} m^{-2}}$ is the Stefan-Boltmann constant. For example, a typical human body has a skin surface area of about $A = 1.8\ \mathrm{m}^2$. The normal body temperature $T_h = 36.8\,°\mathrm{C}$ or $T_h = 273 + 36.8 = 309.8$ K (Kelvin) in a room with a constant room temperature $T_0 = 20\,°\mathrm{C}$ or $T_0 = 273 + 20 = 297$ K. The skin temperature T_s in contact with clothes will be about $T_s \approx (T_h + T_0)/2 = (36.8 + 20)/2 = 28.4\,°\mathrm{C}$ or

$$T_s = 273 + 28.4 = 301.4 \text{ K}.$$

Thus, total energy per unit time radiated by an adult human body is

$$E = A(\sigma T_s^4 - \sigma T_0^4) = A\sigma(T_s^4 - T_0^4)$$
$$= 1.8 \times 5.67 \times 10^{-8} \times (301.4^4 - 297^4) \approx 90 \text{ J/s} = 90 \text{ Watts}.$$

This is equivalent to the power of three 30-watt light bulbs.

2.3 POLYNOMIALS

The functions we have discussed so far are a special case of polynomials. In particular, a quadratic function $ax^2 + bx + c$ is a special case of polynomials. In general, a polynomial $p(x)$ is an explicit expression, often written as a sum or linear combination, which consists of multiples of non-negative integer powers of x. Therefore, we often write a polynomial in the descending order (in terms of the power x^n)

$$p(x) = a_n x^n + a_{n-1} x^{n-1} + ... + a_1 x + a_0, \quad (n \geq 0), \tag{2.17}$$

where a_i $(i = 0, 1, ..., n)$ are known constants. The coefficient a_n is often called the leading coefficient, and the highest power n of x is called the degree of the polynomial. Each partial expression such as $a_n x^n$ and $a_{n-1} x^{n-1}$ is called a term. The term a_0 is called the constant term. Obviously, the simplest polynomial is a constant (degree 0).

Example 2.6

For example, the quadratic function $3x^2 - 4x + 5$ is a polynomial of degree 2 with a leading coefficient 3. It has three terms.

The expression

$$-2x^5 - 3x^4 + 5x^2 + 1$$

is a polynomial of the fifth degree with a leading coefficient -2. It has only four terms as the terms x^3 and x do not appear in the expression (their coefficients are zero).

When adding (or subtracting) two polynomials, we have to add (or subtract) the corresponding terms (with the same power n of x), and the degree of resulting polynomial may have any degree from 0 to the highest degree of two polynomials.

However, when multiplying two polynomials, the degree of the resulting polynomial is the sum of the degrees of the two polynomials. For example,

$$(a_n x^n + a_{n-1}x^{n-1} + ... + a_0)(b_m x^m + b_{m-1}x^{m-1} + ... + b_0)$$
$$= a_n b_m x^{n+m} + (a_n b_{m-1} + a_{n-1}b_m)x^{n+m-1} + ... + a_0 b_0. \tag{2.18}$$

Example 2.7

If we add $2x^3 + 3x^2 + 1$ with $2x^3 - 3x^2 + 4x - 5$, we have

$$(2x^3 + 3x^2 + 1) + (2x^3 - 3x^2 + 4x - 5)$$
$$= (2x^3 + 3x^2 + 0x + 1) + (2x^3 - 3x^3 + 4x - 5)$$
$$= (2+2)x^3 + (3-3)x^2 + (0+4)x + (1-5)$$
$$= 4x^3 + 0x^2 + 4x - 4 = 4x^3 + 4x - 4,$$

which is a cubic polynomial. However, if we subtract the second polynomial from the first, we have

$$(2x^3 + 3x^2 + 1) - (2x^3 - 3x^2 + 4x - 5)$$
$$= (2-2)x^3 + [3-(-3)]x^2 + (0-4)x + [1-(-5)]$$
$$= 0x^3 + 6x^2 - 4x + 6 = 6x^2 - 4x + 6,$$

which is a quadratic.

Similarly, the product of $3x^4 - x - 1$ and $-x^5 - x^4 + 2$ is a polynomial of degree 9 because

$$(3x^4 - x - 1)(-x^5 - x^4 + 2) = -3x^9 - 3x^8 + x^6 + 2x^5 + 7x^4 - 2x - 2,$$

has 9 as the exponent in the highest-order term (i.e., x^9).

2.4 ROOTS

For a given polynomial such as

$$f(x) = x^3 - x = 0, \tag{2.19}$$

the value x_* that satisfies the above equation is called a root. We use $*$ to highlight this value is a special value of x. That is

$$x_*^3 - x_* = 0. \tag{2.20}$$

Obviously, for this simple equation, we can guess that $x_* = 1$ is a solution or root because $1^3 - 1 = 0$. It is also easy to check that $x_* = 0$ is also a root. Now the question is: have we found all the roots for this simple problem? If not, how many are left? If yes, how can we be sure?

If we look at this problem more carefully, we realize that $x_* = -1$ is also a root because $(-1)^3 - (-1) = -1 + 1 = 0$. Now there are three roots:

$$x_* = 0, \quad x_* = +1, \quad x_* = -1. \tag{2.21}$$

If we try even harder, it seems that we cannot find another root.

Another way to look this problem is to factorize the polynomial. Since $f(x) = x^3 - x = x(x^2 - 1) = x(x+1)(x-1)$, we have

$$x(x+1)(x-1) = 0. \tag{2.22}$$

In order to solve the above equation, we can see that the right-hand side is zero, and the left-hand side consists of three factors $x, x+1, x-1$. Therefore, if any of the three factors is zero, the equation will be true, thus we can have

$$x = 0, \tag{2.23}$$

or

$$x + 1 = 0, \tag{2.24}$$

or

$$x - 1 = 0. \tag{2.25}$$

Thus, we finally have

$$x = 0, \quad \text{or} \quad x = -1, \quad \text{or} \quad x = +1, \tag{2.26}$$

which are the same as the three roots we obtained earlier. In this example, the order of the polynomial is 3, and thus three roots exist.

Now a natural question is: how to find the roots of a polynomial in general? We know that there exists an explicit formula for finding the roots of a quadratic. From the discussion in the previous chapter about the standard quadratic equation

$$ax^2 + bx + c = 0, \quad (a \neq 0), \tag{2.27}$$

we know that its roots can be obtained analytically by

$$x = \frac{-b \pm \sqrt{b^2 - 4ac}}{2a}. \tag{2.28}$$

This formula is valid for

$$b^2 - 4ac \geq 0.$$

Otherwise, there is no root in the real-number domain.

Analytical forms are also possible for a cubic, and a quartic function (polynomial of degree 4), though quite complicated. In some special cases, it may be possible to find a factor via factorization or even by educated guess; then some of the solutions can be found. For example, for $f(x) = x^6 - 1$, we can have

$$f(x) = x^6 - 1 = (x - 1)(x + 1)(x^2 + x + 1)(x^2 - x + 1) = 0, \tag{2.29}$$

but to figure out such factors is not easy. In general, factorization of high-order polynomials is not always possible.

Example 2.8

Let us try to solve the following cubic equation

$$f(x) = x^3 - 2x^2 + 3x - 6 = 0,$$

which can be factorized as

$$x^3 - 2x^2 + 3x - 6 = (x - 2)(x^2 + 3) = 0.$$

So the solution should be either

$$x - 2 = 0,$$

or

$$x^2 + 3 = 0.$$

Therefore, the only real solution is $x = 2$, and the other condition is impossible to satisfy for any real number unless in context of complex numbers, to be discussed later. The graph of $f(x)$ is shown in Fig. 2.1.

Now we introduce a generic method for finding the roots of a general cubic polynomial

$$f(x) = ax^3 + bx^2 + cx + d = 0, \tag{2.30}$$

where a, b, c, d are real numbers, and $a \neq 0$. We can use a simple change of variable

$$z = x + \frac{b}{3a}. \tag{2.31}$$

First, we can write the above equation as

$$f(x) = ax^3 + 3\beta x^3 + 3\gamma x + d = 0, \tag{2.32}$$

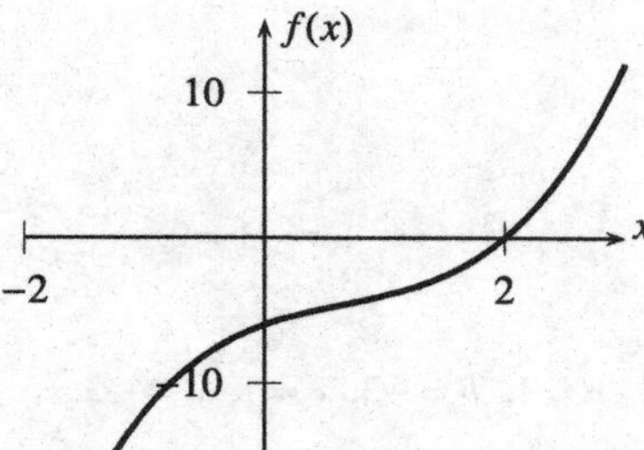

FIGURE 2.1 Cubic function $f(x) = x^3 - 2x^2 + 3x - 6$.

where

$$\beta = \frac{b}{3}, \qquad \gamma = \frac{c}{3}. \tag{2.33}$$

Substituting the new variable $z = x + \frac{\beta}{a}$ or $x = z - \frac{\beta}{a}$, we have

$$a(z - \frac{\beta}{a})^3 + 3\beta(z - \frac{\beta}{a})^2 + 3\gamma(z - \frac{\beta}{a}) + d = 0. \tag{2.34}$$

Using the formula $(x+y)^3 = x^3 + 3x^2y + 3xy^2 + y^3$ and expanding the above equation, after some algebraic manipulations we have

$$z^3 + \frac{3h}{a^2}z + \frac{g}{a^3} = 0, \tag{2.35}$$

where

$$h = a\gamma - \beta^2, \qquad g = a^2 d - 3a\beta\gamma + 2\beta^3. \tag{2.36}$$

This essentially transforms the original equation into a reduced or depressed cubic equation. By defining a constant

$$\begin{aligned}\delta = f(-b/(3a)) &= a(-\frac{b}{3a})^3 + b(-\frac{b}{3a})^2 + c(-\frac{b}{3a}) + d \\ &= \frac{2b^3}{27a^2} - \frac{bc}{3a} + d = \frac{2\beta^3 - 3a\beta\gamma + a^2 d}{a^2} = \frac{g}{a^2},\end{aligned} \tag{2.37}$$

and the Cardan's discriminant $\Delta = g^2 + 4h^3$, we have

$$\Delta = a^2(a^2d^2 - 6a\beta\gamma d + 4a\gamma^3 + 4\beta^3 d - 3\beta^2\gamma^2) = a^4(\delta^2 - \epsilon^2), \tag{2.38}$$

or

$$\epsilon^2 = \delta^2 - \frac{\Delta}{a^4}. \tag{2.39}$$

In the case of $\Delta > 0$, the reduced cubic equation has one real root. The real root can be calculated by

$$z = \sqrt[3]{\frac{1}{2a}(-\delta + \sqrt{\delta^2 - \epsilon^2})} + \sqrt[3]{\frac{1}{2a}(-\delta - \sqrt{\delta^2 - \epsilon^2})}, \tag{2.40}$$

or

$$x = z - \frac{b}{3a} = -\frac{b}{3a} + \sqrt[3]{\frac{1}{2a}(-\delta + \frac{1}{a^2}\sqrt{\Delta})} + \sqrt[3]{\frac{1}{2a}(-\delta - \frac{1}{a^2}\sqrt{\Delta})}. \tag{2.41}$$

If $\Delta = 0$, it has three real roots. If $\epsilon \neq 0$, it gives two equal roots, and the three roots are

$$z = \zeta, \ \zeta, \ -2\zeta, \tag{2.42}$$

where ζ is determined by $\zeta^2 = -h/a^2$. Thus, the signs of ζ and δ are important (for details of derivations, please refer to Nickalls RWD, *Mathematical Gazzette*, **77**, 354–359, 1993). In modern scientific applications, we rarely use these formulas, as we can easily obtain the solutions using a computer. Here we just use this case to illustrate how complicated the process can become even for cubic polynomials.

So let us look at an example.

Example 2.9

For a cubic equation

$$x^3 - 3x^2 + x - 3 = 0,$$

we know that

$$a = 1, \ b = -3, \ c = 1, \ d = -3,$$

and

$$\beta = b/3 = -1, \quad \gamma = c/3 = 1/3.$$

So we have

$$h = a\gamma - \beta^2 = 1 \times (1/3) - (-1)^2 = -2/3,$$
$$g = a^2 d - 3a\beta\gamma + 2\beta^2 = 1^2 \times (-3) - 3 \times 1 \times (-1) \times (1/3) + 2 \times (-1)^3 = -4,$$

and

$$\delta = \frac{g}{a^2} = \frac{-4}{1^2} = -4.$$

This gives that

$$\Delta = g^2 + 4h^3 = (-4)^2 + 4 \times (-2/3)^3 = 400/27 > 0,$$

which means that there is only one real root that is given by

$$x = -\frac{b}{3a} + \sqrt[3]{\frac{1}{2}(-\delta + \frac{\sqrt{\Delta}}{a^2})} + \sqrt[3]{\frac{1}{2}(-\delta - \frac{\sqrt{\Delta}}{a^2})}$$
$$= -\frac{-3}{3 \times 1} + \sqrt[3]{\frac{1}{2}(-(-4) + \frac{\sqrt{400/27}}{1^2})} + \sqrt[3]{\frac{1}{2}(-(-4) - \frac{\sqrt{400/27}}{1^2})}$$
$$\approx 1 + 1.577350269 + 0.422649731 \approx 3.0,$$

which is still not a straightforward task. In fact, we can rewrite the equation by factoring as

$$x^3 - 3x^2 + x - 3 = (x - 3)(x^2 + 1) = 0.$$

In the domain of real numbers, $x^2 + 1 > 0$, so $x - 3 = 0$ or $x = 3$, which is the only real root. This example shows that an analytical formula may be good for mathematical analysis, but it does not necessarily simplify the task for numerical calculations.

In general, there is no analytical formula for the roots of polynomials higher than the fifth order. The roots of a quintic function can still have analytical forms, though they become very complicated.

Therefore, for practical applications, the roots of most polynomials have to be estimated numerically, and the root-finding algorithms such as Newton's method become very useful. We will introduce such algorithms in later chapters.

EXERCISES

2.1. Simplify the following expressions:
- $3^2 \times 2^3/6^2$
- $2^{-3} \times 2^5 \times 100^0$
- $\sqrt{3}(\sqrt{7})^3/\sqrt{21}$
- $2^8 2^{-6} - 3^3 3^{-2} + \pi^0$

2.2. Show that $x^5 - 1 = (x - 1)(x^4 + x^3 + x^2 + x + 1)$.

Hint. Expand the right-hand side first and simplify.

2.3. Show that $x^4 - 16 = (x + 2)(x - 2)(x^2 + 4)$ and find the roots of $x^4 - 16 = 0$.

2.4. Solve $x^3 - 2x^2 - 2x + 1 = 0$. How many roots can you find?

Hint. Factorize it to $(x+1)(x^2 - 3x + 1) = 0$. Thus, $x = -1$ and $x = (3 \pm \sqrt{5})/2$.

2.5. Suppose k is an integer so that $x^3 - kx = 0$ has three roots. What are the values of k to ensure that all three roots are also integers. Can k be negative?

Hint. Since $x^3 - kx = x(x^2 - k)$, the roots are 0 and $\pm\sqrt{k}$. So $k = 1, 4, 9, 16, ...$ will lead to integer roots.

Chapter 3

Binomial Theorem and Expansions

Chapter Points

- Binomial theorem will be introduced to discuss higher-order expansions and their relations to expansion coefficients and Pascal's triangle.
- Factorials are presented, together with various identities related to binomial expansions.

After reviewing the fundamentals of the polynomials and index notations, we are now ready to introduce more important topics in mathematics. In the notations of indices, we have tried to simplify certain expressions such as $(a^4b^2)^5/a^{12}$, and we can also expand

$$(a+b)^2 = (a+b)(a+b) = a^2 + 2ab + b^2, \tag{3.1}$$

or

$$(a-b)^2 = (a-b)(a-b) = a^2 - 2ab + b^2. \tag{3.2}$$

However, we never actually dealt with the factors such as $(a+b)^n$ where $n \geq 2$ is an integer. This requires the introduction of the binomial theorem, which is an important mathematical technique. Binomial theorem is a basis for mathematical analysis of probability as commonly used in statistics and data modeling. Sum of series is crucial for numerical analysis and approximation techniques.

3.1 BINOMIAL EXPANSIONS

From the discussions in the previous chapters, we know $(a+b)^0 = 1$ and $(a+b)^1 = (a+b) = a+b$. We also know that

$$(a+b)^2 = (a+b)(a+b) = a(a+b) + b(a+b) = a^2 + 2ab + b^2. \tag{3.3}$$

If we follow a similar procedure to expand, we can deal with higher-order expansions. For example,

$$\begin{aligned}(a+b)^3 &= (a+b)(a+b)^2 = (a+b)(a^2+2ab+b^2),\\ &= a(a^2+2ab+b^2) + b(a^2+2ab+b^2) = a^3 + 3a^2b + 3ab^2 + b^3.\end{aligned} \tag{3.4}$$

Similarly, we have

$$(a+b)^4 = a^4 + 4a^3b + 6a^2b^2 + 4ab^3 + b^4, \tag{3.5}$$

$$(a+b)^5 = a^5 + 5a^4b + 10a^3b^2 + 10a^2b^3 + 5ab^4 + b^5. \tag{3.6}$$

Example 3.1

In a special case when $a = x$ and $b = 1$, we have

$$\begin{aligned}(a+b)^5 = (x+1)^5 &= x^5 + 5x^4 \times 1 + 10x^3 \times 1^2 + 10x^2 \times 1^3 + 5x \times 1^4 + 1^5\\ &= x^5 + 5x^4 + 10x^3 + 10x^2 + 5x + 1.\end{aligned}$$

Similarly, for $a = x$ and $b = -1$, we have

$$(x-1)^5 = x^5 + 5x^4 \times (-1) + 10x^3 \times (-1)^2 + 10x^2 \times (-1)^3 + 5x \times (-1)^4 + (-1)^5$$
$$= x^5 - 5x^4 + 10x^3 - 10x^2 + 5x - 1.$$

These expressions become longer and longer, and now the question is whether a general expression exists for $(a+b)^n$ for any integer $n > 0$? The answer is yes, and this is the binomial theorem.

3.2 FACTORIALS

Before we introduce the binomial theorem, we have to introduce some other relevant concepts such as factorials.

For any positive integer n, the factorial n (or n factorial), denoted by $n!$, is the product of all the n natural numbers from 1 to n. That is

$$n! = n \times (n-1) \times (n-2) \times \ldots \times 1. \tag{3.7}$$

For example, we have $1! = 1$, $2! = 2 \times 1 = 2$, $3! = 3 \times 2 \times 1 = 6$, and $5! = 5 \times 4 \times 3 \times 2 \times 1 = 120$. Similarly, $10! = 10 \times 9 \times \ldots \times 2 \times 1 = 3628800$. From the above definition, we can easily obtain a recursive formula

$$(n+1)! = n! \times (n+1), \qquad (n = 1, 2, 3, \ldots). \tag{3.8}$$

In fact, the above equation is still valid for $n = 0$. Now you may wonder what is the value of $0!$? For consistency and simplicity in writing mathematical expressions, it requires that we define $0! = 1$. In fact, it is also possible to define the factorial of negative integers, but it involves more difficult concepts such as Γ function and Γ integrals. Anyway, factorials of negative numbers such as $(-5)!$ are rarely used in sciences and engineering, so we do not need to study them further.

The combinatorial coefficient or binomial coefficient is often written in the form

$$\binom{n}{r} \equiv {}^nC_r \equiv \frac{n!}{r!(n-r)!}, \tag{3.9}$$

where $n, r \geq 0$ are integers. Here the symbol '$\equiv$' means 'it is defined as' or 'exactly the same as'. In some literature, the notation ${}^nC_r$ is also widely used. To write the expression explicitly, we have

$$\binom{n}{r} = \frac{n(n-1)(n-2)\ldots(n-r+1)}{1 \times 2 \times \ldots \times (r-1) \times r}. \tag{3.10}$$

Example 3.2

For example, we can calculate $\binom{10}{5}$ by using either

$$\binom{10}{5} = \frac{10 \times 9 \times 8 \times 7 \times 6}{1 \times 2 \times 3 \times 4 \times 5} = \frac{30240}{120} = 252,$$

or

$$\binom{10}{5} = \frac{10!}{5!(10-5)!} = \frac{3628800}{120 \times 120} = 252.$$

This number also represents the number of choices or combinations of choosing (for example) 5 among 10 candidates to form a team.

Special cases are when $r = 0$ and $r = n$, we have $\binom{n}{0} = 1$, and $\binom{n}{n} = 1$. From the definition of a factorial and (3.8), we have

$$\binom{n}{m+1} = \frac{n!}{(m+1)!(n-m-1)!}$$

$$= \frac{n! \times (n-m)}{m! \times (m+1) \times \underbrace{(n-m-1)! \times (n-m)}_{=(n-m)!}}$$

$$= \frac{n!}{m!(n-m)!} \times \frac{(n-m)}{m+1} = \frac{(n-m)}{(m+1)} \binom{n}{m}, \tag{3.11}$$

which is valid for $m = 0, 1, 2, ...$ and $n \geq m$.

Example 3.3

Since we already know that $\binom{10}{5} = 252$, we can calculate $\binom{10}{6}$ easily. We know $n = 10$ and $m = 5$, and we have

$$\binom{10}{6} = \frac{(n-m)}{(m+1)} \binom{n}{m} = \frac{(10-5)}{(5+1)} \binom{10}{5} = \frac{5}{6} \times 252 = 210.$$

Similarly, we have

$$\binom{10}{7} = \frac{(10-6)}{(6+1)} \binom{10}{6} = \frac{4}{7} \times 210 = 120.$$

3.3 BINOMIAL THEOREM AND PASCAL'S TRIANGLE

Now we can state the binomial theorem. For any natural number n, the binomial expansion can be written as

$$(a+b)^n = \binom{n}{0} a^n + \binom{n}{1} a^{n-1} b + \binom{n}{2} a^{n-2} b^2 + ... + \binom{n}{n-1} a b^{n-1} + \binom{n}{n} b^n.$$

Example 3.4

In the case when $n = 5$, the binomial theorem gives

$$(a+b)^5 = \binom{5}{0} a^5 + \binom{5}{1} a^4 b + \binom{5}{2} a^3 b^2 + \binom{5}{3} a^2 b^4 + \binom{5}{4} a b^4 + \binom{5}{5} b^5.$$

It is straightforward to verify that

$$\binom{5}{0} = 1, \binom{5}{1} = 5, \binom{5}{2} = 10, \binom{5}{3} = 10, \binom{5}{4} = 5, \binom{5}{5} = 1,$$

and we have

$$(a+b)^5 = a^5 + 5a^4 b + 10a^3 b^2 + 10a^2 b^3 + 5ab^4 + b^5.$$

In this example, it is worth pointing out that there is some symmetry in terms of the coefficients. We can see that

$$\binom{5}{0} = \binom{5}{5} = 1, \quad \binom{5}{1} = \binom{5}{4} = 5, \quad \binom{5}{2} = \binom{5}{3} = 10. \tag{3.12}$$

In fact, such symmetry of coefficients is universal, and the coefficients form the famous pattern – Pascal's triangle.

$$\begin{array}{ccccccccccc} & & & & & 1 & & & & & \\ & & & & 1 & & 1 & & & & \\ & & & 1 & & 2 & & 1 & & & \\ & & 1 & & 3 & & 3 & & 1 & & \\ & 1 & & \underbrace{4 \quad 6} & & & & \underbrace{4 \quad 1} & & & \\ 1 & & 5 & & 10 & & 10 & & 5 & & 1 \\ & & & & & ... ... & & & & & \end{array}$$

The coefficients on the extreme left (and right) are always 1, and the coefficient of the next row is the sum of the two coefficients on its 'shoulders' in the previous row. For example, 10 in the 5th row is the sum of 4 and 6 in the 4th row, so is $5 = 4 + 1$.

In general, we have the recursive relationship

$$\binom{n}{r} = \binom{n-1}{r-1} + \binom{n-1}{r}, \qquad (n = 1, 2, 3, ...). \tag{3.13}$$

Example 3.5

For the special case of $n = 4$ and $r = 2$, we have

$$\binom{4}{2} = \binom{4-1}{2-1} + \binom{4-1}{2} = \binom{3}{1} + \binom{3}{2}.$$

Since $\binom{3}{1} = 3$ and $\binom{3}{2} = 3$, we have

$$\binom{4}{2} = 3 + 3 = 6.$$

We now use (3.11) to prove the above relationship. We know that

$$\binom{n-1}{r} = \binom{n-1}{(r-1)+1} = \frac{(n-1)-(r-1)}{(r-1)+1}\binom{n-1}{r-1} = \frac{n-r}{r}\binom{n-1}{r-1}, \tag{3.14}$$

we then have

$$\begin{aligned}\binom{n-1}{r-1} + \binom{n-1}{r} &= \binom{n-1}{r-1} + \frac{(n-r)}{r}\binom{n-1}{r-1} \\ &= \frac{r+(n-r)}{r}\binom{n-1}{r-1} = \frac{n}{r}\binom{n-1}{r-1} \\ &= \frac{n}{r} \times \frac{(n-1)\times(n-2)\times ... \times((n-1)-(r-1)+1)}{1 \times 2 \times ... \times (r-1)} \\ &= \frac{n \times (n-1) \times ... \times (n-r+1)}{1 \times 2 \times ... \times (r-1) \times r} = \binom{n}{r}.\end{aligned} \tag{3.15}$$

This is relatively abstract. Now let us use some actual numbers to see the relevance of this relationship.

Example 3.6

For the case of $n = 5$ and $r = 4$, we have

$$\binom{n}{r} = \binom{5}{4} = \frac{5!}{4!(5-1)!} = \frac{120}{24 \times 1} = 5,$$

$$\binom{n-1}{r-1} = \binom{4}{3} = \frac{4!}{3!(4-3)!} = \frac{24}{6 \times 1} = 4,$$

$$\binom{n-1}{r} = \binom{4}{4} = \frac{4!}{4!(4-4)!} = \frac{24}{24 \times 1} = 1,$$

so indeed the following holds

$$\binom{n}{r} = 5 = \binom{n-1}{r-1} + \binom{n-1}{r} = 4 + 1 = 5.$$

An interesting issue here is that you may wonder what happens if we extend the integer n to fractions and even negative numbers, say $(a+b)^{-1/2}$? This will lead to the generalized binomial theorems which involve infinite series, which will be discussed briefly in later chapters. The above formula provides a way to construct the coefficients from $n = 1$. In essence, this formula describes a sequence, which is the main topics of the next chapter.

EXERCISES

3.1. Calculate the values of the following expressions:

- $7!/6!$
- $5! - 4! - 3! - 2! - 1!$
- $\binom{5}{2} - \binom{4}{2} - \binom{3}{2}$
- $(1! - 0!)^2 + 2!$

3.2. Show that $\binom{6}{4} = \binom{5}{3} + \binom{5}{4}$.

3.3. Show that $(x+1)^3 - (x-1)^3 = 6x^2 + 2$.

Hint. Expand the left-hand side.

3.4. If two positive integers $n, k < 5$ such that $\binom{n}{k} = 6$, what are the possible values of n and k? Is it true that $n = 4$ and $k = 2$?

3.5. Show that

$$\binom{n}{0} + \binom{n}{1} + \binom{n}{2} + \ldots + \binom{n}{n-1} + \binom{n}{n} = 2^n,$$

for any positive integer n.

Hint. Set $a = b = 1$ in the binomial theorem.

Chapter 4

Sequences

Chapter Points

- Sequences will be discussed in detail, including both arithmetic progression and geometric progression.
- The difference equation for Fibonacci sequences and its solution techniques will be explained in detail. The link with the golden ratio will also be highlighted.
- Show the ways of calculating the sums of both arithmetic and geometric series, especially the sum of infinite geometrical series.

4.1 SIMPLE SEQUENCES

The simplest sequence is probably the natural numbers

$$1, \ 2, \ 3, \ 4, \ 5, \ ... \tag{4.1}$$

where we know that the next numbers should be 6, 7 and so on and so forth. Each isolated number is called a term. So 1 is the first term, and 5 is the 5th term. Mathematically speaking, a sequence is a row of numbers that obey certain rules, and such rules can be either a formula, a recursive relationship (or equation) between consecutive terms, or any other deterministic or inductive relationship.

In general, we have a sequence

$$a_1, \ a_2, \ a_3, \ ..., \ a_n, ... \tag{4.2}$$

where a_1 is the first term and the nth term is a_n. It is worth pointing out that it is sometimes more convenient to write the sequence as $a_0, a_1, a_2, a_3, ..., a_n, ...$, depending on the context. In this case, we have to be careful to name the term. The first term should be a_0 and the nth term is a_{n-1}.

Example 4.1

All positive even integers

$$2, 4, 6, 8, 10, ...,$$

form a simple sequence, which can be expressed as

$$a_n = 2n, \quad (n = 1, 2, 3, ...).$$

In addition, all the positive odd integers

$$1, 3, 5, 7, 9, ...,$$

can be written as

$$b_n = 2n - 1, \quad (n = 1, 2, 3, ...).$$

It is interesting to note that the differences between any two consecutive terms for both sequences are 2, that is $a_2 - a_1 = a_3 - a_2 = a_4 - a_3 = 2$ and $b_2 - b_1 = b_3 - b_2 = b_4 - b_3 = 2$. In general, we have

$$a_{n+1} - a_n = 2, \quad b_{n+1} - b_n = 2,$$

or

$$a_{n+1} = a_n + 2, \quad b_{n+1} = b_n + 2.$$

The terms in the above example are all linear in terms of n. However, the term (nth term) can in general be nonlinear. For example, the sequence

$$2, 5, 10, 17, 26, ..., \tag{4.3}$$

can be described by the formula for the nth term

$$a_n = n^2 + 1, \quad (n = 1, 2, 3, ...).$$

Example 4.2

The sequence $-1, 1, 3, 5, 7, ...$ corresponds to the formula $a_n = 2n - 3$ for $(n = 1, 2, 3, ...)$. It can also be described by the recursive formula

$$a_{n+1} = a_n + 2, \quad a_1 = -1, \quad (n = 1, 2, 3, ...).$$

There is a constant difference between consecutive terms a_n and a_{n+1}.

On the other hand, the sequence

$$2, 1, \frac{1}{2}, \frac{1}{4}, \frac{1}{8}, ...$$

has a constant ratio $r = 1/2$. The general formula is

$$a_n = 2r^{n-1}, \quad (n = 1, 2, 3, ...),$$

which is also equivalent to the recursive relationship

$$a_{n+1} = a_n r, \quad a_1 = 2, \ (n = 1, 2, 3, ...).$$

There are two special classes of widely-used sequences. They are arithmetic sequences and geometric sequences. In the previous example, the formula $a_n = 2n - 3$ $(n = 1, 2, 3, ...)$ is an arithmetic sequence because there is a common difference $d = 2$ between a_{n+1} and a_n. On the other hand, the formula $a_n = 2r^{n-1}$ describes a geometric sequence with a common ratio r.

4.1.1 Arithmetic Sequence

In general, an arithmetic sequence, also called arithmetic progression, can be written as

$$a_n = a + (n - 1)d, \tag{4.4}$$

where a is a known constant (often the first term), and d is the common difference (also called the step). We often assume $d \neq 0$, otherwise we have a trivial sequence with every term being the same number.

Example 4.3

For example, the formula $a_n = 2n - 3$ for the sequence $-1, 1, 3, 5, ...$ can be written as

$$a_n = 2n - 3 = -1 + 2(n - 1), \tag{4.5}$$

which gives $a = -1$ and $d = 2$.

Let us try to figure out the formula for the following sequence:

$$2, \frac{7}{2}, 5, \frac{13}{2}, 8, \frac{19}{2}, 11, ...$$

First, it is easy to check that the difference between the first two terms is $7/2-2=3/2$. This difference is the same between any two consecutive terms. Thus, we can conclude that $d=3/2$.

We also know that the first term is 2, so we can set $a=2$. From the above general formula (4.4), we have

$$a_n = \frac{3}{2}(n-1)+2, \quad (n=1,2,3,4,...).$$

4.1.2 Geometric Sequence

On the other hand, a geometric sequence, or geometric progression, is defined by $a_1 = a$ and $a_{n+1} = ra_n (n = 1, 2, 3, ...)$, where the common ratio $r \neq 0$ or 1. This definition gives

$$a_1 = a,\ a_2 = ar,\ a_3 = ar^2,\ ...,\ a_n = ar^{n-1},\ ..., \tag{4.6}$$

whose nth term is simply

$$a_n = ar^{n-1}, \quad (n=1,2,3,...). \tag{4.7}$$

Example 4.4

Let us try to figure out the formula for the following sequence:

$$2, \frac{2}{3}, \frac{2}{9}, \frac{2}{27}, \frac{2}{81}, ...$$

It is easy to see that there is a factor of 2 in each term. Thus, we can be sure that $a=2$. The ratio between the second term to the first term is

$$r = \frac{2}{3}/2 = \frac{1}{3}.$$

Similarly, the ratio of the third term to the second term is

$$r = \frac{2}{9}/\frac{2}{3} = \frac{1}{3},$$

which is the same ratio for all the other two consecutive terms. Thus, from the above general formula, we can write

$$a_n = ar^n = 2\Big(\frac{1}{3}\Big)^n, \quad (n=1,2,3,...).$$

But we should check if this formula can indeed produce the original sequence. For $n=1$, we get $a_1 = 2/3$, which is the second term, not the first term. Thus, there is something wrong with the formula. If we look carefully at the sequence, we should start with $n=0$ instead of $n=1$. With this minor modification, we have a correct formula

$$a_n = 2\Big(\frac{1}{3}\Big)^n, \quad (n=0,1,2,...).$$

Alternatively, if we insist to start with $n=1$, we can modify the formula as

$$a_n = 2\Big(\frac{1}{3}\Big)^{n-1}, \quad (n=1,2,3,...).$$

Obviously, r can be negative if necessary. For example, for $a=1$ and $r=-\frac{2}{3}$, we have

$$1, -\frac{2}{3}, \frac{4}{9}, -\frac{9}{27}, \frac{16}{81}, ..., \tag{4.8}$$

whose nth term is

$$a_n = (-\frac{2}{3})^{n-1}, \quad (n=1,2,3,...). \tag{4.9}$$

4.2 FIBONACCI SEQUENCE

Now the question is which form of formula we should use? In general, we should use the formula for the nth term because we can easily calculate the actual number for each term. If we try to study the relationship between terms (often among consecutive terms, though not always), we should use the recursive relationship. It is worth pointing out that the relationship can be complicated in some cases. Let us look at a classic example.

Example 4.5

A famous classic sequence is the Fibonacci sequence

$$0,\ 1,\ 1,\ 2,\ 3,\ 5,\ 8,\ 13,\ 21, ...$$

whose recursive relationship (or equation) is

$$a_{n+1} = a_n + a_{n-1}, \qquad (n = 1, 2, 3, ...),$$

with $a_0 = 0$ and $a_1 = 1$. The formula for the nth term is not simple at all, and its construction often involves the solution to a recurrence equation.

In order to get the general expression of nth term for the Fibonacci sequence, we have to solve its corresponding recurrence equation

$$a_{n+1} = a_n + a_{n-1}. \tag{4.10}$$

This is a second-order equation because it involves the relationship among three consecutive terms. Mathematically speaking, this equation is a linear recursive equation or linear recurrent equation.

First, we can assume that the general term consists of a generic factor $a_n = B\lambda^n$ where λ is the unknown value to be determined, and B is an arbitrary constant (readers can refer to more advanced textbooks such as Yang XS, *Mathematical Modelling for Earth Sciences*, Dunedin Academic, 2008, and its bibliography).

Substituting $a_n = B\lambda^n$, equation (4.10) becomes

$$B\lambda^{n+1} = B\lambda^n + B\lambda^{n-1}. \tag{4.11}$$

By dividing $B\lambda^{n-1}$ (assuming it is not zero, otherwise the solution is trivial, that is zero), we have a characteristic equation

$$\lambda^2 - \lambda - 1 = 0. \tag{4.12}$$

This is a quadratic equation for λ, and its solution can be obtained easily. We have

$$\lambda_1 = \frac{1+\sqrt{5}}{2}, \qquad \lambda_2 = \frac{1-\sqrt{5}}{2}. \tag{4.13}$$

It is obvious to see that λ_1 is the golden ratio ϕ, while λ_2 is the golden ratio conjugate Φ as discussed in Chapter 1.

The trick here is again to assume that the generic form of the solution is a linear combination of the two possible basic solutions

$$a_n = \alpha\lambda_1^n + \beta\lambda_2^n = \alpha\Big(\frac{1+\sqrt{5}}{2}\Big)^2 + \beta\Big(\frac{1-\sqrt{5}}{2}\Big)^n, \tag{4.14}$$

where α and β are undetermined coefficients. Now we have to determine α and β using the initial two terms $a_0 = 0$ and $a_1 = 1$.

For $n = 0$ (the first term), we have

$$a_0 = 0 = \alpha + \beta, \tag{4.15}$$

which means $\beta = -\alpha$. For $n = 1$ (the second term), we have

$$a_1 = 1 = \alpha\Big(\frac{1+\sqrt{5}}{2}\Big) + \beta\Big(\frac{1-\sqrt{5}}{2}\Big) = \alpha\Big(\frac{1+\sqrt{5}}{2}\Big) - \alpha\Big(\frac{1-\sqrt{5}}{2}\Big). \tag{4.16}$$

Its solution is simply $\alpha = 1/\sqrt{5}$. Subsequently, we have $\beta = -\alpha = -1/\sqrt{5}$. Therefore, the general formula for the nth term in the Fibonacci sequence is

$$a_n = \frac{1}{\sqrt{5}}\Big[\Big(\frac{1+\sqrt{5}}{2}\Big)^n - \Big(\frac{1-\sqrt{5}}{2}\Big)^n\Big]. \tag{4.17}$$

This formula is almost impossible to guess from the relationship of the sequence itself. It leaves us with the task of verifying that this formula indeed provides the right terms for the Fibonacci sequence.

Example 4.6

For the general formula (4.17), we can now check if the first three terms are indeed 0, 1, 1. For $n = 0$, we have $[(1+\sqrt{5})/2]^0 = 1$ and $[(1-\sqrt{5})/2]^0 = 1$, so we get

$$a_0 = \frac{1}{\sqrt{5}}[1-1] = 0.$$

For $n = 1$, we have

$$a_1 = \frac{1}{\sqrt{5}}\Big[\frac{1+\sqrt{5}}{2} - \frac{1-\sqrt{5}}{2}\Big] = \frac{1}{\sqrt{5}}\Big[\frac{2\sqrt{5}}{2}\Big] = 1.$$

For $n = 2$, we have

$$\begin{aligned} a_2 &= \frac{1}{\sqrt{5}}\Big[\Big(\frac{1+\sqrt{5}}{2}\Big)^2 - \Big(\frac{1-\sqrt{5}}{2}\Big)^2\Big] \\ &= \frac{1}{\sqrt{5}}\Big[\frac{1+2\sqrt{5}+(\sqrt{5})^2}{4} - \frac{1-2\sqrt{5}+(\sqrt{5})^2}{4}\Big] \\ &= \frac{1}{\sqrt{5}}\Big[\frac{6+2\sqrt{5}}{4} - \frac{6-2\sqrt{5}}{4}\Big] = \frac{1}{\sqrt{5}}[\frac{4\sqrt{5}}{4}] = \frac{1}{\sqrt{5}}[\sqrt{5}] = 1. \end{aligned}$$

4.3 SUM OF A SERIES

Sometimes, we have to calculate the sum of all the terms of a sequence. For example, when German mathematician Carl F. Gauss was a child, he was able to add all the natural numbers from 1 to 100 in an amazingly short time

$$1 + 2 + 3 ... + 99 + 100 = 5050. \tag{4.18}$$

This is because the sum of all the numbers from 1 to n is

$$1 + 2 + ... + n = \frac{n(n+1)}{2}. \tag{4.19}$$

In this case, we are in fact dealing with the sum of all the terms of a sequence. Traditionally, we often refer to a sequence as a series when we are dealing with the sum of all the terms (usually, a finite number of terms).

Conventionally, we use the sigma notation, $\sum$ from the Greek capital letter S, to express the sum of all the terms in a series

$$\sum_{i=1}^{n} a_i = a_1 + a_2 + ... + a_n, \tag{4.20}$$

where a_i denotes the general term, and the sum starts from $i = 1$ until the last term $i = n$.

For the sum of an arithmetic series with the ith term of $a_i = a + (i-1)d$, we have

$$\begin{aligned} S = \sum_{i=1}^{n} a_i = a_1 + a_2 + ... + a_n &= \underbrace{a}_{=a_1} + (a+d) + (a+2d) + ... + \underbrace{(a+(n-1)d)}_{=a_n} \\ &= a_1 + (a_1+d) + (a_1+2d) + ... + (a_n - d) + a_n. \end{aligned} \tag{4.21}$$

We can also write this sum from the back (last term a_n) to the front (first term $a_1 = a$), and we have

$$\begin{aligned} S &= a_n + (a_n - d) + (a_n - 2d) + ... + (a_n - (n-2)d) + a \\ &= a_n + (a_n - d) + (a_n - 2d) + ... + (a_1 + 2d) + (a_1 + d) + a_1. \end{aligned} \tag{4.22}$$

Adding these above two expressions and grouping the similar terms, we have

$$\begin{aligned} 2S &= (a_1 + a_n) + (a_1 + d + a_n - d) + ... + (a_n - d + a_1 + d) + (a_n + a_1) \\ &= (a_1 + a_n) + (a_1 + a_n) + ... + (a_1 + a_n) = n(a_1 + a_n). \end{aligned} \tag{4.23}$$

Therefore, the sum S is

$$S = \frac{n}{2}(a_1 + a_n). \tag{4.24}$$

As $a_1 = a$ and $a_n = a + (n-1)d$, we also have

$$S = \frac{n}{2}[2a + (n-1)d] = n[a + \frac{(n-1)}{2}d]. \tag{4.25}$$

Example 4.7

In the case of the sum of all natural numbers $1+2+3+...+n$, we have $a = 1$ and $d = 1$, so the sum is

$$S = 1+2+3+...+n = \frac{n}{2}[2 \times 1 + (n-1) \times 1] = \frac{n(n+1)}{2}.$$

For the sum of $S = 1+3+5+...+(2n-1)$, we have $a = 1$, $d = 2$ and $a_n = 2n-1$. Now we have

$$S = 1+3+5+...+(2n-1) = \frac{n}{2}[1 + (2n-1)] = n^2.$$

For the sum of a geometric series, we have

$$S = \sum_{i=1}^{n} a_i = a + \overbrace{ar + ar^2 + ... + ar^{n-2} + ar^{n-1}}. \tag{4.26}$$

Multiplying the above equation by r on both sides, we have

$$rS = \overbrace{ar + ar^2 + ar^3 + ... + ar^{n-2} + ar^{n-1}} + ar^n. \tag{4.27}$$

Now if we subtract the first from the second equation, we have

$$rS - S = ar^n - a, \tag{4.28}$$

or

$$(r-1)S = a(r^n - 1), \tag{4.29}$$

which leads to

$$S = \frac{a(r^n - 1)}{(r-1)} = \frac{a(1 - r^n)}{(1-r)}. \tag{4.30}$$

It is worth pointing out that it may cause problems when $r = 1$ becomes the division by zero. In fact, when $r = 1$, all the terms in the series are the same (as the first term a), so the sum is now $a + a + ... + a = na$.

Another special case is when $r = -1$; then we have an oscillatory series $a, -a, +a, -a, ...$, and the sum will be $S = 0$ when n is even, or $S = a$ when n is odd. Both $r = 1$ and $r = -1$ are thus trivial and we will not discuss them any further.

Example 4.8

To calculate the sum of a geometric series

$$1+\frac{1}{2}+\frac{1}{4}+\frac{1}{8}+\frac{1}{16}+\frac{1}{32}+\frac{1}{64}+\cdots+\frac{1}{1024},$$

we first have to figure out the ratio. It is easy to check that the ratio is $r=1/2$. In fact, the above sum of the series can be written as

$$1+\frac{1}{2}+(\frac{1}{2})^2+(\frac{1}{2})^3+(\frac{1}{2})^4+(\frac{1}{2})^5+\cdots(\frac{1}{2})^{10}.$$

Thus, we have $a=1$ and $r=1/2$. However, $n=11$ (not 10) because there are 11 terms in the sum. So the sum can be calculated by

$$S=\frac{a(1-r^n)}{1-r}=\frac{1\times[1-(\frac{1}{2})^{11}]}{1-\frac{1}{2}}=\frac{[1-\frac{1}{2048}]}{\frac{1}{2}}=2\times\frac{2048-1}{2048}=\frac{2\times 2047}{2\times 1024}=\frac{2047}{1024}.$$

Where $r<0$ is negative, we can do the same, but care should be taken to avoid any mistake. Let us look at an example.

Example 4.9

For the sum of the geometric series

$$S=1-\frac{2}{3}+\frac{4}{9}-\frac{8}{27}+...+\frac{(-1)^{n-1}2^{n-1}}{3^{n-1}},$$

which can be written as

$$S=1+(\frac{-2}{3})+(\frac{-2}{3})^2+...+(\frac{-2}{3})^{n-1}.$$

Thus, we have $a=1$ and $r=-2/3$, so we have

$$S=\frac{1\times[1-(\frac{-2}{3})^n]}{1-(-\frac{2}{3})}=\frac{1-(\frac{-2}{3})^n}{\frac{5}{3}}.$$

For example, when $n=10$, we have

$$S=1-\frac{2}{3}+\frac{4}{9}-\frac{9}{27}+...-\frac{512}{19683}=\frac{1-(-2/3)^{10}}{5/3}=\frac{11605}{19683}.$$

Because the addition of any two numbers is interchangeable, that is $a+b=b+a$, it is easy to verify that the sum of two series follows the addition rule

$$\sum_{i=1}^{n}(a_i+b_i)=(a_1+b_1)+(a_2+b_2)+...+(a_n+b_n)$$
$$=(a_1+...+a_n)+(b_1+...+b_n)=\sum_{i=1}^{n}a_i+\sum_{i=1}^{n}b_i. \quad (4.31)$$

Similarly, for any non-zero constant β, we have

$$\sum_{i=1}^{n}\beta a_i=\beta a_1+\beta a_2+...+\beta a_n=\beta(a_1+...+a_n)=\beta\sum_{i=1}^{n}a_i. \quad (4.32)$$

These properties become useful in calculating the sum of complicated series. Let us look at an example.

Example 4.10

We now try to calculate the sum of the series

$$S = 5 + 10 + 18 + 32 + 58 + 108 + ... + 1556.$$

The general formula to generate this sequence is

$$a_n = 2n + 3 \times 2^{n-1}.$$

The sum can be written as

$$S = \sum_{i=1}^{n} a_i = \sum_{i=1}^{n} (2i + 3 \times 2^{i-1}) = 2\sum_{i=1}^{n} i + 3\sum_{i=1}^{n} 2^{i-1}.$$

The first part is twice the sum of an arithmetic series, while the second part is the sum of a geometric series multiplied by a factor of 3. Using the sum formula for arithmetic and geometrical series, we have the sum

$$S = 2 \times \frac{n(n+1)}{2} + 3 \times \frac{1 \times (2^n - 1)}{(2-1)} = n(n+1) + 3(2^n - 1).$$

In the case of $n = 10$, we have

$$S = 5 + 10 + 18 + 32 + ... + 1556 = 10 \times (10 + 1) + 3 \times (2^{10} - 1) = 3179.$$

4.4 INFINITE SERIES

For an arithmetic series, the sum $S = n[a + \frac{n-1}{2}d]$ will increase as n increases for $d > 0$, and this increase will be unbounded (to any possible large number).

On the other hand, the sum of a geometrical series

$$S = \frac{a(1 - r^n)}{1 - r}, \tag{4.33}$$

will have interesting properties as n increases to very large values. For $r > 1$, r^n will increase indefinitely as n tends to infinity, and the magnitude or the absolute value $|S|$ of the sum S will also increase indefinitely. In this case, we say the sum diverges to infinity. Similarly, for $r < -1$, r^n also increase indefinitely (to large negative when n is odd, or to large positive when n is even). As mentioned earlier, both $r = 1$ and $r = -1$ are trivial cases. So we are now only interested in the case $-1 < r < 1$, or $|r| < 1$ where $|r|$ is the absolute value of r. For example, $|-0.5| = 0.5$ and $|0.5| = 0.5$.

As n increases and tends to infinity, we use the notation $n \to \infty$. In the case of $-1 < r < 1$, the interesting feature is that r^n becomes smaller and smaller, approaching zero. We denote this by $r^n \to 0$ as $n \to \infty$. As the number of terms in the series increases to infinity, we call the series an infinite series. Therefore, the sum of an infinite geometrical series with $|r| < 1$ will tend to a finite value (the limit), and we will use the following notation

$$\begin{aligned} S &= a + ar + ar^2 + ... + ar^{i-1} + ... \\ &= \lim_{n\to\infty} \sum_{i=1}^{n} a_i = \lim_{n\to\infty} \frac{a(1 - r^n)}{1 - r} = \frac{a}{1 - r}. \end{aligned} \tag{4.34}$$

Let us look at some applications.

Example 4.11

The classic example is the sum of

$$S = 1 + \frac{1}{2} + \frac{1}{4} + \frac{1}{8} + \frac{1}{16} + ...,$$

where $a = 1$ and $r = 1/2$. So we have

$$S = \frac{a}{1-r} = \frac{1}{1-1/2} = 2.$$

For the infinite series

$$S = 3 - 3 \cdot \frac{1}{5} + 3 \cdot \frac{1}{25} - 3 \cdot \frac{1}{125} + ...,$$

we have to change the series into

$$S = 3 \times [1 - \frac{1}{5^1} + \frac{1}{5^2} - \frac{1}{5^3} + ...] = 3 \sum_{i=1}^{n} (\frac{-1}{5})^{i-1} = 3 \times \frac{1}{1-(\frac{-1}{5})} = \frac{5}{2}.$$

Series and infinite series will become useful when approximating a complex function using Taylor series which will be introduced in later chapters concerning advanced calculus.

EXERCISES

4.1. Find the general terms for the following sequences:

- $1, 8, 15, 22, 29, 36, 43, 50, \cdots$.
- $5, 6, 9, 14, 21, 30, 41, 54, \cdots$.
- $2, 6, 12, 20, 30, 42, 56, 72, \cdots$.
- $4, 2, 1, \frac{1}{2}, \frac{1}{4}, \frac{1}{8}, \frac{1}{16}, \cdots$.

Hint. They partially contain $7(n-1)$, $(n-1)^2$, n^2, and $1/2^n$, respectively.

4.2. The famous mathematician Euler once claimed that $n^2 + n + 41$ will give all prime numbers for $n = 0, 1, 2, 3, \cdots$. Verify if it is always true. Will $n^2 - n + 41$ also give all prime numbers?

Hint. Check up to $n = 40$.

4.3. Find the sum of the following series:

- $\frac{7}{10} + \frac{7}{100} + \frac{7}{1000} + \frac{7}{10000} + \frac{7}{100000} + \cdots$.
- $1 + \frac{1}{4} + \frac{1}{16} + \frac{1}{64} + \frac{1}{256} + \cdots$.
- $1 + 0.09 + 0.0009 + 0.000009 + 0.00000009 + \cdots$.
- $12 + 8 + \frac{16}{3} + \frac{32}{9} + \cdots$.
- $64 - 32 + 16 - 8 + 4 - 2 + 1 + \cdots$.

Hint. Answers: $9/7, 4/3, 12/11, 36, 128/3$.

4.4. Discuss and find the sum of the series:

$$1 + (\frac{1}{2} + \frac{1}{3}) + (\frac{1}{4} + \frac{1}{9}) + (\frac{1}{8} + \frac{1}{27}) + (\frac{1}{16} + \frac{1}{81}) + \cdots,$$

and

$$1 + (\frac{1}{2} - \frac{1}{3}) + (\frac{1}{4} - \frac{1}{9}) + (\frac{1}{8} - \frac{1}{27}) + (\frac{1}{16} - \frac{1}{81}) + \cdots.$$

Hint. Try to write into the sum of two independent series, then find the sum for each series. Answers are $2\frac{1}{2}$ and $1\frac{1}{2}$, respectively.

Chapter 5

Exponentials and Logarithms

Chapter Points

- Exponential functions will be introduced, together with their links to logarithmic functions.
- Logarithmic functions are explained in depth using different bases. Natural logarithms are also explained in detail.
- Mathematical properties of both exponential functions and logarithms are demonstrated using examples and the changes of logarithmic bases are also given briefly using examples.

When we discussed geometric sequences, we frequently used the nth term ar^{n-1} where $a \neq 0$ and the ratio r are given constants. By varying the positive integer n, we can calculate the value of any term in the sequence. If we extend the values of n from integers to real numbers x, r^x is in fact an exponential function.

5.1 EXPONENTIAL FUNCTION

An exponential function is in the generic form b^x

$$f(x) = b^x, \tag{5.1}$$

where $b > 0$ is a known or given real constant, called the base, while the independent variable x is called the exponent. In the context of real numbers, we have to limit our discussion here to $b > 0$. When $b < 0$, for some values of x, b^x will be meaningless. For example, if $b = -3$ and $x = 1/2$, then $(-3)^{1/2} = \sqrt{-3}$ would be meaningless. Similarly, say, when $b = 0$ and $x = -5$, then $b^x = 0^{-5} = 1/0^5$ would also be meaningless.

In addition, a trivial case is when $b = 1$, then $f(x) = 1^x = 1$ everywhere. It becomes a single constant, not a function in the standard sense. Three functions $f(x) = 2^x$, $(\frac{1}{2})^x$ and $(\frac{1}{4})^x$ are shown in Fig. 5.1 where all the functions will go though the points $(0, 1)$ where $x = 0$ and $f(0) = 1$. If $b > 0$, then $b^x > 0$ for all x. This is because if $x \geq 0$, $b^x \geq 1 > 0$, while $x < 0$, $b^x = b^{-|x|} = \frac{1}{b^{|x|}} > 0$. We can also see that the reflection in the y-axis of 2^x is $(1/2)^x$. In fact, both b^x and $(1/b)^x$ are the reflection of each other in the y-axis.

Example 5.1

Suppose you have £1000 to save in a bank for 5 years with a yearly interest of $r = 4\%$. What is the value you can expect to get after 5 years?

The amount of money you can get at the end of nth year is

$$A = 1000(1 + r)^n,$$

where r is the yearly interest rate and in the present case $r = 0.04$. So we have

$$A = 1000(1 + 0.04)^5 = 1216.65,$$

which is about £216 more than your initial saving. Obviously, if you save it much longer for 10 years with the same interest rate, you will get

$$1000(1 + 0.04)^{10} = 1480.24.$$

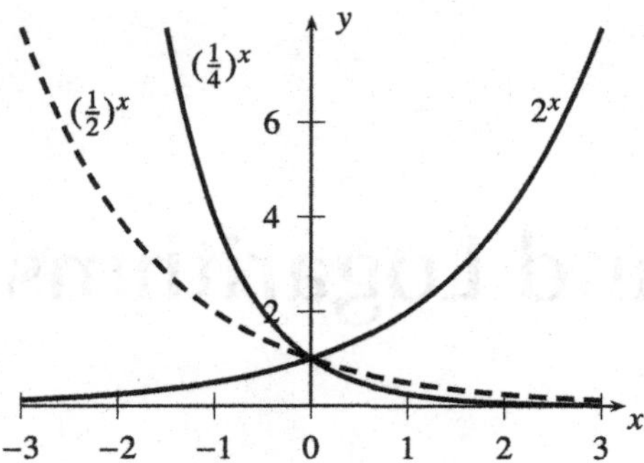

FIGURE 5.1 Exponential functions.

The above calculations have been based on the fixed interest rate. Imagine that the interest rate after 6 years changes to 7%, then the interest rate changes again to 3% for the final 2 years. In this case, we have

$$A = \underbrace{1000(1+0.04)^6}_{\text{first 6 years}} \cdot \underbrace{(1+0.07)^2}_{\text{7th and 8th years}} \cdot \underbrace{(1+0.03)^2}_{\text{final 2 years}}$$
$$= 1000 \cdot 1.2653 \cdot 1.1449 \cdot 1.0609 \approx 1536.89.$$

An interesting extension for $f(x) = a^x$ is that

$$f(-x) = a^{-x} = \frac{1}{a^x} = (\frac{1}{a})^x. \tag{5.2}$$

Therefore, the ratio of two exponential functions a^x and b^x is

$$\frac{a^x}{b^x} = (\frac{a}{b})^x. \tag{5.3}$$

Example 5.2

The numeric value of an exponential function can increase very quickly when $a > 1$. For example, when $a = 2$, we have

$$2^1 = 2, \ 2^{10} = 1024, \ 2^{20} = 1048576, \ 2^{40} = 1099511627776,$$
$$2^{100} = 1267650600228229401496703205376 \approx 1.267 \times 10^{30}.$$

As a well-known example is the so-called wheat grains on a chessboard. If one puts one grain on the first square, 2 on the second square, 4 on the third square, and so on and so forth. This essentially doubles until all the squares are filled. As there are 64 squares (labeled as 0, 1, 2, ..., 63), the total number of grains is

$$N = 2^0 + 2^1 + 2^2 + 2^3 + ... + 2^{63} = \sum_{i=0}^{63} 2^i$$
$$= 2^{64} - 1 = 18446744073709551615 \approx 1.8 \times 10^{19},$$

which is a huge number and it is much more than all the grains in the world.

But for the negative exponents, the values can also approach to zero very quickly. For example, we have

$$2^{-1} = \frac{1}{2} = 0.5, \quad 2^{-10} = \frac{1}{1024} = 0.0009765625,$$
$$2^{-40} \approx 0.90947 \times 10^{-12}, \ 2^{-100} \approx 7.8886 \times 10^{-31}.$$

5.2 LOGARITHM

Now suppose we know the value of $b > 0$ and the value of the function $y = b^x$, the question is if we can determine x given b and y. This is essentially to calculate the inverse of the exponential function so as to find x that satisfies

$$b^x = y. \tag{5.4}$$

The notation for such inverse is

$$x = \log_b y, \tag{5.5}$$

where b is again the base. That is to say, x is the logarithm of y to the base b. As discussed earlier, $y = b^x > 0$ is always positive for $b > 0$, the logarithm is only possible for $y > 0$ and $b > 0$. Since $b^0 = 1$, we have $0 = \log_b 1$. From $b^1 = b$, we have $1 = \log_b b$. In addition, we can also combine the above two equations, and we have

$$b^{\log_b y} = y. \tag{5.6}$$

Example 5.3

Claude Shannon's information theory laid the foundation for modern digital communications. The key ideas are Shannon's *bit* and entropy. In order to transmit a series of 0s and 1s, it is useful to know the information contents they contain. For example, with a simple 'yes' (1) or 'no' (0), each digit has a 50–50 chance to appear. That is, 1 may appear $p = 1/2$ of times, so is 0. Shannon's information is defined as

$$I = -\log_2 p.$$

Here, the base is 2.

If you ask your friend a question whether he or she will come to a party, the information content of a simple answer 'yes' (or 'no') is exactly

$$I = -\log_2 \frac{1}{2} = 1 \text{ (bit)}.$$

For simplicity, consider a message written in English without punctuations. We can also ignore the case difference (i.e., the same information whether capital letters or not), so there are 27 different characters in the message (26 letters plus a space). Then, the information content of each letter can be estimated by

$$I = -\log_2(1/27) \approx 4.75488 \text{ bits}.$$

In general, for a series of n characters with each individual character appearing with a probability p_i, the Shannon entropy is given by

$$H = -\sum_{i=1}^{n} p_i \log p_i. \tag{5.7}$$

We will explain this in more detail when we discuss probabilities in later chapters of this book.

It is worth pointing out that the input argument of a logarithm function cannot be zero or negative. That is, the domain of

$$y = \log_b(x), \quad b > 0, \tag{5.8}$$

is $x > 0$. For $x > b$, $y > 1$. When $x = 1$, then $y = \log_b(1) = 0$. In the case $0 < x < 1$, then $y < 0$. The range of y is thus $-\infty < y < +\infty$.

Example 5.4

For a function

$$y = \log_2(x + 1),$$

its domain can be derived from the condition $x + 1 > 0$. That is $x > -1$.

Similarly, for the following function

$$y = \log_b(x^2 - 4), \quad b > 0,$$

it requires that

$$x^2 - 4 > 0,$$

which leads to

$$x > 2, \quad \text{or } x < -2.$$

Thus, the domain of this function is $(-\infty, -2) \cup (2, +\infty)$.

Since, for any real numbers p, q and $b > 0$, we can define two new variables $u = b^p$, $v = b^q$, and

$$y = b^p b^q = uv = b^{p+q}. \tag{5.9}$$

From the definition of the logarithm, we now have

$$p + q = \log_b y = \log_b (uv). \tag{5.10}$$

Since $p = \log_b u$ and $q = \log_b v$, we finally obtain

$$\log_b (uv) = p + q = \log_b u + \log_b v. \tag{5.11}$$

This is the multiplication rule. If we replace q by $-q$ so that $b^p b^{-q} = b^p / b^q = u/v$, we can easily verify the division rule

$$\log_b \frac{u}{v} = \log_b u - \log_b v. \tag{5.12}$$

When $p = q$ (so $u = v$), we have $\log_b (u^2) = 2 \log_b u$ which can easily be extended to any x

$$\log_b u^x = x \log_b u. \tag{5.13}$$

A special case when $x = 1/n$, we have

$$\log_b u^{1/n} = \log_b \sqrt[n]{u} = \frac{1}{n} \log_b u. \tag{5.14}$$

As these rules are valid for any $b > 0$, it is sometimes convenient to simply write the above rule without explicitly stating the base b. This means that we more often use

$$\log(uv) = \log u + \log v. \tag{5.15}$$

Even though any base $b > 0$ is valid, some bases are convenient for calculating logarithms and thus more widely used than others. For historical reasons and because of the decimal systems, base $b = 10$ is widely used and we write $\log_{10} u$.

Another special base is the base e for natural or Napierian logarithms where

$$e = 2.7182818284..., \tag{5.16}$$

and in this case, we simply write the logarithm as ln, using a special notation to distinguish from the common logarithms (log)

$$\ln u \equiv \log_e u. \tag{5.17}$$

The reasons why $e = 2.71828...$ is a natural choice for the base are many, and readers can refer to some textbooks about the history of mathematics. One of the reasons is that its corresponding exponential function

$$y = e^x, \tag{5.18}$$

and its various forms such as $\exp[-x]$ or $\exp[-x^2]$ appear in a wide range of real-world applications.

It is worth pointing out that sometimes we use simple notations if no confusion arise. Like we use $\sin^2(x) = (\sin(x))^2 = (\sin x)^2$ in trigonometry, so we use here

$$\ln^2(x) = \ln^2 x = \Big(\ln(x)\Big)^2 = (\ln x)^2. \tag{5.19}$$

All these notations mean the same thing, so it is purely for convenience and simplicity.

Example 5.5

There is a rule of 70 (or rule of 69) about the calculations related to the interest rate. If the interest rate is r, how long does it take for an investment m or saving to double its value? The value V will increase with time T in the following way:

$$V = m(1+r)^T,$$

which gives

$$\frac{V}{m} = (1+r)^T.$$

If the value doubles (or $V/m = 2$), we have

$$2 = (1+r)^T.$$

Taking the logarithm of both sides, we have

$$\ln 2 = \ln(1+r)^T = T\ln(1+r).$$

So

$$T = \frac{\ln 2}{\ln(1+r)}.$$

When r is very small, we have $\ln(1+r) \approx r$ (see the Taylor series to be introduced in later chapters), and thus we have

$$T = \frac{\ln 2}{\ln(1+r)} \approx \frac{\ln 2}{r} = \frac{0.693147}{r} \approx \frac{0.69}{r}.$$

Since the interest rate is mostly expressed in percentage $R\%$, we have $r = R/100$, thus we finally have

$$T = \frac{0.69}{R/100} \approx \frac{0.69 \times 100}{R} \approx \frac{69}{R} \approx \frac{70}{R}.$$

This is the so-called rule of 69 or rule of 70.

For example, if the interest rate is 2% per year (or $r = 0.02 = 2\%$), then $T \approx 70/2 = 35$, which means it will take about 35 years to double the value. If r rise to 7% per year ($r = 0.07 = 7\%$), then $T \approx 70/7 = 10$ years. If we use the exact formula, we have

$$T = \frac{\ln 2}{\ln(1+0.07)} \approx \frac{0.693147}{0.0676586} \approx 10.24,$$

which is quite close to 10. Obviously, as r increases, the error will also increase.

Logarithms, especially the natural logarithms, can have a wider range of applications in science and engineering.

Half-life is a term used often in describing decay such as radiative materials and drug delivery. Most decays obey the so-called exponential law in the form

$$N(t) = N_0 e^{-\lambda t} = N_0 \exp[-\lambda t],$$

where N_0 is the initial quantity of the substance and $N(t)$ the remaining quantity of the same material at time t. The parameter λ is the so-called decay constant, and $1/\lambda$ represents a time constant for this system.

The so-called half-life $t_{1/2}$ is the time taken for the substance to reduce to half of its initial quantity, that is

$$\frac{1}{2}=\frac{N(t_{1/2})}{N_0}=\exp[-\lambda t_{1/2}].$$

By taking the logarithm of both sides, we have

$$\ln(1/2)=\ln\{\exp[-\lambda t_{1/2}]\}=-\lambda t_{1/2},$$

which gives

$$\lambda=\frac{\ln 2}{t_{1/2}},$$

or

$$t_{1/2}=\frac{\ln 2}{\lambda}.$$

For example, the half-life of carbon-14 (^{14}C) is about 5730 years, which forms the basis of carbon dating technology.

The fundamental idea of radiocarbon dating or carbon dating is that all living organisms contain some minor fraction of carbon-14. When the organism is alive, the ratio of carbon-14 to the normal carbon (carbon-12) is fixed, depending the living environment. From the time the organism is dead and becomes a fossil, the radioactive carbon isotope (carbon-14) will start to decay and the amount of carbon-14 will start to decrease, while the normal carbon will remain about the same. Thus, the ratio of carbon-14 to carbon-12 will provide some information about the age of the fossil.

Example 5.6

For example, if the ratio of carbon-14 to carbon-12 of a found fossil is about 20% of the ratio in the same type of living things, what is the approximate age of the fossil?

Let T be the age of the fossil. We have

$$N(T)=N_0\exp[-\lambda T]=N_0\exp[-T\ln 2/t_{1/2}],$$

where we have used $\lambda=\ln 2/t_{1/2}$. Since $N(T)/N_0=20\%=0.2$ is given, we have

$$\frac{N(T)}{N_0}=0.2=\exp[-T\ln 2/t_{1/2}].$$

Taking the natural logarithm of both sides, we get

$$\ln(0.2)=-\frac{T\ln 2}{t_{1/2}}.$$

Thus, we have

$$T=\frac{-\ln(0.2)t_{1/2}}{\ln 2}.$$

Since $\ln 2=0.693$, $-\ln(0.2)=1.609$ and $t_{1/2}=5730$ years, we finally obtain

$$T=\frac{1.609t_{1/2}}{0.693}=\frac{1.609\times 5730}{0.693}\approx 13304 \text{ years}.$$

This becomes a very useful technology for archeology.

Interestingly, the biological half-life of water in a human body varies from about 7 days to 14 days, depending on the rate of metabolism.

5.3 CHANGE OF BASE FOR LOGARITHM

The calculations of logarithms are not straightforward, so they were commonly listed in various mathematical tables before computers became widely used.

Sometimes, the calculation of logarithm becomes simpler if we can change its base. The so-called change-of-base formula can be written as

$$\log_b(x) = \frac{\log_k(x)}{\log_k(b)}, \tag{5.20}$$

where $k > 0$ is the new base. This formula changes the calculation of $\log_b(x)$ from the base b to base k. In fact, we can use any base that is convenient to our calculation. Thus, we have

$$\log_b(x) = \frac{\log_2(x)}{\log_2(b)} = \frac{\log_{10}(x)}{\log_{10}(b)} = \frac{\ln(x)}{\ln b}. \tag{5.21}$$

Example 5.7

For example, to calculate $\log_3(7)$, we can change it to base e, and we have

$$\log_3(7) = \frac{\ln(7)}{\ln(3)} = \frac{1.9459}{1.0981} \approx 1.7712,$$

where we have used a calculator to estimate ln(3) and ln(7).

Sometimes, it may be useful to use different forms of mathematical formulas, depending on the purpose of application. Thus, certain transformation may be useful to rewrite formulas in different forms. This point can be illustrated by an example.

Example 5.8

According to a study by the NASA's near-Earth asteroids (NEA) team, the frequency of near-Earth asteroids potentially impacting the Earth obeys approximately the following relationship

$$\log_{10} N_{>D} = 1.568 - 2.70 \log_{10} D,$$

where D is the diameter in meters of the colliding bodies, and $N_{>D}$ is the number or frequency per year for all the asteroids with a diameter greater than D (in meters). If we use this equation to do calculations, it is not straightforward. So it is better to re-arrange the above equation slightly. By taking the power of both sides, we have

$$10^{\log_{10} N_{>D}} = 10^{1.568 - 2.70 \log_{10} D} = 10^{1.568} 10^{\log_{10} D^{-2.70}} = 10^{1.568} D^{-2.70}.$$

Since $10^{\log_{10} N_{>D}} = N_{>D}$, we have

$$N_{>D} = 10^{1.568} D^{-2.70} \approx 36.98 D^{-2.70}.$$

For example, if $D = 2$ m, we have

$$N_{>2} = 36.98 \times 2^{-2.70} \approx 6,$$

which means that the total number of objects with a diameter greater than 2 m colliding with the Earth is about 6 per year. Usually, these small NEAs will break up and explode in the upper atmosphere without causing any noticeable damage.

However, on the other hand, for $D = 100$ meters, we have

$$N_{>100} \approx 36.98 \times 100^{-2.70} \approx 0.00015,$$

which is about 0.00015 per year. This means that it takes about $1/0.00015 \approx 6700$ years to potentially see one event. So such larger impacts are rare.

In general, the calculations of logarithms can be carried out by numerical methods and computers. Nowadays, even a pocket calculator can do logarithms and very advanced computation.

EXERCISES

5.1. Explain which of the following is a power function, an exponential function?

$$7^x, x^{5.5}, 2^{x^2}, \pi^x, x^\pi, e^{-x}, x^x.$$

5.2. First find out the domain of the following functions and then plot out these function using any software:

$$3^{-x}, e^{-2x}, 2x^2 - 3, x^{-2}, x^{3/2}.$$

Hint. Check if it is valid at $x = 0$.

5.3. Find the domain and range of the following functions:

$$\ln(x+5), \log_2(2^x), \ln^2(x^2+1), \log_{10}(1/x), \log(\frac{1}{x-1}).$$

5.4. Simplify the following expressions:

- $\log_2(3) + \log_2(9) - \log_2(18)$.
- $\ln(2) + \ln(5) + \ln(10) - \ln(100) + \log_2(2)$.
- $2^2 + 3^{-2}\log_3(81) + \ln(e^2)$.
- $\frac{\ln(6)}{\ln e} - \ln 3 + \ln 4 + \log_{10}(1000)$.

5.5. Find roots of the following equation

$$x^x = x^2, \quad x > 0.$$

Hint. Try $x = 1$ and $x = 2$. Then, use graphs to see if there are more roots.

Chapter 6

Trigonometry

Chapter Points

- Introduce the concepts of angle, degrees and radians and their relationships.
- The formulas for adding angles, double angle and half angle are explained.
- All basic trigonometric functions are introduced in this chapter and some of the identities of trigonometric functions will be proved briefly.
- The inverse functions of some trigonometrical functions are explained so as to calculate the angle when its function value is given.
- Both sine and cosine rules are discussed and their applications are shown using examples.

6.1 ANGLE

Let us first define angles in trigonometry. An angle is the measure of the space between two intersecting lines or surfaces. It is denoted as positive when measured in an anti-clockwise direction from the x-axis, while a negative angle is measured in a clockwise direction from the x-axis. The angles are often expressed either in degrees (°) or in radians. A whole circle is often divided into 360°, and a right angle is 90°.

A radian is defined as the angle whose arc length is r which is also the radius (see Fig. 6.1). As the total arc length or circumference of a circle is $2\pi r$, so we have

$$2\pi \text{ in radians} = 360° \text{ in degrees}, \tag{6.1}$$

which means that a right angle is $\pi/2$. By dividing both sides of the above equation by 2π, we have

$$1 \text{ radian} = \frac{360°}{2 \times 3.14159} = 57.296° \text{ degrees}. \tag{6.2}$$

Conversely, we have

$$1° = \frac{2\pi}{360} = 0.017453 \text{ radians}. \tag{6.3}$$

In some engineering applications, we have to convert quantities using different units. For example, the rotation speed of a motor is often given in round per minute (rpm), but we often have to deal with angular velocity in radian/second. Another example is that the density of a material can be given in either gram/cm^3 or kg/m^3.

Example 6.1

From the specifications of an electric motor, we know that it has a maximum angular velocity of 3000 round per minute (rpm). What is this spin speed in radian per second?

We know that 1 round is 360° or 2π radians (i.e., 1 round = 2π radians) and 1 minute has 60 seconds, so

$$1 \text{ round/minute} = 1 \frac{\text{round}}{\text{minute}} = \frac{2\pi \text{ radian}}{60 \text{ second}} = \frac{\pi}{30} \text{ radian/second}.$$

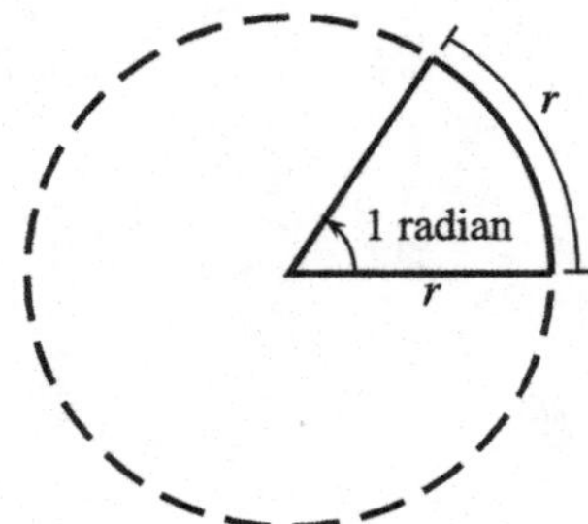

FIGURE 6.1 Definition of a radian.

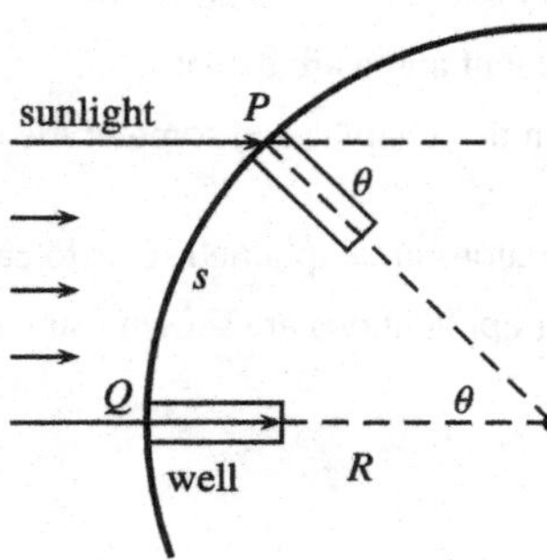

FIGURE 6.2 Eratosthenes' method of measuring the Earth's size.

For the motor, we have

$$3000 \text{ rpm} = 3000 \cdot \frac{\pi}{30} = 100\pi \text{ rad/s} \approx 314 \text{ rad/s}.$$

Trigonometry has many applications. For example, the ancient Greek geographer, Eratosthenes, was probably the first person who tried to measure the size of the Earth quite accurately using two vertical wells (see Fig. 6.2).

Example 6.2

Let us look at the methodology used by Eratosthenes more closely. In one well Q at Syene, the sunlight happened to shine down the well at noon exactly during the Summer Solstice, while the sunlight had an angle θ with its vertical axis in another well P at Alexandria. The distance d between P and Q is known. We know the circumference L of the Earth is related to its radius R by $L = 2\pi R$. From simple trigonometry, we know that

$$\frac{d}{L} = \frac{\theta}{360^\circ},$$

which leads to

$$L = \frac{360d}{\theta}.$$

Eratosthenes measured that $\theta \approx 7.2^\circ$ and $d = 790$ km, so the circumference of the Earth was estimated to be

$$L \approx \frac{360 \times 790}{7.2} \approx 39{,}500 \text{ km},$$

which is very close to the modern value of $L = 40{,}040$ km.

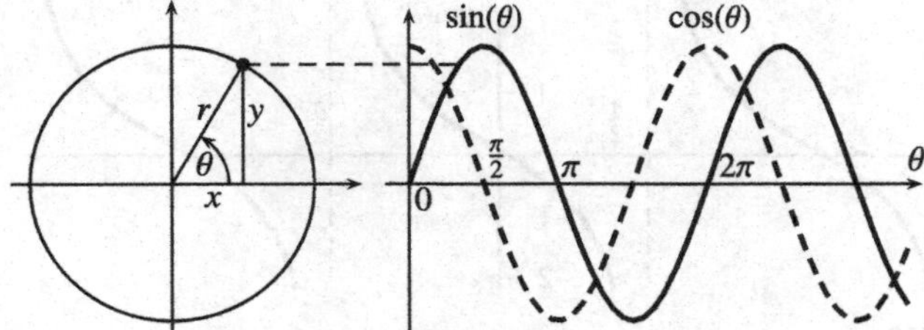

FIGURE 6.3 A sine function sin(θ) (solid) and a cosine function cos(θ) (dashed).

6.2 TRIGONOMETRICAL FUNCTIONS

Referring to the variables in Fig. 6.3, the basic sine and cosine functions are defined as

$$\sin\theta = \frac{y}{r}, \qquad \cos\theta = \frac{x}{r}. \tag{6.4}$$

Since $|x| \leq r$ and $|y| \leq r$ for any triangle, we have

$$-1 \leq \sin\theta \leq 1, \qquad 1 \leq \cos\theta \leq 1. \tag{6.5}$$

From the graphs of $\sin\theta$ and $\cos\theta$ in Fig. 6.3, we know that

$$\sin 0 = \sin\pi = 0, \;\; \cos 0 = \cos 2\pi = 1, \;\; \sin\frac{\pi}{2} = 1, \tag{6.6}$$

and

$$\cos\pi = -1, \;\; \sin\frac{3\pi}{2} = -1. \tag{6.7}$$

In fact, the sine curve becomes a cosine curve if you shift to the left by $\pi/2$. That is

$$\sin(\theta + \frac{\pi}{2}) = \cos\theta. \tag{6.8}$$

Similarly, we have $\cos(\theta + \frac{\pi}{2}) = -\sin\theta$. If we add 2π or 360° to the angle θ, we will reach to the same point, we then have

$$\sin(2\pi + \theta) = \sin(\theta), \qquad \cos(2\pi + \theta) = \cos\theta. \tag{6.9}$$

This is equivalent to saying that both $\sin\theta$ and $\cos\theta$ functions have a period of 2π or 360°.

In addition, if we replace θ by $-\theta$, then from the definition $\sin\theta = y/r$, the point at (x, y) becomes $(x, -y)$ with the same r. This means that

$$\sin(-\theta) = \frac{-y}{r} = -\sin(\theta), \tag{6.10}$$

which suggests that $\sin\theta$ is any odd function. Similar argument suggests that $\cos(-\theta) = \cos(\theta)$, which is an even function.

Other trigonometrical functions can be defined using the basic sin and cos functions. For example, the $\tan\theta$ can be defined as

$$\tan\theta = \frac{y}{x} = \frac{\sin\theta}{\cos\theta}, \tag{6.11}$$

whose graph is shown in Fig. 6.4. Similarly, we have

$$\cot\theta = \frac{1}{\tan\theta} = \frac{\cos\theta}{\sin\theta}, \qquad \text{cosec}\;\theta = \frac{1}{\sin\theta}, \qquad \sec\theta = \frac{1}{\cos\theta}. \tag{6.12}$$

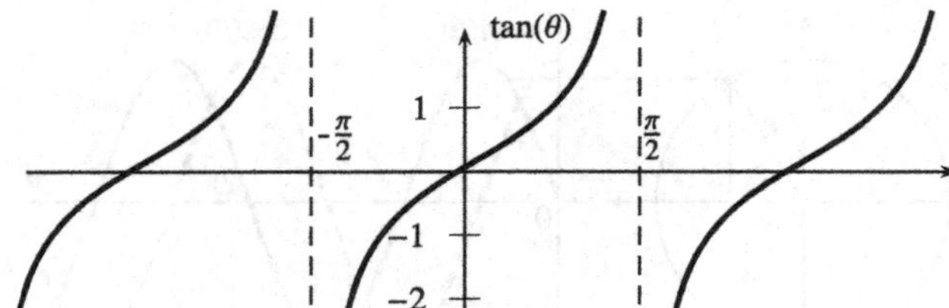

FIGURE 6.4 Graph of the $\tan(\theta)$ function.

In trigonometry, we often write either $\sin(\theta)$ or $\sin\theta$ as long as there is no ambiguity. We also write the power as $\sin^n\theta \equiv (\sin\theta)^n$ for clarity and simplicity.

Example 6.3

Using the Pythagoras' theorem $x^2 + y^2 = r^2$, we have

$$\sin^2\theta + \cos^2\theta = (\frac{y}{r})^2 + (\frac{x}{r})^2 = \frac{y^2 + x^2}{r^2} = \frac{r^2}{r^2} = 1. \tag{6.13}$$

This equality is true for any angle θ, so we call it an identity. In some mathematical texts, they often use the $\equiv$ symbol to emphasize the fact that it is an identity and write it as

$$\sin^2\theta + \cos^2\theta \equiv 1. \tag{6.14}$$

Again this suggests that $|\sin\theta| \le 1$, and $|\cos\theta| \le 1$.

6.2.1 Identities

Since $\tan\theta = \sin\theta/\cos\theta$ and

$$1 + \frac{\sin^2\theta}{\cos^2\theta} = \frac{\cos^2\theta + \sin^2\theta}{\cos^2\theta} = \frac{1}{\cos^2\theta}, \tag{6.15}$$

we have the following identity

$$1 + \tan^2\theta \equiv \frac{1}{\cos^2\theta} = \sec^2\theta. \tag{6.16}$$

Other identities can be proved in the similar manner.

Example 6.4

Prove that the following identity

$$\frac{1}{\tan^2\theta + 1} + \frac{1}{\cot^2\theta + 1} \equiv 1.$$

Since $\tan\theta = \sin\theta/\cos\theta$ and $\cot\theta = \cos\theta/\sin\theta$, we have

$$\frac{1}{\tan^2\theta + 1} + \frac{1}{\cot^2\theta + 1} = \frac{1}{\frac{\sin^2\theta}{\cos^2\theta} + 1} + \frac{1}{\frac{\cos^2\theta}{\sin^2\theta} + 1}$$
$$= \frac{1}{\frac{\sin^2\theta + \cos^2\theta}{\cos^2\theta}} + \frac{1}{\frac{\cos^2\theta + \sin^2\theta}{\sin^2\theta}} = \frac{1}{\frac{1}{\cos^2\theta}} + \frac{1}{\frac{1}{\sin^2\theta}} = \cos^2\theta + \sin^2\theta = 1,$$

which becomes the original identity.

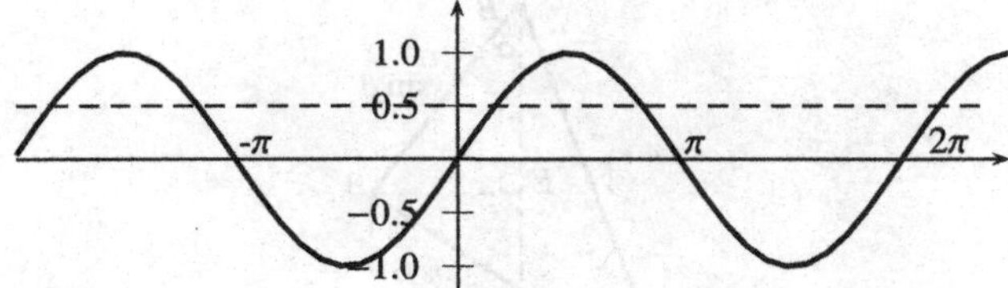

FIGURE 6.5 Inverse of $\sin\theta = 0.5$.

6.2.2 Inverse

The inverse of a sine function can be obtained by finding the value of θ once the value of $y = \sin\theta$ is given. That is

$$\sin\theta = y, \qquad \text{or} \qquad \theta = \sin^{-1}(y). \tag{6.17}$$

In some literature, $\sin^{-1}$ is also written as arcsin. For example, in order to find the values θ so that $\sin\theta = 1/2$, we can plot $\sin(\theta)$ for various values of θ (solid curve) and then draw a straight line shown in Fig. 6.5 where we can see that there are multiple solutions at $x = -210°, 30°, 150°, ...$ (or $x = -7\pi/6, \pi/6, 5\pi/6, ...$). In fact, there are infinitely many solutions in the form of

$$x = \frac{\pi}{6} + 2n\pi, \qquad x = \frac{5\pi}{6} + 2n\pi, \tag{6.18}$$

where n is an integer ($n = 0, \pm 1, \pm 2, ...$). In order to avoid any possible ambiguity, the angles in radians in the interval $[-\frac{\pi}{2}, \frac{\pi}{2}]$ are commonly defined as the principal values of the $\sin^{-1}$ function.

Similarly, the inverse of $y = \cos\theta$ is denoted by $\cos^{-1}$, and its principal values are in the interval of $[0, \pi]$. That is

$$\theta = \cos^{-1} y = \cos^{-1}(y), \quad \theta \in [0, \pi], \tag{6.19}$$

where () is purely for ease of reading or to avoid confusion.

Furthermore, the inverse of $y = \tan\theta$ is

$$\theta = \tan^{-1}(y), \tag{6.20}$$

whose principal values are between $-\pi/2$ and $\pi/2$. That is $\theta \in (-\pi/2, \pi/2)$.

It is worth noting that, since $-1 \leq \sin\theta \leq 1$, the expression such as $\sin^{-1}(2)$ is not valid as there exists no angle such that $\sin\theta = 2$. Similarly, $\cos^{-1}(2)$ is not valid either.

In some textbooks and computing, $\sin^{-1}$ is called asin or arcsin, while $\cos^{-1}$ is often called acos or arccos. Similarly, $\tan^{-1}$ is also referred to as atan or arctan. Thus, we have

$$\theta = \tan^{-1}(y) = \text{atan}(y) = \arctan(y). \tag{6.21}$$

6.2.3 Trigonometrical Functions of Two Angles

For the sine function of the addition of the two angles α and β (see Fig. 6.6), we let the length of OB be 1 or $OB = 1$. Now we have

$$\sin(\alpha + \beta) = \frac{BD}{OB} = BD = BE + ED. \tag{6.22}$$

Since $OA = \cos\beta$, $AB = \sin\beta$ and $AE = \sin\alpha \sin\beta$, we have

$$AC = ED = OA \sin\alpha = \cos\beta \sin\alpha, \quad BE = AB\cos\alpha = \sin\beta\cos\alpha,$$

which leads to

$$\sin(\alpha + \beta) = \sin\alpha\cos\beta + \cos\alpha\sin\beta. \tag{6.23}$$

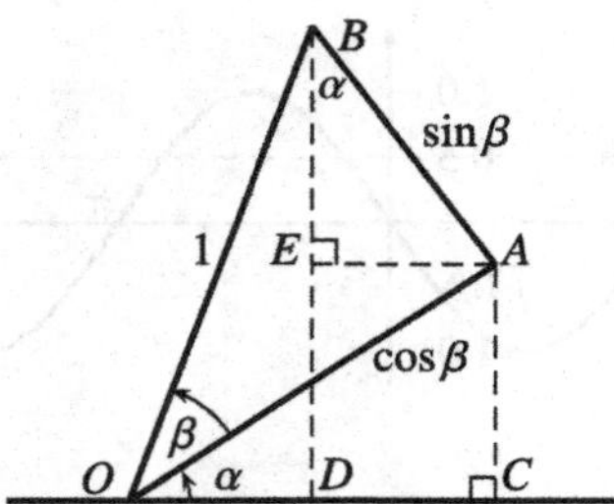

FIGURE 6.6 Derivation of $\sin(\alpha+\beta)$.

In the special case when $\alpha=\beta$, we have

$$\sin 2\alpha = \sin\alpha\cos\alpha + \cos\alpha\sin\alpha = 2\sin\alpha\cos\alpha. \tag{6.24}$$

Similarly, we have

$$\cos(\alpha+\beta) = \frac{OD}{OB} = OD = OC - DC. \tag{6.25}$$

From $OC = OA\cos\alpha = \cos\beta\cos\alpha$ and $DC = EA = AB\sin\alpha = \sin\beta\sin\alpha$, we finally have

$$\cos(\alpha+\beta) = \cos\alpha\cos\beta - \sin\alpha\sin\beta. \tag{6.26}$$

Replacing β by $-\beta$ and using $\sin(-\beta) = -\sin\beta$ and $\cos(-\beta) = \cos\beta$, we have

$$\sin(\alpha-\beta) = \sin\alpha\cos\beta - \cos\alpha\sin\beta, \tag{6.27}$$

and

$$\cos(\alpha-\beta) = \cos\alpha\cos\beta + \sin\alpha\sin\beta. \tag{6.28}$$

From equation (6.26), if $\alpha=\beta$, we have

$$\cos 2\alpha = \cos\alpha\cos\alpha - \sin\alpha\sin\alpha = \cos^2\alpha - \sin^2\alpha.$$

Since $\sin^2\alpha + \cos^2\alpha = 1$ or $\sin^2\alpha = 1 - \cos^2\alpha$, we have

$$\cos 2\alpha = \cos^2\alpha - \sin^2\alpha = 2\cos^2\alpha - 1 = 1 - 2\sin^2\alpha. \tag{6.29}$$

Example 6.5

Prove the identity

$$\tan 2\alpha = \frac{2\tan\alpha}{1-\tan^2\alpha}.$$

By combining (6.24) and (6.29), we have

$$\tan 2\alpha = \frac{\sin 2\alpha}{\cos 2\alpha} = \frac{2\sin\alpha\cos\alpha}{\cos^2\alpha - \sin^2\alpha}.$$

If we divide both the numerator and denominator by $\cos^2\alpha$, we have

$$\tan 2\alpha = \frac{\frac{2\sin\alpha}{\cos\alpha}}{1 - \frac{\sin^2\alpha}{\cos^2\alpha}} = \frac{2\tan\alpha}{1-\tan^2\alpha}.$$

Here, we have implicitly assumed that $\cos(\alpha) \neq 0$ (otherwise, we cannot do the division). However, the above identity is also valid for $\cos(\alpha) = 0$ (i.e., $\alpha = \pm\pi/2, \pm 3\pi/2, \cdots$), which is straightforward to verify.

By subtracting (6.27) from (6.23), we have

$$\sin(\alpha+\beta) - \sin(\alpha-\beta) = \sin\alpha\cos\beta + \cos\alpha\sin\beta - (\sin\alpha\cos\beta - \cos\alpha\sin\beta)$$
$$= 2\cos\alpha\sin\beta. \tag{6.30}$$

Similarly, by adding the two identities (6.23) and (6.27), we have

$$\sin(\alpha+\beta) + \sin(\alpha-\beta) = 2\sin\alpha\cos\beta. \tag{6.31}$$

By introducing new notations $A = \alpha + \beta$, and $B = \alpha - \beta$, we have after simple addition and subtraction

$$\alpha = \frac{A+B}{2}, \qquad \beta = \frac{A-B}{2}. \tag{6.32}$$

The above identities (6.30) and (6.31) now become

$$\sin A - \sin B = 2\cos\frac{A+B}{2}\sin\frac{A-B}{2}, \tag{6.33}$$

and

$$\sin A + \sin B = 2\sin\frac{A+B}{2}\cos\frac{A-B}{2}. \tag{6.34}$$

Example 6.6

The formula for the addition of two angles can be used to derive a general expression

$$a\cos\theta + a\sin\theta = R\cos(\theta - \psi), \tag{6.35}$$

for any θ and constants a and b. From (6.28), we know that

$$\cos(\theta - \psi) = \cos\theta\cos\psi + \sin\theta\sin\psi. \tag{6.36}$$

Multiplying both sides by R, we have

$$R\cos(\theta - \psi) = (R\cos\psi)\cos\theta + (R\sin\psi)\sin\theta. \tag{6.37}$$

By comparing with the original equation (6.35), we have

$$a\cos\theta + b\sin\theta = (R\cos\psi)\cos\theta + (R\sin\psi)\sin\theta. \tag{6.38}$$

This must be true for any θ, so we have

$$a = R\cos\psi, \qquad b = R\sin\psi. \tag{6.39}$$

Taking squares and using $\cos^2\psi + \sin^2\psi = 1$, we have

$$a^2 + b^2 = R^2\cos^2\psi + R^2\sin^2\psi = R^2(\cos^2\psi + \sin^2\psi) = R^2, \tag{6.40}$$

which gives

$$R = \sqrt{a^2 + b^2}. \tag{6.41}$$

Here we have only used the positive root as R is conventionally considered as the amplitude of the variations.

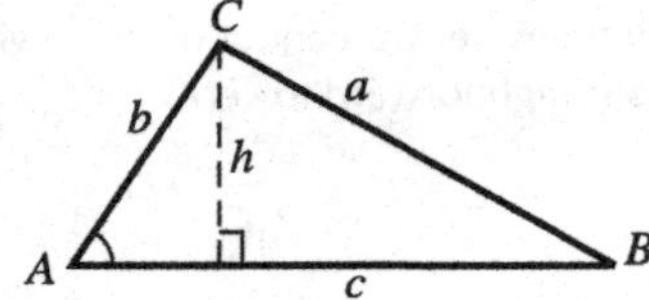

FIGURE 6.7 Derivation of the sine rule.

By taking the ratio of b and a, we have

$$\frac{b}{a} = \frac{R\sin\psi}{R\cos\psi} = \tan\psi, \tag{6.42}$$

or

$$\psi = \tan^{-1}\frac{b}{a}. \tag{6.43}$$

For example, in mechanical vibrations, two waves with the same angular frequency ω or period, their superimposition can always be written in a single sinusoidal form as

$$a\cos(\omega t) + b\sin(\omega t) = R\cos(\omega t - \psi),$$

where ψ is the phase shift.

6.3 SINE RULE

The sine rule is important in trigonometry as it provides the relationships between the three sides and their corresponding angles of a triangle. The height h in the triangle shown in Fig. 6.7 can be calculated using

$$h = b\sin A, \quad \text{or } h = a\sin B, \tag{6.44}$$

depending on which right-angled smaller triangle you are using. The combination of the above two formulas leads to

$$b\sin A = a\sin B, \quad \text{or} \quad \frac{a}{\sin A} = \frac{b}{\sin B}. \tag{6.45}$$

Applying the same argument to other perpendicular heights, we can obtain

$$\frac{b}{\sin B} = \frac{c}{\sin C}. \tag{6.46}$$

Combing the above equations, we have the sine rule

$$\frac{a}{\sin A} = \frac{b}{\sin B} = \frac{c}{\sin C}. \tag{6.47}$$

This rule makes it easier to determine all other two sides very easily once the length of a side and two angles are given in a triangle.

On the other hand, if one angle and its two sides are given, it is easier to use the cosine rule to determine the rest quantities of the given triangle.

6.4 COSINE RULE

In order to derive the cosine rule, we refer to the same triangle in deriving the sine rule, but now we put it in a Cartesian coordinate system shown in Fig. 6.8. Now the coordinates are $(0, 0)$, $(c, 0)$ and $(b\cos A, b\sin A)$. The length a is simply the Cartesian distance between points A and B. We now have

$$\begin{aligned} a^2 &= (b\cos A - c)^2 + (b\sin A - 0)^2 = b^2\cos^2 A + c^2 - 2bc\cos A + b^2\sin^2 A \\ &= b^2(\cos^2 A + \sin^2 A) + c^2 - 2bc\cos A. \end{aligned} \tag{6.48}$$

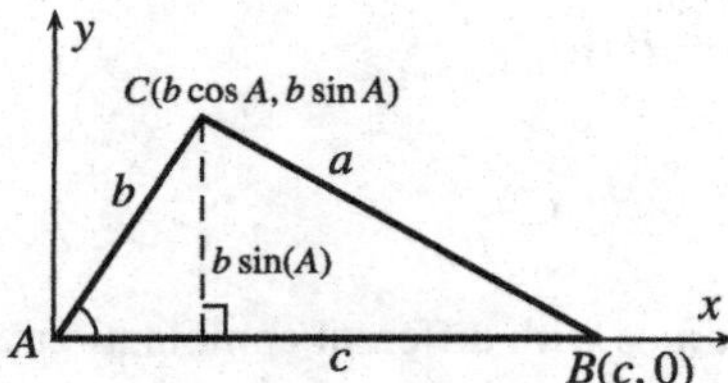

FIGURE 6.8 Derivation of the cosine rule.

By using $\sin^2 A + \cos^2 A = 1$, we finally obtain the cosine rule

$$a^2 = b^2 + c^2 - 2bc\cos A. \tag{6.49}$$

Similarly, by proper permutation using the same argument, we have

$$b^2 = c^2 + a^2 - 2ca\cos B, \qquad c^2 = a^2 + b^2 - 2ab\cos C. \tag{6.50}$$

By rearranging the cosine rule, we can obtain the formulas to determine the angles of a triangle once the lengths of the three sides are given. We have

$$\cos A = \frac{b^2 + c^2 - a^2}{2bc}, \qquad \cos B = \frac{c^2 + a^2 - b^2}{2ca}, \tag{6.51}$$

and

$$\cos C = \frac{a^2 + b^2 - c^2}{2ab}. \tag{6.52}$$

Example 6.7

In the triangle shown in Fig. 6.7, we have the angle $A = 60°$, and two sides $b = 2$ and $c = 4$. Now we try to determine the other two angles and a. First, using the cosine rule, we have

$$a^2 = b^2 + c^2 - 2bc\cos A = 2^2 + 4^2 - 2 \times 2 \times 4\cos 60° = 4 + 16 - 16\frac{1}{2} = 12,$$

or $a = \sqrt{12}$. In order to determine the angle B, we can use (6.51)

$$\cos B = \frac{c^2 + a^2 - b^2}{2ca} = \frac{4^2 + (\sqrt{12})^2 - 2^2}{2 \times 4 \times \sqrt{12}} = \frac{3}{2\sqrt{3}} = \frac{\sqrt{3}}{2},$$

which gives

$$B = \cos^{-1}\frac{\sqrt{3}}{2} = 30°.$$

Thus, the other angle is $C = 180° - A - B = 90°$. This means that it is a right-angled triangle.

EXERCISES

6.1. An electric motor has a final speed of 9000 round per minute, what is its speed in radian per second. If the motor rotates for 10 seconds, how many degrees does it turn?

6.2. A triangle has one angle of 45° and the lengths of the two adjacent sides of this angle are 2 and $2\sqrt{2}$, respectively. What is the length of the third side? What are the rest two angles?

Hint. Try to use the cosine rule.

6.3. Prove the following trigonometrical identities:

- $\sin A \sin B = \frac{1}{2}[\cos(A - B) - \cos(A + B)]$
- $\sin(3A) = 3\sin A - 4\sin^3 A$
- $\cos(3A) = 4\cos^3 A - 3\cos A$
- $\sin(4A) = \sin A\ (8\cos^3 A - 4\cos A)$

Hint. Use the sum of two angles a few times with different combinations.

6.4. Analyse the scenario of a total solar eclipse and show that the ratio of the distance from the Sun to the Earth to the distance from the Moon to the Earth is approximately the same ratio of the Sun's diameter to the Moon's diameter. In addition, can you design a way to measure the Sun's (or the Moon's) diameter using a pinhole image technique with a ruler and a cardboard?

6.5. An ellipse can be described as an equation in a two-dimensional Cartesian coordinate system (x,y) in the following form

$$\frac{x^2}{a^2} + \frac{y^2}{b^2} = 1,$$

where $a, b > 0$. The eccentricity (e) of an ellipse is defined as $e = \sqrt{1 - (\frac{b}{a})^2}$. The same ellipse can also be described by a set of two parametric equations

$$x = a\cos(t), \quad y = b\cos(t),$$

where t is a parameter. Discuss the domain of t and the links between these two representations. Use any programming language to show that the two representations can indeed give the same ellipse.

Hint. First plot out the ellipse for $a = 2$ and $b = 1$ using two different representations.

Part II

Complex Numbers

Chapter 7

Complex Numbers

Chapter Points

- Basic concepts of complex numbers are introduced, and complex algebraic operations are also explained, including modulus, Argand diagram and their links to vectors.
- The well-known Euler's formula is explained and examples are given to show how to use this formula in its proper context.
- Hyperbolic functions are introduced, and their links with trigonometrical functions are also discussed, including some identities.

7.1 WHY DO NEED COMPLEX NUMBERS?

You may wonder why complex numbers are necessary. There are many reasons and situations that complex numbers are needed. From the mathematical point of view, complex numbers make the number system complete. For example, when we solve quadratic equations, we say the square root of a negative number, for example $\sqrt{-1}$, does not exist because there is no number whose square is negative (for all real numbers x, $x^2 \geq 0$). Well, this is only true in the context of real numbers. Such limitations mean that the system of real numbers is incomplete, as the mathematical operations of numbers could lead to something which does not belong to the number system.

On the other hand, from the engineering point of view, it is much easier to study the behavior of certain systems (such as control systems) using Laplace transforms, which requires complex numbers. For example, circuit theory and the modeling of power engineering can rely on the complex models, and complex numbers can make such models simpler.

7.2 COMPLEX NUMBERS

A significant extension is to introduce imaginary numbers by defining an imaginary unit

$$i = \sqrt{-1}, \qquad i^2 = (\sqrt{-1})^2 = -1. \tag{7.1}$$

This is a seemingly simple step but it may have many profound consequences. It is worth pointing out that i is a special notation, but you cannot use $i^2 = (\sqrt{-1})^2 = \sqrt{(-1)^2} = \sqrt{1} = \pm 1$ because this may lead to some confusion. To avoid such possible confusion, it is better to think of i as $\sqrt{-}$ or the imaginary unit, so for any real number $a > 0$

$$\sqrt{-a} = \sqrt{(-1) \times a} = \sqrt{-1}\sqrt{a} = i\sqrt{a}. \tag{7.2}$$

For example, $\sqrt{-2} = i\sqrt{2}$ and $\sqrt{-25} = i5 = 5i$ (we prefer to write numbers first followed by i).

Example 7.1

The imaginary number i follows the same rules for mathematical functions defined in the real-number system. For example, we can calculate the following

$$i^3 = i^2 i = -1i = -i, \; i^4 = (i^2)^2 = (-1)^2 = 1, \; i^5 = i^4 i = i, \; i^6 = i^4 \cdot i^2 = -1,$$
$$i^7 = i^4 \cdot i^2 \cdot i = -i, \; i^8 = (i^4)^2 = 1, \; i^9 = i^8 \cdot i = i, \; i^{10} = i^{4\times2+2} = (i^4)^2 \cdot i^2 = -1.$$

Thus

$$i^{50} = i^{4\times12+2} = (i^4)^{12} \times i^2 = -1, \; i^{101} = i^{4\times25+1} = i,$$

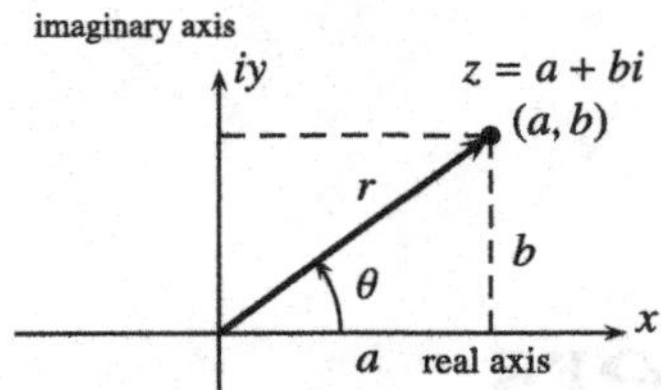

FIGURE 7.1 Complex plane, angle and modulus.

and

$$i^{1001} = i^{4\times 250+1} = i, \ i^{12345} = i^{12344+1} = i^{4\times 3086+1} = i.$$

So the best way to estimate such an expression is to try to write exponents in terms of the multiples of 4 (as close as possible). To simply use odd or even exponents is not enough, and often leads to incorrect results.

It is worth pointing out that the notation j (instead of i) is often used in engineering textbooks. One of the reasons is that i is often associated with a current in circuit theory. Whatever the notation may be, it is just $j \equiv i = \sqrt{-1}$.

You might ask what $\sqrt{i}$ or $i^{1/2}$ is? Obviously, we can calculate it, but we first have to define the algebraic rule for complex numbers. Before we proceed to do this, let us try to represent complex numbers geometrically.

Another way of thinking of a complex number is to represent it as a pair of two numbers x and y on a complex plane (see Fig. 7.1), similar to the Cartesian coordinates (x, y) where each point corresponds uniquely to an ordered pair of two coordinates (real numbers), but with a significant difference. Here the vertical axis is no longer a real axis; it becomes the imaginary axis iy-axis. Similarly, a point on the complex plane corresponds also uniquely to a complex number

$$z = a + bi, \tag{7.3}$$

which consists of a real part a and an imaginary part bi. This also means that it corresponds to an ordered pair of real numbers a and b. For a given $z = a + bi$, the real part is denoted by $a = \Re(z)$ and imaginary part is denoted by $b = \Im(z)$. The length or modulus of a complex number z is defined as

$$|z| = \sqrt{a^2 + b^2}, \tag{7.4}$$

which is always non-negative (i.e., $|z| \geq 0$).

7.3 COMPLEX ALGEBRA

Now we can define the mathematical operations of any two complex numbers $z_1 = a + bi$ and $z_2 = c + di$. The addition of two complex numbers is carried out by adding their real parts and imaginary parts, respectively. That is

$$(a + bi) + (c + di) = (a + c) + (b + d)i. \tag{7.5}$$

Similarly, the subtraction of two complex numbers is defined by

$$(a + bi) - (c + di) = (a - c) + (b - d)i. \tag{7.6}$$

Two complex numbers are equal if, and only if, their real parts and imaginary parts are equal, respectively. That is, $z_1 = z_2$ if, and only if, $a = c$ and $b = d$.

The multiplication of two complex numbers is carried out similarly to expanding an expression, using $i^2 = -1$ when necessary

$$\begin{aligned}(a + bi) \cdot (c + di) &= a \cdot (c + di) + bi \cdot (c + di) \\ &= ac + adi + bci + bdi^2 = (ac - bd) + (bc + ad)i.\end{aligned} \tag{7.7}$$

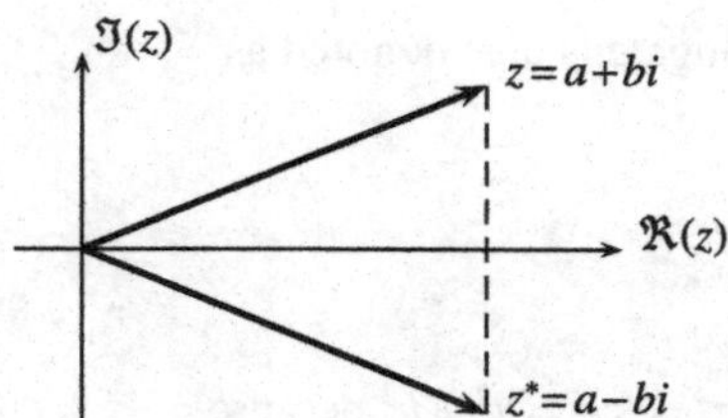

FIGURE 7.2 Complex conjugate and reflection.

The division of two complex numbers

$$\frac{a+bi}{c+di} = \frac{(a+bi)\cdot(c-di)}{(c+di)\cdot(c-di)}$$
$$= \frac{ac - adi + bci - bdi^2}{c^2 - d^2 i^2} = \frac{(ac+bd)}{(c^2+d^2)} + \frac{(bc-ad)}{(c^2+d^2)} i, \tag{7.8}$$

where we have used the equality $(x+y)(x-y) = x^2 - y^2$.

Example 7.2

Now let us try to find $\sqrt{i}$. Let $z = a + bi$ and $z^2 = i$, we have

$$(a+bi)^2 = (a^2 - b^2) + 2abi = i = 0 + i.$$

As two complex numbers, here $(a^2 - b^2) + 2abi$ and $0 + i$, must be equal on both sides, we have

$$a^2 - b^2 = 0, \qquad 2ab = 1.$$

The first condition gives $a = \pm b$. In case of $a = b$, the second condition gives

$$a^2 = b^2 = \frac{1}{2},$$

or

$$a = b = \pm\frac{1}{\sqrt{2}} = \pm\frac{\sqrt{2}}{2}.$$

In case of $a = -b$, the second condition leads to

$$-2a^2 = 1,$$

which does not have any solution for any real number a.

Therefore, the only possibility is that $a = b$ and thus $\sqrt{i}$ has two distinct roots

$$\sqrt{i} = \frac{\sqrt{2}}{2} + \frac{\sqrt{2}}{2} i,$$

and

$$\sqrt{i} = -\frac{\sqrt{2}}{2} - \frac{\sqrt{2}}{2} i.$$

The complex conjugate z^* of a complex number $z = a + bi$ is defined by changing the sign of the imaginary part

$$z^* = a - bi, \tag{7.9}$$

which is the reflection in the real axis of the original z (see Fig. 7.2). The definition also leads to

$$(z^*)^* = (a - bi)^* = a + bi = z. \tag{7.10}$$

It is worth pointing out that the complex conjugate is also denoted as

$$\bar{z} = z^*. \tag{7.11}$$

Example 7.3

For any number $z = a + bi$, we can prove that $zz^* = |z|^2 = a^2 + b^2$ because

$$zz^* = (a+bi)(a-bi) = a^2 + \underline{abi} - \underline{abi} - b^2i^2 = a^2 - b^2(-1) = a^2 + b^2.$$

It also straight forward to show that

$$|z|^2 = |z^2|.$$

From $|z| = \sqrt{a^2+b^2}$, we get $|z|^2 = a^2 + b^2$.

Since $z^2 = (a+bi)^2 = a^2 + 2abi + b^2i^2 = (a^2 - b^2) + 2abi$, we have

$$|z^2| = \sqrt{(a^2-b^2)^2 + (2ab)^2} = \sqrt{a^4 - 2a^2b^2 + b^4 + 4a^2b^2}$$
$$= \sqrt{a^4 + 2a^2b^2 + b^4} = \sqrt{(a^2+b^2)^2} = (a^2+b^2),$$

which means that $|z^2| = |z^2|$.

From Fig. 7.1, we can also represent a complex number in the polar form in terms of an ordered pair (r, θ). From basic trigonometry, we know that

$$\sin(\theta) = \frac{b}{r}, \qquad \cos(\theta) = \frac{a}{r}, \tag{7.12}$$

or

$$a = r\cos(\theta), \qquad b = r\sin(\theta), \tag{7.13}$$

where θ is the argument or the angle of z, and r is the modulus or magnitude of the complex number $z = a + bi$, which can be obtained using Pythagoras' theorem

$$r = |z| = \sqrt{a^2 + b^2}. \tag{7.14}$$

For any given $z = a + bi$, the angle θ is given by

$$\theta = \arg(z) = \tan^{-1}\left(\frac{b}{a}\right), \tag{7.15}$$

where we only use the range $-\pi < \theta \le \pi$, called the principal values. In some textbooks, the principal values are defined as $0 \le \theta < 2\pi$, i.e., $[0, 2\pi)$.

Thus, the same complex number can be expressed as the polar form

$$z = a + bi = r\cos(\theta) + ir\sin(\theta). \tag{7.16}$$

Sometimes, it is also conveniently written as the exponential form

$$z = re^{i\theta} = r[\cos(\theta) + i\sin(\theta)], \tag{7.17}$$

which requires Euler's formula

$$e^{i\theta} = \cos(\theta) + i\sin(\theta). \tag{7.18}$$

The proof of this formula usually involves the expansion of infinite power series, and you can find the detail in later chapters [see Eq. (10.39) in Section 10.3].

7.4 EULER'S FORMULA

An interesting extension is to replace θ by $-\theta$ in Euler's formula; we have

$$e^{-i\theta} = \cos(-\theta) + i\sin(-\theta) = \cos(\theta) - i\sin(\theta). \tag{7.19}$$

Adding this to the original formula (7.18), we have

$$e^{i\theta} + e^{-i\theta} = [\cos(\theta) + \cos(\theta)] + i[\sin(\theta) - \sin(\theta)] = 2\cos(\theta), \tag{7.20}$$

or

$$\cos(\theta) = \frac{e^{i\theta} + e^{-i\theta}}{2}. \tag{7.21}$$

If we follow the same procedure, but subtract these two formulas, we have

$$e^{i\theta} - e^{-i\theta} = [\cos(\theta) - \cos(\theta)] + 2\sin(\theta)i, \;\text{ or } \sin(\theta) = \frac{e^{i\theta} - e^{-i\theta}}{2i}. \tag{7.22}$$

The polar form is especially convenient for multiplication, division, exponential manipulations and other mathematical manipulations. For example, the complex conjugate $z = re^{i\theta}$ is simply $z^* = re^{-i\theta}$.

Example 7.4

For example, for two complex numbers $z_1 = r_1 e^{i\theta_1}$ and $z_2 = r_2 e^{i\theta_2}$, their product is simply

$$z_1 z_2 = r_1 e^{i\theta_1} \times r_2 e^{i\theta_2} = r_1 r_2 e^{i(\theta_1+\theta_2)}.$$

Their ratio is

$$\frac{z_1}{z_2} = \frac{r_1 e^{i\theta_1}}{r_2 e^{i\theta_2}} = \frac{r_1}{r_2} e^{i(\theta_1-\theta_2)}.$$

Furthermore, for $z = re^{i\theta}$, we have

$$z^n = (re^{i\theta})^n = r^n (e^{i\theta})^n = r^n[\cos(\theta) + i\sin(\theta)]^n.$$

Also using Euler's formula, we have

$$z^n = r^n (e^{i\theta})^n = r^n e^{in\theta} = r^n[\cos(n\theta) + i\sin(n\theta)].$$

Combining the above two equations, we have

$$[\cos(\theta) + i\sin(\theta)]^n = \cos(n\theta) + i\sin(n\theta),$$

which is the famous de Moivre's formula.

If we now revise the previous example, we have $n = 1/2$

$$\sqrt{i} = i^{1/2} = [\cos(\frac{\pi}{2}) + i\sin(\frac{\pi}{2})]^{1/2} = \cos\frac{\pi}{4} + i\sin(\frac{\pi}{4}) = \frac{\sqrt{2}}{2} + \frac{\sqrt{2}}{2}i,$$

which is indeed one of the two roots we obtained earlier. As the imaginary axis i forms an angle of 90° (or $\pi/2$) from the real axis, the square root takes the half angle $\pi/4$. However, a cautious reader may ask what happens to the other root? What choice of angle can influence the results. The imaginary axis i can also be considered to have angle of $360° + 90° = 450°$ (or $2\pi + \pi/2 = 5\pi/2$) from the real axis, another square root will have an angle of $5\pi/4$. Thus, the other root is

$$\sqrt{i} = \cos\frac{5\pi}{4} + i\sin\frac{5\pi}{4} = -\frac{\sqrt{2}}{2}(1+i),$$

which is indeed the other root we obtained earlier.

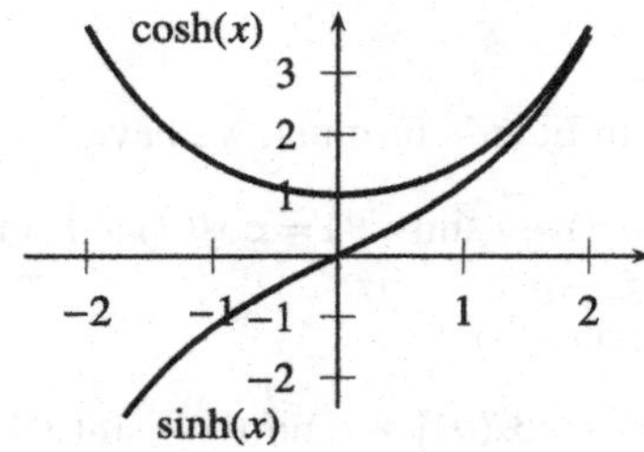

FIGURE 7.3 Graphs of sinh(x) and cosh(x).

The real functions such as logarithms can be extended directly to the functions of complex numbers. For example, we know from Euler's formula when $\theta = \pi$

$$e^{i\pi} = \cos(\pi) + i\sin(\pi) = -1 + 0i = -1. \tag{7.23}$$

If we use the logarithm definition, we should have

$$i\pi = \log_e(-1) = \ln(-1). \tag{7.24}$$

In the standard $\log(x)$, we usually assume that $x > 0$. Now we can see that $x < 0$ is also valid, though its logarithm is a complex number.

7.5 HYPERBOLIC FUNCTIONS

Hyperbolic functions have many similarities to trigonometrical functions, and are commonly used in engineering and science.

7.5.1 Hyperbolic Sine and Cosine

The hyperbolic sine and cosine functions are defined as

$$\sinh x = \frac{e^x - e^{-x}}{2}, \qquad \cosh x = \frac{e^x + e^{-x}}{2}, \tag{7.25}$$

where x is a real number. If we replace x by $-x$, and substitute it into the above definitions, we have

$$\sinh(-x) = \frac{e^{-x} - e^{-(-x)}}{2} = \frac{-[e^x - e^{-x}]}{2} = -\sinh x, \tag{7.26}$$

and

$$\cosh(-x) = \frac{e^{-x} + e^{-(-x)}}{2} = \frac{e^{-x} + e^x}{2} = \cosh x. \tag{7.27}$$

This suggests that $\sinh x$ is an odd function while $\cosh x$ is an even function. Their graphs are shown in Fig. 7.3. It is worth pointing out that the domain for both sinh and cosh are all the real numbers $(-\infty < x < +\infty)$.

Example 7.5

It is easy to check that

$$\sinh(0) = \frac{e^0 - e^{-0}}{2} = \frac{1-1}{2} = 0,$$

and

$$\cosh(0) = \frac{e^0 + e^{-0}}{2} = \frac{1+1}{2} = 1.$$

Since both $\exp[x]$ and $\exp[-x]$ are always positive for any real number x, we can conclude that $\cosh(x)$ is always positive. In fact, $\cosh(x) \geq 1$. Therefore, the range of $\cosh(x)$ is $[1, +\infty)$, while the range of $\sinh(x)$ is the whole range of real numbers.

Example 7.6

The speed (v) of water waves in the ocean can be estimated by

$$v = \sqrt{\frac{g\lambda}{2\pi} \tanh\left(\frac{2\pi h}{\lambda}\right)},$$

where h is the depth of the water and λ is the wavelength. Here $g = 9.8$ m/s^2 is the acceleration due to gravity.

This formula indicates that waves with longer wavelengths travel faster than waves with shorter wavelengths. Waves travel faster in deeper waters than in shallow waters. When we say deep waters, it means that the wavelength λ is much smaller than the depth h and we write this as $\lambda \ll h$. In this case, we have $2\pi h/\lambda$ becomes very large (so we write $2\pi h/\lambda \to \infty$).

For any $x > 0$, we have

$$\tanh(x) = \frac{\sinh(x)}{\cosh(x)} = \frac{e^x - e^{-x}}{e^x + e^{-x}}.$$

If x is very large and positive, we have $e^{-x} \approx 0$. So we have

$$\tanh(x) \approx \frac{e^x - 0}{e^x + 0} \approx \frac{e^x}{e^x} \approx 1, \quad \text{if } x \text{ is large.}$$

Setting $x = 2\pi h/\lambda$ and using the above result, we have

$$v = \sqrt{\frac{g\lambda}{2\pi} \tanh\left(\frac{2\pi h}{\lambda}\right)} \approx \sqrt{\frac{g\lambda}{2\pi}}.$$

A tsunami is a giant water wave whose wavelength and speed are constantly changing as it travels towards the shore. In deep ocean waters, its wavelength is about 25 km to 50 km. So for $\lambda = 25$ km $= 25 \times 1000 = 25000$ m, we have

$$v = \sqrt{\frac{g\lambda}{2\pi}} = \sqrt{\frac{9.8 \times 25000}{2\pi}} \approx 197 \text{ m/s},$$

which is about 440 mph. For longer wavelength $\lambda = 50$ km, its speed is about 630 mph. Obviously, as they travel towards shores, their wavelengths will gradually reduce to the range of about 1.5 km to 5 km and their speed can also reduce to the range of about 200 mph to 25 mph, depending on the structures of the shores. However, the height of such waves can reach up to 30 meters, which can cause significant damage onshore.

If we replace x by ix in the hyperbolic cosine function, we know from (7.21) in our earlier discussion that

$$\cosh(ix) = \frac{e^{ix} + e^{-ix}}{2} = \cos x, \ \sinh(ix) = \frac{e^{ix} - e^{-ix}}{2} = i \sin x, \tag{7.28}$$

where we have used $\sin x = (e^{ix} - e^{-ix})/2$ from (7.22).

Other hyperbolic functions are defined in a similar manner to the ratio of the basic hyperbolic sine and cosine functions. For example, we have the hyperbolic tangent

$$\tanh x = \frac{\sinh x}{\cosh x} = \frac{e^x - e^{-x}}{e^x + e^{-x}}, \tag{7.29}$$

and $\coth x = 1/\tanh x$, $\text{sech}\ x = 1/\cosh x$, and $\text{cosech}\ x = 1/\sinh x$.

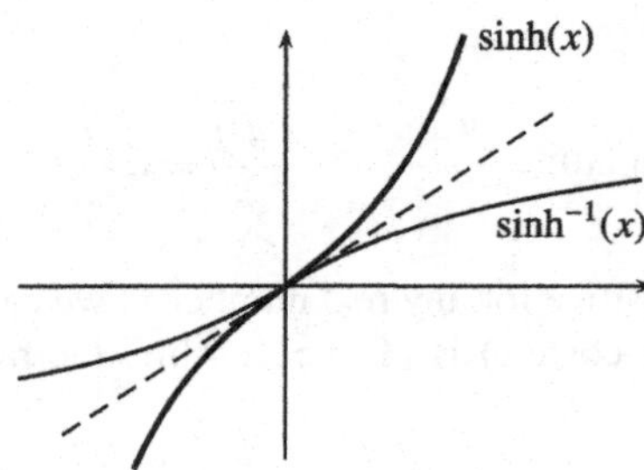

FIGURE 7.4 Graphs of $\sinh(x)$ and $\sinh^{-1}(x)$.

7.5.2 Hyperbolic Identities

The hyperbolic functions also have similar identities to their trigonometrical counterparts.

Example 7.7

In order to prove $\cosh^2 x - \sinh^2 x = 1$, we start from

$$\begin{aligned}\cosh^2 x - \sinh^2 &= (\cosh x + \sinh x)(\cosh x - \sinh x)\\ &= (\frac{e^x + e^{-x}}{2} + \frac{e^x - e^{-x}}{2})(\frac{e^x + e^{-x}}{2} - \frac{e^x - e^{-x}}{2}) = (\frac{2e^x}{2})(\frac{2e^{-x}}{2}) = 1.\end{aligned}$$

This is similar to $\cos^2 x + \sin^2 x = 1$.

There is a quick way to obtain the corresponding identities from those identities for trigonometrical functions. Taking the squares of both sides of (7.28), we have

$$\cos^2 x = \cosh^2(ix), \qquad \sin^2 x = \frac{1}{i^2}\sinh^2(ix) = -\sinh^2(ix). \tag{7.30}$$

This implies that we can replace $\cos^2 x$ by $\cosh^2 x$ and $\sin^2 x$ by $-\sinh^2 x$ in the identity $\cos^2 x + \sin^2 x = 1$; we will get the identity $\cosh^2 x - \sinh^2 x = 1$ for hyperbolic functions. That is Osborn's rule for converting identities where we only need to change the sign of any terms containing the squares of a sine function including tangent (tan) and cotangent (cot). For example, we can change

$$\tan 2\theta = \frac{2\tan\theta}{1 - \tan^2\theta}, \quad \text{into} \quad \tanh 2x = \frac{2\tanh x}{1 + \tanh^2 x}. \tag{7.31}$$

In fact, we can even extend Osborn's rule further to include $\cos \to \cosh$ and $\sin \to i\sinh$. For example, from the identity $\cos(\alpha + \beta) = \cos\alpha\cos\beta - \sin\alpha\sin\beta$, we have

$$\begin{aligned}\cosh(\alpha + \beta) &= \cosh\alpha\cosh\beta - i\sinh\alpha \times i\sinh\beta\\ &= \cosh\alpha\cosh\beta + \sinh\alpha\sinh\beta,\end{aligned} \tag{7.32}$$

where we have used $i^2 = -1$.

7.5.3 Inverse Hyperbolic Functions

The inverse of $\sinh x$ can easily be obtained (graphically shown in Fig. 7.4) by simply reflecting the graph of $\sinh x$ in the line of $y = x$ (dashed line). Mathematically speaking, we want to find $y = \sinh^{-1} x$ such that $\sinh y = x$. From the identity $\cosh^2 y - \sinh^2 y = 1$, we have

$$\cosh^2 y = 1 + \sinh^2 y = 1 + x^2. \tag{7.33}$$

Since $\cosh y = (e^y + e^{-y})/2 \geq 1$ (see Fig. 7.3), the above equation becomes

$$\cosh y = \sqrt{1 + x^2}. \tag{7.34}$$

In addition, from the previous example, we know that $\cosh y + \sinh y = \frac{e^y+e^{-y}}{2} + \frac{e^y-e^{-y}}{2} = e^y$. After combining the above two equations, we now have

$$\sqrt{1+x^2} + \sinh y = e^y. \tag{7.35}$$

Using $\sinh y = x$, we have $\sqrt{1+x^2} + x = e^y$. Taking the logarithm of both sides, we finally obtain

$$y = \sinh^{-1} x = \ln[x + \sqrt{1+x^2}]. \tag{7.36}$$

The inverse of other functions can be obtained in a similar manner.

Example 7.8

In order to get $y = \cosh^{-1} x$ or $\cosh y = x$, we use

$$\cosh^2 y - \sinh^2 y = 1,$$

so that $\sinh^2 y = \cosh^2 y - 1 = x^2 - 1$, which gives $\sinh y = \sqrt{x^2 - 1}$ where $(x \geq 1)$. Again using the identity $\cosh y + \sinh y = e^y$, we have

$$x + \sinh y = x + \sqrt{x^2 - 1} = e^y.$$

Taking the logarithms, we now have

$$y = \cosh^{-1} x = \ln[x + \sqrt{x^2 - 1}]$$

where $x \geq 1$. It is worth pointing out that there are two branches of $\cosh^{-1} x$, and we have assumed here that $\cosh^{-1} x > 0$.

Complex numbers have a wide range of applications as many mathematical techniques become simpler and more powerful using complex numbers. For example, in the discrete Fourier series to be introduced later in this book, formulas are simpler when written in the form of complex numbers.

EXERCISES

7.1. Simplify the following expressions:

- $i^{100} + i^{10} + i + 1$
- $e^{-\frac{\pi i}{2}} + e^{-i\pi} + e^{-2\pi i}$
- $\sinh(1) + \cosh(1) - \sinh(0) - \cosh(0)$
- $\cos(\pi) + i\sin(\pi) - \sqrt{2i}$
- $e^{i\pi} e^{-i\pi/2}/(2i) - i$

7.2. Show that $\cosh(2x) = \cosh^2(x) + \sinh^2(x)$.

7.3. Use any programming language to plot out the basic hyperbolic functions $\sinh(x)$, $\sinh(2x)$, $\cosh(x)$, $\cosh(2x)$ and $\tanh(x)$ on the same graph.

7.4. Find a possible use of $\sinh(x)$

7.5. A hyperbola in a two-dimensional Cartesian coordinate system (x, y) can be described as

$$\frac{x^2}{a^2} - \frac{y^2}{b^2} = 1,$$

where $a, b > 0$. This can also be written as a set of two parametric equations $x = a\cosh(t)$ and $y = b\sinh(t)$ where t is a varying parameter, called hyperbolic angle. Discuss the links between these two representations and then try to plot out a hyperbola for $a = 1$ and $b = 2$. What is the proper domain of t?

Part III

Vectors and Matrices

Chapter 8

Vectors and Vector Algebra

Chapter Points

- Basic concepts of vectors are explained, together with vector algebra such as the addition and subtraction of vectors.
- Vector products are introduced, including the dot product, cross product and triple vector product.
- The meaning of vector products is also explained and demonstrated with examples related to force, area and volume.

8.1 VECTORS

Suppose we travel from a point P at (x_1, y_1) to another point Q at (x_2, y_2), we have a displacement vector $\mathbf{d} = \overrightarrow{PQ}$ (see Fig. 8.1).

For the displacement vector, we need the magnitude (or the length or distance) between P and Q, and also the direction or angle θ to determine the vector uniquely. Since the coordinates of two points P and Q are given, the distance between P and Q can be calculated using the Cartesian distance. The length or magnitude of $\mathbf{d}$ can conveniently be written as $d = |\mathbf{d}|$, and we have

$$PQ = |\overrightarrow{PQ}| = |\mathbf{d}| = \sqrt{(x_2 - x_1)^2 + (y_2 - y_1)^2}. \tag{8.1}$$

The magnitude of a vector is also called the modulus of the vector. Here we follow the conventions of using a single letter in italic form to denote the magnitude while using the same letter in bold type to denote the vector itself.

The direction of the vector is represented by the angle θ from the x-axis. We have

$$\tan\theta = \frac{y_2 - y_1}{x_2 - x_1}. \tag{8.2}$$

Conventionally, we often write a vector using bold font $\mathbf{d}$, rather than d. In many books, vectors are also written in the overhead arrow form such as $\overrightarrow{PQ}$ or simply $\vec{d}$. The notation $\overrightarrow{PQ}$ signifies that the vector is pointing from P to Q. Here we will use the bold-type notations as they are more popularly used in mathematics. The components of the vector $\mathbf{d}$ are $x_2 - x_1$ along the x-axis and $y_2 - y_1$ along the y-axis. This provides a way to write the vector as

$$\mathbf{d} = \overrightarrow{PQ} = \begin{pmatrix} x_2 - x_1 \\ y_2 - y_1 \end{pmatrix}. \tag{8.3}$$

Here we write the vector as a column, called a column vector.

Example 8.1

Reading from the graph shown in Fig. 8.1, we know that P is at $(1, 1)$ and Q is at $(5, 4)$. The displacement can be represented in the mathematical form as

$$\mathbf{d} = \begin{pmatrix} x_2 - x_1 \\ y_2 - y_1 \end{pmatrix} = \begin{pmatrix} 5 - 1 \\ 4 - 1 \end{pmatrix} = \begin{pmatrix} 4 \\ 3 \end{pmatrix}.$$

Therefore, the distance PQ or the magnitude d of the displacement $\mathbf{d}$ is

$$d = |\mathbf{d}| = \sqrt{(5-1)^2 + (4-1)^2} = \sqrt{4^2 + 3^2} = 5.$$

Engineering Mathematics with Examples and Applications

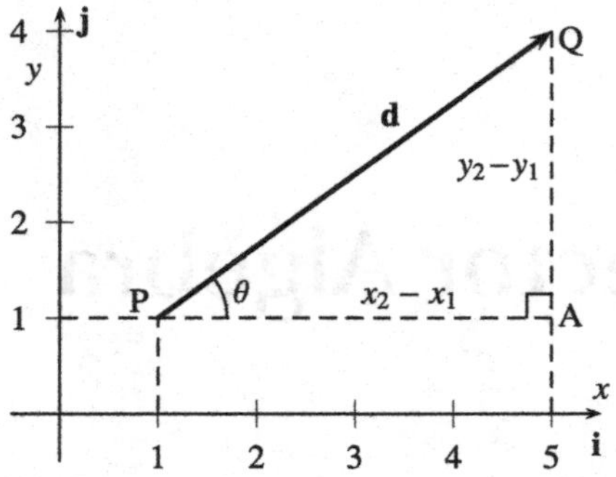

FIGURE 8.1 The displacement from $P(x_1, y_1)$ to point $Q(x_2, y_2)$.

The angle θ is given by

$$\tan\theta = \frac{4-1}{5-1} = \frac{3}{4} = 0.75,$$

or

$$\theta = \tan^{-1} = 0.75 \approx 36.87°.$$

For real-world problems, the unit of length must be given, either in meters or any other suitable units.

It is worth pointing out that the vector $\overrightarrow{QP}$ is pointing the opposite direction $\overrightarrow{PQ}$, and we thus have

$$\overrightarrow{QP} = \begin{pmatrix} x_1 - x_2 \\ y_1 - y_2 \end{pmatrix} = \begin{pmatrix} (-1)(x_2 - x_1) \\ (-1)(y_2 - y_1) \end{pmatrix} = -\begin{pmatrix} x_2 - x_1 \\ y_2 - y_1 \end{pmatrix} = -\overrightarrow{PQ} = -\mathbf{d}.$$

In general for any real number $\beta \neq 0$ and a vector $\mathbf{v} = \begin{pmatrix} a \\ b \end{pmatrix}$, we have

$$\beta\mathbf{v} = \beta\begin{pmatrix} a \\ b \end{pmatrix} = \begin{pmatrix} \beta a \\ \beta b \end{pmatrix}. \tag{8.4}$$

A vector whose magnitude is 1 is called a unit vector. So all the following vectors are unit vectors

$$\mathbf{i} = \begin{pmatrix} 1 \\ 0 \end{pmatrix}, \quad \mathbf{j} = \begin{pmatrix} 0 \\ 1 \end{pmatrix}, \quad \mathbf{w} = \begin{pmatrix} \cos\theta \\ \sin\theta \end{pmatrix}. \tag{8.5}$$

For any θ, we know $|\mathbf{w}| = \sqrt{\cos^2\theta + \sin^2\theta} = 1$ due to the identity $\sin^2\theta + \cos^2\theta = 1$. The vectors $\mathbf{i}$ and $\mathbf{j}$ are the unit vectors along x-axis and y-axis directions, respectively.

8.2 VECTOR ALGEBRA

Since a vector has a magnitude and a direction, any two vectors with the same magnitude and direction should be equal since there is no other constraint. This means that we can shift and move both ends of a vector by any same amount in any direction, and we still have the same vector. In other words, if two vectors are equal, they must have the same magnitude and direction. Mathematically, their corresponding components must be equal.

If no physical barrier is our concern, then we can reach point Q from point P in an infinite number of ways. We can go along the x-axis direction to the right for a distance $x_2 - x_1$ to the point A, and then go upward along the y-direction for a distance $y_2 - y_1$. This is equivalent to saying that $\mathbf{d}$ is the sum of two vectors $(x_2 - x_1)\mathbf{i}$ and $(y_2 - y_1)\mathbf{j}$. We have

$$\mathbf{d} = \begin{pmatrix} x_2 - x_1 \\ y_2 - y_1 \end{pmatrix} = (x_2 - x_1)\mathbf{i} + (y_2 - y_1)\mathbf{j}. \tag{8.6}$$

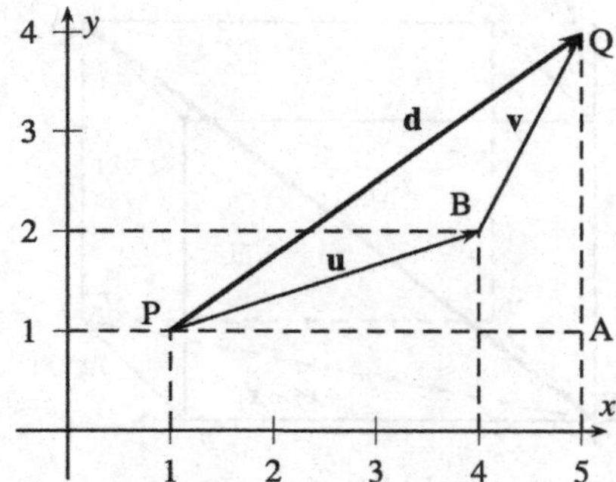

FIGURE 8.2 The displacement from point P(1,1) to point Q(5,4).

Example 8.2

Similarly, we can first go along the direction of PB and then along BQ (see Fig. 8.2). This also suggests that

$$\overrightarrow{PQ} = \overrightarrow{PB} + \overrightarrow{BQ}, \quad \text{or} \quad \mathbf{d} = \mathbf{u} + \mathbf{v}. \tag{8.7}$$

Now the point B is at (4, 2). So we have

$$\mathbf{u} = \begin{pmatrix} 4-1 \\ 2-1 \end{pmatrix} = \begin{pmatrix} 3 \\ 1 \end{pmatrix}, \quad \mathbf{v} = \begin{pmatrix} 5-4 \\ 4-2 \end{pmatrix} = \begin{pmatrix} 1 \\ 2 \end{pmatrix}, \quad \mathbf{d} = \begin{pmatrix} 4 \\ 3 \end{pmatrix}, \tag{8.8}$$

which suggests that

$$\mathbf{u} + \mathbf{v} = \begin{pmatrix} 3 \\ 1 \end{pmatrix} + \begin{pmatrix} 1 \\ 2 \end{pmatrix} = \begin{pmatrix} 3+1 \\ 1+2 \end{pmatrix} = \begin{pmatrix} 4 \\ 3 \end{pmatrix} = \mathbf{d}. \tag{8.9}$$

The addition of two vectors is a vector whose components are simply the addition of their corresponding components. If we define the subtraction of any two vectors **u** and **v** as

$$\mathbf{u} - \mathbf{v} = \mathbf{u} + (-\mathbf{v}), \tag{8.10}$$

where $-\mathbf{v}$ is obtained by flipping **v** by 180°. In general, we have

$$\mathbf{v}_1 \pm \mathbf{v}_2 = \begin{pmatrix} a_1 \\ b_1 \end{pmatrix} \pm \begin{pmatrix} a_2 \\ b_2 \end{pmatrix} = \begin{pmatrix} a_1 \pm a_2 \\ b_1 \pm b_2 \end{pmatrix}. \tag{8.11}$$

The addition of any two vectors **u** and **v** is commutative, that is

$$\mathbf{v}_1 + \mathbf{v}_2 = \mathbf{v}_2 + \mathbf{v}_1. \tag{8.12}$$

This is because each of its components is commutative: $a_1 + a_2 = a_2 + a_1$ and $b_1 + b_2 = b_2 + b_1$. Similarly, as the addition of scalars is associative (i.e., $a_1 + (a_2 + a_3) = (a_1 + a_2) + a_3$), then the addition of vectors is associative as well. That is

$$\mathbf{v}_1 + (\mathbf{v}_2 + \mathbf{v}_3) = (\mathbf{v}_1 + \mathbf{v}_2) + \mathbf{v}_3. \tag{8.13}$$

So far we have only focused on the vectors in a two-dimensional plane; we can easily extend our discussion to 3D vectors or higher-dimensional vectors. For the 3D vector shown in Fig. 8.3, we have

$$\mathbf{d} = \overrightarrow{PQ} = \begin{pmatrix} x_2 - x_1 \\ y_2 - y_1 \\ z_2 - z_1 \end{pmatrix}. \tag{8.14}$$

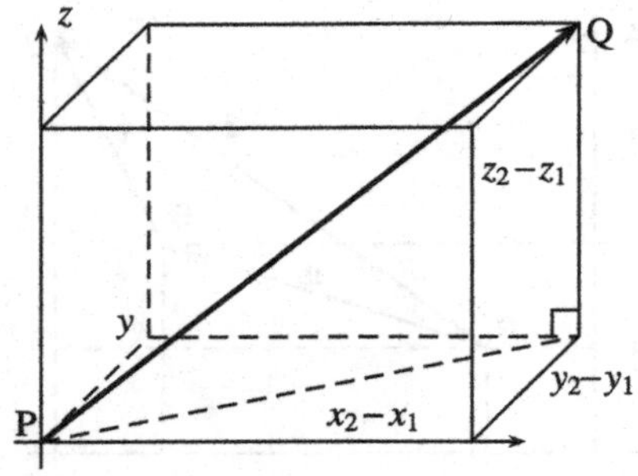

FIGURE 8.3 The 3D displacement vector from $P(x_1, y_1, z_1)$ to point $Q(x_2, y_2, z_2)$.

If we define the unit vectors as

$$\mathbf{i} = \begin{pmatrix} 1 \\ 0 \\ 0 \end{pmatrix}, \quad \mathbf{j} = \begin{pmatrix} 0 \\ 1 \\ 0 \end{pmatrix}, \quad \mathbf{k} = \begin{pmatrix} 0 \\ 0 \\ 1 \end{pmatrix}, \tag{8.15}$$

for the three perpendicular directions, we can write **d** as

$$\mathbf{d} = \begin{pmatrix} x_2 - x_1 \\ y_2 - y_1 \\ z_2 - z_1 \end{pmatrix} = (x_2 - x_1)\mathbf{i} + (y_2 - y_1)\mathbf{j} + (z_2 - z_1)\mathbf{k}. \tag{8.16}$$

The addition and subtraction of any two vectors now becomes

$$\mathbf{v}_1 \pm \mathbf{v}_2 = \begin{pmatrix} a_1 \\ b_1 \\ c_1 \end{pmatrix} \pm \begin{pmatrix} a_2 \\ b_2 \\ c_2 \end{pmatrix} = \begin{pmatrix} a_1 \pm a_2 \\ b_1 \pm b_2 \\ c_1 \pm c_2 \end{pmatrix}. \tag{8.17}$$

For any two vectors **u** and **v**, we have an inequality in terms of their lengths

$$|\mathbf{u} + \mathbf{v}| \leq |\mathbf{u}| + |\mathbf{v}|. \tag{8.18}$$

Example 8.3

For two 3D vectors

$$\mathbf{u} = \begin{pmatrix} 1 \\ 2 \\ -2 \end{pmatrix}, \quad \mathbf{v} = \begin{pmatrix} -2 \\ -1 \\ 3 \end{pmatrix},$$

we have

$$\mathbf{u} + \mathbf{v} = \begin{pmatrix} 1 \\ 2 \\ -2 \end{pmatrix} + \begin{pmatrix} -2 \\ -1 \\ 3 \end{pmatrix} = \begin{pmatrix} 1 + (-2) \\ 2 + (-1) \\ -2 + 3 \end{pmatrix} = \begin{pmatrix} -1 \\ 1 \\ 1 \end{pmatrix},$$

and

$$\mathbf{u} - \mathbf{v} = \begin{pmatrix} 1 \\ 2 \\ -2 \end{pmatrix} - \begin{pmatrix} -2 \\ -1 \\ 3 \end{pmatrix} = \begin{pmatrix} 1 - (-2) \\ 2 - (-1) \\ -2 - 3 \end{pmatrix} = \begin{pmatrix} 3 \\ 3 \\ -5 \end{pmatrix}.$$

The lengths of two vectors are

$$|\mathbf{u}| = \sqrt{1^2 + 2^2 + (-2)^2} = \sqrt{9} = 3,$$

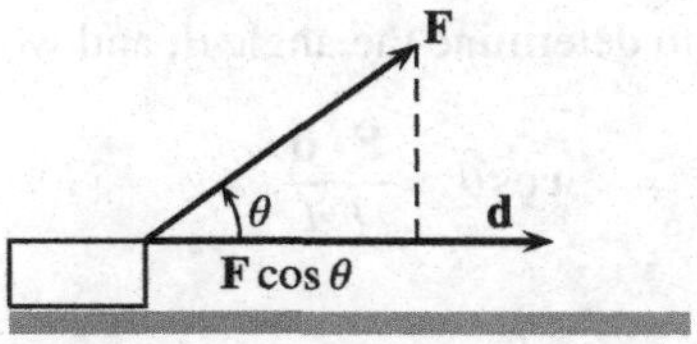

FIGURE 8.4 Work done W by a force $\mathbf{F}$ to move an object in the direction of displacement $\mathbf{d}$ is $W = \mathbf{F} \cdot \mathbf{d} = Fd\cos\theta$.

and

$$|\mathbf{v}| = \sqrt{(-2)^2 + (-1)^2 + 3^2} = \sqrt{14}.$$

Since $|\mathbf{u}+\mathbf{v}| = \sqrt{(-1)^2 + 1^2 + 1^2} = \sqrt{3}$, we indeed have

$$|\mathbf{u}| + |\mathbf{v}| = 3 + \sqrt{14} > |\mathbf{u}+\mathbf{v}| = \sqrt{3}.$$

These formulas can be extended to the addition and subtraction of multiple vectors.

8.3 VECTOR PRODUCTS

The product of two vectors can be either a scalar or a vector, depending on the way we carry out the multiplications. The scalar product of two vectors $\mathbf{F}$ and $\mathbf{d}$ is defined as

$$\mathbf{F} \cdot \mathbf{d} = Fd\cos\theta. \tag{8.19}$$

This rather odd definition has some physical meaning. For example, we know that the work W done by a force f to move an object a distance s, is simply $W = fs$ on the condition that the force is applied along the direction of movement. If a force $\mathbf{F}$ is applied at an angle θ related to the displacement $\mathbf{d}$ (see Fig. 8.4), we first have to decompose or project the force $\mathbf{F}$ onto the displacement direction so that the component actually acts on the object along the direction of $\mathbf{d}$ is $\mathbf{F}_{\|} = F\cos\theta$. So the actual work done becomes

$$W = \mathbf{F}_{\|} d = Fd\cos\theta, \tag{8.20}$$

which means that the amount of work W is the scalar product

$$W = \mathbf{F} \cdot \mathbf{d}. \tag{8.21}$$

Here the $\cdot$ symbol denotes such a scalar product. From such notations, the scalar product of two vectors is also called the dot product or inner product.

If we intend to compute in terms of their components

$$\mathbf{F} = \begin{pmatrix} f_1 \\ f_2 \\ f_3 \end{pmatrix}, \qquad \mathbf{d} = \begin{pmatrix} d_1 \\ d_2 \\ d_3 \end{pmatrix}, \tag{8.22}$$

the dot product can be calculated by

$$\mathbf{F} \cdot \mathbf{d} = f_1 d_1 + f_2 d_2 + f_3 d_3. \tag{8.23}$$

Since $\cos 90° = 0$, when the scalar product is zero, it suggests that the two vectors are perpendicular to each other; sometimes we also say they are orthogonal. So for the unit vectors $\mathbf{i}$, $\mathbf{j}$, and $\mathbf{k}$, we have

$$\mathbf{i}\cdot\mathbf{i} = \mathbf{j}\cdot\mathbf{j} = \mathbf{k}\cdot\mathbf{k} = 1, \quad \text{or} \quad \mathbf{i}\cdot\mathbf{j} = \mathbf{j}\cdot\mathbf{k} = \mathbf{k}\cdot\mathbf{i} = 0. \tag{8.24}$$

These basic properties can easily be verified by using the formula (8.23).

If we know the dot product, we can use it to determine the angle θ, and we have

$$\cos\theta = \frac{\mathbf{F} \cdot \mathbf{d}}{Fd}. \tag{8.25}$$

Example 8.4

To test if the following vectors are orthogonal or not

$$\mathbf{u} = \begin{pmatrix} 1 \\ 2 \\ 3 \end{pmatrix}, \quad \mathbf{v} = \begin{pmatrix} 2 \\ 2 \\ -2 \end{pmatrix},$$

we can calculate their dot product

$$\mathbf{u} \cdot \mathbf{v} = 1 \cdot 2 + 2 \cdot 2 + 3 \cdot (-2) = 2 + 4 - 6 = 0.$$

As their dot product is zero, these two vectors are perpendicular to each other.

The dot product has some interesting properties. From its definition, it is easy to see that $\mathbf{F} \cdot \mathbf{d} = \mathbf{d} \cdot \mathbf{F}$. Another interesting property is the distributive law:

$$\mathbf{F} \cdot (\mathbf{d} + \mathbf{s}) = \mathbf{F} \cdot \mathbf{d} + \mathbf{F} \cdot \mathbf{s}. \tag{8.26}$$

Now let us prove the above distributive law. Using

$$\mathbf{s} = \begin{pmatrix} s_1 \\ s_2 \\ s_3 \end{pmatrix}, \tag{8.27}$$

we have

$$\mathbf{F} \cdot \mathbf{d} = f_1 d_1 + f_2 d_2 + f_3 d_3, \qquad \mathbf{F} \cdot \mathbf{s} = f_1 s_1 + f_2 s_2 + f_3 s_3. \tag{8.28}$$

Now we have

$$\begin{aligned} \mathbf{F} \cdot (\mathbf{d} + \mathbf{s}) &= \mathbf{F} \cdot \left[\begin{pmatrix} d_1 \\ d_2 \\ d_3 \end{pmatrix} + \begin{pmatrix} s_1 \\ s_2 \\ s_3 \end{pmatrix} \right] = \mathbf{F} \cdot \begin{pmatrix} d_1 + s_1 \\ d_2 + s_2 \\ d_3 + s_3 \end{pmatrix} \\ &= f_1(d_1 + s_1) + f_2(d_2 + s_2) + f_3(d_3 + s_3) \\ &= (f_1 d_1 + f_2 d_2 + f_3 d_3) + (f_1 s_1 + f_2 s_2 + f_3 s_3) = \mathbf{F} \cdot \mathbf{d} + \mathbf{F} \cdot \mathbf{s}, \end{aligned} \tag{8.29}$$

which is the distributive law.

Example 8.5

If a horse is towing a boat with a similar configuration as shown in Fig. 8.4, the force is 500 N with an angle of 30° degree of the traveling direction. The boat travels for 1 km = 1000 m at the speed of 3.6 km/hour. What is the work done? What is the output power of the horse?

The work done W is

$$W = \mathbf{F} \cdot \mathbf{d} = 500 \times 1000 \cos 30° = 5 \times 10^5 \times \frac{\sqrt{3}}{2} \approx 4.3 \times 10^5 \text{ Joules}.$$

The speed of the boat (and the horse) is 3.6 km/hour or $\mathbf{v} = 3.6 \times 1000/3600 = 1$ m/s. Thus, the output power of the horse is

$$P = \mathbf{F} \cdot \mathbf{v} = 500 \times 1 \cos 30° = 500 \times \frac{\sqrt{3}}{2} \approx 433 \text{ watts}.$$

FIGURE 8.5 The direction of $\mathbf{u} \times \mathbf{v}$ points to the direction along $\mathbf{n}$ while the magnitude $\mathbf{Q} = |\mathbf{u} \times \mathbf{v}| = |u||v| \sin\theta$ is the area of the shaded region.

The vector product, also called the cross product or outer product, of two vectors $\mathbf{u}$ and $\mathbf{v}$ forms another vector $\mathbf{Q}$. The definition can be written as

$$\mathbf{u} \times \mathbf{v} = uv \sin\theta \, \mathbf{n}, \tag{8.30}$$

where $\mathbf{n}$ is the unit vector, and $\mathbf{Q}$ points to the direction of $\mathbf{n}$ which is perpendicular to both vectors $\mathbf{u}$ and $\mathbf{v}$, forming a right-handed system (see Fig. 8.5). In addition, θ is the angle between $\mathbf{u}$ and $\mathbf{v}$, and u and v are the magnitudes of $\mathbf{u}$ and $\mathbf{v}$, respectively.

In many books, the notation $\mathbf{u} \wedge \mathbf{v}$ is also used, that is

$$\mathbf{u} \wedge \mathbf{v} \equiv \mathbf{u} \times \mathbf{v}. \tag{8.31}$$

The right-handed system suggests that, if we change the order of the product, there is a sign change. That is, $\mathbf{v} \times \mathbf{u} = -\mathbf{u} \times \mathbf{v}$.

Though the vector product is a vector; however, its magnitude has a geometrical meaning. That is, the magnitude is the area of the shaded parallelogram shown in Fig. 8.5.

Using $\mathbf{u} = \begin{pmatrix} u_1 & u_2 & u_3 \end{pmatrix}^T$, and $\mathbf{v} = \begin{pmatrix} v_1 & v_2 & v_3 \end{pmatrix}^T$ where the superscript T means the transpose which turns a column vector into a row vector or vice versa, we can write the vector product in terms of their components

$$\mathbf{u} \times \mathbf{v} = \begin{vmatrix} \mathbf{i} & \mathbf{j} & \mathbf{k} \\ u_1 & u_2 & u_3 \\ v_1 & v_2 & v_3 \end{vmatrix} = \begin{pmatrix} u_2 v_3 - u_3 v_2 \\ u_3 v_1 - u_1 v_3 \\ u_1 v_2 - u_2 v_1 \end{pmatrix}$$
$$= (u_2 v_3 - u_3 v_2)\mathbf{i} + (u_3 v_1 - u_1 v_3)\mathbf{j} + (u_1 v_2 - u_2 v_1)\mathbf{k}. \tag{8.32}$$

Here, we have used $| \cdot |$ as a notation, though it is the determinant of a matrix, and we will introduce this concept in the next chapter.

Example 8.6

A robotic arm rotates around an axis with an angular velocity $\omega = (1, 0, 1)^T$, what is the linear velocity at the position $\mathbf{r} = (2, 2, 1)^T$?

The velocity $\mathbf{v}$ can be calculated by

$$\mathbf{v} = \omega \times \mathbf{r} = \begin{pmatrix} 1 \\ 0 \\ 1 \end{pmatrix} \times \begin{pmatrix} 2 \\ 2 \\ 1 \end{pmatrix} = \begin{vmatrix} \mathbf{i} & \mathbf{j} & \mathbf{k} \\ 1 & 0 & 1 \\ 2 & 2 & 1 \end{vmatrix}$$
$$= (0 \times 1 - 1 \times 2)\mathbf{i} + (1 \times 2 - 1 \times 1)\mathbf{j} + (1 \times 2 - 0 \times 2)\mathbf{k} = -2\mathbf{i} + \mathbf{j} + 2\mathbf{k} = \begin{pmatrix} -2 \\ 1 \\ 2 \end{pmatrix}.$$

So the velocity points towards the $(-2, 1, 2)^T$ direction.

8.4 TRIPLE PRODUCT OF VECTORS

For three different vectors $\mathbf{u}$, $\mathbf{v}$ and $\mathbf{w}$, it is possible to form a triple product with some physical meaning.

For example, the product $\mathbf{u} \cdot (\mathbf{v} \times \mathbf{w})$ is the volume V of a parallelepiped with sides $\mathbf{u}$, $\mathbf{v}$ and $\mathbf{w}$. That is

$$V = \mathbf{u} \cdot (\mathbf{v} \times \mathbf{w}). \tag{8.33}$$

Since $|\mathbf{v} \times \mathbf{w}|$ is the area of the (base) parallelogram formed by $\mathbf{v}$ and $\mathbf{w}$, and the component of $\mathbf{u}$ in the direction of $\mathbf{v} \times \mathbf{w}$ is the height, the triple product is indeed the volume.

Careful readers may already wonder why we choose **v** and **w** to form the base parallelogram? In fact, there are two other ways to calculate the triple product with the same result. We have

$$\mathbf{u} \cdot (\mathbf{v} \times \mathbf{w}) = \mathbf{w} \cdot (\mathbf{u} \times \mathbf{v}) = \mathbf{v} \cdot (\mathbf{w} \times \mathbf{u}), \tag{8.34}$$

which is often referred to as the cyclic permutation.

Example 8.7

Let us calculate the volume of the parallelepiped formed by the following three vectors

$$\mathbf{u} = \begin{pmatrix} 1 \\ 0 \\ 0 \end{pmatrix}, \quad \mathbf{v} = \begin{pmatrix} 1 \\ 1 \\ 0 \end{pmatrix}, \quad \mathbf{w} = \begin{pmatrix} 2 \\ 3 \\ 5 \end{pmatrix}.$$

First, we form the cross product

$$\mathbf{u} \times \mathbf{v} = \begin{pmatrix} u_2v_3 - u_3v_2 \\ u_3v_1 - u_1v_3 \\ u_1v_2 - u_2v_1 \end{pmatrix} = \begin{pmatrix} 0 \times 0 - 0 \times 1 \\ 0 \times 1 - 1 \times 0 \\ 1 \times 1 - 0 \times 1 \end{pmatrix} = \begin{pmatrix} 0 \\ 0 \\ 1 \end{pmatrix}.$$

The volume is the triple product

$$\mathbf{w} \cdot (\mathbf{u} \times \mathbf{v}) = \begin{pmatrix} 2 \\ 3 \\ 5 \end{pmatrix} \cdot \begin{pmatrix} 0 \\ 0 \\ 1 \end{pmatrix} = 2 \times 0 + 3 \times 0 + 5 \times 1 = 5.$$

Here we have only covered the very basics of the vector algebra and there are more complex manipulations of vectors including other triple products, differentiation and integrals. We will introduce such more advanced concepts in later chapters.

EXERCISES

8.1. For three vectors: $\mathbf{u} = \begin{pmatrix} 1 \\ 2 \end{pmatrix}$, $\mathbf{v} = \begin{pmatrix} 2 \\ -1 \end{pmatrix}$, and $\mathbf{w} = \begin{pmatrix} 2 \\ 3 \end{pmatrix}$, show that $|\mathbf{u} + \mathbf{v} + \mathbf{w}| \le |\mathbf{u}| + |\mathbf{v}| + |\mathbf{w}|$.

8.2. For the same vectors as before, what is the angle between **u** and **v**? What is angle between **v** and **w**? Is the sum of the previous two angles the same as the angle between **u** and **w**?

Hint. Try to compute the modulus of each vector, and then compute their dot products as well as angles.

8.3. A balloon first flies in the northwest direction for 2 km, then it flies to the south for 500 meter, followed by the move towards the east for 1.5 km. Find out how far the balloon is from its starting location?

8.4. A person tries to swim across a river of 10 meter wide and the water in the river flows at 2 m/s downstream. If the person has an average swimming speed of 1.5 m/s, at what angle should the swimmer aim so as to cross the river with the fastest time (or the shortest swimming path)?

8.5. An object is sliding, at a constant velocity of 3 m/s, under its gravity down a slope of 30° above the horizontal direction. What is the friction coefficient between the object and surface?

Hint. Try to draw the directions of the frictional force, gravity and then balance all the forces. The friction coefficient should be independent of the weight.

Chapter 9

Matrices

Chapter Points

- Introduce all the fundamentals of matrices, including the definition of matrices, their addition and subtraction and products.
- Both the inverse of a square matrix and the transformation matrix are explained, together with the calculation of determinant of a square matrix.
- A linear system of simultaneous equations and its matrix representation are discussed in detail.
- Eigenvalues are associated with many applications and the ways of calculating eigenvalues are explained in detail, together with the explanation of constructing their corresponding eigenvectors.

9.1 MATRICES

A matrix is a rectangular array of numbers. For example, a coffee shop sells four different type of coffees, and the sales in terms of the numbers of cups for three consecutive days are recorded as follows:

$$\begin{array}{ccccc} & A & B & C & D \\ \text{Day 1} & 210 & 256 & 197 & 207 \\ \text{Day 2} & 242 & 250 & 205 & 199 \\ \text{Day 3} & 192 & 249 & 220 & 215 \end{array}, \tag{9.1}$$

where different products form a row, and different days for the same product form a column. This can be written as the sale matrix with 3 rows and 4 columns

$$\mathbf{S} = \begin{pmatrix} 210 & 256 & 197 & 207 \\ 242 & 250 & 205 & 199 \\ 192 & 249 & 220 & 215 \end{pmatrix}. \tag{9.2}$$

Each item of the numbers is called an entry or element of the matrix. We usually use a bold-type upper case to denote a matrix, and we use the lower case to denote its elements. Therefore, we have

$$\mathbf{S} = [s_{ij}], \qquad (i = 1, 2, 3, \text{and } j = 1, 2, 3, 4). \tag{9.3}$$

The element on the second row and the third column is $s_{23} = 205$.

The transpose of a matrix $\mathbf{S}$ can be obtained by interchanging its rows and columns, and is denoted by $\mathbf{S}^T$. We have

$$\mathbf{S}^T = \begin{pmatrix} 210 & 256 & 197 & 207 \\ 242 & 250 & 205 & 199 \\ 192 & 249 & 220 & 215 \end{pmatrix}^T = \begin{pmatrix} 210 & 242 & 192 \\ 256 & 250 & 249 \\ 197 & 205 & 220 \\ 207 & 199 & 215 \end{pmatrix}. \tag{9.4}$$

The same coffee shop owner owns another coffee shop on a different street, selling the same products. The same three-day sales are

$$\mathbf{Q} = \begin{pmatrix} 191 & 229 & 170 & 240 \\ 195 & 209 & 199 & 214 \\ 207 & 272 & 149 & 190 \end{pmatrix}. \tag{9.5}$$

The total sales of both shops are obtained by the addition of their corresponding entries

$$\mathbf{S}+\mathbf{Q}=\begin{pmatrix} 210 & 256 & 197 & 207 \\ 242 & 250 & 205 & 199 \\ 192 & 249 & 220 & 215 \end{pmatrix}+\begin{pmatrix} 191 & 229 & 170 & 240 \\ 195 & 209 & 199 & 214 \\ 207 & 272 & 149 & 190 \end{pmatrix}$$
$$=\begin{pmatrix} 210+191 & 256+229 & 197+170 & 207+240 \\ 242+195 & 250+209 & 205+199 & 199+214 \\ 192+207 & 249+272 & 220+149 & 215+190 \end{pmatrix}$$
$$=\begin{pmatrix} 401 & 485 & 367 & 447 \\ 437 & 459 & 404 & 413 \\ 399 & 521 & 369 & 405 \end{pmatrix}. \tag{9.6}$$

Their sales differences are

$$\mathbf{S}-\mathbf{Q}=\begin{pmatrix} 210 & 256 & 197 & 207 \\ 242 & 250 & 205 & 199 \\ 192 & 249 & 220 & 215 \end{pmatrix}-\begin{pmatrix} 191 & 229 & 170 & 240 \\ 195 & 209 & 199 & 214 \\ 207 & 272 & 149 & 190 \end{pmatrix}$$
$$=\begin{pmatrix} 210-191 & 256-229 & 197-170 & 207-240 \\ 242-195 & 250-209 & 205-199 & 199-214 \\ 192-207 & 249-272 & 220-149 & 215-190 \end{pmatrix}$$
$$=\begin{pmatrix} 19 & 27 & 27 & -33 \\ 47 & 41 & 6 & -15 \\ -15 & -23 & 71 & 25 \end{pmatrix}. \tag{9.7}$$

We can see here that the addition and subtraction of the matrices are carried out entry by entry. It is only possible to carry out addition and subtraction if and only if the matrices **S** and **Q** have the same numbers of rows and columns.

Suppose that the prices for each product are: GBP 0.99 for A, GBP 1.50 for B, GBP 1.15 for C, and GBP 0.90 for D. This can be written as a vector

$$\mathbf{P}=(\ 0.99, 1.50, 1.15, 0.90\ )^T, \tag{9.8}$$

or more formally as a column matrix or a column vector

$$\mathbf{P}=\begin{pmatrix} 0.99 \\ 1.50 \\ 1.15 \\ 0.90 \end{pmatrix}. \tag{9.9}$$

The total sales income for each day is given by the multiplication of **S** and **p**.

$$\mathbf{I}_1=\mathbf{SP}=\begin{pmatrix} 210 & 256 & 197 & 207 \\ 242 & 250 & 205 & 199 \\ 192 & 249 & 220 & 215 \end{pmatrix}\begin{pmatrix} 0.99 \\ 1.50 \\ 1.15 \\ 0.90 \end{pmatrix}$$
$$=\begin{pmatrix} 210\times 0.99+256\times 1.50+197\times 1.15+207\times 0.90 \\ 242\times 0.99+250\times 1.50+205\times 1.15+199\times 0.90 \\ 192\times 0.99+249\times 1.50+220\times 1.15+215\times 0.90 \end{pmatrix}$$
$$=\begin{pmatrix} 1004.75 \\ 1029.43 \\ 1010.08 \end{pmatrix}, \tag{9.10}$$

which means that the total incomes for the three days are GBP 1004.75, GBP 1029.43 and GBP 1010.08, respectively. In general, the multiplication of two matrices is possible if and only if the number of columns of the first matrix on the left (**S**) is the same as the number of rows of the second matrix on the right (**p**).

Mathematical speaking, a matrix is a set of numbers arranged in a rectangular array in the form:

$$\mathbf{A} = \begin{pmatrix} a_{11} & a_{12} & \cdots & a_{1n} \\ a_{21} & a_{22} & \cdots & a_{2n} \\ \vdots & \vdots & \ddots & \vdots \\ a_{m1} & a_{m2} & \cdots & a_{mn} \end{pmatrix}, \tag{9.11}$$

where $a_{ij} (i = 1, 2, ..., m;\ j = 1, 2, ..., n)$ are numbers. m is the number of rows and n is the number of columns. In this case, the matrix $\mathbf{A}$ is said to have a size of $m \times n$. In mathematics, we usually write the matrix in bold font such as $\mathbf{A}$ and write its element in low case such as a_{11}. Sometimes, in order to show its elements, we may have to write

$$\mathbf{A} = [a_{ij}], \quad (i = 1, 2, ..., m;\ j = 1, 2, ..., n), \tag{9.12}$$

where we use $[a_{ij}]$ to explicitly show the elements of $\mathbf{A}$. If the number of rows (m) is equal to the number of columns n (i.e., $m = n$), we say the matrix is a square matrix.

The transpose of a matrix can be formally defined as

$$\mathbf{A}^T = [a_{ij}]^T = [a_{ji}], \text{ for all } (i, j). \tag{9.13}$$

Sometimes, it makes a statement simpler if we write

$$\forall (i, j), \tag{9.14}$$

to mean for all values of $(i = 1, 2, ..., m;\ j = 1, 2, ..., n)$.

If a matrix is a square matrix satisfying

$$\mathbf{A}^T = \mathbf{A}, \text{ or } a_{ij} = a_{ji}, \quad \forall (i, j), \tag{9.15}$$

it is called a symmetric matrix.

Example 9.1

For matrices

$$\mathbf{A} = \begin{pmatrix} 1 & 2 & 3 \\ 4 & 5 & 6 \end{pmatrix}, \quad \mathbf{B} = \begin{pmatrix} 2 & 3 \\ 3 & 1 \end{pmatrix},$$

their sizes are 2 by 3, and 2 by 2, respectively. So $\mathbf{B}$ is a square matrix.

Their transposes are

$$\mathbf{A}^T = \begin{pmatrix} 1 & 4 \\ 2 & 5 \\ 3 & 6 \end{pmatrix}, \quad \mathbf{B}^T = \begin{pmatrix} 2 & 3 \\ 3 & 1 \end{pmatrix}.$$

Since $\mathbf{B}^T = \mathbf{B}$, matrix $\mathbf{B}$ is symmetric.

Each element a_{ij} of a matrix $\mathbf{A}$ can be real or complex. If all elements are real numbers, the matrix is called a real matrix. If at least one element contains a complex number, the matrix is called complex.

In case of a complex matrix $\mathbf{C} = [c_{ij}]$, its conjugate matrix $\bar{\mathbf{C}}$ is obtained by taking the complex conjugate of each element. That is

$$\bar{\mathbf{C}} = [\bar{c}_{ij}], \quad \forall (i, j), \tag{9.16}$$

where we use the notation $(\bar{\ })$ to show the conjugate. In addition, the so-called conjugate transpose is denoted by $\mathbf{C}^*$ or $\mathbf{C}^H$ and can be obtained by

$$\mathbf{C}^* = \mathbf{C}^H = [\bar{c}_{ij}]^T = \overline{\mathbf{C}^T} = (\bar{\mathbf{C}})^T, \tag{9.17}$$

where the order of transpose or conjugate is not important. In mathematics, the conjugate transpose of a matrix is also called Hermitian conjugate, or adjoint matrix. Obviously, if a matrix is real, the Hermitian conjugate becomes the normal transpose.

Example 9.2

For a matrix

$$\mathbf{C} = \begin{pmatrix} 1 & 2-3i \\ 4-5i & 6i \\ 7+8i & 9-i \end{pmatrix},$$

its transpose is simply

$$\mathbf{C}^T = \begin{pmatrix} 1 & 4-5i & 7+8i \\ 2-3i & 6i & 9-i \end{pmatrix}.$$

The conjugate of $\mathbf{C}$ is

$$\bar{\mathbf{C}} = \begin{pmatrix} 1 & 2+3i \\ 4+5i & -6i \\ 7-8i & 9+i \end{pmatrix},$$

and the conjugate transpose of $\mathbf{C}$ is

$$\mathbf{C}^* = \mathbf{C}^H = \begin{pmatrix} 1 & 4+5i & 7-8i \\ 2+3i & -6i & 9+i \end{pmatrix}.$$

It is easy to check that the result is the same whether the conjugate or transpose is carried out first.

9.2 MATRIX ADDITION AND MULTIPLICATION

Matrix addition and subtraction are possible only if both matrices are of the same size. If $\mathbf{A} = [a_{ij}]$ and $\mathbf{G} = [g_{ij}]$ are the same size of $m \times n$, their sum $\mathbf{S} = [s_{ij}]$ can be obtained by adding their corresponding entries. That is

$$s_{ij} = a_{ij} + g_{ij}, \quad \forall(i, j), \tag{9.18}$$

where $\mathbf{S}$ has the same size of $m \times n$. Their differences also form a matrix $\mathbf{M} = [m_{ij}]$

$$m_{ij} = a_{ij} - g_{ij}, \tag{9.19}$$

which also has the same size. If a matrix can be multiplied by a scalar $\beta \neq 0$, we have

$$\beta\mathbf{A} = [\beta a_{ij}], \quad \forall(i, j). \tag{9.20}$$

Example 9.3

For $\mathbf{A} = \begin{pmatrix} 1 & 2 & 3 \\ 4 & 5 & 6 \end{pmatrix}$ and $\mathbf{G} = \begin{pmatrix} 1 & 1 & 7 \\ 2 & -2 & 1 \end{pmatrix}$, we have

$$\mathbf{A}+\mathbf{G} = \begin{pmatrix} 1 & 2 & 3 \\ 4 & 5 & 6 \end{pmatrix} + \begin{pmatrix} 1 & 1 & 7 \\ 2 & -2 & 1 \end{pmatrix}$$
$$= \begin{pmatrix} 1+1 & 2+1 & 3+7 \\ 4+2 & 5+(-2) & 6+1 \end{pmatrix} = \begin{pmatrix} 2 & 3 & 10 \\ 6 & 3 & 7 \end{pmatrix},$$

and

$$\mathbf{A}-\mathbf{G} = \begin{pmatrix} 1 & 2 & 3 \\ 4 & 5 & 6 \end{pmatrix} - \begin{pmatrix} 1 & 1 & 7 \\ 2 & -2 & 1 \end{pmatrix}$$

$$= \begin{pmatrix} 1-1 & 2-1 & 3-7 \\ 4-2 & 5-(-2) & 6-1 \end{pmatrix} = \begin{pmatrix} 0 & 1 & -4 \\ 2 & 7 & 5 \end{pmatrix}.$$

In addition, we have

$$3\mathbf{A} = 3\begin{pmatrix} 1 & 2 & 3 \\ 4 & 5 & 6 \end{pmatrix} = \begin{pmatrix} 3\times 1 & 3\times 2 & 3\times 3 \\ 3\times 4 & 3\times 5 & 3\times 6 \end{pmatrix} = \begin{pmatrix} 3 & 6 & 9 \\ 12 & 15 & 18 \end{pmatrix}.$$

Matrix multiplication requires that the number of columns of the first matrix is equal to the number of rows of the second matrix. If $\mathbf{A} = [a_{ij}]$ is an $m \times n$ matrix, and $\mathbf{B} = [b_{jk}]$ is an $n \times p$ matrix, then $\mathbf{C} = \mathbf{AB}$ is an $m \times p$ matrix. We have

$$\begin{aligned} \mathbf{C} = [c_{ik}] &= \begin{pmatrix} \boxed{c_{11}} & c_{12} & \cdots & c_{1p} \\ c_{21} & c_{22} & \cdots & c_{2p} \\ \vdots & \vdots & \ddots & \vdots \\ c_{m1} & c_{m2} & \cdots & c_{mp} \end{pmatrix} \\ &= \mathbf{AB} = [a_{ij}][b_{jk}] \\ &= \begin{pmatrix} \boxed{a_{11}} & \boxed{a_{12}} & \boxed{\cdots} & \boxed{a_{1n}} \\ a_{21} & a_{22} & \cdots & a_{2n} \\ \vdots & \vdots & \ddots & \vdots \\ a_{m1} & a_{m2} & \cdots & a_{mn} \end{pmatrix} \begin{pmatrix} \boxed{b_{11}} & b_{12} & \cdots & b_{1p} \\ \boxed{b_{21}} & b_{22} & \cdots & b_{2p} \\ \boxed{\vdots} & \vdots & \ddots & \vdots \\ \boxed{b_{n1}} & b_{n2} & \cdots & b_{np} \end{pmatrix}, \end{aligned} \tag{9.21}$$

where

$$c_{ik} = \sum_{j=1}^{n} a_{ij} b_{jk}. \tag{9.22}$$

Here, the entry c_{ik} is obtained by the sum of multiplying the ith row by the corresponding entry in the kth column. For example, we have

$$c_{11} = \sum_{j=1}^{n} a_{1j} b_{j1} = a_{11}b_{11} + a_{12}b_{21} + \ldots + a_{1n}b_{n1}, \tag{9.23}$$

and these entries are highlighted in boxes as Eq. (9.21). Similarly, we have

$$c_{23} = \sum_{j=1}^{n} a_{2j} b_{j3} = a_{21}b_{13} + a_{22}b_{23} + \ldots + a_{2n}b_{n3}, \tag{9.24}$$

and so on. Let us look at an example.

Example 9.4

For two matrices

$$\mathbf{A} = \begin{pmatrix} a_{11} & a_{12} \\ a_{21} & a_{22} \end{pmatrix} = \begin{pmatrix} 1 & 2 \\ 3 & 4 \end{pmatrix}, \quad \mathbf{B} = \begin{pmatrix} b_{11} & b_{12} \\ b_{21} & b_{22} \end{pmatrix} = \begin{pmatrix} 5 & 6 \\ 7 & 8 \end{pmatrix},$$

where $a_{11} = 1$, $b_{11} = 5$, ..., $b_{22} = 8$. For the multiplication of these two matrices, we have

$$\mathbf{C} = \begin{pmatrix} c_{11} & c_{12} \\ c_{21} & c_{22} \end{pmatrix} = \mathbf{AB} = \begin{pmatrix} a_{11} & a_{12} \\ a_{21} & a_{22} \end{pmatrix} \begin{pmatrix} b_{11} & b_{12} \\ b_{21} & b_{22} \end{pmatrix}.$$

So we have

$$c_{11}=\sum_{j=1}^{2}a_{1j}b_{j1}=a_{11}b_{11}+a_{12}b_{21}=1\times5+2\times7=19,$$

$$c_{12}=\sum_{j=1}^{2}a_{1j}b_{j2}=a_{11}b_{12}+a_{12}b_{22}=1\times6+2\times8=22,$$

$$c_{21}=\sum_{j=1}^{2}a_{2j}b_{j1}=a_{21}b_{11}+a_{22}b_{21}=3\times5+4\times7=43,$$

$$c_{22}=\sum_{j=1}^{2}a_{2j}b_{j2}=a_{21}b_{12}+a_{22}b_{22}=3\times6+4\times8=50.$$

Therefore, we have

$$\mathbf{C}=\mathbf{AB}=\begin{pmatrix}1&2\\3&4\end{pmatrix}\begin{pmatrix}5&6\\7&8\end{pmatrix}=\begin{pmatrix}19&22\\43&50\end{pmatrix}.$$

When a scalar α multiplies a matrix $\mathbf{A}$, the result is the matrix with each of $\mathbf{A}$'s elements multiplying by α. For example,

$$\alpha\mathbf{A}=\alpha\begin{pmatrix}a&b\\c&d\\e&f\end{pmatrix}=\begin{pmatrix}\alpha a&\alpha b\\\alpha c&\alpha d\\\alpha e&\alpha f\end{pmatrix}. \tag{9.25}$$

Generally speaking, the addition of two matrices is commutative

$$\mathbf{P}+\mathbf{Q}=\mathbf{Q}+\mathbf{P}. \tag{9.26}$$

The addition of three matrices is associative, that is

$$(\mathbf{P}+\mathbf{Q})+\mathbf{A}=\mathbf{P}+(\mathbf{Q}+\mathbf{A}). \tag{9.27}$$

However, matrices multiplication is not commutative. That is

$$\mathbf{AB}\neq\mathbf{BA}. \tag{9.28}$$

Example 9.5

For example, for two matrices

$$\mathbf{A}=\begin{pmatrix}1&2\\3&4\end{pmatrix},\quad \mathbf{B}=\begin{pmatrix}5&6\\7&8\end{pmatrix},$$

we have

$$\mathbf{D}=\begin{pmatrix}d_{11}&d_{12}\\d_{21}&d_{22}\end{pmatrix}=\mathbf{BA}=\begin{pmatrix}b_{11}&b_{12}\\b_{21}&b_{22}\end{pmatrix}\begin{pmatrix}a_{11}&a_{12}\\a_{21}&a_{22}\end{pmatrix}.$$

Following the same procedure as in the previous example, we have

$$d_{11}=\sum_{j=1}^{2}b_{1j}a_{j1}=b_{11}a_{11}+b_{12}\times a_{21}=5\times1+6\times3=23.$$

Similarly, we have

$$d_{12}=34,\quad d_{21}=31,\quad d_{22}=46.$$

Thus, we have

$$\mathbf{D} = \mathbf{BA} = \begin{pmatrix} 23 & 34 \\ 31 & 46 \end{pmatrix}.$$

From the previous example, we know that

$$\mathbf{AB} = \begin{pmatrix} 19 & 22 \\ 43 & 50 \end{pmatrix},$$

we can conclude that

$$\mathbf{AB} \neq \mathbf{BA}.$$

There are two special matrices: the zero matrix and the identity matrix. A zero matrix is a matrix whose every element is zero. We have 1×4 zero matrix as $\mathbf{O} = \begin{pmatrix} 0 & 0 & 0 & 0 \end{pmatrix}$. If the number of rows of a matrix is the same as the number columns, that is $m = n$, the matrix is called a square matrix. If the diagonal elements are 1's and all the other elements are zeros, it is called a unit matrix or an identity matrix. For example, the 3×3 unit matrix can be written as

$$\mathbf{I} = \begin{pmatrix} 1 & 0 & 0 \\ 0 & 1 & 0 \\ 0 & 0 & 1 \end{pmatrix}. \tag{9.29}$$

If a matrix $\mathbf{A}$ is the same size as the unit matrix, then it is commutative with $\mathbf{I}$. That is

$$\mathbf{IA} = \mathbf{AI} = \mathbf{A}. \tag{9.30}$$

Example 9.6

For the matrix $\mathbf{A} = \begin{pmatrix} 1 & 2 \\ 3 & 4 \end{pmatrix}$ given earlier, we can check that

$$\mathbf{AI} = \begin{pmatrix} 1 & 2 \\ 3 & 4 \end{pmatrix}\begin{pmatrix} 1 & 0 \\ 0 & 1 \end{pmatrix} = \begin{pmatrix} 1 & 2 \\ 3 & 4 \end{pmatrix}.$$

Similarly, we have

$$\mathbf{IA} = \begin{pmatrix} 1 & 0 \\ 0 & 1 \end{pmatrix}\begin{pmatrix} 1 & 2 \\ 3 & 4 \end{pmatrix} = \begin{pmatrix} 1 & 2 \\ 3 & 4 \end{pmatrix},$$

which indeed means that

$$\mathbf{AI} = \mathbf{IA}.$$

9.3 TRANSFORMATION AND INVERSE

When a point $P(x, y)$ (say, a point on a robotic arm) is rotated by an angle θ, it becomes its corresponding point $P'(x', y')$ (see Fig. 9.1). The relationship between the old coordinates (x, y) and the new coordinates (x', y') can be derived by using trigonometry.

The new coordinates at the new location $P'(x', y')$ can be written in terms of the coordinates at the original location $P(x, y)$ and the angle θ. From basic geometry, we know that $\angle A'P'Q = \theta = \angle A'OW$ Since $x = OA = OA'$, $y = AP = A'P'$, and $x' = OS = OW - SW = OW - QA'$, we have

$$x' = OW - QA' = x\cos\theta - y\sin\theta. \tag{9.31}$$

Similarly, we have

$$y' = SP' = SQ + QP' = WA' + QP' = x\sin\theta + y\cos\theta. \tag{9.32}$$

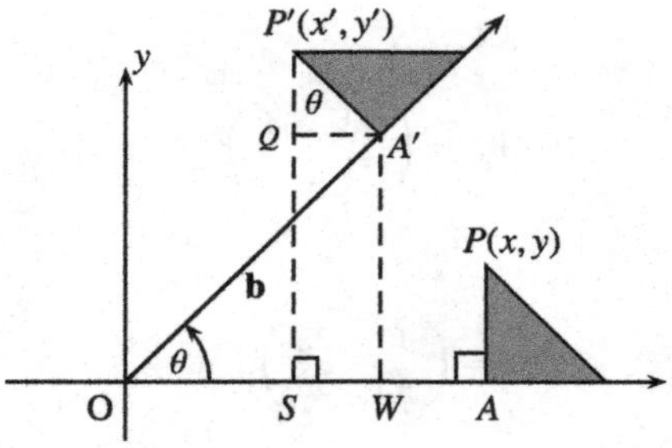

FIGURE 9.1 The rotational transformation.

The above two equations can be written in a compact form

$$\begin{pmatrix} x' \\ y' \end{pmatrix} = \begin{pmatrix} \cos\theta & -\sin\theta \\ \sin\theta & \cos\theta \end{pmatrix} \begin{pmatrix} x \\ y \end{pmatrix}, \tag{9.33}$$

where the matrix for the rotation or transformation is

$$\mathbf{R}_\theta = \begin{pmatrix} \cos\theta & -\sin\theta \\ \sin\theta & \cos\theta \end{pmatrix}. \tag{9.34}$$

Therefore, for any point (x, y), its new coordinates after rotating by an angle θ can be obtained by

$$\begin{pmatrix} x' \\ y' \end{pmatrix} = \mathbf{R}_\theta \begin{pmatrix} x \\ y \end{pmatrix}. \tag{9.35}$$

Example 9.7

For a vector **u** linking point A = (1, 2) to B = (2, 1), both points A and B rotate 45° anticlockwise in the Cartesian coordinate system. What are the new coordinates of A and B? What is the new vector **u**′?

The rotation matrix is

$$\mathbf{R}_{45} = \begin{pmatrix} \cos 45^\circ & -\sin 45^\circ \\ \sin 45^\circ & \cos 45^\circ \end{pmatrix} = \begin{pmatrix} \sqrt{2}/2 & -\sqrt{2}/2 \\ \sqrt{2}/2 & \sqrt{2}/2 \end{pmatrix} = \frac{\sqrt{2}}{2}\begin{pmatrix} 1 & -1 \\ 1 & 1 \end{pmatrix}.$$

The new point A' after rotating from A is

$$\frac{\sqrt{2}}{2}\begin{pmatrix} 1 & -1 \\ 1 & 1 \end{pmatrix}\begin{pmatrix} 1 \\ 2 \end{pmatrix} = \frac{\sqrt{2}}{2}\begin{pmatrix} 1\times 1+(-1)\times 2 \\ 1\times 1+1\times 2 \end{pmatrix} = \frac{\sqrt{2}}{2}\begin{pmatrix} -1 \\ 3 \end{pmatrix}.$$

Point B will becomes point B' at

$$\frac{\sqrt{2}}{2}\begin{pmatrix} 1 & -1 \\ 1 & 1 \end{pmatrix}\begin{pmatrix} 2 \\ 1 \end{pmatrix} = \frac{\sqrt{2}}{2}\begin{pmatrix} 1\times 2+(-1)\times 1 \\ 1\times 2+1\times 1 \end{pmatrix} = \frac{\sqrt{2}}{2}\begin{pmatrix} 1 \\ 3 \end{pmatrix}.$$

Before the rotation, vector **u** is

$$\mathbf{u} = \overrightarrow{AB} = \begin{pmatrix} 2 \\ 1 \end{pmatrix} - \begin{pmatrix} 1 \\ 2 \end{pmatrix} = \begin{pmatrix} 1 \\ -1 \end{pmatrix},$$

which points at 45°downwards to the right. Now the new vector **u**′ after the rotation becomes

$$\mathbf{u}' = \overrightarrow{A'B'} = \frac{\sqrt{2}}{2}\begin{pmatrix} 1 \\ 3 \end{pmatrix} - \frac{\sqrt{2}}{2}\begin{pmatrix} -1 \\ 3 \end{pmatrix} = \frac{\sqrt{2}}{2}\begin{pmatrix} 2 \\ 0 \end{pmatrix},$$

which points to the right horizontally.

If point P is rotated by $\theta + \psi$, we have

$$\begin{pmatrix} x'' \\ y'' \end{pmatrix} = \begin{pmatrix} \cos(\theta+\psi) & -\sin(\theta+\psi) \\ \sin(\theta+\psi) & \cos(\theta+\psi) \end{pmatrix} = \mathbf{R}_{\theta+\psi} \begin{pmatrix} x \\ y \end{pmatrix}, \tag{9.36}$$

which can also be achieved by two steps: first by rotating θ to get $P'(x', y')$ and then rotating by ψ from $P'(x', y')$ to $P''(x'', y'')$. This is to say

$$\begin{pmatrix} x'' \\ y'' \end{pmatrix} = \mathbf{R}_{\psi} \begin{pmatrix} x' \\ y' \end{pmatrix} = \mathbf{R}_{\psi}\mathbf{R}_{\theta} \begin{pmatrix} x \\ y \end{pmatrix}. \tag{9.37}$$

Combining with (9.36), we have

$$\mathbf{R}_{\theta+\psi} = \mathbf{R}_{\psi}\mathbf{R}_{\theta}, \tag{9.38}$$

or

$$\begin{aligned} \begin{pmatrix} \cos(\theta+\psi) & -\sin(\theta+\psi) \\ \sin(\theta+\psi) & \cos(\theta+\psi) \end{pmatrix} &= \begin{pmatrix} \cos\psi & -\sin\psi \\ \sin\psi & \cos\psi \end{pmatrix} \begin{pmatrix} \cos\theta & -\sin\theta \\ \sin\theta & \cos\theta \end{pmatrix} \\ &= \begin{pmatrix} \cos\psi\cos\theta - \sin\psi\sin\theta & -[\cos\psi\sin\theta + \sin\psi\cos\theta] \\ \sin\psi\cos\theta + \cos\psi\sin\theta & \cos\psi\cos\theta - \sin\psi\sin\theta \end{pmatrix}, \end{aligned} \tag{9.39}$$

which is another way of deriving the sine and cosine of the addition of two angles.

In a special case when first rotating by θ, followed by rotating back by $-\theta$, a point $P(x, y)$ should reach its original point. That is

$$\begin{pmatrix} x \\ y \end{pmatrix} = \begin{pmatrix} 1 & 0 \\ 0 & 1 \end{pmatrix} \begin{pmatrix} x \\ y \end{pmatrix} = \mathbf{R}_{-\theta}\mathbf{R}_{\theta} \begin{pmatrix} x \\ y \end{pmatrix}, \tag{9.40}$$

which means that

$$\mathbf{R}_{-\theta}\mathbf{R}_{\theta} = \begin{pmatrix} 1 & 0 \\ 0 & 1 \end{pmatrix} = \mathbf{I}. \tag{9.41}$$

In other words, $\mathbf{R}_{-\theta}$ is the inverse of $\mathbf{R}_{\theta}$. That is to say

$$\mathbf{R}_{-\theta} = \begin{pmatrix} \cos(-\theta) & -\sin(-\theta) \\ \sin(-\theta) & \cos(-\theta) \end{pmatrix} = \begin{pmatrix} \cos\theta & \sin\theta \\ -\sin\theta & \cos\theta \end{pmatrix}, \tag{9.42}$$

is the inverse of

$$\mathbf{R}_{\theta} = \begin{pmatrix} \cos\theta & -\sin\theta \\ \sin\theta & \cos\theta \end{pmatrix}. \tag{9.43}$$

In general, the inverse $\mathbf{A}^{-1}$ of a square matrix $\mathbf{A}$, if it exists, is defined by

$$\mathbf{A}^{-1}\mathbf{A} = \mathbf{A}\mathbf{A}^{-1} = \mathbf{I}, \tag{9.44}$$

where $\mathbf{I}$ is a unit matrix which is the same size as $\mathbf{A}$.

Example 9.8

For example, a 2×2 matrix $\mathbf{A}$ and its inverse $\mathbf{A}^{-1}$

$$\mathbf{A} = \begin{pmatrix} a & b \\ c & d \end{pmatrix}, \quad \mathbf{A}^{-1} = \begin{pmatrix} \alpha & \beta \\ \gamma & \kappa \end{pmatrix},$$

can be related by

$$\mathbf{A}\mathbf{A}^{-1} = \begin{pmatrix} a & b \\ c & d \end{pmatrix} \begin{pmatrix} \alpha & \beta \\ \gamma & \kappa \end{pmatrix} = \begin{pmatrix} a\alpha + b\gamma & a\beta + b\kappa \\ c\alpha + d\gamma & c\beta + d\kappa \end{pmatrix} = \begin{pmatrix} 1 & 0 \\ 0 & 1 \end{pmatrix} = \mathbf{I}.$$

This means that

$$a\alpha + b\gamma = 1,\ a\beta + b\kappa = 0,\ c\alpha + d\gamma = 0,\ c\gamma + d\kappa = 1.$$

These four equations will solve the four unknowns α, β, γ and κ. After some simple rearrangement and calculations, we have

$$\alpha = \frac{d}{\Delta},\ \beta = \frac{-b}{\Delta},\ \gamma = \frac{-c}{\Delta},\ \kappa = \frac{a}{\Delta},$$

where $\Delta = ad - bc$ is the determinant of $\mathbf{A}$. Therefore, the inverse becomes

$$\mathbf{A}^{-1} = \frac{1}{ad - bc}\begin{pmatrix} d & -b \\ -c & a \end{pmatrix}.$$

It is straightforward to verify that $\mathbf{A}^{-1}\mathbf{A} = \begin{pmatrix} 1 & 0 \\ 0 & 1 \end{pmatrix}$.

In the case of

$$\mathbf{R}_\theta = \begin{pmatrix} \cos\theta & -\sin\theta \\ \sin\theta & \cos\theta \end{pmatrix},$$

we have

$$\mathbf{R}^{-1} = \frac{1}{\cos\theta\cos\theta - (-\sin\theta)\sin\theta}\begin{pmatrix} \cos\theta & \sin\theta \\ -\sin\theta & \cos\theta \end{pmatrix} = \begin{pmatrix} \cos\theta & \sin\theta \\ -\sin\theta & \cos\theta \end{pmatrix},$$

where we have used $\cos^2\theta + \sin^2\theta = 1$. This is the same as (9.42).

We have seen that some special combinations of the elements such as the determinant $\Delta = ad - bc$ is very important. We now try to define it more generally.

The determinant of an $n \times n$ square matrix $\mathbf{A} = [a_{ij}]$ is a number which can be obtained by a cofactor expansion either by row or by column

$$\det(\mathbf{A}) \equiv |A| = \sum_{j=1}^{n}(-1)^{i+j} a_{ij}|\mathbf{M}_{ij}|, \tag{9.45}$$

where $|\mathbf{M}_{ij}|$ is the cofactor or the determinant of a minor matrix $\mathbf{M}$ of $\mathbf{A}$, obtained by deleting row i and column j. This is a recursive relationship. For example, $\mathbf{M}_{12}$ of a 3×3 matrix is obtained by deleting the first row and the second column

$$\begin{vmatrix} \not{a}_{11} & -\not{a}_{12} & \not{a}_{13} \\ a_{21} & a_{22} & a_{23} \\ a_{31} & a_{32} & a_{33} \end{vmatrix} \Longrightarrow |\mathbf{M}|_{12} = \begin{vmatrix} a_{21} & a_{23} \\ a_{31} & a_{33} \end{vmatrix}. \tag{9.46}$$

Obviously, the determinant of a 1×1 matrix $|a_{11}| = a_{11}$ is the number itself. The determinant of a 2×2 matrix

$$\det(\mathbf{A}) = \begin{vmatrix} a_{11} & a_{12} \\ a_{21} & a_{22} \end{vmatrix} = a_{11}a_{22} - a_{12}a_{21}. \tag{9.47}$$

The determinant of a 3×3 matrix is given by $\det(\mathbf{A})$ or

$$\begin{aligned} \begin{vmatrix} a_{11} & a_{12} & a_{23} \\ a_{21} & a_{22} & a_{23} \\ a_{31} & a_{32} & a_{33} \end{vmatrix} &= (-1)^{1+1}a_{11}\begin{vmatrix} a_{22} & a_{23} \\ a_{32} & a_{33} \end{vmatrix} + (-1)^{1+2}a_{12}\begin{vmatrix} a_{21} & a_{23} \\ a_{31} & a_{33} \end{vmatrix} \\ &+ (-1)^{1+3}a_{13}\begin{vmatrix} a_{21} & a_{22} \\ a_{31} & a_{32} \end{vmatrix} = a_{11}(a_{22}a_{33} - a_{32}a_{23}) \\ &- a_{12}(a_{21}a_{33} - a_{31}a_{23}) + a_{13}(a_{21}a_{32} - a_{31}a_{22}). \end{aligned} \tag{9.48}$$

Here we used the expansion along the first row $i = 1$. We can also expand it along any other rows or columns, and the results are the same. As the determinant of a matrix is a scalar or a simple number, it is not difficult to understand the following properties

$$\det(\mathbf{AB}) = \det(\mathbf{A})\det(\mathbf{B}), \qquad \det(\mathbf{A}^T) = \det(\mathbf{A}). \tag{9.49}$$

There are many applications of determinants. For example, if $\det(\mathbf{A}) = 0$, the square matrix is called singular, and the inverse of such a matrix does not exist. The inverse of a matrix exists only if $\det(\mathbf{A}) \neq 0$. Here we will use it to calculate the inverse $\mathbf{A}^{-1}$ using

$$\mathbf{A}^{-1} = \frac{\text{adj}(\mathbf{A})}{\det(\mathbf{A})} = \frac{1}{\det(\mathbf{A})}\mathbf{B}^T, \qquad \mathbf{B} = \Big[(-1)^{i+j}|M_{ij}|\Big], \tag{9.50}$$

where the matrix $\mathbf{B}^T$ is called the adjoint of matrix $\mathbf{A}$ with the same size as $\mathbf{A}$, and $i, j = 1, ..., n$. Each of the element $\mathbf{B}$ is expressed in terms of a cofactor so that $b_{ij} = (-1)^{i+j}|M_{ij}|$. $\mathbf{B}$ itself is called the cofactor matrix, while $\text{adj}(\mathbf{A}) = \mathbf{B}^T$ is sometimes used to denote the adjoint matrix. This seems too complicated, and let us compute the inverse of a 3×3 matrix as an example.

Example 9.9

In order to compute the inverse of

$$\mathbf{A} = \begin{pmatrix} 1 & 1 & -2 \\ 1 & 0 & 2 \\ 2 & 1 & 1 \end{pmatrix},$$

we first construct its adjoint matrix $\mathbf{B}^T$ with

$$\mathbf{B} = [b_{ij}] = \Big[(-1)^{i+j}|M_{ij}|\Big].$$

The first element b_{11} can be obtained by

$$b_{11} = (-1)^{1+1}\begin{vmatrix} 0 & 2 \\ 1 & 1 \end{vmatrix} = (-1)^2 \times (0 \times 1 - 2 \times 1) = -2.$$

The element b_{12} is

$$b_{12} = (-1)^{1+2}\begin{vmatrix} 1 & 2 \\ 2 & 1 \end{vmatrix} = -1 \times (1 \times 1 - 2 \times 2) = 3,$$

while the element b_{21} is

$$b_{21} = (-1)^{2+1}\begin{vmatrix} 1 & -2 \\ 1 & 1 \end{vmatrix} = (-1)^3 \times (1 \times 1 - 1 \times (-2)) = -3.$$

Following a similar procedure, we have $\mathbf{B}$ and its transpose $\mathbf{B}^T$ as

$$\mathbf{B} = \begin{pmatrix} -2 & 3 & 1 \\ -3 & 5 & 1 \\ 2 & -4 & -1 \end{pmatrix}, \quad \text{or} \quad \mathbf{B}^T = \begin{pmatrix} -2 & -3 & 2 \\ 3 & 5 & -4 \\ 1 & 1 & -1 \end{pmatrix}.$$

Then, the determinant of $\mathbf{A}$ is

$$\det(\mathbf{A}) = \begin{vmatrix} 1 & 1 & -2 \\ 1 & 0 & 2 \\ 2 & 1 & 1 \end{vmatrix} = 1 \times \begin{vmatrix} 0 & 2 \\ 1 & 1 \end{vmatrix} - 1 \times \begin{vmatrix} 1 & 2 \\ 2 & 1 \end{vmatrix} + (-2) \times \begin{vmatrix} 1 & 0 \\ 2 & 1 \end{vmatrix}$$

$$= 1 \times (0 \times 1 - 2 \times 1) - 1 \times (1 \times 1 - 2 \times 2) - 2 \times (1 \times 1 - 2 \times 0)$$

$$= 1 \times (-2) - 1 \times (-3) - 2 \times 1 = -1.$$

Finally, the inverse becomes

$$\mathbf{A}^{-1} = \frac{\mathbf{B}^T}{\det(\mathbf{A})} = \frac{1}{-1}\begin{pmatrix} -2 & -3 & 2 \\ 3 & 5 & -4 \\ 1 & 1 & -1 \end{pmatrix} = \begin{pmatrix} 2 & 3 & -2 \\ -3 & -5 & 4 \\ -1 & -1 & 1 \end{pmatrix}.$$

This result will be used in the next example.

9.4 SYSTEM OF LINEAR EQUATIONS

A system of linear equations can be written as a large matrix equation, and the solution of such a linear system will become straightforward if the inverse of a square matrix is used. Let us demonstrate this by an example. For a linear system consisting of three simultaneous equations, we have

$$a_{11}x + a_{12}y + a_{13}z = b_1,$$
$$a_{21}x + a_{22}y + a_{23}z = b_2,$$
$$a_{31}x + a_{32}y + a_{33}z = b_3,$$

which can be written compactly as

$$\begin{pmatrix} a_{11} & a_{12} & a_{13} \\ a_{21} & a_{22} & a_{23} \\ a_{31} & a_{32} & a_{33} \end{pmatrix}\begin{pmatrix} x \\ y \\ z \end{pmatrix} = \begin{pmatrix} b_1 \\ b_2 \\ b_3 \end{pmatrix}, \tag{9.51}$$

or more compactly as

$$\mathbf{Au} = \mathbf{b}, \tag{9.52}$$

where $\mathbf{u} = \begin{pmatrix} x & y & z \end{pmatrix}^T$. By multiplying $\mathbf{A}^{-1}$ on both sides, we have

$$\mathbf{A}^{-1}\mathbf{Au} = \mathbf{A}^{-1}\mathbf{b}, \tag{9.53}$$

or

$$\mathbf{Iu} = \mathbf{A}^{-1}\mathbf{b}. \tag{9.54}$$

Therefore, its solution can be written as

$$\mathbf{u} = \mathbf{A}^{-1}\mathbf{b}.$$

Now let us look at an example to solve a simple linear system.

Example 9.10

In order to solve the following system

$$x + y - 2z = -6,$$
$$x + 2z = 8,$$
$$2x + y + z = 5,$$

we first write it as $\mathbf{Au} = \mathbf{b}$, or

$$\begin{pmatrix} 1 & 1 & -2 \\ 1 & 0 & 2 \\ 2 & 1 & 1 \end{pmatrix}\begin{pmatrix} x \\ y \\ z \end{pmatrix} = \begin{pmatrix} -6 \\ 8 \\ 5 \end{pmatrix}.$$

We know from the earlier example that the inverse of $\mathbf{A}^{-1}$ is

$$\mathbf{A}^{-1} = \begin{pmatrix} 2 & 3 & -2 \\ -3 & -5 & 4 \\ -1 & -1 & 1 \end{pmatrix},$$

we now have $\mathbf{u} = \mathbf{A}^{-1}\mathbf{b}$ or

$$\begin{pmatrix} x \\ y \\ z \end{pmatrix} = \begin{pmatrix} 2 & 3 & -2 \\ -3 & -5 & 4 \\ -1 & -1 & 1 \end{pmatrix} \begin{pmatrix} -6 \\ 8 \\ 5 \end{pmatrix} = \begin{pmatrix} 2 \times (-6) + 3 \times 8 + (-2) \times 5 \\ -3 \times (-6) + (-5) \times 8 + 4 \times 5 \\ -1 \times (-6) + (-1) \times 8 + 1 \times 5 \end{pmatrix} = \begin{pmatrix} 2 \\ -2 \\ 3 \end{pmatrix},$$

which gives a unique set of solutions $x = 2$, $y = -2$ and $z = 3$.

In general, a linear system of n equations for n unknowns can be written in the compact form as

$$\begin{pmatrix} a_{11} & a_{12} & \dots & a_{1n} \\ a_{21} & a_{22} & \dots & a_{2n} \\ \vdots & \vdots & & \\ a_{n1} & a_{n2} & \dots & a_{nn} \end{pmatrix} \begin{pmatrix} u_1 \\ u_2 \\ \vdots \\ u_n \end{pmatrix} = \begin{pmatrix} b_1 \\ b_2 \\ \vdots \\ b_n \end{pmatrix}, \tag{9.55}$$

or simply

$$\mathbf{Au} = \mathbf{b}. \tag{9.56}$$

Its solution can be obtained by inverse

$$\mathbf{u} = \mathbf{A}^{-1}\mathbf{b}. \tag{9.57}$$

You may wonder how you can get the inverse $\mathbf{A}^{-1}$ of a larger matrix $\mathbf{A}$ more efficiently. For large systems, direct inverse is not a good option. There are many other more efficient methods to obtain the solutions, including the powerful Gauss-Jordan elimination, matrix decomposition, and iteration methods. Interested readers can refer to more advanced literature for details.

9.5 EIGENVALUES AND EIGENVECTORS

A special case of a linear system $\mathbf{Au} = \mathbf{b}$ is when $\mathbf{b} = \lambda\mathbf{u}$, and this becomes an eigenvalue problem.

An eigenvalue λ and corresponding eigenvector $\mathbf{u}$ of a square matrix $\mathbf{A}$ satisfy

$$\mathbf{Au} = \lambda\mathbf{u}. \tag{9.58}$$

Obviously, we can multiply both sides by an identity matrix $\mathbf{I}$ of the same size as that of $\mathbf{A}$, and we have

$$\mathbf{IAu} = \mathbf{I}\lambda\mathbf{u}, \tag{9.59}$$

or

$$\mathbf{Au} = \lambda\mathbf{Iu}, \tag{9.60}$$

where we have used $\mathbf{IA} = \mathbf{A}$ and $\mathbf{I}\lambda = \lambda\mathbf{I}$. The above equation can also be written as

$$(\mathbf{A} - \lambda\mathbf{I})\mathbf{u} = 0. \tag{9.61}$$

In order to satisfy this equation, we can either have $\mathbf{u} = 0$ (a trivial solution) or

$$\mathbf{A} - \lambda\mathbf{I} = 0.$$

More formally, any nontrivial solution requires that

$$\det|\mathbf{A}-\lambda\mathbf{I}|=0. \tag{9.62}$$

Let us look at a simple example.

Example 9.11

For a 2×2 matrix

$$\mathbf{A}=\begin{pmatrix}1 & 2\\ 2 & 1\end{pmatrix},$$

its eigenvalues can be obtained by solving

$$\det|\mathbf{A}-\lambda\mathbf{I}|=\begin{vmatrix}1-\lambda & 2\\ 2 & 1-\lambda\end{vmatrix}$$
$$=(1-\lambda)(1-\lambda)-2\times 2=(1-\lambda)^2-4=0,$$

which can be written as

$$(1-\lambda)^2=2^2.$$

Taking the square root of both sides, we have

$$1-\lambda=\pm 2,$$

which gives two eigenvalues

$$\lambda_1=-1,\quad \lambda_2=3.$$

Here, we have seen that a 2×2 matrix can have two eigenvalues. In general, for a real matrix $\mathbf{A}$ of size $n\times n$, we have

$$\mathbf{A}=\begin{pmatrix}a_{11} & a_{12} & \cdots & a_{1n}\\ a_{21} & a_{22} & \cdots & a_{2n}\\ \vdots & \vdots & \ddots & \vdots\\ a_{n1} & a_{n2} & \cdots & a_{nn}\end{pmatrix}, \tag{9.63}$$

its eigenvalues can be determined by

$$\begin{vmatrix}a_{11}-\lambda & a_{12} & \ldots & a_{1n}\\ a_{21} & a_{22}-\lambda & \ldots & a_{2n}\\ \vdots & \vdots & \ddots & \vdots\\ a_{n1} & a_{n2} & \ldots & a_{nn}-\lambda\end{vmatrix}=0, \tag{9.64}$$

which is equivalent to

$$\lambda^n+\alpha_{n-1}\lambda^{n-1}+\ldots+\alpha_0=(\lambda-\lambda_1)(\lambda-\lambda_2)\ldots(\lambda-\lambda_n)=0. \tag{9.65}$$

In general, the characteristic equation has n solutions ($\lambda_i, i=1,2,\ldots n$), though it is usually not easy to find them.

Example 9.12

For a simple 2×2 matrix

$$\mathbf{A}=\begin{pmatrix}1 & 5\\ 2 & 4\end{pmatrix},$$

its eigenvalues can be determined by

$$\begin{vmatrix} 1-\lambda & 5 \\ 2 & 4-\lambda \end{vmatrix} = 0,$$

or

$$(1-\lambda)(4-\lambda) - 2 \times 5 = 0,$$

which is equivalent to

$$(\lambda+1)(\lambda-6) = 0.$$

Thus, the eigenvalues are $\lambda_1 = -1$ and $\lambda_2 = 6$. The trace of $\mathbf{A}$ is $\mathrm{tr}(\mathbf{A}) = a_{11} + a_{22} = 1 + 4 = 5 = \lambda_1 + \lambda_2$. In this relationship can be generalized as we will see later in this chapter.

For an eigenvalue λ, there is an eigenvector $\mathbf{u}$ that corresponds to it so that they satisfy

$$\mathbf{Au} = \lambda \mathbf{u}. \tag{9.66}$$

Interestingly, we can multiply a non-zero scalar or number $\beta \neq 0$ and we have

$$\mathbf{Au}\beta = \lambda \mathbf{u}\beta, \tag{9.67}$$

which can be written as

$$\mathbf{A}(\beta \mathbf{u}) = \lambda(\beta \mathbf{u}). \tag{9.68}$$

This means that $\beta \mathbf{u}$ is also an eigenvector that corresponds to the same eigenvalue β. That is to say that there could be an infinite number of eigenvectors if we choose different values of β. In the other words, we can only determine the direction of the eigenvector $\mathbf{u}$, not its length uniquely.

Therefore, it is customary to impose an extra condition that the length of the eigenvector is unity, and in this case, the eigenvector can be determined uniquely. Even with the unity requirement, there are still two vectors $\mathbf{u}$ and $-\mathbf{u}$ (or the same direction but with opposite unit vectors) that correspond to the same eigenvalue and some care should be taken.

Example 9.13

In order to obtain the eigenvector for each eigenvalue in the previous example, we assume

$$\mathbf{v} = \begin{pmatrix} v_1 \\ v_2 \end{pmatrix}.$$

For the eigenvalue $\lambda_1 = -1$, we plug this into

$$(\mathbf{A} - \lambda \mathbf{I})\mathbf{v} = 0,$$

and we have

$$\begin{pmatrix} 1-(-1) & 5 \\ 2 & 4-(-1) \end{pmatrix} \begin{pmatrix} v_1 \\ v_2 \end{pmatrix} = 0, \quad \begin{pmatrix} 2 & 5 \\ 2 & 5 \end{pmatrix} \begin{pmatrix} v_1 \\ v_2 \end{pmatrix} = 0,$$

which is equivalent to

$$2v_1 + 5v_2 = 0, \quad \text{or} \quad v_1 = -\frac{5}{2} v_2.$$

This equation has infinite solutions; each corresponds to the vector parallel to the unit eigenvector. As the eigenvector should be normalized so that its modulus is unity, this additional condition requires

$$v_1^2 + v_2^2 = 1,$$

which means

$$(\frac{-5v_2}{2})^2 + v_2^2 = 1.$$

This leads to

$$v_2^2 = \frac{4}{29},$$

which gives two solutions

$$v_2 = \pm\frac{2}{\sqrt{29}}.$$

If we choose the positive root, we have $v_2 = 2/\sqrt{29}$ and $v_1 = -5/\sqrt{29}$. Thus, we have the first set of eigenvalue and eigenvector

$$\lambda_1 = -1, \qquad \mathbf{v}_1 = \begin{pmatrix} -\frac{5}{\sqrt{29}} \\ \frac{2}{\sqrt{29}} \end{pmatrix}. \tag{9.69}$$

Similarly, the second eigenvalue $\lambda_2 = 6$ gives

$$\mathbf{v}_2 = \begin{pmatrix} \frac{\sqrt{2}}{2} \\ \frac{\sqrt{2}}{2} \end{pmatrix}.$$

Eigenvalues have interesting connections with their original matrix. The trace of any square matrix is defined as the sum of its diagonal elements, i.e.,

$$\text{tr}(\mathbf{A}) = \sum_{i=1}^{n} a_{ii} = a_{11} + a_{22} + ... + a_{nn}. \tag{9.70}$$

The sum of all the eigenvalues of a square matrix **A** is equivalent to the trace of **A**. That is

$$\text{tr}(\mathbf{A}) = a_{11} + a_{22} + ... + a_{nn} = \sum_{i=1}^{n} \lambda_i = \lambda_1 + \lambda_2 + ... + \lambda_n. \tag{9.71}$$

In addition, the eigenvalues are also related to the determinant by

$$\det(\mathbf{A}) = \prod_{i=1}^{n} \lambda_i. \tag{9.72}$$

Example 9.14

For the previous example, we know that

$$\mathbf{A} = \begin{pmatrix} 1 & 5 \\ 2 & 4 \end{pmatrix},$$

its trace and determinant of **A** are

$$\text{tr}(\mathbf{A}) = 1 + 4 = 5, \quad \det(\mathbf{A}) = 1 \times 4 - 2 \times 5 = -6.$$

The sum of the eigenvalues is

$$\sum_{i=1}^{2} \lambda_i = -1 + 6 = 5 = \text{tr}(\mathbf{A}),$$

while the product of the eigenvalues is

$$\prod_{i=1}^{2} \lambda_i = -1 \times 6 = -6 = \det(\mathbf{A}).$$

Indeed, the above relationships about eigenvalues are true.

When a real square matrix is symmetric, its eigenvalues are all real and distinct. In addition, the eigenvectors (say, $\mathbf{u}_i$ and $\mathbf{u}_j$) that correspond to different eigenvalues (say, λ_i and λ_j) of a real symmetric matrix are orthogonal to each other. When we say two vectors are orthogonal, their dot product is zero. That is $\mathbf{u}_i \cdot \mathbf{u}_j = 0$ for $\lambda_i \neq \lambda_j$. Let us see a simple example.

Example 9.15

From an earlier example (Example 9.11), we know that matrix

$$\mathbf{A} = \begin{pmatrix} 1 & 2 \\ 2 & 1 \end{pmatrix},$$

has two eigenvalues

$$\lambda_1 = -1, \quad \lambda_2 = 3.$$

In order to determine the eigenvector

$$\mathbf{u} = \begin{pmatrix} u_1 \\ u_2 \end{pmatrix},$$

for the first eigenvalue $\lambda_1 = -1$, we have

$$(\mathbf{A} - \lambda \mathbf{I})\mathbf{u} = 0,$$

which gives

$$\begin{pmatrix} 1-(-1) & 2 \\ 2 & 1-(-1) \end{pmatrix} \begin{pmatrix} u_1 \\ u_2 \end{pmatrix} = 0,$$

or

$$\begin{pmatrix} 2 & 2 \\ 2 & 2 \end{pmatrix} \begin{pmatrix} u_1 \\ u_2 \end{pmatrix} = 0.$$

This leads to two identical equations

$$2u_1 + 2u_2 = 0, \quad 2u_1 + 2u_2 = 0.$$

This means $u_2 = -u_1$ (or $u_1 = -u_2$).

In order to determine the eigenvector uniquely, it requires that the length is 1. That is

$$\sqrt{u_1^2 + u_2^2} = 1,$$

or

$$u_1^2 + u_2^2 = 1.$$

From $u_2 = -u_1$, we get

$$u_1^2 + (-u_1)^2 = 1,$$

which gives

$$u_1 = +\frac{\sqrt{2}}{2}, \quad \text{or} \quad u_1 = -\frac{\sqrt{2}}{2}.$$

Therefore, the eigenvector becomes

$$\mathbf{u}_1 = \frac{\sqrt{2}}{2}\begin{pmatrix} 1 \\ -1 \end{pmatrix},$$

or

$$\mathbf{u}_2 = \frac{\sqrt{2}}{2}\begin{pmatrix} -1 \\ 1 \end{pmatrix},$$

which is simply $\mathbf{u}_1$ in the reverse direction (multiplied by -1). This means that we can only determined the general direction of an eigenvector uniquely.

Following the same procedure for $\lambda_2 = 3$, we have

$$(\mathbf{A} - \lambda\mathbf{I})\mathbf{v} = \begin{pmatrix} 1-3 & 2 \\ 2 & 1-3 \end{pmatrix}\begin{pmatrix} v_1 \\ v_2 \end{pmatrix} = 0,$$

or

$$\begin{pmatrix} -2 & 2 \\ 2 & -2 \end{pmatrix}\begin{pmatrix} v_1 \\ v_2 \end{pmatrix} = 0.$$

Again, this gives two equations

$$-2v_1 + 2v_2 = 0, \quad 2v_1 - 2v_2 = 0,$$

which are essentially the same as $v_1 = v_2$. With the additional condition $\sqrt{v_1^2 + v_2^2} = 1$, we obtain

$$\mathbf{v}_1 = \frac{\sqrt{2}}{2}\begin{pmatrix} 1 \\ 1 \end{pmatrix}, \quad \text{or} \quad \mathbf{v}_2 = -\frac{\sqrt{2}}{2}\begin{pmatrix} 1 \\ 1 \end{pmatrix},$$

which is simply $\mathbf{v}_1$ multiplied by -1. It is easy to check that the dot product of $\mathbf{u}$ and $\mathbf{v}$ is zero. That is

$$\mathbf{u}_1 \cdot \mathbf{v}_1 = \frac{\sqrt{2}}{2}\frac{\sqrt{2}}{2}\Big[1 \times 1 + (-1) \times 1\Big] = 0,$$

which means that the eigenvectors (corresponding to the eigenvalues -1 and 3) are indeed orthogonal to each other.

9.5.1 Distribution of Eigenvalues

For an $n \times n$ matrix $\mathbf{A} = [a_{ij}]$, there is an important theorem, called Gerschgorin theorem, concerning the locations of all the eigenvalues λ_i of $\mathbf{A}$.

Let us first define a number (or radius) r_i by

$$r_i \equiv \sum_{j=1, j\neq i}^{n} |a_{ij}| = \sum_{j=1}^{n} |a_{ij}| - |a_{ii}|, \qquad (i = 1, 2, ..., n), \tag{9.73}$$

and then denote Ω_i as the circle, $|z - a_{ii}| \leq r_i$, centered at a_{ii} with a radius r_i in the complex plane $z \in \mathcal{C}$. Such circles are often called Gerschgorin's circles or discs.

Since the eigenvalues λ_i (counting the multiplicity of roots) and their corresponding eigenvectors $\mathbf{u}^{(i)}$ are determined by

$$\mathbf{A}\mathbf{u}^{(i)} = \lambda_i \mathbf{u}^{(i)}, \tag{9.74}$$

for all $i = 1, 2, ..., n$, we have each component $u_k^{(i)}$ $(k = 1, 2, ..., n)$ satisfying

$$\sum_{j=1}^{n} a_{kj} u_j^{(i)} = \lambda_k u_k^{(i)}, \tag{9.75}$$

where $\mathbf{u}^{(i)} = (u_1^{(i)}, u_2^{(i)}, ..., u_n^{(i)})^T$ and $u_j^{(i)}$ is the jth component of the vector $\mathbf{u}^{(i)}$. Furthermore, we also define the largest absolute component of $\mathbf{u}^{(i)}$ (or its infinity norm) as

$$|\mathbf{u}^{(i)}| = \|\mathbf{u}^{(i)}\|_\infty = \max_{1 \le j \le n} \{|u_j^{(i)}|\}. \tag{9.76}$$

As the length of an eigenvector is not zero, we get $|\mathbf{u}^{(i)}| > 0$. We now have

$$a_{kk} u_k^{(i)} + \sum_{j \neq k} a_{kj} u_j^{(i)} = \lambda_k u_k^{(i)}, \tag{9.77}$$

whose norm leads to

$$|\lambda_k - a_{kk}| = \left| \frac{\sum_{j \neq k} a_{kj} u_j^{(i)}}{u_k^{(i)}} \right| \le \frac{\sum_{j \neq k} |a_{kj}| |\mathbf{u}^{(i)}|}{|\mathbf{u}^{(i)}|} = \sum_{j \neq k} |a_{kj}| = r_k.$$

This is equivalent to the following simple inequality (for any eigenvalue λ)

$$|\lambda - a_{ii}| \le r_i, \tag{9.78}$$

which is essentially the Gerschgorin circle theorem. Geometrically speaking, this important theorem states that the eigenvalues λ_i of $\mathbf{A}$ must be inside one of these circles Ω_i.

In addition, if p of these circles form a connected set $\mathcal{S}$ which is disjointed from the remaining $n - p$ circles, it can be proven that there are exactly p of the eigenvalues are inside the set $\mathcal{S}$, here we have to count the multiplicity of roots. Furthermore, if $\mathbf{A}$ is symmetric and real (or $\mathbf{A} = \mathbf{A}^T$), all the eigenvalues are real, and thus they all fall on the real axis.

First, let us look at a simple example.

Example 9.16

For example, the matrix $\mathbf{A}$

$$\mathbf{A} = \begin{pmatrix} 2 & 2 & 0 \\ 2 & -2 & 4 \\ 4 & 0 & 0 \end{pmatrix},$$

has three eigenvalues λ. These eigenvalues should satisfy

$$|\lambda - 2| \le r_1 = |2| + |0| = 2,$$

$$|\lambda - (-2)| \le r_2 = |2| + |4| = 6,$$

$$|\lambda - 0| \le |4| + |0| = 4.$$

These circles are shown in Fig. 9.2.

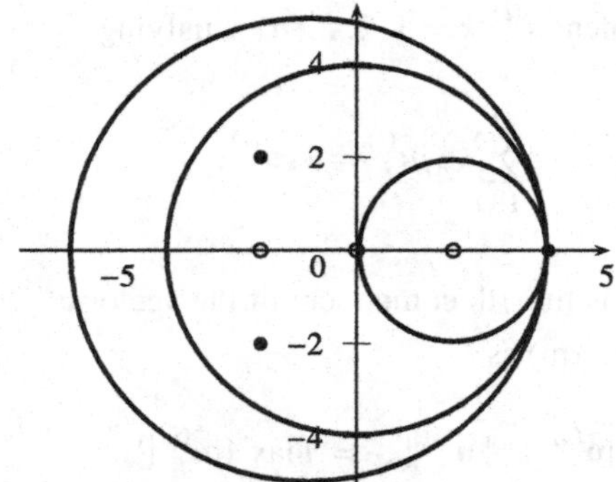

FIGURE 9.2 Gerschgorin circles of eigenvalues.

Following the method of finding the eigenvalues discussed earlier in this chapter, we have the eigenvalues of **A**

$$\lambda_i = 4, \ -2+2i, \ -2-2i,$$

which are marked as solid dots in the same figure. We can see that all these eigenvalues λ_i are within the union of all the Greschgorin discs.

Tridiagonal matrices are important in many applications. For example, the finite difference scheme in an equally-spaced grid often results in a simple tridiagonal matrix

$$\mathbf{A} = \begin{pmatrix} b & a & & & 0 \\ a & b & a & & \\ & a & b & \ddots & \\ & & \ddots & \ddots & a \\ 0 & & & a & b \end{pmatrix}, \tag{9.79}$$

where a and b are constants. In this case, the radius r_i becomes

$$r_1 = r_n = |a|,$$

$$r_i = |a| + |a| = 2|a|, \qquad (i = 2, 3, ..., n-1). \tag{9.80}$$

All the eigenvalues of **A** will satisfy

$$|\lambda - b| \leq r_i, \tag{9.81}$$

or

$$-2|a| + b \leq \lambda \leq b + 2|a|, \tag{9.82}$$

where we have used that $r_1 = r_n = |a| \leq 2|a|$.

In many applications, we also have to use the inverse $\mathbf{A}^{-1}$ of **A**. The eigenvalues Λ of the inverse $\mathbf{A}^{-1}$ are simply the reciprocals of the eigenvalues λ of **A**. That is, $\Lambda = 1/\lambda$.

Example 9.17

In our present case, we have

$$-2|a| + b \leq \frac{1}{\Lambda} \leq b + 2|a|, \tag{9.83}$$

or

$$\frac{1}{b+2|a|} \le \Lambda \le \frac{1}{b-2|a|}. \tag{9.84}$$

For example, if $a=1$ and $b=4$, we have $1/6 \le \Lambda \le 1/2$. However, if $a=-1$ and $b=-4$, we have $-1/2 \le \Lambda \le -1/6$.

9.5.2 Definiteness of a Matrix

A square symmetric matrix **A** is said to be positive definite if all its eigenvalues are strictly positive ($\lambda_i > 0$ where $i = 1, 2, ..., n$). By multiplying $\mathbf{Au} = \lambda \mathbf{u}$ by $\mathbf{u}^T$, we have

$$\mathbf{u}^T \mathbf{Au} = \mathbf{u}^T \lambda \mathbf{u} = \lambda \mathbf{u}^T \mathbf{u}, \tag{9.85}$$

which leads to

$$\lambda = \frac{\mathbf{u}^T \mathbf{Au}}{\mathbf{u}^T \mathbf{u}}. \tag{9.86}$$

This means that

$$\mathbf{u}^T \mathbf{Au} > 0, \qquad \text{if } \lambda > 0. \tag{9.87}$$

In fact, for any vector **v**, the following relationship holds

$$\mathbf{v}^T \mathbf{Av} > 0. \tag{9.88}$$

For **v** can be a unit vector, thus all the diagonal elements of **A** should be strictly positive as well.

If all the eigenvalues are non-negative or $\lambda_i \ge 0$, then the matrix is called positive semi-definite. In general, an indefinite matrix can have both positive and negative eigenvalues. Furthermore, the inverse of a positive definite matrix is also positive definite.

There are many applications of definiteness. For example, for a linear system $\mathbf{Au} = \mathbf{f}$, if **A** is positive definite, the system can be solved more efficiently by matrix decomposition methods.

Example 9.18

In order to determine the definiteness of a 2×2 symmetric matrix **A**

$$\mathbf{A} = \begin{pmatrix} \alpha & \beta \\ \beta & \alpha \end{pmatrix},$$

we first have to determine its eigenvalues. From $|\mathbf{A} - \lambda \mathbf{I}| = 0$, we have

$$\det \begin{pmatrix} \alpha - \lambda & \beta \\ \beta & \alpha - \lambda \end{pmatrix} = (\alpha - \lambda)^2 - \beta^2 = 0,$$

or

$$\lambda = \alpha \pm \beta.$$

Their corresponding eigenvectors are

$$\mathbf{v}_1 = \frac{1}{\sqrt{2}} \begin{pmatrix} 1 \\ 1 \end{pmatrix}, \qquad \mathbf{v}_2 = \frac{1}{\sqrt{2}} \begin{pmatrix} 1 \\ -1 \end{pmatrix}.$$

Eigenvectors associated with distinct eigenvalues of a symmetric square matrix are orthogonal. Indeed, they are orthogonal since we have

$$\mathbf{v}_1^T\mathbf{v}_2 = \begin{pmatrix} 1/\sqrt{2} & 1/\sqrt{2} \end{pmatrix}\begin{pmatrix} 1/\sqrt{2} \\ -1/\sqrt{2} \end{pmatrix}$$
$$= \frac{1}{\sqrt{2}} \times \frac{1}{\sqrt{2}} + \frac{1}{\sqrt{2}} \times (\frac{-1}{\sqrt{2}}) = 0.$$

The matrix $\mathbf{A}$ will be positive definite if

$$\alpha \pm \beta > 0,$$

which means $\alpha > 0$ and $\alpha > \max(+\beta, -\beta)$. The inverse

$$\mathbf{A}^{-1} = \frac{1}{\alpha^2 - \beta^2}\begin{pmatrix} \alpha & -\beta \\ -\beta & \alpha \end{pmatrix},$$

will also be positive definite. For example,

$$\mathbf{A} = \begin{pmatrix} 3 & -2 \\ -2 & 3 \end{pmatrix}, \quad \mathbf{B} = \begin{pmatrix} 10 & 9 \\ 9 & 10 \end{pmatrix},$$

are positive definite because all the eigenvalues are positive, while

$$\mathbf{D} = \begin{pmatrix} 2 & 3 \\ 3 & 2 \end{pmatrix},$$

is indefinite because one eigenvalue 5 is positive and one eigenvalue -1 is negative.

Eigenvalues and eigenvectors can have many real-world applications. For example, in engineering, the vibration frequencies of a structure are related to the so-called eigenfrequencies, and the vibration modes are often related to eigenvectors. Principal stresses and strains in engineering are also closely linked to eigenvalues and eigenvectors. In addition, the energy levels of atoms can be obtained by solving Schrödinger equation in quantum mechanics in terms of its eigenvalues.

EXERCISES

The exercises use the following matrices:

$$\mathbf{A} = \begin{pmatrix} 2 & 3 \\ 3 & 2 \end{pmatrix}, \mathbf{B} = \begin{pmatrix} 1 & -2 \\ -3 & 5 \end{pmatrix}, \mathbf{C} = \begin{pmatrix} 1 & i \\ 2 & 1 \end{pmatrix}.$$

9.1. Calculate $\mathbf{A}^T$, $\mathbf{B}^T$, $\bar{\mathbf{C}}$ and $\mathbf{C}^H$.
9.2. Compute **AB**, **BA** and **BC**. What is **AB** – **BA**?
9.3. Compute det(**A**) and det(**C**).
9.4. Compute $\mathbf{A}^{-1}$ and $\mathbf{B}^{-1}$.

Hint. $\begin{pmatrix} -2/5 & 3/5 \\ 3/5 & -2/5 \end{pmatrix}$ and $\begin{pmatrix} -5 & -2 \\ -3 & -1 \end{pmatrix}$.

9.5. Find the eigenvalues of **A** and their corresponding eigenvectors.
9.6. Show that

$$(\mathbf{AB})^T = \mathbf{B}^T\mathbf{A}^T.$$

Part IV

Calculus

Chapter 10

Differentiation

Chapter Points

- Introduce all the important concepts concerning gradients, derivatives and higher-order derivatives.
- Differentiation and differentiation rules are first explained and then shown by examples.
- Functions can be represented by series expansions, which can be useful for approximations. Series expansions and Taylor series in general are explained with the support of many examples.

10.1 GRADIENT AND DERIVATIVE

The gradient of a curve at any point P is the rate of change at that point, and it is the gradient of the tangent to the curve at that point. We know geometrically what a tangent means, but it is not easy to draw it accurately without calculations. In order to find the true gradient and the tangent, we normally use some adjustment points and try to use the ratio of the small change in y, $\delta y = y_Q - y_P$ to the change in x, $\delta x = x_Q - x_P$ to approximate the gradient. That is to say, we use

$$\frac{\delta y}{\delta x} = \frac{y_Q - y_P}{x_Q - x_P}, \tag{10.1}$$

to estimate the gradient. As Q is closer than R to P, thus the gradient estimated using P and Q is more accurate than that using P and R (see Fig. 10.1). We hope that the estimate will become the true gradient as the point Q becomes very, very close to P. How do we describe such closeness? A simple way is to use the distance $h = \delta x = PS = x_Q - x_P$ and let h tend to zero. That is to say, $h \to 0$.

Since we know that the point $x_P = a$, its coordinates for the curve x^3 are (a, a^3). The adjustment point Q now has coordinates $(a + h, (a + h)^3)$. Therefore, the gradient at point P is approximately

$$\frac{\delta y}{\delta x} = \frac{(a+h)^3 - a^3}{h}. \tag{10.2}$$

Since $(a + h)^3 = (a^3 + 3a^2h + 3ah^2 + h^3)$, we now have

$$\frac{\delta y}{\delta x} = \frac{(a^3 + 3a^2h + 3ah^2 + h^3) - a^3}{h} = 3a^2 + 3ah + h^2. \tag{10.3}$$

Since $h \to 0$, and also $h^2 \to 0$, both terms $3ah$ and h^2 will tend to zero. This means that the true gradient of x^3 at $x = a$ is $3a^2$. Since $x = a$ is any point, so in general the gradient of x^3 is $3x^2$. That is

$$(x^3)' = 3x^2, \tag{10.4}$$

where we use the $'$ notation to denote the gradient.

Now let us introduce a more formal notation for the gradient of any curve $y = f(x)$. We define the gradient as

$$f'(x) \equiv \frac{dy}{dx} \equiv \frac{df(x)}{dx} = \lim_{h \to 0} \frac{f(x+h) - f(x)}{h}. \tag{10.5}$$

The gradient is also called the first derivative. The three notations $f'(x)$, dy/dx and $df(x)/dx$ are interchangeable.

Conventionally, the notation dy/dx is called Leibnitz's notation, while the prime notation $'$ is called Lagrange's notation. Newton's dot notation $\dot{y} = dy/dt$ is now exclusively used for time derivatives. The choice of such notations is purely for clarity, convention and/or personal preference.

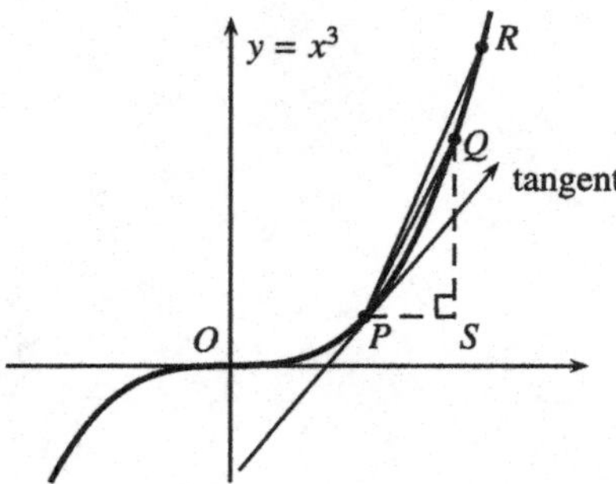

FIGURE 10.1 The tangent and gradient of a curve with $h = \text{PS} = x_Q - x_P$.

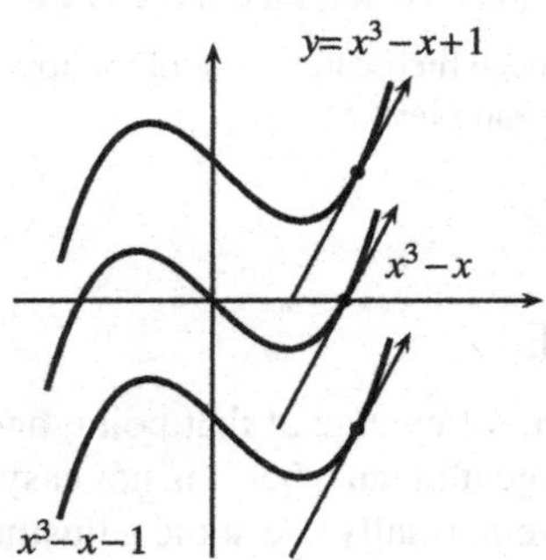

FIGURE 10.2 The gradients of a family of curves $y = x^3 - x + k$ (where $k = 0, \pm 1$) at $x = 1$ are the same $3x^2 - 1$.

In addition, here the notation 'lim' is used, as it is associated with the fact that $h \to 0$. The limit can be understood that for any $0 < h \leq 1$ where h is much less than 1 and infinitesimally small, then there always exists another even smaller number $0 < \epsilon < h$ so that we can find a better estimate for the gradient using ϵ.

Using this standard notation, our earlier example becomes

$$(x^3)' \equiv \frac{dx^3}{dx} = 3x^2. \tag{10.6}$$

From the basic definition of the derivative, we can verify that such differentiation manipulation is a linear operator. That is to say that for any two functions $f(x)$, $g(x)$ and two constants α and β, the derivative or gradient of a linear combination of the two functions can be obtained by differentiating the combination term by term. We have

$$[\alpha f(x) + \beta g(x)]' = \alpha f'(x) + \beta g'(x), \tag{10.7}$$

which can easily be extended to multiple terms. Furthermore, if k is a real constant, then its derivative is zero. For example, if we have $f(x) = x^3 - x + k$, we have

$$\begin{aligned} \frac{df(x)}{dx} &= \frac{d(x^3 - x + k)}{dx} \\ &= \frac{d(x^3)}{dx} - \frac{dx}{dx} + \frac{dk}{dx} \\ &= 3x^2 - 1 + 0 = 3x^2 - 1, \end{aligned} \tag{10.8}$$

which means that for a family of curves shifted by a constant k, the gradients at any same point (say $x = 1$) are all the same, and the tangent lines are parallel to each other (see Fig. 10.2).

Example 10.1

To find the gradient $f(x) = x^n$ where $n = 1, 2, 3, 4, \ldots$ is a positive integer, we use

$$f'(x) = \lim_{h \to 0} \frac{f(x+h) - f(x)}{h} = \lim_{h \to 0} \frac{(x+h)^n - x^n}{h}.$$

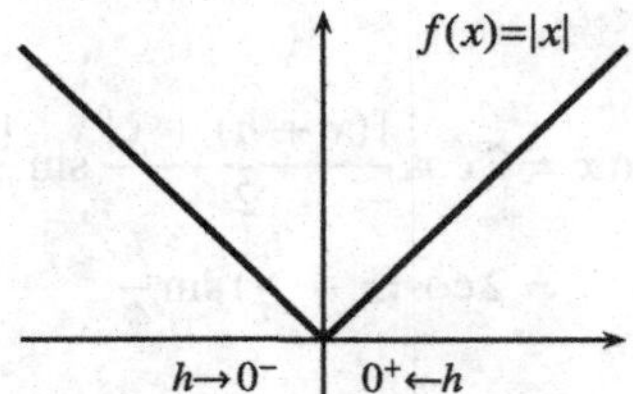

FIGURE 10.3 Limit and thus the derivative at $x = 0$ is not defined.

Using the binomial theorem

$$(x+h)^n = \binom{n}{0}x^n + \binom{n}{1}x^{n-1}h + \binom{n}{2}x^{n-2}h^2 + \ldots + \binom{n}{n}h^n,$$

and the fact that

$$\binom{n}{0} = 1, \binom{n}{1} = n, \binom{n}{2} = \frac{n(n-1)}{2!}, \binom{n}{k} = \frac{n!}{k!(n-k)!}, \binom{n}{n} = 1,$$

we have

$$(x+h)^n - x^n = (x^n + nx^{n-1}h + \frac{n(n-1)}{2!}x^{n-2}h^2 + \ldots + h^n) - x^n$$
$$= nx^{n-1}h + \frac{n(n-1)}{2!}x^{n-2}h^2 + \ldots + h^n.$$

Therefore, we get

$$f'(x) = \lim_{h\to 0} \frac{nx^{n-1}h + \frac{n(n-1)}{2!}x^{n-2}h^2 + \ldots + h^n}{h}$$
$$= \lim_{h\to 0}\Big[nx^{n-1} + \frac{n(n-1)}{2!}x^{n-2}h + \ldots + h^{n-1}\Big].$$

Since $h \to 0$, all the terms except the first (nx^{n-1}) will tend to zero. We finally have

$$f'(x) = nx^{n-1}.$$

In fact, this formula is valid for any integer ($n = 0, \pm 1, \pm 2, \ldots$).

It is worth pointing out that the limit $\lim_{h\to 0}$ must exist. If such a limit does not exist, then the derivative or the gradient is not defined. For example, the gradient of a rather seemingly simple function $f(x) = |x|$ where $|x|$ means the absolute value of $f'(x)$ does not exist at $x = 0$ (see Fig. 10.3). This is because we will get different values depending on how we approach $x = 0$. If we approach $h \to 0$ from the right ($h > 0$) and use the notation $h \to 0^+$, we have

$$f'(0^+) = \lim_{h\to 0^+} \frac{|h| - |0|}{h} = \lim_{h\to 0^+} \frac{h}{h} = \lim_{h\to 0^+} 1 = 1. \tag{10.9}$$

However, if we approach from the left ($h < 0$), we have

$$f'(0^-) = \lim_{h\to 0^-} \frac{|h| - |0|}{h} = \lim_{h\to 0^-} \frac{-h}{h} = \lim_{h\to 0^-} -1 = -1. \tag{10.10}$$

The correct definition of the derivative should be independent of the way we approach the point using $h \to 0$. In this case, we say, the gradient or derivative at $x = 0$ does not exist.

We can obtain the derivative of any function from first principles. For example, we now try to derive $d(\sin x)/dx$.

In order to obtain the derivative of $f(x) = \sin(x)$, we now use the definition

$$f'(x) = \lim_{h\to 0} \frac{\sin(x+h) - \sin x}{h}.$$

From trigonometry, we know that

$$\sin(x+h) - \sin x = 2\cos\frac{[(x+h)+x]}{2}\sin\frac{[(x+h)-x]}{2}$$
$$= 2\cos(x+\frac{h}{2})\sin\frac{h}{2}.$$

We have

$$f'(x) = \lim_{h\to 0}\frac{2\cos(x+\frac{h}{2})\sin\frac{h}{2}}{h} = \lim_{h\to 0}\cos(x+\frac{h}{2})\frac{\sin\frac{h}{2}}{\frac{h}{2}}.$$

Using

$$\lim_{h\to 0}\frac{\sin\frac{h}{2}}{\frac{h}{2}} = 1, \qquad \lim_{h\to 0}\cos(x+\frac{h}{2}) = \cos x,$$

we finally have

$$f'(x) = \frac{d(\sin x)}{dx} = \cos x.$$

Following the same procedure, we can also derive that

$$\frac{d(\cos x)}{dx} = -\sin x,$$

which is left as an exercise.

This example demonstrates that even if we are able to calculate the derivative using first principles, it is usually not the best way to do so in practice. The good thing is that we only have to do it once to understand how it works. For more complicated functions, we should use certain rules such as the chain rule and product rule. In addition, we can sometimes use tables or mathematical software packages. Now let us introduce some differentiation rules.

10.2 DIFFERENTIATION RULES

We know how to differentiate x^3 and $\sin(x)$, and a natural question is how we can differentiate $\sin(x^3)$? Here we need the chain rule. The trick to derive the chain rule is to consider dx and dy (and other similar quantities) as infinitesimal but non-zero quantities or increments δx and δy so that we can divide and multiply them as necessary. In the limit of $\delta x \to 0$, we have $\delta x \to dx$ and $\delta y/\delta x \to dy/dx$. If $f(u)$ is a function of u, and u is in turn a function of x, we want to calculate dy/dx. We then have

$$\frac{dy}{dx} = \frac{dy}{du}\cdot\frac{du}{dx}, \tag{10.11}$$

or

$$\frac{df[u(x)]}{dx} = \frac{df(u)}{du}\cdot\frac{du(x)}{dx}. \tag{10.12}$$

This is the well-known chain rule.

Example 10.2

Now we come back to our original problem of differentiating $\sin(x^3)$. First we let $u = x^3$, we then

$$\frac{d(\sin(x^3))}{dx} = \frac{d(\sin u)}{du}\cdot\frac{dx^3}{dx} = \cos u \cdot 3x^2 = 3x^2\cos(x^3).$$

As we practice more, we can write the derivatives more compactly and quickly.

For example,

$$\frac{d(\sin^n(x))}{dx} = n\sin^{n-1}(x)\cdot\cos x,$$

where we consider $\sin^n(x)$ to be in the form of u^n and $u = \sin(x)$.

A further question is if we can differentiate $x^5\sin(x)$ easily, since we already know how to differentiate x^5 and $\sin x$ separately? For this, we need the differentiation rule for products. Let $y = uv$ where $u(x)$ and $v(x)$ are (often simpler) functions of x. From the definition, we have

$$\frac{dy}{dx} = \frac{d(uv)}{dx} = \lim_{h\to 0}\frac{u(x+h)v(x+h) - u(x)v(x)}{h}. \tag{10.13}$$

Since adding a term and deducting the same term does not change an expression, we have

$$\begin{aligned}&\frac{u(x+h)v(x+h) - u(x)v(x)}{h}\\ &= \frac{u(x+h)v(x+h) \overbrace{- u(x+h)v(x) + u(x+h)v(x)}^{=0} - u(x)v(x)}{h}\\ &= u(x+h)\frac{[v(x+h)-v(x)]}{h} + v(x)\frac{[u(x+h)-u(x)]}{h}.\end{aligned} \tag{10.14}$$

In addition, $u(x+h) \to u(x)$ as $h \to 0$, we finally have

$$\begin{aligned}\frac{d(uv)}{dx} &= \lim_{h\to 0}\left\{u(x+h)\frac{[v(x+h)-v(x)]}{h} + v(x)\frac{[u(x+h)-u(x)]}{h}\right\}\\ &= u\frac{dv}{dx} + v\frac{du}{dx},\end{aligned} \tag{10.15}$$

or simply

$$(uv)' = uv' + vu'. \tag{10.16}$$

This is the formula for products or product rule. If we replace v by $1/v = v^{-1}$ and apply the chain rule

$$\frac{d(v^{-1})}{dx} = -1\times v^{-1-1}\times\frac{dv}{dx} = -\frac{1}{v^2}\frac{dv}{dx}, \tag{10.17}$$

we have the formula for quotients or the quotient rule

$$\frac{d(\frac{u}{v})}{dx} = \frac{d(uv^{-1})}{dx} = u(\frac{-1}{v^2})\frac{dv}{dx} + v^{-1}\frac{du}{dx} = \frac{v\frac{du}{dx} - u\frac{dv}{dx}}{v^2}. \tag{10.18}$$

Now let us apply these rules to differentiating $\tan x$.

Example 10.3

Since $\tan x = \sin x/\cos x$, we have $u = \sin x$ and $v = \cos x$. Therefore, we get

$$\begin{aligned}\frac{d(\tan x)}{dx} &= \frac{\cos x\frac{d\sin x}{dx} - \sin x\frac{d\cos x}{dx}}{\cos^2 x} = \frac{\cos x\cos x - \sin x\times(-\sin x)}{\cos^2 x}\\ &= \frac{\cos^2 x + \sin^2 x}{\cos^2 x} = \frac{1}{\cos^2 x}.\end{aligned}$$

Sometimes, it is not easy to find the derivative dy/dx directly. In this case, it is usually a good idea to try to find dx/dy since

$$\frac{dy}{dx} = 1/\frac{dx}{dy}, \tag{10.19}$$

TABLE 10.1 First derivatives of common functions.

$f(x)$	$f'(x)$	$f(x)$	$f'(x)$
x^n	nx^{n-1}	e^x	e^x
a^x	$a^x \ln a \ (a>0)$	$\ln x$	$\frac{1}{x} \ (x>0)$
$\sin x$	$\cos x$	$\cos x$	$-\sin x$
$\tan x$	$1+\tan^2 x$	$\sec x$	$\sec x \tan x$
$\log_a x$	$\frac{1}{x \ln a}$	$\tanh x$	$\mathrm{sech}^2 x$
$\sinh x$	$\cosh x$	$\cosh x$	$\sinh x$
$\sin^{-1} x$	$\frac{1}{\sqrt{1-x^2}}$	$\cos^{-1} x$	$\frac{-1}{\sqrt{1-x^2}}$
$\tan^{-1} x$	$\frac{1}{1+x^2}$	$\sinh^{-1} x$	$\frac{1}{\sqrt{x^2+1}}$
$\tanh^{-1} x$	$\frac{1}{1-x^2}$	$\cosh^{-1} x$	$\frac{1}{\sqrt{x^2-1}}$

or carry out the derivatives term-by-term and then find dy/dx. This is especially the case for implicit functions. Let us demonstrate this with an example.

Example 10.4

To find the gradient of the curve $\sin^2 x + y^4 - 2y = e^{3x}$, at the point (0, 1), we first differentiate each term with respect to x, and we have

$$\frac{d\sin^2 x}{dx} + \frac{dy^4}{dx} - \frac{d(2y)}{dx} = \frac{de^{3x}}{dx},$$

or

$$2\sin x\cos(x) + 4y^3\frac{dy}{dx} - 2\frac{dy}{dx} = 3e^{3x},$$

where we have used the chain rule

$$\frac{du}{dx} = \frac{du}{dy}\cdot\frac{dy}{dx},$$

so that $u = y^4$ gives

$$\frac{dy^4}{dx} = 4y^3\frac{dy}{dx}.$$

After some rearrangement, we have

$$(4y^3 - 2)\frac{dy}{dx} = 3e^{3x} - 2\sin x\cos x,$$

or

$$\frac{dy}{dx} = \frac{3e^{3x} - 2\sin x\cos x}{(4y^3 - 2)}.$$

Therefore, the gradient at $x = 0$ and $y = 1$ becomes

$$\frac{dy}{dx} = \frac{3e^{3\times 0} - 2\times\sin 0\times\cos 0}{4\times 1^3 - 2} = \frac{3}{2} = 1.5.$$

The first derivatives of commonly used mathematical functions are listed in Table 10.1, and the derivatives of other functions can be obtained using the differentiation rules in combination with this table.

The derivative we discussed so far is the gradient or the first derivative of a function $f(x)$. If we want to see how the gradient itself varies, we may need to take the gradient of the gradient of a function. In this case, we are in fact calculating the second derivative. We write

$$\frac{d^2 f(x)}{dx^2} \equiv f''(x) \equiv \frac{d}{dx}\left(\frac{df(x)}{dx}\right). \tag{10.20}$$

Following a similar procedure, we can write any higher-order derivative as

$$\frac{d^3 f(x)}{dx^3} = f'''(x) = \frac{d(f''(x))}{dx}, \ldots, \frac{d^n f(x)}{dx^n} = f^{(n)}(x). \tag{10.21}$$

Let us look some examples.

Example 10.5

For example, $f(x) = x^3$, we have

$$f'(x) = 3x^2, \quad f''(x) = d(3x^2)/dx = 3 \times 2x = 6x, \quad f'''(x) = 6$$

and

$$f''''(x) = 0.$$

For $g(x) = e^x \sin(x^2)$, we have

$$\begin{aligned} g' &= [e^x \sin(x^2)]' = (e^x)' \sin(x^2) + e^x (\sin(x^2))' = e^x \sin(x^2) + e^x \cos(x^2)2x. \\ g'' &= [e^x \sin(x^2)]' + [e^x \cos(x^2)2x]' \\ &= [e^x \sin(x^2) + 2xe^x \cos(x^2)] + [e^x \cos(x^2)2x + e^x(-\sin(x^2)(2x)(2x) + e^x \cos(x^2)2] \\ &= e^x[\sin(x^2) + 4x\cos(x^2) - 4x^2 \sin(x^2) + 2\cos(x^2)]. \end{aligned}$$

Following the same procedure, we have

$$g''' = e^x[\sin(x^2) - 12x^2 \sin(x^2) - 12x \sin(x^2) - 8x^3 \cos(x^2) + 6x\cos(x^2) + 6\cos(x^2)].$$

The expressions can get quite lengthy for higher-order derivatives.

It is worth pointing out that notation for the second derivative is $\frac{d^2y}{dx^2}$, not $\frac{d^2y}{d^2x}$ or $\frac{dy^2}{dx^2}$ (which is wrong).

10.3 SERIES EXPANSIONS AND TAYLOR SERIES

In numerical methods and some mathematical analysis, series expansions make some calculations easier. For example, we can write the exponential function e^x as a series about $x_0 = 0$

$$e^x = \alpha_0 + \alpha_1 x + \alpha_2 x^2 + \alpha_3 x^3 + \ldots + \alpha_n x^n. \tag{10.22}$$

Now let us try to determine these coefficients. At $x = 0$, we have

$$e^0 = 1 = \alpha_0 + \alpha_1 \times 0 + \alpha_2 \times 0^2 + \ldots \alpha_n \times 0^n = \alpha_0, \tag{10.23}$$

which gives $\alpha_0 = 1$. In order to reduce the power or order of the expansion so that we can determine the next coefficient, we first differentiate both sides of (10.22) once; we have

$$e^x = \alpha_1 + 2\alpha_2 x + 3\alpha_3 x^2 + \ldots + n\alpha_n x^{n-1}. \tag{10.24}$$

By setting again $x = 0$, we have

$$e^0 = 1 = \alpha_1 + 2\alpha_2 \times 0 + \ldots + n\alpha_n \times 0^{n-1} = \alpha_1, \tag{10.25}$$

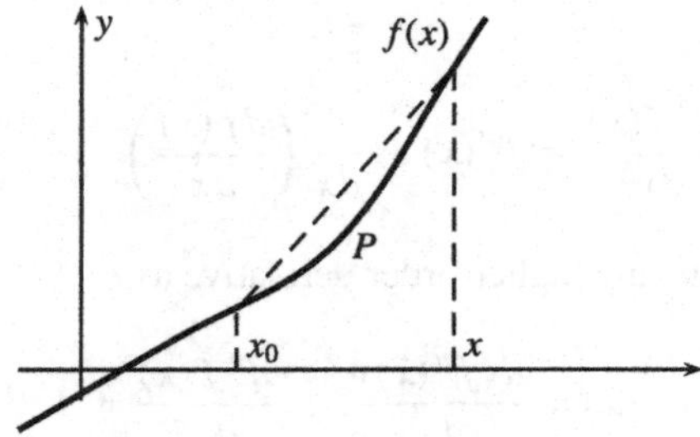

FIGURE 10.4 Expansion and approximations for $f(x) = f(x_0 + h)$ where $h = x - x_0$.

which gives $\alpha_1 = 1$. Similarly, differentiating it again, we have

$$e^x = (2 \times 1) \times \alpha_2 + 3 \times 2\alpha_3 x + ... + n(n-1)\alpha_n x^{n-2}. \tag{10.26}$$

At $x = 0$, we get

$$e^0 = (2 \times 1) \times \alpha_2 + 3 \times 2\alpha_3 \times 0 + ... + n(n-1)\alpha_n \times 0^{n-2} = 2\alpha_2, \tag{10.27}$$

or $\alpha_2 = 1/(2 \times 1) = 1/2!$. Following the same procedure and differentiating it n times, we have

$$e^x = n!\alpha_n, \tag{10.28}$$

and $x = 0$ leads to $\alpha_n = 1/n!$. Therefore, the final series expansion can be written as

$$e^x = 1 + x + \frac{1}{2!}x^2 + \frac{1}{3!}x^3 + ... + \frac{1}{n!}x^n. \tag{10.29}$$

Obviously, we can follow a similar process to expand other functions. We have seen here the importance of differentiation and derivatives.

If we know the value of $f(x)$ at x_0, we know that the gradient $f'(x)$ in a small interval $h = x - x_0$ can be approximated by (see Fig. 10.4)

$$f'(x) = \frac{f(x_0 + h) - f(x_0)}{h}, \tag{10.30}$$

which is equivalent to saying that the adjacent value of $f(x_0 + h)$ can be approximated by

$$f(x_0 + h) \approx f(x_0) + hf'(x). \tag{10.31}$$

Then as a first approximation we can use $f'(x_0)$ to approximate $f'(x)$ in the interval. However, if we know higher derivatives $f''(x_0)$ at x_0, we can use an even better approximation to estimate $f'(x)$

$$f'(x_0 + h) = f'(x_0) + h\frac{f''(x)}{2}. \tag{10.32}$$

This is equivalent to stating that

$$f(x_0 + h) = f(x_0) + hf'(x_0) + \frac{h^2}{2}f''(x). \tag{10.33}$$

If we again approximate $f''(x)$ using higher-order derivatives, we can obtain the Taylor series expansions

$$f(x_0 + h) = f(x_0) + f'(x_0)h + \frac{f''(x_0)}{2!}h^2 + \frac{f'''(x_0)}{3!}h^3 + ... + \frac{f^{(n)}(x_0)}{n!}h^n + R_{n+1}(h), \tag{10.34}$$

where $f^{(n)}$ is the nth derivative of $f(x)$ and $R_{n+1}(h)$ is the error estimation for this series. In theory, we can use as many terms as possible, but in practice, the series converges very quickly and only a few terms are sufficient. It is straightforward to verify that the exponential series for e^x is identical to the results given earlier. Now let us look at other examples.

Example 10.6

Let us expand $f(x) = \sin x$ about $x_0 = 0$. We know that

$$f'(x) = \cos x, \quad f''(x) = -\sin x, \quad f'''(x) = -\cos x, \quad ...,$$

or $f'(0) = 1, \; f''(0) = 0, \; f'''(0) = -1, \; f''''(0) = 0, ...$, which means that

$$\sin x = \sin 0 + x f'(0) + \frac{f''(0)}{2!}x^2 + \frac{f'''(0)}{3!}x^3 + ... = x - \frac{x^3}{3!} + \frac{x^5}{5!} + ...,$$

where the angle x is in radians.

For example, we know that $\sin 30° = \sin\frac{\pi}{6} = 1/2$. We now use the expansion to estimate it for $x = \pi/3 = 0.523598$

$$\sin\frac{\pi}{6} \approx \frac{\pi}{6} - \frac{(\pi/6)^3}{3!} + \frac{(\pi/6)^5}{5!}$$
$$= 0.523599 - 0.02392 + 0.0000328 \approx 0.5000021326,$$

which is very close to the true value 1/2.

If we continue the process to infinity, we then reach the infinite power series and the error $\lim_{n\to\infty} R_{n+1} \to 0$ if the series converges. The end results are the Maclaurin series. For example, some common series are

$$e^x = 1 + x + \frac{x^2}{2!} + ... + \frac{x^n}{n!} + ..., \quad (x \in \mathcal{R}), \tag{10.35}$$

$$\sin x = x - \frac{x^3}{3!} + \frac{x^5}{5!} - ..., \quad (x \in \mathcal{R}), \tag{10.36}$$

$$\cos x = 1 - \frac{x^2}{2!} + \frac{x^4}{4!} - ..., \quad (x \in \mathcal{R}), \tag{10.37}$$

and

$$\ln(1 + x) = x - \frac{x^2}{2} + \frac{x^3}{3} - \frac{x^4}{4} + \frac{x^5}{5} - ..., \quad x \in (-1, 1]. \tag{10.38}$$

It leaves as an exercise to prove the above series.

Example 10.7

Let us use the above Maclaurin series to prove the Euler formula:

$$e^{i\theta} = \cos(\theta) + i\sin(\theta). \tag{10.39}$$

Using $x = i\theta$, $i^2 = -1$ and

$$e^x = 1 + x + \frac{x^2}{2!} + \frac{x^3}{3!} + \frac{x^4}{4!} + \frac{x^5}{5!} + ...$$

we have

$$e^x = e^{i\theta} = 1 + i\theta + \frac{(i\theta)^2}{2!} + \frac{(i\theta)^3}{3!} + \frac{(i\theta)^4}{4!} + \frac{(i\theta)^5}{5!} + ...$$
$$1 + i\theta - \frac{\theta^2}{2!} - i\frac{\theta^3}{3!} + \frac{\theta^4}{4!} + i\frac{\theta^5}{5!} + ...$$

$$= [1 - \frac{\theta^2}{2!} + \frac{\theta^4}{4!} + ...] + i[\theta - \frac{\theta^3}{3!} + \frac{\theta^5}{5!} + ...]$$
$$= \cos\theta + i\sin\theta,$$

where we have used Eqs. (10.36) and (10.37).

Differentiation is just one important part of calculus, and another important part is integration, which will be introduced in the next chapter.

EXERCISES

10.1. Find the first derivatives of the following expressions:

$$f(x) = x^3 + 2x^2 + 3x, \quad g(x) = xe^{-x}, \quad h(x) = x^3 + \cos(x^2) + x\ln x.$$

10.2. Find the first derivative of $f(x) = x^x$ for $x > 0$.
10.3. Find $y'(x)$ from the following equation:

$$y(x) - \sin(x)e^{-x} + x^2 y(x) = 0.$$

10.4. Find the derivatives (up to the fifth derivative) of $xe^{-x} + \sin(x^2)$.
10.5. Expand $\exp(-x)$ into a series.
10.6. Expand $x^2\exp(-x^2)$ into a series.

Chapter 11

Integration

Chapter Points

- Integration is explained in great detail, including both infinite integral, definite integrals and their basic rules.
- Techniques are explained and demonstrated to obtain the integrals of basic functions, including integration by substitution and integration by parts.
- Many examples are used to show how to apply these basic integration techniques in necessary combinations.

11.1 INTEGRATION

Differentiation is used to find the gradient for a given function. Now a natural question is how to find the original function for a given gradient. This is the integration process, which can be considered as the reverse of the differentiation process. Since we know that

$$\frac{d \sin x}{dx} = \cos x, \tag{11.1}$$

that is the gradient of $\sin x$ is $\cos x$, we can easily say that the original function is $\sin x$ if we know its gradient is $\cos x$. We can write

$$\int \cos x \, dx = \sin x + C, \tag{11.2}$$

where C is the constant of integration. Here $\int dx$ is the standard notation showing the integration is with respect to x, and we usually call this the integral. The function $\cos x$ is called the integrand.

The integration constant comes from the fact that a family of curves shifted by a constant will have the same gradient at their corresponding points (see Fig. 10.2). This means that the integration can be determined up to an arbitrary constant. For this reason, we call it the indefinite integral.

Integration is more complicated than differentiation in general. Even when we know the derivative of a function, we have to be careful. For example, we know that $(x^{n+1})' = (n+1)x^n$ or

$$\Big(\frac{1}{n+1}x^{n+1}\Big)' = x^n$$

for any n integers, so we can write

$$\int x^n dx = \frac{1}{n+1}x^{n+1} + C. \tag{11.3}$$

However, there is a possible problem when $n=-1$ because $1/(n+1)$ will become $1/0$. In fact, the above integral is valid for any n except for $n=-1$. When $n=-1$, we have

$$\int \frac{1}{x} dx = \ln x + C. \tag{11.4}$$

If we know that the gradient of a function $F(x)$ is $f(x)$ or $F'(x) = f(x)$, it is possible and sometimes useful to express where the integration starts and ends, and we often write

$$\int_a^b f(x)dx = \Big[F(x)\Big]_a^b$$

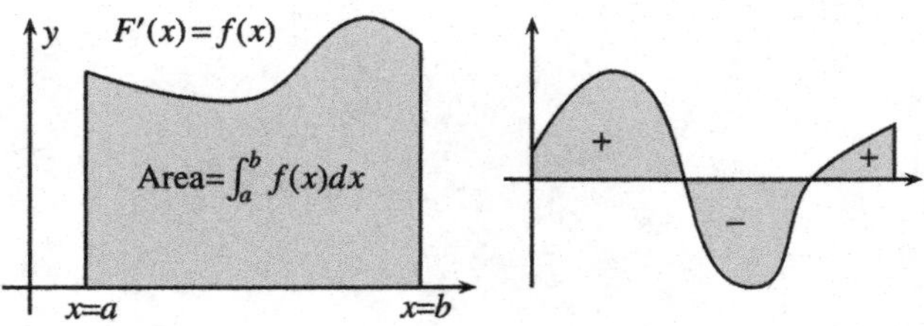

FIGURE 11.1 Integration and geometrical interpretation.

$$= F(x)\Big|_a^b = F(b) - F(a). \tag{11.5}$$

Here a is called the lower limit of the integration, while b is the upper limit of the integration. In this case, the constant of integration has dropped out because the integral can be determined accurately. The integral becomes a definite integral and it corresponds to the area under a curve $f(x)$ from a to $b \geq a$ (see Fig. 11.1).

It is worth pointing out that the area in Fig. 11.1 may have different signs depending on whether the curve $f(x)$ is above or below the x-axis. If $f(x) > 0$, then that section of area is positive; otherwise, it is negative.

Example 11.1

To calculate the integral $\int_0^1 \sqrt{x}dx$, we know that $n = 1/2$, so we have

$$\int_0^1 \sqrt{x}\,dx = \Big[\frac{1}{\frac{1}{2}+1}x^{\frac{1}{2}+1}\Big]_0^1 = \Big[\frac{2}{3}x^{\frac{3}{2}}\Big]_0^1 = \frac{2}{3}1^{\frac{3}{2}} - \frac{2}{3}0^{\frac{3}{2}} = \frac{2}{3}.$$

From (11.5), if we interchange the limits of the integral, we have

$$\int_b^a f(x)dx = \Big[F(x)\Big]_b^a = F(a) - F(b) = -(F(b) - F(a)) = -\int_a^b f(x)dx. \tag{11.6}$$

This means the interchange of the integration limits is equivalent to multiplying the integral by -1.

Example 11.2

Since we have already obtained that

$$\int_0^1 \sqrt{x}\,dx = \frac{2}{3},$$

we can easily conclude that

$$\int_1^0 \sqrt{x}\,dx = -\frac{2}{3}.$$

In addition, since differentiation is linear, that is $[\alpha f(x) + \beta g(x)]' = \alpha f'(x) + \beta g'(x)$ where α and β are two real constants, it is easy to understand that integration is also a linear operator. We have

$$\int_a^b [\alpha f(x) + \beta g(x)]dx = \alpha \int_a^b f(x)dx + \beta \int_a^b g(x)dx. \tag{11.7}$$

Also if $a \leq c \leq b$, we have

$$\int_a^b f(x)dx = \int_a^c f(x)dx + \int_c^b f(x)dx. \tag{11.8}$$

This may become useful when dealing with certain integrals, especially if we do not want the areas (positive and negative) to simply add together. (See Fig. 11.2.)

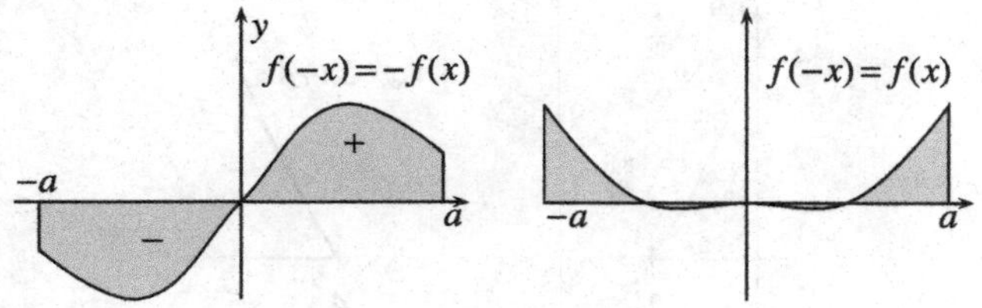

FIGURE 11.2 Integrals of odd and even functions.

In the special case of an odd function $f(-x) = -f(x)$, we have

$$\int_{-a}^{a} f(x)dx = \int_{-a}^{0} f(x)dx + \int_{0}^{a} f(x)dx = -\int_{0}^{-a} f(x)dx + \int_{0}^{a} f(x)dx$$
$$= \int_{0}^{a} f(-x)dx + \int_{0}^{a} f(x)dx = \int_{0}^{a} [-f(x) + f(x)]dx = 0. \quad (11.9)$$

The other way of understanding this is that the area for $x > 0$ is opposite in sign to the area for $x < 0$. So the total area or the integral is zero.

Similarly, for an even function $f(-x) = f(x)$, the areas for both $x > 0$ and $x < 0$ are the same, so we have

$$\int_{-a}^{a} f(x)dx = 2\int_{0}^{a} f(x)dx. \quad (11.10)$$

The odd and even properties can become useful in simplifying some complex integrals, especially when the integrand is odd and the integration limits are symmetric.

Example 11.3

For example, to evaluate the integral

$$I = \int_{-\pi}^{\pi} \sin(x)e^{-x^{100}} dx,$$

may be difficult because there may not be any explicit form for the integral. However, we know that x^{100} is an even function and $\exp(-x^{100})$ is also an even function. We also know that $\sin(x)$ is an odd function, thus the integrand $f(x) = \sin(x)e^{-x^{100}}$ is an odd function. Alternatively, we know that

$$f(-x) = \sin(-x)e^{-(-x)^{100}} = -\sin(x)e^{-x^{100}} = -f(x),$$

which also confirms that $f(x)$ is an odd function.

Therefore, we have

$$\int_{-\pi}^{0} \sin(x)e^{-x^{100}} dx = -\int_{0}^{\pi} \sin(x)e^{-x^{100}} dx,$$

which means that

$$\int_{-\pi}^{\pi} \sin(x)e^{-x^{100}} dx = 0.$$

On the other hand, an even integrand can still make the calculation slightly easier. To calculate $\int_{-\pi/2}^{\pi/2} \cos(x)dx$, we know that $\cos(x)$ is an even function, and the integration limits are symmetric, so we have

$$\int_{-\pi/2}^{0} \cos(x)dx = \int_{0}^{\pi/2} \cos(x)dx.$$

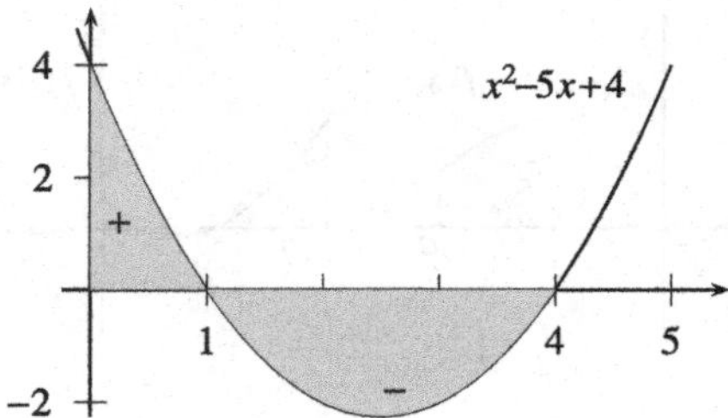

FIGURE 11.3 Computing the total area of the shaded regions.

Thus, we have

$$\int_{-\pi/2}^{\pi/2} \cos(x)dx = 2\int_0^{\pi/2} \cos(x)dx = 2\sin(x)\Big|_0^{\pi/2} = 2[\sin(\frac{\pi}{2}) - \sin(0)] = 2[1-0] = 2.$$

Sometimes, the integral that we evaluate can have precise physical meaning. In this case, some care should be taken when the value of an integral is negative. Let us look at an example.

Example 11.4

Let us now estimate the total area of the shaded regions under the curve $f(x) = x^2 - 5x + 4$ shown in Fig. 11.3.

First, we can simply say that the area is obviously the definite integral

$$A = \int_0^4 f(x)dx = \int_0^4 (x^2 - 5x + 4)dx.$$

From the basic linear properties of the integration, we have

$$A = \int_0^4 x^2 dx - 5\int_0^4 xdx + 4\int_0^4 dx = \Big[\frac{1}{3}x^3\Big]_0^4 - 5\Big[\frac{1}{2}x^2\Big]_0^4 + 4\Big[x\Big]_0^4$$
$$= \frac{1}{3} \times [4^3 - 0^3] - \frac{5}{2} \times [4^2 - 0^2] + 4 \times [4 - 0] = \frac{64}{3} - 40 + 16 = -\frac{8}{3}.$$

The total area is negative; this is not what we expected. The area should be positive, right? The geometrical meaning of the integral is the total area, adding the positive areas (above the x-axis) and the negative (below the x-axis). However, what we wanted to calculate is simply the total area of the two shaded regions, not to include its signs. In this case, we have to divide the integral into two intervals $[0, 1]$ and $[1, 4]$. The area of the first region is

$$\int_0^1 f(x)dx = \frac{1}{3} \times [1^3 - 0^3] - \frac{5}{2} \times [1^2 - 0^2] + 4 \times [1 - 0] = \frac{11}{6}.$$

Similarly, the area of the second region is

$$\int_1^4 f(x)dx = \frac{1}{3} \times [4^3 - 1^3] - \frac{5}{2} \times [4^2 - 1^2] + 4 \times [4 - 1]$$
$$= \frac{63}{3} - \frac{75}{2} + 12 = -\frac{9}{2}.$$

Here we only want the value of the area, so the area of the second region is $9/2$. Therefore, the total area of the shaded regions is $A = \frac{11}{6} + \frac{9}{2} = 6\frac{1}{3}$.

The integrals of commonly used functions are listed in Table 11.1. For many simple functions, we can easily write their integrals using the table of differentiation. For more complicated functions, what are the general techniques to obtain integrals? There are many techniques such as integration by parts, method of substitution, complex integrals and others. In the next section, we will briefly introduce integration by parts.

It is worth pointing out that some integrals with seemingly simple integrands may not be well defined and can be divergent.

TABLE 11.1 Integrals of common functions.

$f(x)$	$\int f(x)dx$	$f(x)$	$\int f(x)dx$
e^x	e^x	$\frac{1}{x}$	$\ln x$
x^n	$\frac{1}{n+1}x^{n+1}$ $(n \neq -1)$	a^u	$\frac{a^u}{\ln a}$ $(a > 0)$
$\ln x$	$x \ln x - x$	$\log_a x$	$x \log_a x - \frac{x}{\ln a}$
$\frac{1}{a^2+x^2}$	$\frac{1}{a}\tan^{-1}\frac{x}{a}$	$\frac{1}{a^2-x^2}$	$\frac{1}{2a}\ln\frac{a+x}{a-x}$
$\frac{1}{x^2-a^2}$	$\frac{1}{2a}\ln\frac{x-a}{x+a}$	$\frac{1}{\sqrt{a^2-x^2}}$	$\sin^{-1}\frac{x}{a}$
$\frac{1}{\sqrt{x^2+a^2}}$	$\sinh^{-1}\frac{x}{a}$	$\frac{1}{\sqrt{x^2-a^2}}$	$\cosh^{-1}\frac{x}{a}$
$\sin x$	$-\cos x$	$\cos x$	$\sin x$
$\sec^2 x$	$\tan x$	$\csc^2 x$	$-\cot x$
$\sinh x$	$\cosh x$	$\cosh x$	$\sinh x$
$\tanh x$	$\ln\cosh x$	$\coth x$	$\ln\sinh x$

Example 11.5

We know

$$\int \frac{1}{x}dx = \ln(x) + C,$$

so

$$\int_0^1 = \ln(x)\Big|_0^1 = \ln(1) - \ln(0),$$

but $\ln(0)$ is not well defined (or $\ln(0) \to -\infty$), which means that above definite integral is divergent (or does not exist).

When the integral limits involve an infinity, some care should be taken when evaluating the definite integral.

Example 11.6

A very useful definite integral is

$$I = \int_0^\infty e^{-x}dx.$$

We know that $[e^{-x}]' = -e^{-x}$ or $[-e^{-x}]' = e^{-x}$. So we have

$$I = \int_0^\infty e^{-x}dx = \Big[-e^{-x}\Big]_0^\infty = -e^{-\infty} - (-e^{-0}) = 0 - (-1) = 1.$$

Here, we have used $\exp[-\infty] = 0$. However, the integral $\int_0^\infty e^x dx$ does not exist because it diverges (involving the value of infinity).

In fact, many definite integrals with one infinite integration limits may diverge. For example, $\int_{-\infty}^0 e^{-x}dx$, $\int_0^\infty xdx$ and $\int_0^1 \frac{1}{x^2}dx$ will all diverge.

11.2 INTEGRATION BY PARTS

In differentiation, we can easily find the gradient of $x \sin x$ if we know the derivatives of x and $\sin x$ by using the rule of products. In integration, can we find the integral of $x \sin x$ if we know the integrals of x and $\sin x$? The answer is yes, this is the integration by parts.

From the differentiation rule

$$\frac{d(uv)}{dx} = u\frac{dv}{dx} + v\frac{du}{dx}, \tag{11.11}$$

we integrate it with respect to x, we have

$$uv = \int u\frac{dv}{dx}dx + \int v\frac{du}{dx}dx. \tag{11.12}$$

By rearranging, we have

$$\int u\frac{dv}{dx}dx = uv - \int v\frac{du}{dx}dx, \tag{11.13}$$

or

$$\int uv'dx = uv - \int vu'dx. \tag{11.14}$$

This is the well-known formula for the technique known as integration by parts. Since dx appears in both $\frac{du}{dx}$ and dx, it makes sense to write the above as

$$\int udv = uv - \int vdu, \tag{11.15}$$

and this notation is also widely used in many textbooks.

You may wonder where the constant of integration is? You are right, there is a constant of integration for uv, but as we know they exist for indefinite integrals, we simply omit to write it out in the formula, but we have to remember to put it back in the end. Now let us look at an example.

Example 11.7

We know that $\int xdx = \frac{x^2}{2} + C$, and $\int \sin xdx = -\cos x + C$, we now try to find the integral

$$I = \int x\sin xdx. \tag{11.16}$$

Let $u = x$ and $dv/dx = \sin x$, we have

$$\frac{du}{dx} = 1, \qquad v = \int \sin xdx = -\cos x. \tag{11.17}$$

Therefore, we have after substituting into the formula and using integration by parts

$$\begin{aligned} I &= \int x\sin xdx = x(-\cos x) - \int (-\cos x)\frac{du}{dx}dx \\ &= -x\cos x + \int \cos xdx = -x\cos x + \sin x + C. \end{aligned} \tag{11.18}$$

When applying the technique of integration by parts, it might be helpful to pause and look at it from a different perspective, and the solutions sometimes come out more naturally.

Example 11.8

For example, we try to find the integral

$$I = \int e^x \sin xdx. \tag{11.19}$$

If we use $u = e^x$ and $dv/dx = \sin x$, we have $du/dx = e^x$ and $v = -\cos x$. By using integration by parts, we then have

$$\begin{aligned} I &= \int e^x \sin xdx = e^x(-\cos x) - \int (-\cos x)\frac{du}{dx}dx \\ &= -\cos xe^x + \int e^x \cos xdx. \end{aligned} \tag{11.20}$$

It seems that we are stuck here. The integration by parts does not seem to help much, as it only transfers $\int e^x \sin x dx$ to $\int e^x \cos x dx$, which is not an improvement. In this case, we have to pause and think, what happens if we use the integration by parts again for $\int e^x \cos x dx$? We let $u = e^x$ and $dv/dx = \cos x$, which gives $du/dx = e^x$ and $v = \sin x$, and we have

$$\int e^x \cos x dx = e^x \sin x - \int \sin x \frac{du}{dx} dx = e^x \sin x - \int e^x \sin x dx. \tag{11.21}$$

Then, you may think, what is the use, as we have come back to the original integral? But what happens if we combine the above two equations? We have

$$\begin{aligned} I = \int e^x \sin x dx &= -e^x \cos x - \int e^x \cos x dx \\ &= -e^x \cos x + e^x \sin x - \int e^x \sin x = e^x(\sin x - \cos x) - I. \end{aligned} \tag{11.22}$$

We want to find I, so we can solve the above equation by simple rearrangement

$$2I = e^x(\sin x - \cos x), \tag{11.23}$$

or

$$I = \frac{1}{2} e^x(\sin x - \cos x). \tag{11.24}$$

Here we have found the integral as

$$I = \int e^x \sin x dx = \frac{1}{2} e^x(\sin x - \cos x) + C. \tag{11.25}$$

Such a procedure is very useful in deriving reduction formulas for some integrals.

Obviously, the way of treating definite integrals using integration by parts is identical in terms of the procedure. The only difference is that we write out the integration limits explicitly. Thus, the formula for integration by parts become

$$\int_a^b uv' dx = [uv]_a^b - \int_a^b vu' dx. \tag{11.26}$$

Example 11.9

Let us evaluate the integral

$$I = \int_0^{\infty} xe^{-x} dx.$$

We use

$$u = x, \quad v' = e^{-x},$$

which gives

$$du = dx, \quad u' = 1, \quad v = -e^{-x}.$$

Thus we have

$$I = \int_0^{\infty} xe^{-x} dx = \Big[x(-e^{-x})\Big]_0^{\infty} - \int_0^{\infty} 1 \times (-e^{-x}) dx = \Big[-xe^{-x}\Big]_0^{\infty} + \int_0^{\infty} e^{-x} dx.$$

From an earlier example in the previous section, we know that

$$\int_0^{\infty} e^{-x} dx = 1.$$

How do we calculate the term of $Q = xe^{-x}$ as x becomes infinity. One trick as used in many textbooks is that we first let $x = L$ and then take a limit of L. That is

$$Q = \lim_{L\to\infty} Le^{-L}.$$

Since for any value of L, $\exp[-L]$ decreases much faster than the increase rate of L, and their production is very small. Thus, as $L \to \infty$, $Q \to 0$. Using the above results, we finally get

$$I = \lim_{L\to\infty} Le^{-L} - 0e^{-0} + \int_0^\infty e^{-x}dx = 0 - 0 + 1 = 1.$$

Here, we have used the fact that x increases much slower than the decrease of e^{-x} as $x \to \infty$. That is

$$xe^{-x} \to 0, \quad x \to \infty.$$

In fact, we have

$$x^n e^{-x} \to 0, \quad x \to \infty,$$

for all $n > 0$.

It leaves as an exercise to use integration by parts many times to show that

$$\int_0^\infty x^n e^{-x} dx = n!, \tag{11.27}$$

for $n = 1, 2, ...,$ (all positive integer values).

11.3 INTEGRATION BY SUBSTITUTION

Sometimes, it is not easy to carry out the integration directly. However, it might become easier if we use the change of variables or integration by substitution. For example, we want to calculate the integral

$$I = \int f(x)dx. \tag{11.28}$$

We can change the variable x into another variable $u = g(x)$ where $g(x)$ is a known function of x. This means

$$\frac{du}{dx} = g'(x), \tag{11.29}$$

or

$$du = g'(x)dx, \qquad dx = \frac{1}{g'(x)}du. \tag{11.30}$$

This means that

$$I = \int f(x)dx = \int f[g^{-1}(u)]\frac{1}{g'}du, \tag{11.31}$$

where it is usually not necessary to calculate $g^{-1}(u)$ as it is relatively obvious.

Example 11.10

For example, in order to do the integration

$$I = \int xe^{x^2+5}dx, \tag{11.32}$$

we let $u = x^2 + 5$ and we have

$$\frac{du}{dx} = (x^2 + 5)' = 2x, \tag{11.33}$$

or

$$du = 2x dx, \tag{11.34}$$

which means $dx = \frac{1}{2x} du$. Therefore, we have

$$I = \int x e^{x^2+5} dx = \int x e^u \frac{1}{2x} du$$
$$= \frac{1}{2} \int e^u du = \frac{1}{2} e^u + A = e^{x^2+5} + A, \tag{11.35}$$

where A is the constant of integration. Here we have substituted $u = x^2 + 5$ back in the last step.

Sometimes, the integrals involve trigonometrical functions and in many cases a combination of techniques may be required.

Example 11.11

To evaluate the integral $\int_0^\pi \sqrt{1 - \cos^2(x)} dx$, we can use the identity

$$\sin^2(x) + \cos^2(x) = 1, \quad \text{or } 1 - \cos^2(x) = \sin^2(x),$$

thus we have

$$\int_0^\pi \sqrt{1 - \cos^2(x)} dx = \int_0^\pi \sqrt{\sin^2(x)} dx = \int_0^\pi (\pm \sin(x)) dx$$
$$= \Big[\mp \cos(x) \Big]_0^\pi = \mp[\cos(\pi) - \cos(0)] = \pm 2.$$

However, if we try to calculate

$$\int_0^1 \frac{1}{\sqrt{1 - x^2}} dx,$$

what substitution should we use?

If we set $u^2 = 1 - x^2$, we have $2u du = -2x dx$ and

$$dx = -\frac{u}{x} du = -\frac{u}{\sqrt{1 - u^2}} du.$$

In addition, $u = 1$ when $x = 0$, and $u = 0$ when $x = 1$. So we then obtain

$$\int_0^1 \frac{1}{\sqrt{1 - x^2}} dx = \int_1^0 \frac{1}{\sqrt{u^2}} \frac{-u}{\sqrt{1 - u^2}} du$$
$$= -\int_1^0 \frac{1}{\cancel{u}} \frac{\cancel{u}}{\sqrt{1 - u^2}} du = \int_0^1 \frac{1}{\sqrt{1 - u^2}} du,$$

which is exactly the same integral as before. This shows that a proper substitution is very important.

Now let us try $x = \sin(t)$. Using the trigonometric identity $1 - \sin^2(t) = \cos^2(t)$, we have $dx = \cos(t) dt$. In addition, $x = 0$ corresponds to $t = 0$ and $x = 1$ corresponds to $t = \pi/2$ if we only focus on the range $t \in [0, \pi/2]$. Thus, the integral becomes

$$\int_0^1 \frac{1}{\sqrt{1 - x^2}} dx = \int_0^{\pi/2} \frac{1}{\sqrt{1 - \sin^2(t)}} \cos(t) dt$$
$$= \int_0^{\pi/2} \frac{1}{\cos^2(t)} \cos(t) dt = \int_0^{\pi/2} \frac{1}{\cos(t)} \cos(t) dt$$
$$= \int_0^{\pi} 1 dt = t \Big|_0^{\pi/2} = \frac{\pi}{2} - 0 = \frac{\pi}{2}.$$

There are other techniques for integration. However, an interesting fact about calculus is that it is always possible to find the derivatives of complex functions and express them in explicit forms, but it is usually difficult and sometimes impossible to find the explicit form of integrals. Some seemingly simple integrals with simple integrands can be very hard to evaluate. For example, the following integral

$$I = \int_0^x e^{-u^2} du,$$

is not easy to calculate and it is related to the so-called error function:

$$\text{erf}(x) = \frac{2}{\sqrt{\pi}} \int_0^x e^{-u^2} dt. \tag{11.36}$$

On the other hand, the integral

$$K(k) = \int_0^{\pi/2} \frac{d\theta}{\sqrt{1 - k^2 \sin^2(\theta)}}, \tag{11.37}$$

is called the elliptic integral of the first kind, which is again not easy to evaluate. In fact, the evaluations of complex integrals require numerical methods such as the Gaussian integration method and Monte Carlo methods. We will briefly introduce some numerical integration methods in a later chapter in this book.

EXERCISES

11.1. Integrate the following integrals:
- $\int (x - \frac{1}{x}) dx$
- $\int (x^2 + 2x + 1) dx$
- $\int (x + \cos(x) + e^x) dx$
- $\int \ln x dx$
- $\int 2x \cos(x^2) dx$

11.2. Find the values of the following definite integrals:
- $\int_0^1 (x^5 - x^4) dx$
- $\int_0^{\pi/2} \sin(2x) dx$
- $\int_0^{\pi} \cos^2 x dx$
- $\int_0^{\infty} x e^{-x} dx$

11.3. Calculate the area under the curve $|x| e^{-|x|}$.

11.4. A bungee jumper of mass $m = 75$ kg is tied to a rope of length $L = 50$ m with a stiffness constant $k = 750$ N/m. The jumper jumps from a bridge at a height of $h = 100$ m above a river. Discuss the safety issue mathematically and carry out all the necessary calculations.

Hint. For simplicity, assume that $g = 10 \text{ m/s}^2$.

Chapter 12

Ordinary Differential Equations

Chapter Points

- Ordinary differential equations (ODEs) are explained in detail, including the order of a differential equation and its linear forms.
- First-order ODEs are discussed, together with their relevant solution techniques such as direct integration and integrating factors.
- Second-order ODEs are explained in the context of linear equations with constant coefficients with the brief introduction of the basic solution techniques.
- Ways of constructing solutions are presented, including particular integrals and complementary functions as well as characteristic equations.

12.1 DIFFERENTIAL EQUATIONS

In the introduction of basic equations such as $x^3 - x^2 + x - 1 = 0$, we know that the relationship is a function $f(x) = x^3 - x^2 + x - 1$ and the only unknown is x. The aim is to find values of x which satisfy $f(x) = 0$. It is easy to verify that the equation has three solutions $x = 1, \pm i$.

A differential equation, on the other hand, is a relationship that contains functions and their derivatives. For example, the following equation

$$\frac{dy}{dx} = x^3 - x, \tag{12.1}$$

is a differential equation because it provides a relationship between the derivative dy/dx and the function $f(x) = x^3 - x$. The unknown is a function $y(x)$ and the aim is to find a function $y(x)$ (not a simple value) which satisfies the above equation. Here x is the independent variable.

From integration, we know that the gradient dy/dx is $x^3 - x$, and the function $\frac{1}{4}x^4 - \frac{1}{2}x^2$ has the gradient $x^3 - x$. We can say that $y(x) = \frac{x^4}{4} - \frac{x^2}{2}$ satisfies the differential equation (12.1), and thus $y(x)$ is a solution to (12.1). Then the question is: Are there any other solutions to this equation? The answer is an infinite number. We know that there is a family of curves whose gradient is $x^3 - x$. The solution in general should contain an arbitrary constant C. That is to say, the general solution of (12.1) can be written as

$$y(x) = \frac{x^4}{4} - \frac{x^2}{2} + C. \tag{12.2}$$

Any solution that corresponds to a single specific curve is a particular solution. For example, both $x^4/4 - x^2/2$ and $x^4/4 - x^2/2 - 1$ are particular solutions.

Another important concept is the order of a differential equation. The order of a differential equation is the highest derivative of the unknown. For example, the order of (12.1) is 1 as the highest derivative is the gradient, so this equation is called a first-order differential equation. The following equation

$$\frac{d^2y(x)}{dx^2} - 2x\frac{dy(x)}{dx} + y(x) = x^2, \tag{12.3}$$

is the second-order differential equation as the highest derivative is the second derivative d^2y/dx^2.

All the above equations only contain first and/or second derivatives, and there is only a single independent variable x. Such differential equations are called ordinary differential equations (ODE).

12.2 FIRST-ORDER EQUATIONS

We know that $dy/dx = x^3 - x$ is a first-order ordinary differential equation, which can be generalized as

$$\frac{dy(x)}{dx} = f(x), \tag{12.4}$$

where $f(x)$ is a given function of x. Its solution can be obtained by simple integration

$$y(x) = \int f(x)dx + C. \tag{12.5}$$

However, this is a very special case of first-order ODEs.

The integration constant C can be determined by an initial condition at $x = 0$. This initial condition often takes the following form:

$$y(x = 0) = y_0, \quad \text{or} \quad y(0) = y_0, \tag{12.6}$$

where y_0 is a known constant.

Example 12.1

If you have an amount of money y to save in a bank with an interest rate r, the growth of your saving in the bank will obey the following first-order differential equation

$$\frac{dy(t)}{dt} = r\, y(t).$$

Re-arranging the above equation, we have

$$\frac{dy(t)}{y(t)} = rdt.$$

Integrating both side, we have

$$\int \frac{dy(t)}{y(t)} = \int rdt,$$

which gives

$$\ln y(t) = rt + C,$$

where C is the integration constant. Taking exponential on both sides, we have

$$e^{\ln y(t)} = y(t) = e^{rt+C} = Ae^{rt}, \quad A = e^C.$$

Here, the constant A has to be determined by an extra condition or initial condition. Initially, assume that you have $y = y_0$ when $t = 0$; that is

$$y(t = 0) = y_0 = Ae^{r \times 0} = A,$$

which leads to $A = y_0$, so the money grows exponentially as

$$y(t) = y_0 e^{rt}.$$

Suppose you initially have $y_0 = 1000$ pounds and the annual interest rate is $r = 2\% = 0.02$, then your total amount at the end of first year will be

$$y(1) = 1000e^{0.02\times 1} = 1020.2,$$

which means that you earn 20.2 pounds in the first year. At the end of a 5-year period, the total saving will become

$$y(5) = 1000e^{0.02\times 5} = 1105.1.$$

You may wonder why the amount earned in the first year is not $1000 * 0.02 = 20$ pounds, why there is an extra 0.2? The reason is that the interest rate here is a compounding interest rate. If the annual interest rate is r, then the annual growth will be

$$y(t = 1) = y_0(1 + r),$$

however, if the compounding periods is $n = 12$ (monthly), it becomes

$$y(t = 1) = y_0(1 + \frac{r}{n})^n.$$

What happens if we calculate it daily ($n = 365$) or hourly or even every second? As $n \to \infty$, we have

$$y(t = 1) = y_0 \lim_{n\to\infty} (1 + \frac{r}{n})^n = y_0 e^r,$$

where we have used

$$\lim_{n\to\infty} (1 + \frac{1}{n})^n = e.$$

For any t, this becomes

$$y(t) = y_0 e^{rt},$$

which is identical to the expression we obtained by solving the differential equation. Thus, the amount of money is calculated by using a frequency of compounds to infinity. So the effective annual interest rate is essentially

$$r_i = e^r - 1,$$

which is

$$r_i = e^{0.02} - 1 \approx 0.0202,$$

for $r = 0.02$. This means that there is 0.2 extra per year for every 1000 pounds.

In general, the first-order ordinary differential equation can be written as

$$p(x)\frac{dy(x)}{dx} + q(x)y(x) = f(x), \tag{12.7}$$

or

$$p(x)y'(x) + q(x)y(x) = f(x). \tag{12.8}$$

It is obviously that $p(x) \neq 0$; otherwise, the above equation degenerate into an algebraic equation. So we divide both sides by $p(x)$, we have

$$y'(x) + a(x)y(x) = b(x), \quad a(x) = \frac{q(x)}{p(x)}, \quad b(x) = \frac{f(x)}{p(x)}. \tag{12.9}$$

Therefore, a first-order linear differential equation can generally be written as

$$y' + a(x)y = b(x), \tag{12.10}$$

where $a(x)$ and $b(x)$ are known functions of x.

Multiplying both sides of the equation by $\exp[\int a(x)dx]$, called the integrating factor, we have

$$y' e^{\int a(x)dx} + a(x)ye^{\int a(x)dx} = b(x)e^{\int a(x)dx}, \tag{12.11}$$

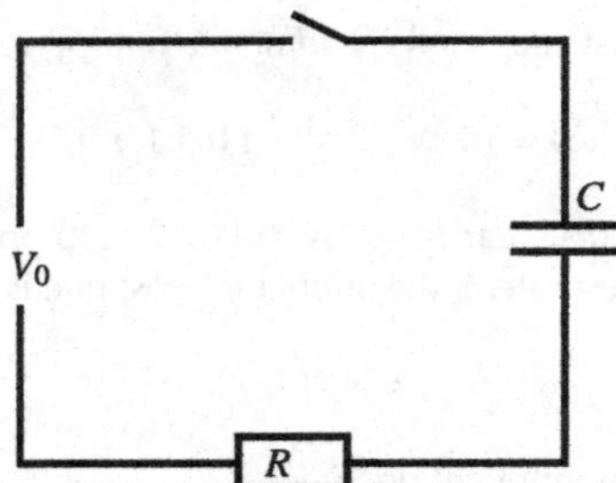

FIGURE 12.1 A simple RC circuit.

which can be written as

$$\left[y e^{\int a(x)dx} \right]' = b(x) e^{\int a(x)dx}. \tag{12.12}$$

By simple integration, we have

$$y e^{\int a(x)dx} = \int b(x) e^{\int a(x)dx} dx + C. \tag{12.13}$$

After dividing the integrating factor $\exp[\int a(x)dx]$, the final solution becomes

$$y(x) = e^{-\int a(x)dx} \int b(x) e^{\int a(x)dx} dx + C e^{-\int a(x)dx}, \tag{12.14}$$

where C is an integration constant to be determined by an initial condition at $x = 0$.

Now let us use first-order differential equations to model a real system. The resistor-capacity (RC) circuit shown in Fig. 12.1 consists of a resistor (R) and a capacitor (C), connected to a voltage source V_0.

Let $V(t)$ be the voltages across the capacitor, we have

$$V(t) + V_R = V_0,$$

where V_R is the voltage across the resistor.

From the basic linear circuit theory, we know that the current through the circuit (when the switch is on) can be given by

$$i = C\frac{dV(t)}{dt}, \quad V_R = iR.$$

So we have

$$V + iR = V + RC\frac{dV}{dt} = V_0,$$

or

$$RC\frac{dV}{dt} + V = V_0.$$

This is essentially a first-order differential equation

$$\tau V'(t) + V(t) = V_0, \quad \tau = RC.$$

Here, τ is the time constant of the RC circuit.

Example 12.2

By comparing with the general form of ODE (12.10), we have $a = 1/\tau$ and $b = V_0/\tau$ (constant). Since the independent variable is now t, the general solution becomes

$$\begin{aligned} y(t) &= e^{-\int \frac{1}{\tau}dt} \int \frac{V_0}{\tau} e^{\int \frac{1}{\tau}dt} dt + Ce^{-\int \frac{1}{\tau}dt} \\ &= \frac{V_0}{\tau} e^{-t/\tau} \int e^{t/\tau} dt + Ce^{-t/\tau} \\ &= \frac{V_0}{\tau} e^{-t/\tau} \left[\tau e^{t/\tau}\right] + Ce^{-t/\tau} = V_0 + Ce^{-t/\tau}. \end{aligned}$$

Before the switch is on, the voltage across the capacitor is zero. That is, $V(t = 0) = 0$. Applying this initial condition to the above equation at $t = 0$, we have

$$y(0) = 0 = V_0 + Ce^{-0/\tau} = V_0 + C,$$

which gives

$$C = -V_0.$$

So the final solution for the voltage variation is

$$V(t) = V_0 - V_0 e^{-t/\tau} = V_0(1 - e^{-t/\tau}).$$

We can see that $V(t) \to V_0$ as $t \to \infty$, which means that this capacitor is fully charged.

In Eq. (12.10), if $a(x)$ is a constant, the equation becomes a linear ordinary differential equation with constant coefficients. In this case, a general way to solve such a linear ODE is to divide the solution process into two steps: first find a general solution $y(t)_c$ to the homogeneous ordinary differential equation by setting $b(x) = 0$ on the right-hand side. Then, the task is to find any particular solution $y(t)_p$ to the full equation (12.10). The general solution $y(t)$ is the sum of these two solutions.

The solution to the homogeneous equation is also called a complementary function, which is the solution to

$$y'(t) + ay(t) = 0. \tag{12.15}$$

Let us revisit the previous example.

Example 12.3

From the equation $\tau V'(t) + V(t) = V_0$, we have

$$V'(t) + \frac{1}{\tau} V(t) = \frac{V_0}{\tau}.$$

Its homogeneous equation is

$$V'(t) + \frac{1}{\tau} V(t) = 0, \quad \text{or} \quad \frac{dV(t)}{dt} + \frac{1}{\tau} V(t) = 0,$$

which can also be written as

$$\frac{dV(t)}{V(t)} = -\frac{1}{\tau}.$$

Integrating both sides, we have

$$\int \frac{dV(t)}{V(t)} dV = -\int \frac{1}{\tau} dt,$$

which gives

$$\ln V(t) = -\frac{1}{\tau}t + C,$$

where C is an integration constant. Taking exponential operations on both sides, we have

$$e^{\ln V(t)} = e^{-\frac{1}{\tau}t+C},$$

or

$$e^{\ln V(t)} = V(t) = e^{-\frac{t}{\tau}}e^{C},$$

which gives

$$V(t) = Ae^{-t/\tau}, \quad A = e^{C}.$$

This is the complementary function.

To find a particular solution V_p, the exact form will depend on the form of the right-hand side. Here, since V_0/τ is a constant, we can try the solution

$$V_p = k\frac{V_0}{\tau},$$

where k is an unknown constant to be determined by substituting into the original equation

$$(k\frac{V_0}{\tau})' + \frac{1}{\tau}k\frac{V_0}{\tau} = \frac{V_0}{\tau},$$

or

$$0 + \frac{kV_0}{\tau^2} = \frac{V_0}{\tau},$$

which gives $k = \tau$. So the particular solution becomes

$$V_p = \tau\frac{V_0}{\tau} = V_0.$$

Therefore, the general solution is the sum of the complementary function and the particular integral. That is

$$V(t) = Ae^{-t/\tau} + V_0,$$

where A is an unknown coefficient to be determined by the initial condition.

From the initial condition $V(t=0) = 0$ at $t = 0$, we have

$$0 = V(t=0) = Ae^{-0/\tau} + V_0 = A + V_0,$$

which gives $A = -V_0$. Thus, the final solution is

$$V(t) = -V_0e^{-t/\tau} + V_0 = V_0(1 - e^{-t/\tau}),$$

which is exactly the solution we obtained in the previous example.

For this simple first-order ODE, this two-step process may seem lengthy. However, this method can be generalized to solve higher-order linear ordinary differential equations.

12.3 SECOND-ORDER EQUATIONS

For second-order ordinary differential equations (ODEs), it is generally more tricky to find their general solutions. However, a special case with significantly practical importance and mathematical simplicity is the second-order linear differential equation with constant coefficients in the following form

$$\frac{d^2y}{dx^2} + b\frac{dy}{dx} + cy(x) = f(x), \tag{12.16}$$

where the coefficients b and c are constants, and $f(x)$ is a known function of x. Obviously, the more general form is

$$a\frac{d^2y}{dx^2} + b\frac{dy}{dx} + cy(x) = f(x). \tag{12.17}$$

However, if we divide both sides by a, we will reach our standard form. Here we assume $a \neq 0$. In a special case of $a = 0$, it reduces to a first-order linear differential equation which has been discussed in the previous section. So we will start our discussion from (12.16).

A differential equation is said to be homogeneous if $f(x) = 0$. For a given generic second-order differential equation (12.16), a function that satisfies the homogeneous equation

$$\frac{d^2y}{dx^2} + b\frac{dy}{dx} + cy(x) = 0, \tag{12.18}$$

is called the complementary function, denoted by $y_c(x)$. Obviously, the complementary function y_c alone cannot satisfy the original equation (12.16) because there is no way to produce the required $f(x)$ on the right-hand side. Therefore, we have to find a specific function, $y_*(x)$ called the particular integral, so that it indeed satisfies the original equation (12.16). The combined general solution

$$y(x) = y_c(x) + y_*(x), \tag{12.19}$$

will automatically satisfy the original equation (12.16). The general solution of (12.16) consists of two parts: the complementary function $y_c(x)$ and the particular integral $y_*(x)$. We can obtain these two parts separately, and simply add them together because the original equation is linear, so their solutions are linear combinations.

First things first, how to obtain the complementary function? The general technique is to assume that it takes the form

$$y_c(x) = Ae^{\lambda x}, \tag{12.20}$$

where A is a constant, and λ is an exponent to be determined. Substituting this assumed form into the homogeneous equation and using both $y_c' = A\lambda e^{\lambda x}$ and $y_c'' = A\lambda^2 e^{\lambda x}$, we have

$$A\lambda^2 e^{\lambda x} + bA\lambda e^{\lambda x} + cAe^{\lambda x} = 0. \tag{12.21}$$

Since $Ae^{\lambda x}$ should not be zero (otherwise, we have a trivial solution $y_c = 0$ everywhere), we can divide all the terms by $Ae^{\lambda x}$, and we have

$$\lambda^2 + b\lambda + c = 0, \tag{12.22}$$

which is the characteristic equation for the homogeneous equation. It is also called the auxiliary equation of the ODE. The solution of λ in this case is simply

$$\lambda = \frac{-b \pm \sqrt{b^2 - 4c}}{2}. \tag{12.23}$$

For simplicity, we can take $A = 1$ as it does not affect the results.

From quadratic equations, we know that there are three possibilities for λ. They are: I) two real distinct roots, II) two identical roots, and III) two complex roots.

In the case of two different roots: $\lambda_1 \neq \lambda_2$. Then, both $e^{\lambda_1 x}$ and $e^{\lambda_2 x}$ satisfy the homogeneous equation, so their linear combination forms the complementary function

$$y_c(x) = Ae^{\lambda_1 x} + Be^{\lambda_2 x}, \tag{12.24}$$

where A and B are constants.

In the special case of identical roots $\lambda_1 = \lambda_2$, or

$$c = \lambda_1^2, \qquad b = -2\lambda_1, \tag{12.25}$$

we cannot simply write

$$y_c(x) = Ae^{\lambda_1 x} + Be^{\lambda_1 x} = (A+B)e^{\lambda_1 x}, \tag{12.26}$$

because it still only one part of the complementary function $y_1 = Ce^{\lambda_1}$ where $C = A + B$ is just another constant. In this case, we should try a different combination, say, $y_2 = xe^{\lambda_1 x}$ to see if it satisfies the homogeneous equation or not. Since $y_2'(x) = e^{\lambda_1 x} + x\lambda_1 e^{\lambda_1 x}$, and $y_2''(x) = \lambda_1 e^{\lambda_1 x} + \lambda_1 e^{\lambda_1 x} + x\lambda_1^2 e^{\lambda_1 x}$, we have

$$\begin{aligned} & y_2''(x) + by_2'(x) + cy_2(x) \\ &= e^{\lambda_1 x}(2\lambda_1 + x\lambda_1^2)e^{\lambda_1 x} + be^{\lambda_1 x}(1 + x\lambda_1) + cxe^{\lambda_1 x} \\ &= e^{\lambda_1 x}[(2\lambda_1 + b)] + xe^{\lambda_1 x}[\lambda_1^2 + b\lambda_1 + c] = 0, \end{aligned} \tag{12.27}$$

where we have used $b + 2\lambda_1 = 0$ (identical roots) and $\lambda_1^2 + b\lambda_1 + c = 0$ (the auxiliary equation). This indeed implies that $xe^{\lambda_1 x}$ also satisfies the homogeneous equation. Therefore, the complementary function for the identical roots is

$$y_c(x) = Ae^{\lambda_1 x} + Bxe^{\lambda_1 x} = (A + Bx)e^{\lambda_1 x}. \tag{12.28}$$

Now let us use this technique to solve a second-order homogeneous ODE.

Example 12.4

The second-order homogeneous equation

$$\frac{d^2y}{dx^2} + 5\frac{dy}{dx} - 6y = 0,$$

has a corresponding auxiliary equation

$$\lambda^2 + 5\lambda - 6 = (\lambda - 2)(\lambda + 3) = 0.$$

It has two real roots

$$\lambda_1 = 2, \qquad \lambda_2 = -3.$$

So the complementary function is

$$y_c(x) = Ae^{2x} + Be^{-3x}.$$

But for the differential equation

$$\frac{d^2y}{dx^2} + 6\frac{dy}{dx} + 9y(x) = 0,$$

its auxiliary equation becomes

$$\lambda^2 + 6\lambda + 9 = 0,$$

which has two identical roots $\lambda_1 = \lambda_2 = -3$. The complementary function in this case can be written as

$$y_c(x) = (A + Bx)e^{-3x}.$$

As complex roots always come in pairs, the case of complex roots would give

$$\lambda_{1,2} = \alpha \pm i\beta, \tag{12.29}$$

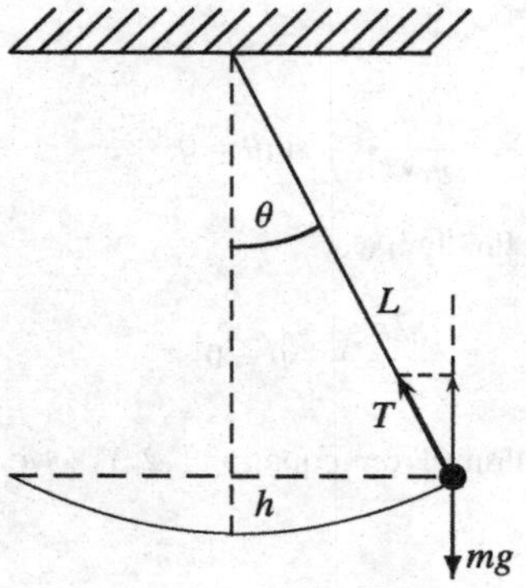

FIGURE 12.2 A simple pendulum and its harmonic motion.

where α and β are real numbers. The complementary function becomes

$$\begin{aligned} y_c(x) &= Ae^{(\alpha+i\beta)x} + Be^{(\alpha-i\beta)x} = Ae^{\alpha x}e^{i\beta x} + Be^{\alpha x}e^{-i\beta x} \\ &= e^{\alpha x}[Ae^{i\beta x} + Be^{-i\beta x}] \\ &= e^{\alpha x}\{A[\cos(\beta x) + i\sin(\beta x)] + B[\cos(-\beta x) + i\sin(-\beta x)]\} \\ &= e^{\alpha x}[(A+B)\cos(\beta x) + i(A-B)\sin(\beta x)] \\ &= e^{\alpha x}[C\cos\beta x + D\sin\beta x], \end{aligned} \tag{12.30}$$

where we have used the Euler formula $e^{\theta i} = \cos\theta + i\sin\theta$ and also absorbed the constants A and B into $C = A + B$ and $D = (A - B)i$.

A special case is when $\alpha = 0$, so the roots are purely imaginary. We have $b = 0$, and $c = \beta^2$. Equation (12.16) in this case becomes

$$\frac{d^2y}{dx^2} + \beta^2 y = 0, \tag{12.31}$$

which is a differential equation for harmonic motions such as the oscillations of a pendulum or a small-amplitude seismic detector. Here β is the angular frequency of the system.

Example 12.5

For a simple pendulum of mass m shown in Fig. 12.2, we now try to derive its equation of oscillations and its period.

Since the motion is circular, the tension or the centripetal force T is thus given by

$$T = m\frac{v^2}{L} = m\dot{\theta}^2 L,$$

where $\dot{\theta} = d\theta/dt$ is the angular velocity.

Forces must be balanced both vertically and horizontally. The component of T in the vertical direction is $T\cos\theta$ which must be equivalent to mg, though in the opposite direction. Here g is the acceleration due to gravity. That is

$$T\cos\theta = mg.$$

Since θ is small or $\theta \ll 1$, we have $\cos\theta \approx 1$. This means that $T \approx mg$.

In the horizontal direction, Newton's second law $F = ma$ implies that the horizontal force $-T\sin\theta$ must be equal to the mass m times the acceleration $L\frac{d^2\theta}{dt^2}$. Now we have

$$m(L\frac{d^2\theta}{dt^2}) = -T\sin\theta \approx -mg\sin\theta.$$

Dividing both sides by mL, we have

$$\frac{d^2\theta}{dt^2} + \frac{g}{L}\sin\theta = 0.$$

Since θ is small, we have $\sin\theta \approx \theta$. Therefore, we finally have

$$\frac{d^2\theta}{dt^2} + \frac{g}{L}\theta = 0.$$

This is the equation of motion for a simple pendulum. From equation (12.31), we know that the angular frequency is $\omega^2 = g/L$ or $\omega = \sqrt{\frac{g}{L}}$. Thus the period of the pendulum is

$$T = \frac{2\pi}{\omega} = 2\pi\sqrt{\frac{L}{g}}. \quad (12.32)$$

We can see that the period is independent of the bob mass. For $L = 1$ m and $g = 9.8$ m/s^2, the period is approximately $T = 2\pi\sqrt{\frac{1}{9.8}} \approx$ 2 seconds.

Up to now, we have found only the complementary function. Now we will try to find the particular integral $y_*(x)$ for the original non-homogeneous equation (12.16). For particular integrals, we do not intend to find the general form; any specific function or integral that satisfies the original equation (12.16) will do. Before we can determine the particular integral, we have to use some trial functions, and such functions will have strong similarity to function $f(x)$. For example, if $f(x)$ is a polynomial such as $x^2 + \alpha x + \beta$, we will try a similar form $y_*(x) = ax^2 + bx + c$ and try to determine the coefficients. Let us demonstrate this by an example.

Example 12.6

In order to solve the differential equation

$$\frac{d^2y}{dx^2} + 5\frac{dy}{dx} - 6y = x - 2,$$

we first find its complementary function. From the earlier example, we know that the complementary function can be written as

$$y_c(x) = Ae^{2x} + Be^{-3x}.$$

For the particular integral, we know that $f(x) = x - 2$, so we try the form

$$y_* = ax + b.$$

Thus $y'_* = a$ and $y''_* = 0$. Substituting them into the original equation, we have

$$0 + 5a - 6(ax + b) = x - 2,$$

or

$$(-6a)x + (5a - 6b) = x - 2.$$

As this equality must be true for any x, so the coefficients of the same power of x on both sides of the equation should be equal. That is

$$-6a = 1, \qquad (5a - 6b) = -2,$$

which gives $a = -\frac{1}{6}$, and $b = \frac{7}{36}$. So the general solution becomes

$$y(x) = Ae^{2x} + Be^{-3x} - \frac{x}{6} + \frac{7}{36}.$$

Similarly, if $f(x) = e^{\alpha x}$, we will try to $y_*(x) = ae^{\alpha x}$ so as to determine a. In addition, for $f(x) = \sin\alpha x$ or $\cos\alpha x$, we will attempt the general form $y_*(x) = a\cos\alpha x + b\sin\alpha x$.

Again let us demonstrate this through an example.

Example 12.7

The motion of a damped pendulum is governed by

$$\frac{d^2y}{dt^2} + 4\frac{dy}{dt} + 5y = 40\cos(3t).$$

Its auxiliary equation becomes

$$\lambda^2 + 4\lambda + 5 = 0,$$

which has two complex solutions

$$\lambda = \frac{-4 \pm \sqrt{4^2 - 4\times 5}}{2} = -2 \pm i.$$

Therefore, the complementary function becomes

$$y_c(x) = e^{-2t}(A\cos t + B\sin t),$$

where A and B are two undetermined constants.

Now we try to get the particular integral. Since $f(t) = 40\cos(3t)$, so we try the similar form

$$y_* = C\cos(3t) + D\sin(3t).$$

Since

$$y_*' = -3C\sin(3t) + 3D\cos(3t), \quad y_*'' = -9C\cos(3t) - 9D\sin(3t),$$

we have

$$-9[C\cos(3t) + D\sin(3t)] + 4[-3C\sin(3t) + 3D\cos(3t)] + 5[C\cos(3t) + D\sin(3t)] = 40\cos(3t),$$

which leads to

$$(-9C + 12D + 5C)\cos(3t) + (-9D - 12C + 5D)\sin(3t) = 40\cos(3t),$$

which requires

$$-9C + 12D + 5C = 40, \quad -9D - 12C + 5D = 0,$$

or

$$12D - 4C = 40, \quad -12C - 4D = 0.$$

This gives

$$C = -1, \quad D = 3.$$

The particular solution is

$$y_* = 3\sin(3t) - \cos(3t).$$

Therefore, the general solution becomes

$$y = e^{-2t}(A\cos x + B\sin x) + 3\sin(3t) - \cos(3t).$$

As time is sufficiently long (i.e., $t \to \infty$), we can see that the first two terms will approach zero (since $e^{-2t} \to 0$). Therefore, the solution will be dominated by the last two terms. That is, the long-term behavior becomes

$$y(r \to \infty) = 3\sin(3t) - \cos(3t).$$

In an earlier chapter about trigonometry, we have shown that the following identity holds

$$a\cos\theta + b\sin\theta = R\cos(\theta - \phi),$$

where

$$R = \sqrt{a^2 + b^2}, \quad \tan\phi = \frac{b}{a}.$$

Therefore, we have

$$y_* = 3\sin(3t) - \cos(3t) = R\cos(t - \phi),$$

where $R = \sqrt{(-1)^2 + 3^2} = \sqrt{10}$ and $\tan\phi = -3$.

12.4 HIGHER-ORDER ODES

Higher order ODEs are more complicated to solve even for linear cases. For the special case of higher-order ODEs where all the coefficients $a_n, ..., a_1, a_0$ are constants,

$$a_n y^{(n)} + ... + a_1 y' + a_0 y = f(x), \tag{12.33}$$

its general solution $y(x)$ consists of two parts: the complementary function $y_c(x)$ and the particular integral or particular solution $y_p^*(x)$. We have

$$y(x) = y_c(x) + y_p^*(x). \tag{12.34}$$

The complementary function which is the solution of the linear homogeneous equation with constant coefficients can be written in a generic form

$$a_n y_c^{(n)} + a_{n-1} y_c^{(n-1)} + ... + a_1 y_c' + a_0 = 0. \tag{12.35}$$

Assuming $y = Ae^{\lambda x}$ where A is a constant, we get the characteristic equation as a polynomial

$$a_n \lambda^n + a_{n-1}\lambda^{(n-1)} + ... + a_1\lambda + a_0 = 0, \tag{12.36}$$

which has n roots in the general case. Then, the solution can be expressed as the summation of various terms

$$y_c(x) = \sum_{k=1}^{n} c_k e^{\lambda_k x}$$

if the polynomial has n distinct zeros $\lambda_1, ...\lambda_n$. Here, $c_k (k = 1, 2, ..., n)$ are undermined coefficients.

For complex roots, and complex roots always occur in pairs $\lambda = r \pm i\omega$, the corresponding linearly independent terms can then be replaced by $e^{rx}[A\cos(\omega x) + B\sin(\omega x)]$.

The particular solution $y_p^*(x)$ is any $y(x)$ that satisfies the original inhomogeneous equation (12.33). Depending on the form of the function $f(x)$, the particular solutions can take various forms. For most of the combinations of basic functions such as $\sin x, \cos x, e^{kx}$, and x^n, the method of the undetermined coefficients is widely used. For $f(x) = \sin(\alpha x)$ or $\cos(\alpha x)$, then we can try $y_p^* = A\sin\alpha x + B\sin\alpha x$. We then substitute it into the original equation (12.33) so that the coefficients A and B can be determined. For a polynomial $f(x) = x^p$ $(p = 0, 1, 2, ..., N)$, we then try $y_p^* = A + Bx + ... + Qx^p$ (polynomial). For $f(x) = e^{kx}x^n$, $y_p^* = (A + Bx + ...Qx^n)e^{kx}$. Similarly, $f(x) = e^{kx}\sin\alpha x$ or $f(x) = e^{kx}\cos\alpha x$, we can use $y_p^* = e^{kx}(A\sin\alpha x + B\cos\alpha x)$. More general cases and their particular solutions can be found in various textbooks.

A very useful technique is to use the method of differential operator D. A differential operator D is defined as

$$D \equiv \frac{d}{dx}. \tag{12.37}$$

Since we know that $De^{\lambda x} = \lambda e^{\lambda x}$ and $D^n e^{\lambda x} = \lambda^n e^{\lambda x}$, so they are equivalent to $D \mapsto \lambda$, and $D^n \mapsto \lambda^n$. Thus, any polynomial $P(D)$ will be mapped to $P(\lambda)$. On the other hand, the integral operator $D^{-1} = \int dx$ is just the inverse of differentiation.

The beauty of the differential operator form is that one can factorize it in the same way as for a polynomial, then solve each factor separately. The differential operator is very useful in finding both the complementary functions and particular integral.

Example 12.8

To find the particular integral for the equation

$$y''''' + 2y = 17e^{2x}, \tag{12.38}$$

we get

$$(D^5 + 2)y_p^* = 17e^{2x}, \quad \text{or} \quad y_p^* = \frac{17}{D^5 + 2}e^{2x}. \tag{12.39}$$

Since $D^5 \mapsto \lambda^5 = 2^5$, we have

$$y_p^* = \frac{17e^{2x}}{2^5 + 2} = \frac{e^{2x}}{2}. \tag{12.40}$$

This method also works for $\sin x$, $\cos x$, $\sinh x$ and others, and this is because they are related to $e^{\lambda x}$ via $\sin\theta = \frac{1}{2i}(e^{i\theta} - e^{-i\theta})$ and $\cosh x = (e^x + e^{-x})/2$.

12.5 SYSTEM OF LINEAR ODES

For a linear ODE of order n (12.35), it can always be written as a linear system

$$\frac{dy}{dx} = y_1, \quad \frac{dy_1}{dx} = y_2, \quad ..., \quad \frac{dy_{n-1}}{dx} = y_{n-1},$$
$$a_n(x)y'_{n-1} = -a_{n-1}(x)y_{n-1} + ... + a_1(x)y_1 + a_0(x)y + \phi(x), \tag{12.41}$$

which is a system for $u = [y \; y_1 \; y_2 \; ... \; y_{n-1}]^T$. If the independent variable x does not appear explicitly in y_i, then the system is said to be autonomous with important properties. For simplicity and in keeping with the convention, we use $t = x$ and $\dot{u} = du/dt$ in our following discussion. A general linear system of n-th order can be written as

$$\begin{pmatrix} \dot{u}_1 \\ \dot{u}_2 \\ \vdots \\ \dot{u}_n \end{pmatrix} = \begin{pmatrix} a_{11} & a_{12} & \dots & a_{1n} \\ a_{21} & a_{22} & \dots & a_{2n} \\ \vdots & & & \vdots \\ a_{n1} & a_{n2} & \dots & a_{nn} \end{pmatrix} \begin{pmatrix} u_1 \\ u_2 \\ \vdots \\ u_n \end{pmatrix}, \tag{12.42}$$

or

$$\dot{\mathbf{u}} = \mathbf{A}\mathbf{u}. \tag{12.43}$$

If $\mathbf{u} = \mathbf{v}\exp(\lambda t)$, then this becomes an eigenvalue problem,

$$(\mathbf{A} - \lambda\mathbf{I})\mathbf{v} = \mathbf{0}, \tag{12.44}$$

which will have a non-null solution only if

$$\det(\mathbf{A} - \lambda\mathbf{I}) = 0. \tag{12.45}$$

The eigenvalues of the above system will control the dynamic behavior of the system. The full mathematical analysis requires the theory of dynamic systems. Interested readers can refer to more advanced literature.

Higher-order differential equations can conveniently be written as a system of differential equations. In fact, an nth-order linear equation can always be written as a linear system of n first-order differential equations. A linear system of ODEs is more suitable for mathematical analysis and numerical integration.

EXERCISES

12.1. Solve differential equation $y'(x) = \sin(x) + x^2$ with the initial condition $y(0) = 1$.

12.2. Solve the following differential equations:

- $y''(x) + 3y'(x) + 2y(x) = 0$.
- $y'(x) + xy(x) = 0$.
- $y''(x) + 4y(x) = 0$.

12.3. Solve $y'''(x) + y''(x) + 4y'(x) + 4y(x) = 0$ with the initial conditions $y(0) = 1$, $y'(0) = 0$ and $y''(0) = 1$. Discuss if this $y''(0) = 1$ is necessary?

12.4. Solve $y'(x) + 2y(x) = 1 - e^{-x}$ with the initial condition $y(0) = 1$.

Chapter 13

Partial Differentiation

Chapter Points

- Partial derivatives are introduced, together with relevant rules for computing partial derivatives, including rules similar to the chain rule and product rule.
- Differentiation of vectors are explained in detail with examples.
- Three basic operators are explained briefly due to their importance in science and engineering.

13.1 PARTIAL DIFFERENTIATION

The derivative defined earlier is for function $f(x)$ which has only one independent variable x, and the gradient will generally depend on the location x. For functions $f(x, y)$ of two variables x and y, their gradient will depend on both x and y in general. In addition, the gradient or rate of change will also depend on the direction (along x-axis or y-axis or any other directions). For example, the function $f(x, y) = x(y - 1)$ has different gradients at $(0, 0)$ along x-axis and y-axis. The gradients along the positive x- and y-directions are called the partial derivatives with respect to x and y, respectively. They are denoted as $\frac{\partial f}{\partial x}$ and $\frac{\partial f}{\partial y}$, respectively.

The partial derivative of $f(x, y)$ with respect to x can be calculated assuming that $y = $ constant. Thus, we have

$$\frac{\partial f(x,y)}{\partial x} \equiv f_x \equiv \frac{\partial f}{\partial x}\Big|_y = \lim_{\Delta x \to 0, y=\text{const}} \frac{f(x+\Delta x, y) - f(x,y)}{\Delta x}. \tag{13.1}$$

Similarly, we have

$$\frac{\partial f(x,y)}{\partial y} \equiv f_y \equiv \frac{\partial f}{\partial y}\Big|_x = \lim_{\Delta y \to 0, x=\text{const}} \frac{f(x, y+\Delta y) - f(x,y)}{\Delta y}. \tag{13.2}$$

The notation $\frac{\partial}{\partial x}\big|_y$ emphasizes that the fact that y is held constant. The subscript notation f_x (or f_y) emphasizes that the derivative is carried out with respect to x (or y).

As the partial derivatives treating all the other variables as constants (except for the one variable of interest), the usual properties and rules for differentiation still apply. That is, for multiple additive terms, you can differentiate each term individually. In addition, both chain rules and product rules also apply here.

Example 13.1

For a function

$$f(x, y) = x^2 + y^3 + 2xy,$$

its partial derivatives are

$$f_x = \frac{\partial f}{\partial x} = \frac{\partial x^2}{\partial x} + \frac{\partial y^3}{\partial x} + \frac{\partial (2xy)}{\partial x} = 2x + 0 + 2y = 2x + 2y,$$
$$f_y = \frac{\partial f}{\partial y} = \frac{\partial x^2}{\partial y} + \frac{\partial y^3}{\partial y} + \frac{\partial (2xy)}{\partial y} = 0 + 3y^2 + 2x = 3y^2 + 2x.$$

Mathematicians like to use the subscript forms as they are simpler notations and can be easily generalized. For example,

$$f_{xx} = \frac{\partial^2 f}{\partial x^2}, \qquad f_{xy} = \frac{\partial^2 f}{\partial x \partial y}. \tag{13.3}$$

Since $\Delta x \Delta y = \Delta y \Delta x$, we have $f_{xy} = f_{yx}$. For any small change $\Delta f = f(x+\Delta x, y+\Delta y) - f(x, y)$ due to Δx and Δy, the total infinitesimal change df can be written as

$$df = \frac{\partial f}{\partial x} dx + \frac{\partial f}{\partial y} dy. \tag{13.4}$$

If x and y are functions of another independent variable ξ, then the above equation leads to the following chain rule

$$\frac{df}{d\xi} = \frac{\partial f}{\partial x}\frac{dx}{d\xi} + \frac{\partial f}{\partial y}\frac{dy}{d\xi}, \tag{13.5}$$

which is very useful in calculating the derivatives in a parametric form or for change of variables.

Example 13.2

For the same function $f = x^2 + y^3 + 2xy$ as in the previous example, we have

$$f_{xx} = \frac{\partial f_x}{\partial x} = \frac{\partial(2x+2y)}{\partial x} = 2 + 0 = 2,$$

$$f_{yy} = \frac{\partial f_y}{\partial y} = \frac{\partial(3y^2+2x)}{\partial y} = 3 \times 2y + 0 = 6y.$$

In addition, we have

$$f_{xy} = \frac{\partial f_y}{\partial x} = \frac{\partial(3y^2+2x)}{\partial x} = 0 + 2 = 2,$$

$$f_{yx} = \frac{\partial f_x}{\partial y} = \frac{\partial(2x+2y)}{\partial y} = 0 + 2 = 0,$$

which indeed shows that

$$f_{xy} = f_{yx}.$$

If a complicated function $f(x)$ can be written in terms of simpler functions u and v so that $f(x) = g(x, u, v)$ where $u(x)$ and $v(x)$ are known functions of x, then we have the generalized chain rule

$$\frac{dg}{dx} = \frac{\partial g}{\partial x} + \frac{\partial g}{\partial u}\frac{du}{dx} + \frac{\partial g}{\partial v}\frac{dv}{dx}. \tag{13.6}$$

Sometimes, a function F can be represented in terms of either (u, v) or (x, y), and these two sets of variables can be considered as two different coordinate systems to represent the same function $F(u, v)$ or $F(x, y)$. In this case, the chain rule can be used to link these two sets of variables. We have

$$\frac{\partial F}{\partial x} = \frac{\partial F}{\partial u}\frac{\partial u}{\partial x} + \frac{\partial F}{\partial v}\frac{\partial v}{\partial x}, \qquad \frac{\partial F}{\partial y} = \frac{\partial F}{\partial u}\frac{\partial u}{\partial y} + \frac{\partial F}{\partial v}\frac{\partial v}{\partial y}, \tag{13.7}$$

$$\frac{\partial F}{\partial u} = \frac{\partial F}{\partial x}\frac{\partial x}{\partial u} + \frac{\partial F}{\partial y}\frac{\partial y}{\partial u}, \qquad \frac{\partial F}{\partial v} = \frac{\partial F}{\partial x}\frac{\partial x}{\partial v} + \frac{\partial F}{\partial y}\frac{\partial y}{\partial v}. \tag{13.8}$$

An example will be given later when we discuss the polar coordinate system.

13.2 DIFFERENTIATION OF VECTORS

The differentiation of a vector is carried out over each component and treating each component as the usual differentiation of a scalar. For example, a position vector of a particle moving in the three-dimensional (3D) space can be written

$$\mathbf{P}(t) = x(t)\mathbf{i} + y(t)\mathbf{j} + z(t)\mathbf{k}, \tag{13.9}$$

where $(x(t), y(t), z(t))$ are its three coordinates and the parameter t is time, which shows that the position is changing with time t.

The velocity of the particle is its vector derivative and we can write its velocity as

$$\mathbf{v} = \frac{d\mathbf{P}}{dt} = \dot{x}(t)\mathbf{i} + \dot{y}(t)\mathbf{j} + \dot{z}(t)\mathbf{k}, \tag{13.10}$$

and acceleration as

$$\mathbf{a} = \frac{d^2\mathbf{P}}{dt^2} = \ddot{x}(t)\mathbf{i} + \ddot{y}(t)\mathbf{j} + \ddot{z}(t)\mathbf{k}, \tag{13.11}$$

where $\dot{()} = d()/dt$. Conversely, the integral of $\mathbf{v}$ is its position

$$\mathbf{P} = \int \mathbf{v} dt + \mathbf{c}, \tag{13.12}$$

where $\mathbf{c}$ is the initial position (a vector constant).

From the basic definition of differentiation, it is easy to check that the differentiation of vectors has the following properties:

$$\frac{d(\alpha\mathbf{a})}{dt} = \alpha\frac{d\mathbf{a}}{dt}, \quad \frac{d(\mathbf{a}\cdot\mathbf{b})}{dt} = \frac{d\mathbf{a}}{dt}\cdot\mathbf{b} + \mathbf{a}\cdot\frac{d\mathbf{b}}{dt}, \tag{13.13}$$

and

$$\frac{d(\mathbf{a}\times\mathbf{b})}{dt} = \frac{d\mathbf{a}}{dt}\times\mathbf{b} + \mathbf{a}\times\frac{d\mathbf{b}}{dt}. \tag{13.14}$$

Example 13.3

For a particle with its position at

$$x(t) = \cos(\omega t), \quad y(t) = \sin(\omega t), \quad z(t) = \frac{1}{2}gt^2,$$

where ω is a constant, what are its velocity and acceleration vectors?

The velocity vector is given by

$$\mathbf{v} = \dot{x}(t)\mathbf{i} + \dot{y}(t)\mathbf{j} + \dot{z}\mathbf{k}$$
$$= -\omega\sin(\omega t)\mathbf{i} + \omega\cos(\omega t)\mathbf{j} + gt\mathbf{k}.$$

Its acceleration is

$$\mathbf{a} = \dot{\mathbf{v}} = \ddot{x}(t)\mathbf{i} + \ddot{y}(t)\mathbf{j} + \ddot{z}(t)\mathbf{k}$$
$$= -\omega^2\cos(\omega t)\mathbf{i} - \omega^2\sin(\omega t)\mathbf{j} + g\mathbf{k}.$$

The orbit is a spiral with the free fall along the z-axis.

13.3 POLAR COORDINATES

A point P on a plane can be represented uniquely by their Cartesian coordinates (x, y). The same point can also be represented using a different coordinate system. For example, in the polar coordinate system, the same point P can be represented by a distance (r) from a reference point (the origin O, often referred to as the pole) and the angle (θ) from the reference direction (the x-axis in this case) as shown in Fig. 13.1.

The distance is also referred to as the radial coordinate, while the angle is also called the polar angle (or azimuth). To ensure the uniqueness of polar coordinates, it requires that $r \geq 0$ and $0 \leq \theta < 2\pi$ (or $\theta \in [0, 360°)$). In some textbook, the angle interval can also be $-\pi < \theta \leq \pi$ (or $\theta \in (-180°, 180°]$). In addition, the pole can be represented by the point $(0, \theta)$ where θ can be any value.

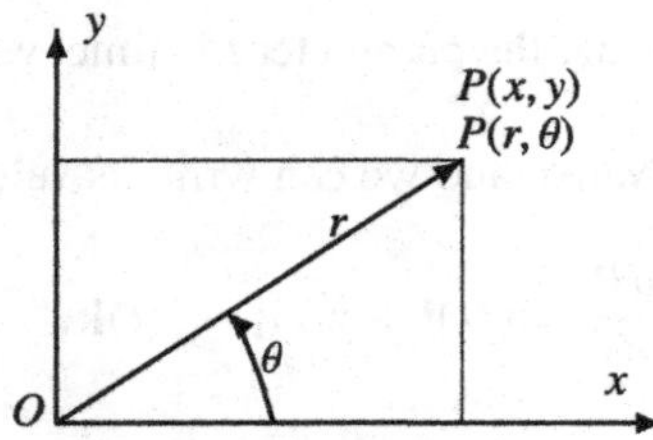

FIGURE 13.1 Polar coordinates and its relationship to Cartesian coordinates.

From Fig. 13.1 and trigonometrical functions, we can easily write down the relationship that links the polar coordinate system to its Cartesian coordinate system

$$x = r\cos\theta, \quad y = r\sin\theta, \tag{13.15}$$

and

$$r = \sqrt{x^2 + y^2}, \quad \tan\theta = \frac{y}{x} \quad [\text{or } \theta = \tan^{-1}(\frac{y}{x})]. \tag{13.16}$$

Here, the angle can also be represented by other trigonometrical functions and we have

$$\cos\theta = \frac{x}{\sqrt{x^2 + y^2}}, \quad \sin\theta = \frac{y}{\sqrt{x^2 + y^2}}. \tag{13.17}$$

Sometimes, there are significant advantages to use the polar coordinate system. For example, a circle of radius a can be represented in the polar coordinate system by a simple equation

$$r = a, \tag{13.18}$$

which is equivalent to

$$x^2 + y^2 = a^2, \tag{13.19}$$

in the Cartesian system. On the other hand, a line in the (x, y) system is in general represented by $y = kx + b$ as we have seen in Chapter 1. In the (r, θ) system, a line can simply be represented by

$$\theta = \theta_0, \tag{13.20}$$

where $\tan\theta_0$ is the slope or gradient of the line.

For a function F, we can use the relationships (13.8) and (13.7) discussed earlier to link two coordinate systems. We have

$$\frac{\partial F}{\partial \theta} = \frac{\partial F}{\partial x}\frac{\partial x}{\partial \theta} + \frac{\partial F}{\partial y}\frac{\partial y}{\partial \theta}. \tag{13.21}$$

From $r = \sqrt{x^2 + y^2}$, $x = r\cos\theta$, and $y = r\sin\theta$ we have

$$\frac{\partial x}{\partial r} = \frac{x}{\sqrt{x^2 + y^2}} = \cos\theta, \quad \frac{\partial y}{\partial r} = \frac{y}{\sqrt{x^2 + y^2}} = \sin\theta, \tag{13.22}$$

and

$$\frac{\partial x}{\partial \theta} = -r\sin\theta, \quad \frac{\partial y}{\partial \theta} = r\cos\theta. \tag{13.23}$$

Using these results, we have

$$\frac{\partial F}{\partial \theta} = \frac{\partial F}{\partial x}\frac{\partial x}{\partial \theta} + \frac{\partial F}{\partial y}\frac{\partial y}{\partial \theta} = -r\sin\theta\frac{\partial F}{\partial x} + r\cos\theta\frac{\partial F}{\partial y} = -y\frac{\partial F}{\partial x} + x\frac{\partial F}{\partial y}. \tag{13.24}$$

Similarly, we also have

$$r\frac{\partial F}{\partial r} = r\cos\theta\frac{\partial F}{\partial x} + r\sin\theta\frac{\partial F}{\partial y} = x\frac{\partial F}{\partial x} + y\frac{\partial F}{\partial y}. \tag{13.25}$$

Here the factor r is for the convenience of derivation. However, it does have mathematical reasons related to the Jacobian.

Let us consider a small element area. The area element in the (x, y) system is simply $dA = dxdy$, but the area element is spanned by an arc element with the arc length of $rd\theta$ and the thickness of dr. Thus, the area element should be $dA = rdrd\theta$ in the polar coordinate system. There is an extra factor r here. This concept becomes clearer when we discuss the multiple integrals in the next chapter.

The Jacobian for transforming from one coordinate system (e.g., (x, y)) to another (e.g., (r, θ)) is the determinant in the following form:

$$J = \frac{\partial(x, y)}{\partial(r, \theta)} = \begin{vmatrix} \frac{\partial x}{\partial r} & \frac{\partial x}{\partial \theta} \\ \frac{\partial y}{\partial r} & \frac{\partial y}{\partial \theta} \end{vmatrix} = \frac{\partial x}{\partial r}\frac{\partial y}{\partial \theta} - \frac{\partial y}{\partial r}\frac{\partial x}{\partial \theta}. \tag{13.26}$$

Example 13.4

For the polar coordinate system (r, θ), we know (from the earlier results)

$$\frac{\partial x}{\partial r} = \cos\theta, \quad \frac{\partial y}{\partial r} = \sin\theta,$$

and

$$\frac{\partial x}{\partial \theta} = -r\sin\theta, \quad \frac{\partial y}{\partial \theta} = r\cos\theta.$$

So the Jacobian becomes

$$J = \begin{vmatrix} \cos\theta & -r\sin\theta \\ \sin\theta & r\cos\theta \end{vmatrix} = \cos\theta \cdot (r\cos\theta) - \sin\theta(-r\sin\theta)$$
$$= r^2(\cos^2\theta + \sin^2\theta) = r.$$

Indeed, the Jacobian gives the factor r as discussed in the above about the area element.

13.4 THREE BASIC OPERATORS

There are three important operators commonly used in vector analysis. These operators are: gradient operator (grad or ∇), the divergence operator (div or $\nabla\cdot$) and the curl operator (curl or $\nabla\times$). They often appear in science and engineering, especially in fluid dynamics, electromagnetic dynamics and solid mechanics.

Sometimes, it is useful to calculate the directional derivative of a function ϕ at the point (x, y, z) in the direction of $\mathbf{n}$

$$\frac{\partial \phi}{\partial \mathbf{n}} = \mathbf{n}\cdot\nabla\phi = \frac{\partial \phi}{\partial x}\cos(\alpha) + \frac{\partial \phi}{\partial y}\cos(\beta) + \frac{\partial \phi}{\partial z}\cos(\gamma), \tag{13.27}$$

where $\mathbf{n} = (\cos\alpha, \cos\beta, \cos\gamma)$ is a unit vector and α, β, γ are the directional angles. Generally speaking, the gradient of any scalar function ϕ of x, y, z can be written in a similar way

$$\text{grad}\,\phi = \nabla\phi = \frac{\partial \phi}{\partial x}\mathbf{i} + \frac{\partial \phi}{\partial y}\mathbf{j} + \frac{\partial \phi}{\partial z}\mathbf{k}. \tag{13.28}$$

This is equivalent to applying the del operator ∇ to the scalar function ϕ

$$\nabla = \frac{\partial}{\partial x}\mathbf{i} + \frac{\partial}{\partial y}\mathbf{j} + \frac{\partial}{\partial z}\mathbf{k}. \tag{13.29}$$

Let us look at an example.

Example 13.5

For a function ϕ that describes a field, we have

$$\phi(x) = (x^2 + y^2)\sin(z).$$

We have

$$\frac{\partial \phi}{\partial x} = 2x\sin(z), \quad \frac{\partial \phi}{\partial y} = 2y\sin(z), \quad \frac{\partial \phi}{\partial z} = (x^2 + y^2)\cos(z).$$

The grad of ϕ can be obtained by

$$\text{grad}\,\phi = \frac{\partial \phi}{\partial x}\mathbf{i} + \frac{\partial \phi}{\partial y}\mathbf{j} + \frac{\partial \phi}{\partial z}\mathbf{k} = 2x\sin(z)\mathbf{i} + 2y\sin(z)\mathbf{j} + (x^2 + y^2)\cos(z)\mathbf{k}.$$

The direction of the gradient operator on a scalar field gives a vector field. The gradient operator has the following properties:

$$\nabla(\alpha\psi + \beta\phi) = \alpha\nabla\psi + \beta\nabla\phi, \qquad \nabla(\psi\phi) = \psi\nabla\phi + \phi\nabla\psi, \tag{13.30}$$

where α, β are constants and ψ, ϕ are scalar functions. In fact, these properties can be proved from the basic rules of differentiation.

For a vector field

$$\mathbf{u}(x, y, z) = u_1(x, y, z)\mathbf{i} + u_2(x, y, z)\mathbf{j} + u_3(x, y, z)\mathbf{k}, \tag{13.31}$$

the application of the operator ∇ can lead to either a scalar field or a vector field, depending on how the del operator applies to the vector field. The divergence of a vector field is the dot product of the del operator ∇ and $\mathbf{u}$

$$\text{div}\,\mathbf{u} = \nabla \cdot \mathbf{u} = \frac{\partial u_1}{\partial x} + \frac{\partial u_2}{\partial y} + \frac{\partial u_3}{\partial z}. \tag{13.32}$$

The divergence of a field can have important physical properties. For example, divergence can be a measure for expansion or compression of a vector field such as stress, strain and fluid flow. Let us look at an example.

Example 13.6

A vector field can be written as

$$\mathbf{u}(x, y, z) = \frac{\mathbf{v}}{r^3},$$

where

$$\mathbf{v} = \begin{pmatrix} x \\ y \\ z \end{pmatrix} = (x, y, z)^T, \quad r = \sqrt{x^2 + y^2 + z^2}.$$

From the formula to compute the divergence as given above, we have

$$u_1 = \frac{x}{r^3}, \quad u_2 = \frac{y}{r^3}, \quad u_3 = \frac{z}{r^3}.$$

It is straightforward to calculate their partial derivatives using the basic rules introduced earlier in this chapter. We have

$$\frac{\partial u_1}{\partial x} = \frac{1}{(x^2 + y^2 + z^2)^{3/2}} - \frac{3x^2}{(x^2 + y^2 + z^2)^{5/2}} = \frac{(x^2 + y^2 + z^2) - 3x^2}{r^5},$$

and

$$\frac{\partial u_2}{\partial y} = \frac{(x^2+y^2+z^2)-3y^2}{r^5}, \quad \frac{\partial u_3}{\partial z} = \frac{(x^2+y^2+z^2)-3z^2}{r^5}.$$

Thus the divergence of the field becomes

$$\nabla \cdot \mathbf{u} = \frac{\partial u_1}{\partial x} + \frac{\partial u_2}{\partial y} + \frac{\partial u_3}{\partial z} = \frac{3(x^2+y^2+z^2)-3x^2-3y^2-3z^2}{r^5} = 0.$$

Furthermore, the curl of $\mathbf{u}$ is the cross product of the del operator and the vector field $\mathbf{u}$

$$\text{curl}\,\mathbf{u} = \nabla \times \mathbf{u} = \begin{vmatrix} \mathbf{i} & \mathbf{j} & \mathbf{k} \\ \frac{\partial}{\partial x} & \frac{\partial}{\partial y} & \frac{\partial}{\partial z} \\ u_1 & u_2 & u_3 \end{vmatrix} = \begin{pmatrix} \frac{\partial}{\partial y} & \frac{\partial}{\partial z} \\ u_2 & u_3 \end{pmatrix}\mathbf{i} - \begin{pmatrix} \frac{\partial}{\partial x} & \frac{\partial}{\partial z} \\ u_1 & u_3 \end{pmatrix}\mathbf{j} + \begin{pmatrix} \frac{\partial}{\partial x} & \frac{\partial}{\partial y} \\ u_1 & u_2 \end{pmatrix}\mathbf{k}$$

$$= (\frac{\partial u_3}{\partial y} - \frac{\partial u_2}{\partial z})\mathbf{i} + (\frac{\partial u_1}{\partial z} - \frac{\partial u_3}{\partial x})\mathbf{j} + (\frac{\partial u_2}{\partial x} - \frac{\partial u_1}{\partial y})\mathbf{k}. \tag{13.33}$$

It is relatively straightforward to verify the following useful identities associated with the ∇ operator:

$$\nabla \cdot \nabla \times \mathbf{u} = 0, \tag{13.34}$$

$$\nabla \times \nabla \psi = 0, \tag{13.35}$$

$$\nabla \times (\psi \mathbf{u}) = \psi \nabla \times \mathbf{u} + (\nabla \psi) \times \mathbf{u}, \tag{13.36}$$

$$\nabla \cdot (\psi \mathbf{u}) = \psi \nabla \cdot \mathbf{u} + (\nabla \psi) \cdot \mathbf{u}. \tag{13.37}$$

We will revisit these identities and provide some proofs when we discuss tensors later in this book.

Example 13.7

For a vector field $\mathbf{u} = xy\mathbf{i} + y\mathbf{j} + z^2\mathbf{k}$, its divergence can be calculated by

$$\nabla \cdot \mathbf{u} = \frac{\partial xy}{\partial x} + \frac{\partial y}{\partial y} + \frac{\partial z^2}{\partial z} = y + 1 + 2z.$$

Its curl is given by

$$\nabla \times \mathbf{u} = (\frac{\partial z^2}{\partial y} - \frac{\partial y}{\partial z})\mathbf{i} + (\frac{\partial (xy)}{\partial z} - \frac{\partial z^2}{\partial x})\mathbf{j} + (\frac{\partial y}{\partial x} - \frac{\partial (xy)}{\partial y})\mathbf{k}$$

$$= (0-0)\mathbf{i} + (0-0)\mathbf{j} + (0-x)\mathbf{k} = -x\mathbf{k}.$$

So

$$\nabla \cdot \nabla \times \mathbf{u} = \nabla \cdot (\nabla \times \mathbf{u}) = \nabla \cdot (-x\mathbf{k}) = \frac{\partial 0}{\partial x} + \frac{\partial 0}{\partial y} + \frac{\partial (-x)}{\partial z} = 0.$$

One of the most common operators in engineering and science is the Laplacian operator is

$$\nabla^2 \Psi = \nabla \cdot (\nabla \Psi)$$

$$= \frac{\partial^2 \Psi}{\partial x^2} + \frac{\partial^2 \Psi}{\partial y^2} + \frac{\partial^2 \Psi}{\partial z^2}, \tag{13.38}$$

for Laplace's equation

$$\Delta \Psi = \nabla^2 \Psi = 0. \tag{13.39}$$

In engineering mathematics, it is sometimes necessary to express the Laplace equation in other coordinates such as cylindrical polar coordinates to be introduced in the next chapter.

Example 13.8

For a potential field of $\Psi = x + y + xyz$, we have its grad

$$\nabla\Psi = \frac{\partial\Psi}{\partial x}\mathbf{i} + \frac{\partial\Psi}{\partial y}\mathbf{j} + \frac{\partial\Psi}{\partial z}\mathbf{k}$$
$$= \frac{\partial(x+y+xyz)}{\partial x}\mathbf{i} + \frac{\partial(x+y+xyz)}{\partial y}\mathbf{j} + \frac{\partial(x+y+xyz)}{\partial z}\mathbf{k}$$
$$= (1+yz)\mathbf{i} + (1+xz)\mathbf{j} + xy\mathbf{k}.$$

In addition, the corresponding Laplacian can be calculated by

$$\nabla^2\Psi = \frac{\partial^2\Psi}{\partial x^2} + \frac{\partial^2\Psi}{\partial y^2} + \frac{\partial^2\Psi}{\partial z^2}$$
$$= \frac{\partial^2(x+y+xyz)}{\partial x^2} + \frac{\partial^2(x+y+xyz)}{\partial y^2} + \frac{\partial^2(x+y+xyz)}{\partial z^2} = 0+0+0=0.$$

Vector calculus has a wide range of applications in almost all disciplines of sciences. Here, we only introduce the very basic part of a very broad subject. For more details, interested readers can refer to more advanced literature.

EXERCISES

13.1. Calculate the first partial derivatives of $f(x, y) = x^2 + y^2 + \sin(x^2y^2)$.
13.2. Calculate the second partial derivatives of $g(x, y) = e^{-xy^2} + xy$.
13.3. What is the divergence of $\mathbf{u} = x^2\mathbf{i} + (y^2 - x^2)\mathbf{j} + z\mathbf{k}$?
13.4. First calculate the curl of $\mathbf{v} = yz^n\mathbf{i} + xy\mathbf{j} + yz^2\mathbf{k}$ where $n \geq 0$. Then, use the result to show that div (curl $\mathbf{v}$) = 0.
13.5. For a so-called potential flow $\Psi = 3x^2y - y^3$, calculate $\nabla^2\Psi$ and then show that $\nabla \times \nabla\Psi = 0$.

Chapter 14

Multiple Integrals and Special Integrals

Chapter Points

- Multivariate integrals involve integration for more than one independent variable and they are useful to calculate areas, volumes and quantities in high dimensions.
- Some specialized methods will be presented to do integration concerning some special integrals such as Gaussian integral and the error function.
- Vector operators in other coordinates are often linked via their Jacobian. Some examples are given to show how they work.
- Order notations are introduced to aid the discussion of approximations and series expansions for some integrals such as the error function.

14.1 LINE INTEGRAL

An important class of integrals in this context is the line integral which integrates along a curve $\mathbf{r}(x, y, z) = x\mathbf{i} + y\mathbf{j} + z\mathbf{k}$. For example, in order to calculate the arc length L of curve $\mathbf{r}$ as shown in Fig. 14.1, we have to use the line integral

$$L = \int_{s_0}^{s} ds = \int_{s_0}^{s} \sqrt{(dx)^2 + (dy)^2} = \int_{x_0}^{x} \Big[\sqrt{\frac{(dx)^2}{(dx)^2} + \frac{(dy)^2}{(dx)^2}}\Big] dx = \int_{x_0}^{x} \sqrt{1 + (\frac{dy}{dx})^2}\, dx. \tag{14.1}$$

Thus, for a curve described by a function $y = f(x)$, its arc length from $x = a$ and $x = b$ can be calculated by

$$L = \int_a^b \sqrt{1 + [f'(x)]^2}\, dx. \tag{14.2}$$

Let us look at a simple example.

Example 14.1

The arc length of the parabola $y(x) = \frac{1}{2}x^2$ from $x = 0$ to $x = 1$ is given by

$$L = \int_0^1 \sqrt{1 + y'^2} dx = \int_0^1 \sqrt{1 + x^2} dx = \frac{1}{2}[x\sqrt{1+x^2} + \ln(x + \sqrt{1+x^2})]\Big|_0^1 = \frac{1}{2}[\sqrt{2} - \ln(\sqrt{2} - 1)] \approx 1.14779.$$

14.2 MULTIPLE INTEGRALS

As the integration of a function $f(x)$ corresponds to the area enclosed under the function between integration limits, this can extend to the double integral and multiple integrals. For a function $f(x, y)$, the double integral is defined as

$$F = \int_{\Omega} f(x, y) dA, \tag{14.3}$$

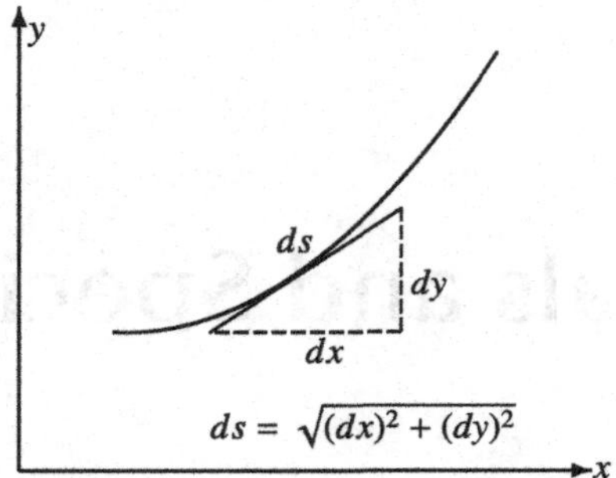

FIGURE 14.1 Arc length along a curve.

where dA is the infinitesimal element of the area, and Ω is the region for integration. The simplest form of dA is $dA = dxdy$ in Cartesian coordinates. In order to emphasize the double integral in this case, the integral is often written as

$$I = \iint_{\Omega} f(x, y)dxdy. \tag{14.4}$$

Since the integration is over the area, the value of the integral corresponds to the volume enclosed under the surface $f(x, y)$. Let us look at an example.

Example 14.2

Calculate the integral

$$I = \iint_{\Omega} x^2 y^3 dxdy,$$

where Ω is a rectangular box with $0 \le x \le 1$ and $0 \le y \le 2$. We have

$$I = \int_0^2 \Big(\int_0^1 x^2 y^3 dx\Big)dx = \int_0^2 \Big[\Big(\frac{x^3}{3}y^3\Big)\Big|_0^1\Big]dy$$

$$= \int_0^2 \Big[\frac{1^3 y^3}{3} - \frac{0^3 y^3}{3}\Big]dy = \int_0^2 \Big(\frac{y^3}{3}\Big)dy = \frac{1}{3}\int_0^2 y^3 dy = \frac{1}{3}\Big(\frac{y^4}{4}\Big|_0^2\Big) = \frac{1}{3}\Big[\frac{2^4}{4} - \frac{0^4}{4}\Big] = \frac{4}{3}.$$

Double integrals are important in a diverse range of engineering applications. For example, in order to estimate the behavior of a structure, we often have to calculate the bending moment and second moment of area.

Example 14.3

The area moment of inertia of a thin rectangular plate, with the width a and the depth b, is defined by

$$I = \iint_{\Omega} y^2 dS = \iint_{\Omega} y^2 dxdy.$$

The plate can be divided into four equal parts, and we have

$$I = 4\int_0^{a/2} \Big[\int_0^{b/2} y^2 dy\Big]dx$$

$$= 4\int_0^{a/2} \frac{1}{3}\Big(\frac{b}{2}\Big)^3 dx = \frac{b^3}{6}\int_0^{a/2} dx = \frac{ab^3}{12}.$$

In describing the behavior of rotation and rigid-body motion, we have to calculate the moment of inertia

$$I = \int_{\Omega} r^2 dm = \int_{\Omega} \rho r^2 dV, \tag{14.5}$$

where $dm = \rho dV$ is the mass of the volume element dV and ρ is the density. The integration must be over the whole body Ω.

Example 14.4

For a thin disc of radius a with a uniform thickness t and density ρ, we have $dV = tdA = trdrd\theta$ and we only need to integrate in two dimensions. Thus, we have

$$I = \int_{\Omega} \rho r^2 dV = \int_0^{2\pi} \int_0^a (\rho t r^2) r dr d\theta = t\rho \int_0^{2\pi} d\theta \int_0^a r^3 dr$$
$$= t\rho(2\pi) \int_0^a r^3 dr = 2\pi\rho t \Big(\frac{r^4}{4}\Big|_0^a\Big) = \pi\rho t \frac{a^4}{2}.$$

Let m be the total mass of the thin disc, we have its volume is $V = \pi a^2 t$ and its density is $\rho = m/V = m/(\pi a^2 t)$. Finally, we have

$$I = \pi t \cdot \frac{m}{\pi a^2 t} \cdot \frac{a^4}{2} = \frac{ma^2}{2}.$$

This is the polar moment of inertia for the thin disc with respect to the axis through the center but perpendicular to the disc surface.

14.3 JACOBIAN

Sometimes it is necessary to change variables when evaluating an integral. For a simple one-dimensional integral, the change of variables from x to a new variable v (say) leads to $x = x(v)$. This is relatively simple as $dx = \frac{dx}{dv} dv$, and we have

$$\int_{x_a}^{x_b} f(x) dx = \int_a^b f(x(v)) \frac{dx}{dv} dv, \tag{14.6}$$

where the integration limits change so that $x(a) = x_a$ and $x(b) = x_b$. Here the extra factor dx/dv in the integrand is referred to as the Jacobian.

For a double integral, it is more complicated. Assuming $x = x(\xi, \eta)$, $y = y(\xi, \eta)$, we have

$$\iint f(x, y) dx dy = \iint f(\xi \eta) |J| d\xi d\eta, \tag{14.7}$$

where J is the Jacobian. That is

$$J \equiv \frac{\partial(x, y)}{\partial(\xi, \eta)} = \begin{vmatrix} \frac{\partial x}{\partial \xi} & \frac{\partial x}{\partial \eta} \\ \frac{\partial y}{\partial \xi} & \frac{\partial y}{\partial \eta} \end{vmatrix} = \begin{vmatrix} \frac{\partial x}{\partial \xi} & \frac{\partial y}{\partial \xi} \\ \frac{\partial x}{\partial \eta} & \frac{\partial y}{\partial \eta} \end{vmatrix}. \tag{14.8}$$

The notation $\partial(x, y)/\partial(\xi, \eta)$ is just a useful shorthand. This is equivalent to saying that the change of the infinitesimal area $dA = dxdy$ becomes

$$dxdy = \Big|\frac{\partial(x, y)}{\partial(\xi, \eta)}\Big| d\xi d\eta = \Big|\frac{\partial x}{\partial \xi}\frac{\partial y}{\partial \eta} - \frac{\partial x}{\partial \eta}\frac{\partial y}{\partial \xi}\Big| d\xi d\eta. \tag{14.9}$$

Example 14.5

When transforming from (x, y) to polar coordinates (r, θ), we have the following relationships

$$x = r\cos\theta, \qquad y = r\sin\theta.$$

Thus, we have

$$\frac{\partial x}{\partial r} = \cos\theta, \quad \frac{\partial x}{\partial \theta} = -r\sin\theta, \quad \frac{\partial y}{\partial r} = \sin\theta, \quad \frac{\partial y}{\partial \theta} = r\cos\theta,$$

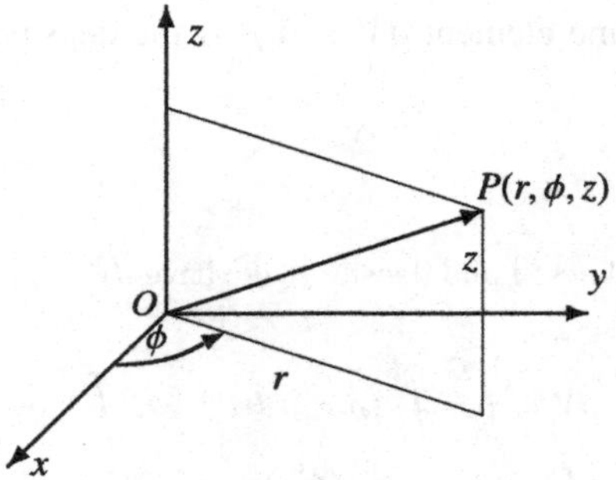

FIGURE 14.2 Cylindrical polar coordinates.

so that the Jacobian is

$$J = \frac{\partial(x, y)}{\partial(r, \theta)} = \frac{\partial x}{\partial r}\frac{\partial y}{\partial \theta} - \frac{\partial x}{\partial \theta}\frac{\partial y}{\partial r}$$
$$= \cos\theta \times r\cos\theta - (-r\sin\theta) \times \sin\theta = r[\cos^2\theta + \sin^2\theta] = r.$$

Thus, an integral in (x, y) will be transformed into

$$\iint \phi(x, y)dxdy = \iint \phi(r\cos\theta, r\sin\theta)rdrd\theta.$$

In a similar fashion, the change of variables in triple integrals gives

$$V = \iiint_{\Omega} \phi(x, y, z)dxdydz = \iiint_{\omega} \psi(\xi, \eta, \zeta)|J|d\xi d\eta d\zeta, \tag{14.10}$$

and

$$J \equiv \frac{\partial(x, y, z)}{\partial(\xi, \eta, \zeta)} = \begin{vmatrix} \frac{\partial x}{\partial \xi} & \frac{\partial y}{\partial \xi} & \frac{\partial z}{\partial \xi} \\ \frac{\partial x}{\partial \eta} & \frac{\partial y}{\partial \eta} & \frac{\partial z}{\partial \eta} \\ \frac{\partial x}{\partial \zeta} & \frac{\partial y}{\partial \zeta} & \frac{\partial z}{\partial \zeta} \end{vmatrix}. \tag{14.11}$$

For cylindrical polar coordinates (r, ϕ, z) as shown in Fig. 14.2, we have

$$x = r\cos\phi, \qquad y = r\sin\phi, \qquad z = z. \tag{14.12}$$

Their derivatives are

$$\frac{\partial x}{\partial r} = \cos\phi, \quad \frac{\partial x}{\partial \phi} = -r\sin\phi, \quad \frac{\partial x}{\partial z} = 0,$$
$$\frac{\partial y}{\partial r} = \sin\phi, \quad \frac{\partial y}{\partial \phi} = r\cos\phi, \quad \frac{\partial z}{\partial z} = 1. \tag{14.13}$$

All the other partial derivatives are zero. Therefore, the Jacobian is

$$J = \frac{\partial(x, y, z)}{\partial(r, \phi, z)} = \begin{vmatrix} \cos\phi & \sin\phi & 0 \\ -r\sin\phi & r\cos\phi & 0 \\ 0 & 0 & 1 \end{vmatrix}$$
$$= 1 \times [\cos\phi\,(r\cos\phi) - (-r\sin\phi)\sin\phi] = r, \tag{14.14}$$

which is the same as that for the polar coordinate system.

In cylindrical polar coordinates (r, ϕ, z), we have

$$\nabla \cdot \mathbf{u} = \frac{1}{r}\frac{\partial(ru_r)}{\partial r} + \frac{1}{r}\frac{\partial u_\phi}{\partial \phi} + \frac{\partial u_z}{\partial z}. \tag{14.15}$$

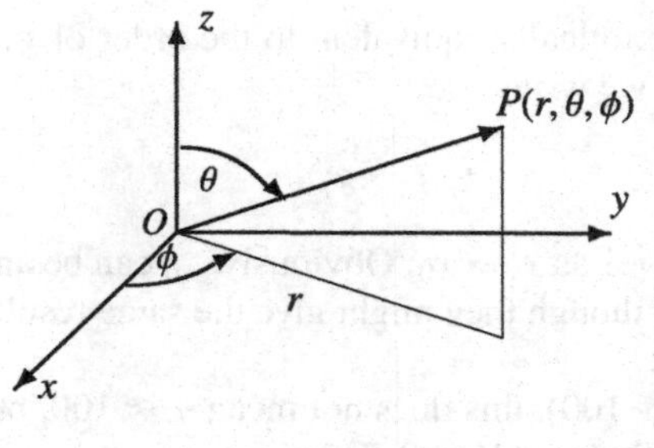

FIGURE 14.3 Spherical polar coordinates.

The Laplace equation becomes

$$\nabla^2 \Psi = \frac{\partial^2 \Psi}{\partial r^2} + \frac{1}{r}\frac{\partial \Psi}{\partial r} + \frac{1}{r^2}\frac{\partial^2 \Psi}{\partial \phi^2} + \frac{\partial^2 \Psi}{\partial z^2}. \tag{14.16}$$

In spherical polar coordinates (r, θ, ϕ), we have

$$\nabla \cdot \mathbf{u} = \frac{1}{r^2}\frac{\partial^2 (r^2 u_r)}{\partial r^2} + \frac{1}{r \sin\theta}\frac{\partial (\sin\theta u_\theta)}{\partial \theta} + \frac{1}{r \sin\theta}\frac{\partial u_\phi}{\partial \phi}. \tag{14.17}$$

The Laplace equation can be written as

$$\nabla^2 \Psi = \frac{1}{r^2}\frac{\partial}{\partial r}[r^2 \frac{\partial \Psi}{\partial r}] + \frac{1}{r^2 \sin\theta}\frac{\partial}{\partial \theta}[\sin\theta \frac{\partial \Psi}{\partial \theta}] + \frac{1}{r^2 \sin^2\theta}\frac{\partial^2 \Psi}{\partial \phi^2}. \tag{14.18}$$

For spherical polar coordinates (r, θ, ϕ) as shown in Fig. 14.3, where θ is the zenithal angle between the z-axis and the position vector $\mathbf{r}$, and ϕ is the azimuthal angle, we have

$$x = r \sin\theta \cos\phi, \qquad y = r \sin\theta \sin\phi, \qquad z = r \cos\theta. \tag{14.19}$$

Therefore, the Jacobian is

$$J = \begin{vmatrix} \sin\theta \cos\phi & \sin\theta \sin\phi & \cos\theta \\ r \cos\theta \cos\phi & r \cos\theta \sin\phi & -r \sin\theta \\ -r \sin\theta \sin\phi & r \sin\theta \cos\phi & 0 \end{vmatrix} = r^2 \sin\theta. \tag{14.20}$$

Thus, the volume element change in the spherical system is

$$dxdydz = r^2 \sin\theta dr d\theta d\phi. \tag{14.21}$$

14.4 SPECIAL INTEGRALS

Some integrals appear so frequently in engineering mathematics that they deserve special attention. Most of these special integrals are also called special functions as they have certain varying parameters or integral limits. We discuss only Gaussian integral and the error function.

14.4.1 Asymptotic Series

Before we discuss any special functions, let us digress first to introduce the asymptotic series and order notations because they will be used to study the characteristics of special functions. Loosely speaking, for two functions $f(x)$ and $g(x)$, if

$$\frac{f(x)}{g(x)} \to K, \qquad x \to x_0, \tag{14.22}$$

where K is a finite, non-zero limit, we write

$$f = O(g). \tag{14.23}$$

The big O notation means that f is asymptotically equivalent to the order of $g(x)$. If the limit is unity or $K = 1$, we say $f(x)$ is order of $g(x)$. In this special case, we write

$$f \sim g, \tag{14.24}$$

which is equivalent to $f/g \to 1$ and $g/f \to 1$ as $x \to x_0$. Obviously, x_0 can be any value, including 0 and ∞. The notation $\sim$ does not necessarily mean $\approx$ in general, though they might give the same results, especially in the case when $x \to 0$. For example, $\sin x \sim x$ and $\sin x \approx x$ if $x \to 0$.

When we say f is order of 100 (or $f \sim 100$), this does not mean $f \approx 100$, but it can mean that f is between about 50 to 150. The small o notation is used if the limit tends to 0. That is

$$\frac{f}{g} \to 0, \qquad x \to x_0, \tag{14.25}$$

or

$$f = o(g). \tag{14.26}$$

If $g > 0$, $f = o(g)$ is equivalent to $f \ll g$. Here, $\ll$ means 'much smaller'. For example, for $\forall x \in \mathcal{R}$, we have $e^x \approx 1 + x + O(x^2) \approx 1 + x + \frac{x^2}{2} + o(x)$.

Another classical example is the Stirling's asymptotic series for factorials

$$n! \sim \sqrt{2\pi n}\,(\frac{n}{e})^n, \qquad n \gg 1. \tag{14.27}$$

Here, $\gg$ mean 'much larger', so $n \gg 1$ means any integer that is much larger than 1. In fact, it can be expanded into more terms

$$n! \sim \sqrt{2\pi n}\,(\frac{n}{e})^n(1 + \frac{1}{12n} + \frac{1}{288n^2} - \frac{139}{51480n^3} - ...). \tag{14.28}$$

This is a good example of asymptotic series. For standard power expansions, the error $R_k(h^k) \to 0$, but for an asymptotic series, the error of the truncated series R_k decreases and gets smaller compared with the leading term [here $\sqrt{2\pi n}(n/e)^n$]. However, R_n does not necessarily tend to zero. In fact, $R_2 = \frac{1}{12n} \cdot \sqrt{2\pi n}(n/e)^n$ is still very large as $R_2 \to \infty$ if $n \gg 1$. For example, for $n = 100$, we have

$$n! = 9.3326 \times 10^{157},$$

while the leading approximation is

$$\sqrt{2\pi n}(n/e)^n = 9.3248 \times 10^{157}.$$

The difference between these two values is 7.7740×10^{154}, which is still very large, though three orders smaller than the leading approximation.

14.4.2 Gaussian Integrals

The Gaussian integral appears in many situations in engineering mathematics and statistics. It can be defined by

$$I(\alpha) = \int_{-\infty}^{\infty} e^{-\alpha x^2} dx. \tag{14.29}$$

In order to evaluate the integral, let us first evaluate I^2. Since the Gaussian integral is a definite integral and must give a constant value, we can change the dummy variable x to anything appropriate (y) as we wish. Thus we have

$$\begin{aligned} I^2 &= [\int_{-\infty}^{\infty} e^{-\alpha x^2} dx]^2 = \int_{-\infty}^{\infty} e^{-\alpha x^2} dx \int_{-\infty}^{\infty} e^{-\alpha y^2} dy \\ &= \int_{-\infty}^{\infty}\int_{-\infty}^{\infty} e^{-\alpha(x^2+y^2)} dx dy. \end{aligned} \tag{14.30}$$

Changing into the polar coordinates (r, θ) and noticing that $r^2 = x^2 + y^2$ and $dxdy = rdrd\theta$, we have

$$I^2 = \int_0^\infty dr \int_0^{2\pi} re^{-\alpha r^2} d\theta = 2\pi \int_0^\infty e^{-\alpha r^2} rdr$$
$$= 2\pi \int_0^\infty \frac{1}{2\alpha} e^{-\alpha r^2} d(\alpha r^2) = \frac{\pi}{\alpha} \int_0^\infty e^{-\alpha r^2} d(\alpha r^2) = \frac{\pi}{\alpha}, \tag{14.31}$$

where we have used

$$\int_0^\infty e^{-\alpha r^2} d(\alpha r^2) = \int_0^\infty e^{-u} du = 1, \quad u = \alpha r^2. \tag{14.32}$$

Therefore, the integral becomes

$$I(\alpha) = \int_{-\infty}^\infty e^{-\alpha x^2} dx = \sqrt{\frac{\pi}{\alpha}}. \tag{14.33}$$

Since α is a parameter, we can differentiate both sides of this equation with respect to α, and we have

$$\int_{-\infty}^\infty x^2 e^{-\alpha x^2} dx = \frac{1}{2\alpha}\sqrt{\frac{\pi}{\alpha}}. \tag{14.34}$$

By differentiating both sides of the Gaussian integral (14.33) n times with respect to α, and we get the generalized Gaussian integral I_n

$$I_n = \int_{-\infty}^\infty x^{2n} e^{-\alpha x^2}$$
$$= \frac{(-1)^n \cdot 1 \cdot 3 \cdot 5 \cdots (2n-1)}{2^n} \sqrt{\frac{\pi}{\alpha^{2n+1}}}, \tag{14.35}$$

where $n > 0$ is an integer.

For a special case when $\alpha = \frac{1}{\sqrt{2\sigma^2}}$ and $n = 0$, Eq. (14.33) can be rearranged as

$$\int_\infty^\infty f(x, \sigma) dx = 1, \tag{14.36}$$

$$f(x) = \frac{1}{\sqrt{2\pi\sigma^2}} e^{-\frac{x^2}{2\sigma^2}}. \tag{14.37}$$

The function $f(x, \sigma)$ is a zero-mean Gaussian probability function. As $\sigma \to 0$, $f(x) \to \delta(x)$ where $\delta(x)$ is the Dirac δ-function which is defined by

$$\delta(x) \neq 0 \ \text{(at } x = 0\text{)}, \ \text{but } \delta(x) = 0, \ \text{for } x \neq 0, \tag{14.38}$$

and

$$\int_\infty^\infty \delta(x) dx = 1. \tag{14.39}$$

It has an interesting property that

$$\int f(x)\delta(x - \beta) dx = f(\beta), \tag{14.40}$$

where $f(x)$ is a function.

14.4.3 Error Functions

The error function, which appears frequently in heat conduction and diffusion problems, is defined by

$$\text{erf}(x) = \frac{2}{\sqrt{\pi}} \int_0^x e^{-\eta^2} d\eta. \tag{14.41}$$

Its complementary error function is defined by

$$\text{erfc}(x) = 1 - \text{erf}(x) = \frac{2}{\sqrt{\pi}} \int_x^{\infty} e^{-t^2} dt. \tag{14.42}$$

The error function is an odd function: $\text{erf}(-x) = -\text{erf}(x)$. Using the results from the Gaussian integral

$$\int_{-\infty}^{\infty} e^{-\eta^2} d\eta = \sqrt{\pi}, \tag{14.43}$$

together with the basic definition, we have $\text{erf}(0) = 0$, and $\text{erf}(\infty) = 1$.

The error function cannot be easily evaluated in a closed form. Using Taylor series for

$$e^x = 1 + x + \frac{x^2}{2!} + \frac{x^3}{3!} + ... + \frac{x^n}{n!} + ..., \tag{14.44}$$

and setting $x = -\eta^2$, we have the integrand

$$e^{-\eta^2} = 1 - \eta^2 + \frac{1}{2}\eta^4 - \frac{1}{6}\eta^6 + ... + \frac{(-1)^n \eta^{2n}}{n!} + ..., \tag{14.45}$$

and integrating term by term, we have

$$\text{erf}(x) = \frac{2}{\sqrt{\pi}} \Big[x - \frac{x^3}{3} + \frac{x^5}{10} - \frac{x^7}{42} + ... + \frac{(-1)^n \eta^{2n+1}}{n!(2n+1)} + ... \Big], \tag{14.46}$$

or

$$\text{erf}(x) = \frac{2}{\sqrt{\pi}} \sum_{n=0}^{\infty} \frac{(-1)^n}{2n+1} \frac{x^{2n+1}}{n!}. \tag{14.47}$$

The integrals of the complementary function are defined by

$$\text{ierfc}(x) = \int_x^{\infty} \text{erfc}(\eta) d\eta, \tag{14.48}$$

and

$$\text{i}^{\text{n}}\text{erfc}(x) = \int_x^{\infty} \text{i}^{\text{n}-1}\text{erfc}(\eta) d\eta. \tag{14.49}$$

Using integration by parts, we can prove the following asymptotic series

$$\text{erf}(x) \sim 1 - \frac{e^{-x^2}}{x\sqrt{\pi}}, \quad (x \to \infty). \tag{14.50}$$

On the other hand, if we replace x in the error function by βx, we have

$$\lim_{\beta \to \infty} \frac{1}{2}[1 + \text{erf}(\beta x)] \to H(x), \tag{14.51}$$

where $H(x)$ is a Heaviside function or a unit step function which is defined by

$$H(x) = 1 \ \ (\text{for } x > 0), \qquad H(x) = 0 \ \ (\text{for } x < 0). \tag{14.52}$$

At $x = 0$, it is discontinuous and it is convention to set $H(0) = 1/2$. Its relationship with the Dirac δ-function is that

$$\frac{dH(x)}{dx} = \delta(x). \tag{14.53}$$

Though these relationships look very complicated, it is relatively easy to compute using any programming language.

EXERCISES

14.1. Compute the arc length of $y = 2x^{3/2}/3$ from $x = 0$ to $x = 2$.

14.2. Compute the volume under the surface $f(x, y) = 5 + x^2 + y$ over a region outlined by $x \in [0, 2]$ and $-x \le y \le x$.

14.3. Calculate the integral $\iint_D e^{-x^2-y^2} dxdy$ in the circular region $x^2 + y^2 \le 1$.

14.4. Use any technique of double integrals to show that the area of an ellipse is πab where a and b are the major and minor radii, respectively.

Hint. Use the equation for an ellipse $\frac{x^2}{a^2} + \frac{y^2}{b^2} = 1$.

Chapter 15

Complex Integrals

Chapter Points

- Analytical functions will be defined, together with the introduction of Cauchy-Riemann conditions.
- For complex integrals, both contour integrals and the residue theorem are introduced and explained with examples.

15.1 ANALYTIC FUNCTIONS

Any function of real variables can be extended to the function of complex variables in the same form while treating the real numbers x as $x + i0$. For example, $f(x) = x^2, x \in \mathbb{R}$ becomes $f(z) = z^2, z \in \mathbb{C}$. Any complex function $f(z)$ can be written as a real part $u = \Re(f(z))$ and an imaginary part $v = \Im(f(z))$. That is

$$f(z) = f(x + iy) = \Re(f(z)) + i\Im(f(z)) = u(x, y) + iv(x, y), \tag{15.1}$$

where $u(x, y)$ and $v(x, y)$ are real-valued functions of two real variables: x and y.

In some textbooks, the domain of $f(z)$ is called z-plane, while the mapping or a function of a point $z = x + iy$ to a point $\omega = u + iv$, and all the points $\omega = f(z) = u + iv$ form a so-called ω-plane.

In many cases, the extension of real functions to complex-valued functions is relatively straightforward. Sometimes, however, care should be taken because a function of complex variables may become a multiple-valued function.

Example 15.1

For a simple function $f(z) = z^2$, we have

$$f(z) = z^2 = (x + iy)^2 = x^2 + 2xyi + (iy)^2 = (x^2 - y^2) + 2xyi,$$

where $u(x, y) = x^2 - y^2$ and $v(x, y) = 2xy$. Similarly, for $f(z) = 1/z$, we have

$$\omega = u + iv = \frac{1}{z} = \frac{1}{x + iy} = \frac{1}{(x + iy)} \cdot \frac{(x - iy)}{(x - iy)} = \frac{x - iy}{x^2 + y^2},$$

where we have used $(x + iy)(x - iy) = x^2 - (iy)^2 = x^2 + y^2$. Thus, we have

$$u = \frac{x}{x^2 + y^2}, \quad v = \frac{-y}{x^2 + y^2}.$$

However, function $g(z) = \sqrt{z}$ is a multiple-valued function because two values $+\sqrt{z}$ and $-\sqrt{z}$ can correspond to the same z. For example, for $z = i$, we know from an earlier chapter that

$$\sqrt{i} = \frac{\sqrt{2}}{2}(1 + i), \quad -\frac{\sqrt{2}}{2}(1 + i).$$

In this case, there are two sets of u and v for the same z.

A function $f(z)$ is called analytic at z_0 if $f'(z)$ exists for all z in some ϵ-neighborhood of z_0 or differentiable in some open disk $|z - z_0| < \epsilon$. If $f(z) = u + iv$ is analytic at every point in a domain Ω, then $u(x, y)$ and $v(x, y)$ satisfies the

Cauchy-Riemann equations

$$\frac{\partial u}{\partial x} = \frac{\partial v}{\partial y}, \qquad \frac{\partial u}{\partial y} = -\frac{\partial v}{\partial x}. \tag{15.2}$$

Conversely, if u and v in $f(z) = u + iv$ satisfy the Cauchy-Riemann equation at all points in a domain, then the complex function $f(z)$ is analytic in the same domain.

Example 15.2

For example, the elementary power function $w = z^n$, $(n > 1)$ is analytic on the whole plane, $w = \rho e^{i\phi}$, $z = re^{i\theta}$, then

$$\rho = r^n, \ \phi = n\theta.$$

The logarithm $\ln z$ is also an elementary function such that

$$\ln z = \ln|z| + i\arg(z) = \ln r + i(\theta + w\pi k),$$

which has infinitely many values, due to the multiple values of θ, with the difference of $2\pi ik$, $(k = 0, \pm 1, \pm 2, ...)$. If we use the principal argument Arg(z), then we have the principal logarithm function

$$\text{Ln}(z) = \ln|z| + i\text{Arg}(z) = \ln r + i\theta.$$

For example, as $-1 = e^{i\pi} = \cos(\pi) - i\sin\pi$, we have

$$\ln(-1) = \ln 1 + i\text{Arg}(-1) = 0 + i\pi = i\pi.$$

This means that the logarithm of a negative number is a pure imaginary number.

If we differentiate the Cauchy-Riemann equations and use $\partial^2 u/\partial x\partial y = \partial^2 u/\partial y\partial x$, we have

$$\frac{\partial^2 u}{\partial x^2} = \frac{\partial^2 v}{\partial x \partial y}, \tag{15.3}$$

and

$$\frac{\partial^2 v}{\partial y^2} = -\frac{\partial^2 v}{\partial y \partial x} = -\frac{\partial^2 v}{\partial x \partial y}. \tag{15.4}$$

Adding these two, we have

$$\frac{\partial^2 u}{\partial x^2} + \frac{\partial^2 u}{\partial y^2} = 0. \tag{15.5}$$

A similar argument for v leads to the following theorem: For a given analytic function $f(z) = u + iv$, both u and v satisfy the Laplace equations

$$\frac{\partial^2 u}{\partial x^2} + \frac{\partial^2 u}{\partial y^2} = 0, \qquad \frac{\partial^2 v}{\partial x^2} + \frac{\partial^2 v}{\partial y^2} = 0. \tag{15.6}$$

This is to say, both real and imaginary parts of an analytic function are harmonic.

Example 15.3

From an earlier example about $f(z) = z^2$, we know that

$$\omega = u + iv = (x^2 - y^2) + 2xyi,$$

or

$$u = x^2 - y^2, \quad v = 2xy.$$

Let us check if they satisfy the Laplace equation.

It is easy to show that

$$\frac{\partial u}{\partial x} = 2x, \quad \frac{\partial^2 u}{\partial x^2} = 2, \quad \frac{\partial u}{\partial y} = -2y, \quad \frac{\partial^2 u}{\partial y^2} = -2,$$

and

$$\frac{\partial v}{\partial x} = 2y, \quad \frac{\partial^2 v}{\partial x^2} = 0, \quad \frac{\partial v}{\partial y} = 2x, \quad \frac{\partial^2 v}{\partial y^2} = 0.$$

Thus, we have

$$\frac{\partial^2 u}{\partial x^2} + \frac{\partial^2 u}{\partial y^2} = 2 + (-2) = 0, \quad \frac{\partial^2 v}{\partial x^2} + \frac{\partial^2 v}{\partial y^2} = 0 + 0 = 0,$$

which imply that they indeed satisfy the Laplace equation. This proves that $f(z) = z^2$ is an analytical function.

15.2 COMPLEX INTEGRALS

For the integral of a real univariate function, the integration is between two limits $x = a$ and $x = b$. That is

$$I = \int_a^b f(x)dx, \tag{15.7}$$

whose interval of integration can be considered as a straight line (or a path) linking $x = a$ and $x = b$. For example, we know that

$$\int_1^2 x dx = \frac{x^2}{2}\Big|_1^2 = \frac{2^2}{2} - \frac{1^2}{2} = 1.5. \tag{15.8}$$

However, if we try to extend this simple integral to the complex domain, we can write

$$I = \int_{z_1}^{z_2} z dz, \tag{15.9}$$

where two integration limits $z = z_1$ and $z = z_2$ corresponding to two points in the plane. Now the question is how to carry out the integral and what path we should take.

For example, let us try to calculate

$$I = \int_1^{2+i} z dz. \tag{15.10}$$

Suppose we naively use the same rule as we did for a real function, we may have

$$\begin{aligned} I &= \int_1^{2+i} z dz = \frac{z^2}{2}\Big|_1^{2+i} = \frac{(2+i)^2}{2} - \frac{1^2}{2} \\ &= \frac{1}{2}[(4 + 4i + i^2) - 1] = \frac{1}{2}[2 + 4i] = 1 + 2i. \end{aligned} \tag{15.11}$$

Is this the right answer? Is this the right way to do it? Before we can answer these two questions, let us look at the integral again from a different perspective.

We know that $z = x + iy$, thus $dz = dx + idy$. Since there are many ways to reach point $z_2 = 2 + i$ from point $z_1 = 1$, we can denote any of such paths as D. Thus we have

$$I = \int_D zdz = \int_D (x+iy)(dx+idy) = \int_D (xdx + ixdy + iydx + i^2 ydy)$$
$$= \int_D (xdx - ydy) + i\int_D (ydx + xdy). \tag{15.12}$$

Now what path should we use? One simple way is to go first along a path D_1 from $z_1 = 1$ to $z_* = 2$ (horizontal with $dy = 0$ and $y = 0$) and then along a path D_2 from $z_* = 2$ to $z_2 = 2 + i$ (vertical with $dx = 0$ and $x = 2$). With this path, we have

$$I = \underbrace{\int_1^2 xdx}_{D_1} + \underbrace{\int_0^i (-y)dy + i\int_0^i 2dy}_{D_2}$$
$$= \frac{x^2}{2}\Big|_1^2 - \frac{y^2}{2}\Big|_0^i + 2iy\Big|_0^i = \frac{3}{2} - \frac{1}{2}(i^2 - 0^2) + 2i(i-0) = \frac{3}{2} + \frac{1}{2} - 2 = 2, \tag{15.13}$$

which is not the answer $(1 + 2i)$ we obtained earlier. On the other hand, if we first go along a path D_3 vertically from $z_1 = 1$ to $z_3 = 1 + i$ (with $dx = 0$ and $x = 1$) and then along a path D_4 horizontally from $z_3 = 1 + i$ to $z_2 = 2 + i$ (with $dy = 0$ and $y = 1$), we now have

$$I = \underbrace{\int_0^i (-y)dy + i\int_0^i 1dy}_{D_3} + \underbrace{\int_1^2 xdx + i\int_1^2 1dx}_{D_4}$$
$$= -\frac{y^2}{2}\Big|_0^i + iy\Big|_0^i + \frac{x^2}{2}\Big|_1^2 + ix\Big|_1^2$$
$$= -\frac{1}{2}(i^2 - 0^2) + i(i-0) + \frac{1}{2}(2^2 - 1^2) + i(2-1) = \frac{1}{2} - 1 + \frac{3}{2} + i = 1 + i, \tag{15.14}$$

which is different from the previous answers $I = 2$ or $I = 1 + 2i$ (by naive integration). It seems that the integral value depends on its integral path or contour. In fact, a specific path has to be specified for complex integration.

Given a function $f(z)$ that is continuous on a piecewise smooth curve Γ, then the integral over Γ is $\int_\Gamma f(z)dz$, called a contour or line integral of $f(z)$. This integral has similar properties to the real integral

$$\int_\Gamma [\alpha f(z) + \beta g(z)]dz = \alpha\int_\Gamma f(z)dz + \beta\int_\Gamma g(z)dz, \tag{15.15}$$

where α and β are two non-zero constants.

15.2.1 Cauchy's Integral Theorem

We say a path is simply closed if its end points and initial points coincide and the curve does not cross itself. For an analytic function $f(z) = u(x, y) + iv(x, y)$, the integral on a simply closed path

$$I = \int_\Gamma (u + iv)(dx + idy)] = \int_\Gamma (udx - vdy) + i\int_\Gamma (vdx + udy). \tag{15.16}$$

However, the formal proof and description of Cauchy's integral theorem requires Green's theorem which states as follows. For a vector field $W(x, y) = P(x, y)dx + Q(x, y)dy$ that is continuously differentiable in a simply-connected domain Ω in the z-plane, a line integral of W along a closed path Γ satisfies

$$\oint_\Gamma [P(x, y)dx + Q(x, y)dy] = \iint_\Omega \Big(\frac{\partial Q}{\partial x} - \frac{\partial P}{\partial y}\Big)dxdy, \tag{15.17}$$

which essentially links a surface integral with a line integral.

By using the Green theorem, this becomes

$$I = \iint_\Omega (-\frac{\partial u}{\partial y} - \frac{\partial v}{\partial x})dxdy + i\iint_\Omega (\frac{\partial u}{\partial x} - \frac{\partial v}{\partial y})dxdy. \tag{15.18}$$

From the Cauchy-Riemann equations, we know that both integrals are zero. Thus, we have Cauchy's Integral Theorem, which states that the integral of any analytic function $f(z)$ on a simply closed path Γ in a simply connected domain Ω is zero. That is

$$\int_\Gamma f(z)dz = 0.$$

It is worth pointing out that the loop (or closed path) must be a positively oriented loop. Here, a positive oriented loop is a simple closed curve or path such that the curve interior is to the left when traveling along the curve. In most cases in practice, we can achieve this by simply integrating anticlockwise along a chosen loop.

This theorem is very important as it has interesting consequences. If the closed path is decomposed into two paths with reverse directions Γ_1 and Γ_2, then Γ_1 and $-\Gamma_2$ form a closed path, which leads to

$$\int_{\Gamma_1} f(z)dz = \int_{\Gamma_2} f(z)dz. \tag{15.19}$$

That is to say that the integral over any curve between two points is independent of the path. This property becomes very useful for evaluation of integrals.

Let us introduce Cauchy's integral formula for complex integration without proof. The integral of $f(z)/(z - z_0)$ over any simply closed path Γ enclosing a point z_0 in the domain Ω is

$$\frac{1}{2\pi i}\oint_\Gamma \frac{f(z)}{z - z_0}dz = f(z_0), \tag{15.20}$$

which is the basic form of the Cauchy integral formula. As we can see that the integrals here are along a closed contour, such integrals are also called contour integrals. Let us look at a simple example.

Example 15.4

To calculate the contour integral of

$$I = \oint_\Gamma \frac{1}{z}dz,$$

along a loop Γ defined by a circle $x^2 + y^2 = 1$, we know that the integrand can be written as

$$\frac{1}{z} = \frac{f(z)}{z - z_0}, \quad f(z) = 1, \ z_0 = 0.$$

Since $f(z) = 1$ is a constant for any z_0, thus $f(z_0) = 1$. Thus, the above equation (15.20) becomes

$$\oint_{x^2+y^2=1} \frac{1}{z}dz = \oint_\Gamma \frac{f(z)}{z - z_0}dz = 2\pi i f(z_0) = 2\pi i.$$

An interesting fact is that the integral along any other loops that encloses the origin $z = 0$ will also give the same integral value. This means that we can choose any loop that makes the integration process easier.

Similar to the above equation, for higher derivatives, we have the generalized Cauchy integral formula:

$$\oint_\Gamma \frac{f(z)}{(z - z_0)^{n+1}}dz = \frac{2\pi i f^{(n)}(z_0)}{n!}, \quad n > 0. \tag{15.21}$$

Example 15.5

For the contour integral

$$I = \oint_{|z|=4} \frac{e^{3z}}{(z - 1)^2}dz,$$

we know that $z_0 = 1$ is within the circle $|z| = 4$ (or $x^2 + y^2 = 2^2$), so we can use the above generalized Cauchy integral formula with $f(z) = e^{3z}$ and $n = 1$ and we have

$$\oint_{|z|=4} \frac{e^{3z}}{(z-1)^2} dz = \frac{2\pi i}{1!} (e^{3z})' \Big|_{z_0=1}$$
$$= 2\pi i \cdot 3e^{3\times 1} = 6\pi i e^3.$$

15.2.2 Residue Theorem

A function $f(z)$ has a pole or singularity of order n at $z = z_0$ if $f(z)$ is not analytic at $z = z_0$ but $(z - z_0)^n f(z)$ is analytic at $z = z_0$. This suggests that $f(z)$ can be expanded as a power series, called Laurant series

$$f(z) = \sum_{k=-n}^{\infty} \alpha_k (z - z_0)^k, \tag{15.22}$$

where α_k are the coefficients. The most important coefficient is probably α_{-1} which is called the residue of $f(z)$ at the pole $z = z_0$. If $f(z)$ has a pole of order N at z_0, the following formula gives a quick way to calculate the residue

$$\text{Res} f(z)|_{z_0} = \frac{1}{(N-1)!} \lim_{z \to z_0} \frac{d^{N-1}[(z - z_0)^N f(z)]}{dz^{N-1}}. \tag{15.23}$$

For any analytic $f(z)$ function in a domain Ω except for isolated singularities at finite points $z_1, z_2, ..., z_N$, the residue theorem states

$$\oint_\Gamma f(z) dz = 2\pi i \sum_{k=1}^{N} \text{Res} f(z)|_{z_k}$$
$$= 2\pi i \Big[\text{Res} f(z_1) + \text{Res} f(z_2) + ... + \text{Res} f(z_N) \Big] \tag{15.24}$$

where Γ is a simple closed path enclosing all these isolated points or poles.

Example 15.6

Let us calculate the contour integral on a unit circle

$$I = \oint_{|z|=1} \frac{e^{2z}}{z^5} dz.$$

Since $n = 4$, $f(z) = e^{2z}$ and $z_0 = 0$, we have

$$I = \oint_{x^2+y^2=1} \frac{e^{2z}}{z^5} dz = \frac{2\pi i}{4!} f''''(z_0) = \frac{2\pi i}{24} \cdot 2^4 e^{2z_0} = \frac{4\pi i}{3} e^0 = \frac{4\pi i}{3}.$$

The residue theorem serves as a powerful tool for calculating some real integrals and summation of series, especially when the integrand is a function of sin and cos that can be changed into complex integrals.

The real integral $\int_{-\infty}^{\infty} \psi(x) dx$ becomes $2\pi i$ multiplying the sum of the residues of $\psi(x)$ at the poles in the upper half-space. Let us look at a more complicated example in detail.

Example 15.7

In order to evaluate the integral

$$I(\alpha) = \int_{-\infty}^{\infty} \frac{e^{i\alpha^2 \zeta}}{\zeta^2 + 1} d\zeta,$$

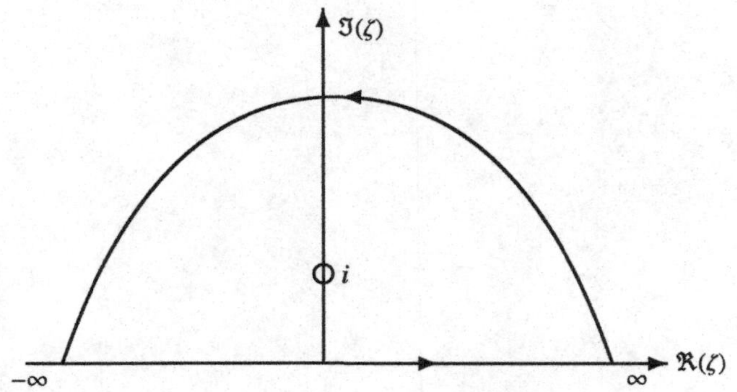

FIGURE 15.1 The contour for the integral $I(\alpha)$.

it is necessary to construct a contour (see Fig. 15.1). As the function

$$\phi = e^{i\alpha^2\zeta}/(1+\zeta^2)$$

has two poles $\zeta = +i$ and $-i$ from $1+\zeta^2=0$, and only one pole $\zeta = +i$ is in the upper half plane, we can construct a contour to encircle the pole at $\zeta = i$ by adding an additional arc at the infinity ($\zeta \to \infty$) on the upper half plane. Combining the arc with the horizontal line from the integral limits from $-\infty$ to ∞ along the ζ-axis, a contour is closed. Hence, we have

$$\phi = \frac{e^{i\alpha^2\zeta}/(\zeta+i)}{\zeta - i} = \frac{f(\zeta)}{\zeta - i},$$

where $f(\zeta) = e^{i\alpha^2\zeta}/(\zeta+i)$. Using the residue theorem, we have

$$I = 2\pi i[f(\zeta = i)] = 2\pi i \frac{e^{-\alpha^2}}{i+i} = \pi e^{-\alpha^2}.$$

In a special case when $\alpha = 0$, we have

$$\int_{-\infty}^{\infty} \frac{1}{\zeta^2+1} d\zeta = \pi.$$

Complex integration and complex analysis in general are indeed a very diverse topic with a wide range of sophisticated techniques. What we have introduced here is just the tip of the iceberg, and interested readers can refer to more specialized literature.

EXERCISES

15.1. For $z = re^{i\theta}$, show that

$$\frac{dz}{z} = \frac{dr}{r} + i d\theta.$$

15.2. Compute $\ln(1+i)$ and $\ln i$.

15.3. Integrate $\oint_C dz/z$ along the half a unit circle from $z = -1$ to $z = +1$. Show that the result depends on the path. Which half?

15.4. Calculate the following contour integral

$$I = \oint_D \frac{e^z}{(z-\pi)^3} dz,$$

where D is the a square path with four corners at $(\pm 10, \pm 10)$.

Part V

Fourier and Laplace Transforms

Chapter 16

Fourier Series and Transform

Chapter Points

- Functions, including non-smooth functions, can be represented by Fourier series. Examples of triangular waves and sawtooth waves are analyzed.
- Fourier transforms are introduced, together with their discrete forms and the fast Fourier transforms.
- The techniques of solving ordinary differential equations using Fourier transforms and inverse Fourier transforms are also explained briefly with two examples.

Mathematical transformation is a method of changing one kind of function and equation into another, often simpler and easier to solve. The Fourier transform maps a function in the time domain such as a signal into another function in the frequency domain, which is commonly used in signal and image processing.

16.1 FOURIER SERIES

From earlier discussions, we know that function e^x can be expanded into a series in terms of a polynomial with terms of x, x^2, ..., x^n. In this case, we are in fact trying to expand it in terms of the basis functions 1, x, x^2, ..., and x^n. There are many other basis functions. For example, the basis functions $\sin(n\pi t)$ and $\cos(n\pi t)$ are more widely used in signal processing. In general, this is essentially about the Fourier series.

16.1.1 Fourier Series

French mathematician Joseph Fourier at the turn of the 1800s was tackling a heat transfer problem and he approximated his functions as a series of trigonometrical functions.

For a function $f(t)$ on an interval $t \in [-T, T]$ where $T > 0$ is a finite constant or half period, the Fourier series is defined as

$$\begin{aligned} f(t) &= \frac{a_0}{2} + a_1 \cos(\frac{\pi t}{T}) + b_1 \sin(\frac{\pi t}{T}) + a_2 \cos(\frac{2\pi t}{T}) + b_2 \sin(\frac{2\pi t}{T}) + \dots \\ &= \frac{a_0}{2} + \sum_{n=1}^{\infty} \Big[a_n \cos(\frac{n\pi t}{T}) + b_n \sin(\frac{n\pi t}{T}) \Big], \end{aligned} \tag{16.1}$$

where

$$a_0 = \frac{1}{T} \int_{-T}^{T} f(t) dt, \qquad a_n = \frac{1}{T} \int_{-T}^{T} f(t) \cos(\frac{n\pi t}{T}) dt, \tag{16.2}$$

and

$$b_n = \frac{1}{T} \int_{-T}^{T} f(t) \sin(\frac{n\pi t}{T}) dt, \quad (n = 1, 2, \dots). \tag{16.3}$$

Here a_n and b_n are the Fourier coefficients of $f(t)$ on $[-T, T]$. It is worth pointing out that an implicit assumption is that $f(t)$ is periodic

$$f(t + 2T) = f(t), \tag{16.4}$$

which can also be seen from the fact that $\cos(n\pi t/T)$ and $\sin(n\pi t/T)$ are also periodic with a period of $2T$.

The function $f(t)$ can be continuous or piecewise continuous with a finite number of jump discontinuities. For a jump discontinuity at $t = t_0$, if both the left value $f'(t_0-)$ and the right value $f'(t_0+)$ exist with $f(t_0-) \neq f(t_0+)$, then the Fourier series converges to their average value. That is

$$f(t_0) = \frac{1}{2}[f(t_0-) + f(t_0+)]. \tag{16.5}$$

In some textbooks, a special case of $T = \pi$ is often used. In this case, the formula becomes simpler and we have

$$f(t) = \frac{a_0}{2} + \sum_{n=1}^{\infty}(a_n \cos nt + b_n \sin nt), \quad (-\pi \leq t \leq \pi), \tag{16.6}$$

with the Fourier coefficients

$$a_0 = \frac{1}{\pi}\int_{-\pi}^{\pi} f(t)dt, \;\; a_n = \frac{1}{\pi}\int_{-\pi}^{\pi} f(t)\cos nt dt, \;\; b_n = \frac{1}{\pi}\int_{-\pi}^{\pi} f(t)\sin nt dt, \tag{16.7}$$

for all positive integers n ($n = 1, 2, ...$).

Example 16.1

The well-known square wave is defined by

$$f(t) = \begin{cases} 1 & \text{if } -\pi < t \leq 0, \\ 0 & \text{if } 0 < t \leq \pi. \end{cases}$$

It has a period of 2π and $f(t + 2\pi) = f(t)$.

It is easy to check that

$$a_0 = \frac{1}{\pi}\int_{-\pi}^{\pi} f(t)dt = \frac{1}{\pi}\int_{-\pi}^{0} 1dt + \frac{1}{\pi}\int_{0}^{\pi} 0dt = \frac{1}{\pi}[0 - (-\pi)] = 1.$$

For $a_n (n \geq 1)$, the coefficient can be obtained by direct integration. We have

$$\begin{aligned} a_n &= \frac{1}{2}\int_{-\pi}^{\pi} f(t)\cos(nt)dt = \frac{1}{\pi}\int_{-\pi}^{0} 1\cos(nt)dt + \frac{1}{\pi}\int_{0}^{\pi} 0\cos(nt)dt \\ &= \frac{1}{\pi}\Big(\frac{\sin nt}{n}\Big)\Big|_{-\pi}^{0} = \frac{1}{n\pi}[\sin 0 - \sin(-n\pi)] = 0, \end{aligned}$$

where we have used $\sin(n\pi) = 0$ when n is an integer.

Similarly, we have

$$\begin{aligned} b_n &= \frac{1}{\pi} f(t)\sin(nt)dt = \frac{1}{\pi}\int_{-\pi}^{0} 1\sin(nt)dt + \frac{1}{\pi}\int_{0}^{\pi} 0\sin(nt)dt = \frac{1}{\pi}\int_{-\pi}^{0} \sin(nt)dt \\ &= \frac{1}{\pi}\Big(-\frac{\cos nt}{n}\Big)\Big|_{-\pi}^{0} = -\frac{1}{n\pi}[\cos 0 - \cos(-n\pi)] \\ &= \frac{-1}{n\pi}[1 - (-1)^n] = \begin{cases} -\frac{2}{n\pi} & \text{for odd } n, \\ 0 & \text{for even } n, \end{cases} \end{aligned}$$

where we have used $\cos(n\pi) = (-1)^n$. To avoid stating that n is odd or even, we can set $n = 2k - 1$ and thus we have

$$b_n \to b_k = -\frac{(-1)^k 2}{(2k-1)\pi}, \quad (k = 1, 2, 3, ...).$$

From the above results, we can see that apart from $a_0 = 1$, all $a_n (n = 1, 2, ...)$ are zero. Therefore, the Fourier series of the square wave becomes

$$f(t) = \frac{1}{2} - \sum_{k=1}^{\infty} \frac{(-1)^k 2}{(2k-1)\pi} \sin((2k-1)t)$$

$$= \frac{1}{2} - \frac{2}{\pi}\Big[\sin t + \frac{\sin 3t}{3} + \frac{\sin 5t}{5} + \frac{\sin 7t}{7} + \frac{\sin 9t}{9} + ...\Big].$$

It seems that the calculations of the Fourier coefficients are quite tedious, but the good news is that for a given function we only need to do it once and the integrals involved are usually easy to obtain.

The Fourier series theory usually guarantees that such expansions are valid. However, it is also sometimes useful to check when appropriate. For example, we know that the above square wave has a discontinuity at $t = 0$ because $f(0-) = 1$ and $f(0+) = 0$. According to (16.5), the series should converge to $(1 + 0)/2 = 1/2$. Let us see if it is the case.

Example 16.2

From the Fourier series obtained in the previous example, we can set $t = 0$ and we have

$$\frac{1}{2} - \frac{2}{\pi}\Big[\sin t + \frac{\sin 3t}{3} + \frac{\sin 5t}{5} + ...\Big] = \frac{1}{2} - \frac{2}{\pi}\Big[\sin 0 + \frac{\sin 0}{3} + \frac{\sin 0}{5} + ...\Big]$$
$$= \frac{1}{2} - \frac{2}{\pi}[0 + 0 + 0 + ...] = \frac{1}{2},$$

where we have used $\sin 0 = 0$. Thus, the series indeed converges to $1/2$ at $t = 0$.

16.1.2 Orthogonality

You may wonder how to derive the formulas for the coefficient a_n and b_n? Before we proceed, let us prove the orthogonality relation

$$J = \int_{-T}^{T} \sin(\frac{n\pi t}{T})\sin(\frac{m\pi t}{T})dt = \begin{cases} 0 & (n \neq m) \\ T & (n = m) \end{cases}, \tag{16.8}$$

where n and m are integers.

From the trigonometrical functions, we know that

$$\cos(A + B) = \cos A \cos B - \sin A \sin B, \tag{16.9}$$

and

$$\cos(A - B) = \cos A \cos B + \sin A \sin B. \tag{16.10}$$

By subtracting, we have

$$\cos(A - B) - \cos(A + B) = 2 \sin A \sin B. \tag{16.11}$$

Now the orthogonality integral becomes

$$J = \int_{-T}^{T} \sin(\frac{n\pi t}{T})\sin(\frac{m\pi t}{T})dt$$
$$= \frac{1}{2}\int_{-T}^{T}\Big\{\cos[\frac{(n-m)\pi t}{T}] - \cos[\frac{(n+m)\pi t}{T}]\Big\}dt. \tag{16.12}$$

If $n \neq m$, we have

$$J = \frac{1}{2}\Big\{\frac{T}{(n-m)\pi}\sin[\frac{(n-m)\pi t}{T}]\Big|_{-T}^{T} - \frac{T}{(n+m)\pi}\sin[\frac{(n+m)\pi t}{T}]\Big|_{-T}^{T}\Big\}$$
$$= \frac{1}{2}[\frac{T}{(n-m)\pi} \times (0 - 0) - \frac{T}{(n+m)\pi} \times (0 - 0)] = 0. \tag{16.13}$$

If $n = m$, we have

$$J = \frac{1}{2}\int_{-T}^{T}\Big\{1 - \cos[\frac{2n\pi t}{T}]\Big\}dt$$
$$= \frac{1}{2}\Big\{t\Big|_{-T}^{T} - \frac{T}{2n\pi}\sin[\frac{2n\pi t}{T}]\Big|_{-T}^{T}\Big\} = \frac{1}{2}[2T - \frac{T}{2n\pi} \times 0] = T, \tag{16.14}$$

which proves the relationship (16.8).

Using similar calculations, we can easily prove the following orthogonality relations

$$\int_{-T}^{T} \cos(\frac{n\pi t}{T})\cos(\frac{m\pi t}{T})dt = \begin{cases} 0 & (n \neq m) \\ T & (n = m) \end{cases}, \tag{16.15}$$

and

$$\int_{-T}^{T} \sin(\frac{n\pi t}{T})\cos(\frac{m\pi t}{T})dt = 0, \qquad \text{for all } n \text{ and } m. \tag{16.16}$$

16.1.3 Determining the Coefficients

Now we can try to derive the expressions for coefficients a_n. Multiplying both sides of the Fourier series (16.1) by $\cos(m\pi t/T)$ and taking the integration from $-T$ to T, we have

$$\int_{-T}^{T} f(t)\cos(\frac{m\pi t}{T})dt = \frac{a_0}{2}\int_{-T}^{T} \cos(\frac{m\pi t}{T})dt$$
$$+\sum_{n=1}^{\infty}\Big\{a_n \int_{-T}^{T} \cos(\frac{n\pi t}{T})\cos(\frac{m\pi t}{T})dt + b_n \int_{-T}^{T} \sin(\frac{n\pi t}{T})\cos(\frac{m\pi t}{T})dt\Big\}.$$

Using the relations (16.15) and (16.16) as well as $\int_{-T}^{T}\cos(m\pi t/T)dt = 0$, we know that the only non-zero integral on the right-hand side is when $n = m$. Therefore, we get

$$\int_{-T}^{T} f(t)\cos(\frac{n\pi t}{T})dt = 0 + [a_n T + b_n \times 0], \tag{16.17}$$

which gives

$$a_n = \frac{1}{T}\int_{-T}^{T} f(t)\cos(\frac{n\pi t}{T})dt, \tag{16.18}$$

where $n = 1, 2, 3, \cdots$. Interestingly, when $n = 0$, it is still valid and becomes a_0 as $\cos 0 = 1$. That is

$$a_0 = \frac{1}{T}\int_{-T}^{T} f(t)dt. \tag{16.19}$$

In fact, $a_0/2$ is the average of $f(t)$ over the period $2T$.

The coefficients b_n can be obtained by multiplying $\sin(m\pi t/T)$ and following similar calculations. We have

$$b_n = \frac{1}{T}\int_{-T}^{T} f(t)\sin(\frac{n\pi t}{T})dt, \quad (n = 1, 2, 3, ...). \tag{16.20}$$

Fourier series in general tends to converge slowly. In order for a function $f(x)$ to be expanded properly, it must satisfy the following Dirichlet conditions:

- A piecewise function $f(x)$ must be periodic with at most a finite number of discontinuities,
- and/or a finite number of minima or maxima within one period.

In addition, the integral of $|f(x)|$ must converge. For example, these conditions suggest that $\ln(x)$ cannot be expanded into a Fourier series in the interval [0, 1] as $\int_0^1 |\ln x|dx$ diverges.

The nth term of the Fourier series,

$$a_n \cos(n\pi t/T) + b_n \sin(n\pi t/T),$$

is called the nth harmonic. The energy of the nth harmonic is defined by $A_n^2 = a_n^2 + b_n^2$, and the sequence of A_n^2 forms the energy or power spectrum of the Fourier series.

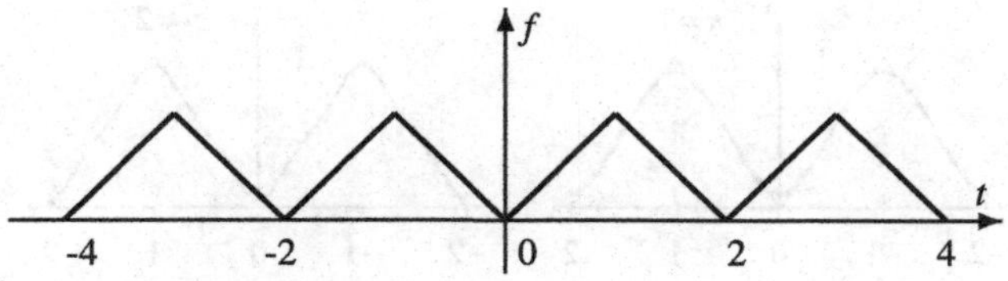

FIGURE 16.1 Triangular wave with a period of 2.

From the coefficient a_n and b_n, we can easily see that $b_n = 0$ for an even function $f(-t) = f(t)$ because $g(t) = f(t)\sin(n\pi t/T)$ is now an odd function $g(-t) = -g(t)$ due to the fact $\sin(2\pi t/T)$ is an odd function. We have

$$b_n = \frac{1}{T}\int_{-T}^{T} f(t)\sin(\frac{n\pi t}{T})dt = \frac{1}{T}\Big[\int_{-T}^{0} g(t)dt + \int_{0}^{T} g(t)dt\Big]$$
$$= \frac{1}{T}\Big[\int_{0}^{T} g(-t)dt + \int_{0}^{T} g(t)dt\Big] = \frac{1}{T}\int_{0}^{T}(-g(t)+g(t))dt = 0. \quad (16.21)$$

Similarly, we have $a_0 = a_n = 0$ for an odd function $f(-t) = -f(t)$. In both cases, only one side $[0, T]$ of the integration is used due to symmetry.

Thus, for even function $f(t)$, we have the Fourier cosine series on $[0, T]$

$$f(t) = \frac{a_0}{2} + \sum_{n=1}^{\infty} a_n \cos(\frac{n\pi t}{T}), \quad \text{if } f(-t) = f(t). \quad (16.22)$$

For odd function $f(t)$ (thus $f(-t) = -f(t)$), we have the sine series

$$f(t) = \sum_{n=1}^{\infty} \sin(\frac{n\pi t}{T}). \quad (16.23)$$

Example 16.3

The triangular wave is defined by $f(t) = |t|$ for $t \in [-1, 1]$ with a period of 2 or $f(t+2) = f(t)$ shown in Fig. 16.1. Using the coefficients of the Fourier series, we have

$$a_0 = \int_{-1}^{1} |t|dt = \int_{-1}^{0}(-t)dt + \int_{0}^{1} tdt = 1.$$

Since both $|t|$ and $\cos(n\pi t)$ are even functions, we have for any $n \geq 1$,

$$a_n = \int_{-1}^{1} |t|\cos(n\pi t)dt = 2\int_{0}^{1} t\cos(n\pi t)dt$$
$$= 2\frac{t}{n\pi}\sin(n\pi t)\Big|_0^1 - \frac{2}{n\pi}\int_0^1 \sin(n\pi t)dt = \frac{2}{n^2\pi^2}[\cos(n\pi) - 1].$$

Because $|t|\sin(n\pi t)$ is an odd function, we have

$$b_n = \int_{-1}^{1} |t|\sin(n\pi t)dt = 0.$$

Hence, the Fourier series for the triangular wave can be written as

$$f(t) = \frac{1}{2} + 2\sum_{n=1}^{\infty} \frac{\cos(n\pi) - 1}{n^2\pi^2}\cos(n\pi t) = \frac{1}{2} + \frac{4}{\pi^2}\sum_{n=1,3,5,...}^{\infty} \frac{(-1)^n}{n^2}\cos(n\pi t),$$

where we have used $\cos(n\pi) = (-1)^n$.

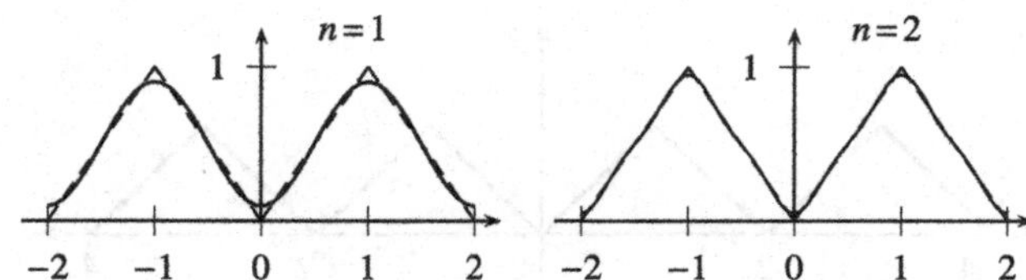

FIGURE 16.2 Fourier series (triangular wave $f(t) = |t|, t \in [-1, 1]$). (Left) first two terms ($n = 1$); (right) first three terms ($n = 2$).

The first few terms, $f_n(t) = 1/2 + 4/\pi^2 \cos(\pi t)$, are shown in Fig. 16.2 where we can see that only a few terms are needed to produce a very good approximation.

Here we can see that the triangular wave with derivative discontinuity can be approximated well by two or three terms. This makes it easy for any mathematical analysis. Fourier series are widely applied in signal and image processing as well as telecommunications.

Sometimes, interesting results can be obtained from the infinite Fourier series when certain appropriate values are used. Let us look at an example.

Example 16.4

From the triangular wave, we know that

$$f(t) = \frac{1}{2} + \frac{4}{\pi^2} \sum_{n=1,3,5,...}^{\infty} \cos(n\pi t) = \frac{1}{2} - \frac{4}{\pi^2}\Big[\cos(\pi t) + \frac{\cos(3\pi t)}{3^2} + \frac{\cos(5\pi t)}{5^2} + ...\Big].$$

For $t = 0$, we know that $f(0) = 0$ and $\cos(n\pi \times 0) = 1$, and we have

$$0 = f(0) = \frac{1}{2} - \frac{4}{\pi^2}\Big[\cos 0 + \frac{\cos 0}{3^2} + \frac{\cos 0}{5^2} + \frac{\cos 0}{7^2} + ...\Big] = \frac{1}{2} - \frac{4}{\pi^2}[1 + \frac{1}{3^2} + \frac{1}{5^2} + \frac{1}{7^2} + ...],$$

which gives

$$1 + \frac{1}{3^2} + \frac{1}{5^2} + \frac{1}{7^2} + ... + \frac{1}{(2n-1)^2} + ... = \frac{\pi^2}{8}.$$

Usually, most Fourier series can converge reasonably quickly and a few terms can often give good approximations. However, care must be taken when computing Fourier series, especially in the case when the so-called Gibbs phenomenon may occur. Loosely speaking, the Gibbs phenomenon refers to the unusual oscillations of the partial sum near the discontinuity or jump of a piecewise periodic function.

Before we proceed, let us look at an example of the sawtooth waves.

Example 16.5

A sawtooth signal is a periodic function with a period of $2a$ where $a > 0$, and it has the following form:

$$f(t) = \begin{cases} t/a & -a \le t < a, \\ 0 & \text{otherwise.} \end{cases}$$

So we have

$$a_n = \frac{1}{a}\int_{-a}^{a} f(t)\cos(\frac{n\pi t}{a})dt = 0, \quad n \ge 0,$$

because the integrand $t\cos(n\pi t/a)$ is an odd function and thus its integral in the symmetrical domain is zero. For b_n, we have

$$b_n = \frac{1}{a}\int_{-a}^{a} \frac{t}{a}\sin(\frac{n\pi t}{a})dt = \frac{2}{a^2}\int_0^a t\sin(\frac{n\pi t}{a})dt$$

FIGURE 16.3 The square wave and its Fourier series. The overshoot is called the Gibbs phenomenon.

$$= -\frac{2}{\pi n}\cos(n\pi) + \frac{2}{\pi^2 n^2}\sin(n\pi) = \frac{2(-1)^{n+1}}{\pi n}, \quad n \geq 1.$$

Here, we have used $\sin(n\pi) = 0$ and $\cos(n\pi) = (-1)^n$ as well as the fact that the integrand is an even function. So the Fourier series of the sawtooth wave is

$$f(t) = \frac{2}{\pi}\sum_{n=1}^{\infty}\frac{(-1)^{n+1}}{n}\sin(\frac{n\pi t}{a}).$$

In the case of $a = \pi$, we have

$$f(t) = \frac{2}{\pi}\sum_{n=1}^{\infty}\frac{(-1)^{n+1}}{n}\sin(nt) = \frac{2}{\pi}[\sin(t) - \frac{1}{2}\sin(2t) + \frac{1}{3}\sin(3t) - \frac{1}{4}\sin(4t) + ...].$$

It seems that the larger the number of terms, the better approximation to the sawtooth signal. However, unlike the previous example on the triangular wave, the approximation is always a few percent overshoot at some points. In fact, no matter how many terms we may use (10, 100 or 1000), there is always an overshoot at the discontinuity at $t = \pi$. This overshoot is known as the Gibbs phenomenon.

Such overshoot does not just occur for the sawtooth wave, and it also occurs in many other cases. For example, it leaves as an exercise to show that a square wave with the function form

$$s(t) = \begin{cases} +1 & 0 \leq t \leq \pi, \\ -1 & -\pi \leq t < 0, \end{cases}$$

has a period of 2π and the following Fourier series

$$s(t) = \frac{4}{\pi}\sum_{n=1,3,5,...}^{\infty}\frac{\sin(nt)}{n} = \frac{4}{\pi}\sum_{k=1}^{\infty}\frac{\sin[(2k-1)t]}{(2k-1)}$$

$$= \frac{4}{\pi}\Big[\sin(t) + \frac{1}{3}\sin(3t) + \frac{1}{5}\sin(5t) + \frac{1}{7}\sin(7t) + ...\Big].$$

Gibbs phenomenon was discovered by J. W. Gibbs in 1899. For the square wave, rigorous mathematical analysis (that is beyond the scope of this book) shows the overshoot is

$$\frac{1}{\pi}\int_0^{\pi}\frac{sin(u)}{u}du - \frac{1}{2} = \frac{1}{\pi}\int_0^{\pi}\text{sinc}(u)du - \frac{1}{2} \approx 0.089490,$$

which is about 9% as shown in Fig. 16.3.

16.2 FOURIER TRANSFORMS

In general, when the period T becomes infinite, the Fourier coefficients of a function defined on the whole real axis $(-\infty, \infty)$ can be written as

$$a(\omega_n) = \int_{-T}^{T} f(t)\cos(\omega_n t)dt, \qquad b(\omega_n) = \int_{-T}^{T} f(t)\sin(\omega_n t)dt, \tag{16.24}$$

where $\omega_n = \frac{n\pi}{T}$ under the limits of $T \to \infty$ and $\omega_n \to 0$. If we further pose the constraint $\int_{\infty}^{\infty} |f(t)| < \infty$, we get $a_0 \to 0$. In this case, the Fourier series becomes the Fourier integral

$$f(t) = \int_0^{\infty} [a(\omega)\cos(\omega t) + b(\omega)\sin(\omega t)]d\omega, \tag{16.25}$$

where

$$a(\omega) = \frac{1}{\pi}\int_{-\infty}^{\infty} f(t)\cos(\omega t)dt, \; b(\omega) = \frac{1}{\pi}\int_{-\infty}^{\infty} f(t)\sin(\omega t)dt. \tag{16.26}$$

Following similar discussions above, even functions lead to Fourier cosine integrals and odd functions lead to Fourier sine integrals.

The Fourier transform $\mathcal{F}[f(t)]$ of $f(t)$ is defined as

$$F(\omega) = \mathcal{F}[f(t)] = \int_{-\infty}^{\infty} f(t)e^{-i\omega t}dt, \tag{16.27}$$

and the inverse Fourier transform can be written as

$$f(t) = \mathcal{F}^{-1}[F(\omega)] = \frac{1}{2\pi}\int_{-\infty}^{\infty} F(\omega)e^{i\omega t}d\omega, \tag{16.28}$$

where we have used

$$\exp[i\omega t] = \cos(\omega t) + i\sin(\omega t). \tag{16.29}$$

Here, we use t as the independent variable and $f(t)$ is a function of t. However, Fourier transforms can also be applied to functions of spatial variable x and other variables. Though there are difference between x and t in terms of physical variables, mathematically they are equivalent as long as their function forms and properties are the same. In addition, it is not required that the function $f(t)$ must be periodic for Fourier transforms. Let us look at an example.

Example 16.6

For a so-called top-hat function that is constant 1 in the range of $[-a, a]$ and zero otherwise, we have

$$f(x) = \begin{cases} 1 & x \in [-a, a], \\ 0 & |x| \geq a, \end{cases}$$

where $a > 0$ is a fixed constant. Its Fourier transform is

$$F(\omega) = \int_{-\infty}^{\infty} f(x)e^{-i\omega x}dx.$$

Since $f(x)$ is non-zero in $[-a, a]$, the integration limits become $-a$ to a. That is

$$F(\omega) = \int_{-a}^{a} 1 \times e^{-i\omega x}dx$$
$$= \frac{1}{-i\omega}[e^{-i\omega a} - e^{-i\omega(-a)}] = \frac{1}{i\omega}[e^{i\omega a} - e^{-i\omega a}].$$

Using

$$e^{i\omega a} = \cos(\omega a) + i\sin(\omega a), \quad e^{-i\omega a} = \cos(\omega a) - i\sin(\omega a),$$

we have

$$e^{i\omega a} - e^{-i\omega a} = i\sin(\omega a) - (-i\sin(\omega a)) = 2i\sin(\omega a).$$

Therefore, the above Fourier transform becomes

$$F(\omega)=\frac{1}{i\omega}[2i\sin(\omega a)]=2a\left[\frac{\sin(\omega a)}{(\omega a)}\right]=2a\,\mathrm{sinc}(\omega a),$$

where $\mathrm{sinc}(x)=\sin(x)/x$ is the 'sinc' function.

Fourier transforms can have important properties such as linearity, shift and scaling. We will state the properties without any formal proof. Interested readers can refer to more advanced textbooks.

- Linearity: $\mathcal{F}[\alpha f(t)+\beta g(t)]=\alpha\mathcal{F}[f(t)]+\beta\mathcal{F}[g(t)]$.
- Scaling: $\mathcal{F}[f(\alpha t)]=\frac{1}{\alpha}F(\omega/\alpha)$.
- Shift: $\mathcal{F}[f(t-a)]=e^{-i\omega a}\mathcal{F}[f(t)]$.
- Derivative: $\mathcal{F}[(-it)^n f(t)]=\frac{d^n F(\omega)}{d\omega^n}$ and $\mathcal{F}[f'(t)]=i\omega\mathcal{F}[f(t)]$.
- Convolution: The convolution of two functions $f(t)$ and $g(t)$ is defined by

$$f(t)*h(t)=\int_{-\infty}^{\infty}f(\tau)g(t-\tau)d\tau. \tag{16.30}$$

Their Fourier transforms obey

$$\mathcal{F}[f(t)*g(t)]=\mathcal{F}[f(t)]\mathcal{F}[g(t)]. \tag{16.31}$$

Some of the properties can be easy to prove by direct calculations.

Example 16.7

For example, by using $u=t-a$, we can show the shift property:

$$\mathcal{F}[f(t-a)]=\int_{-\infty}^{\infty}f(t-a)e^{-i\omega t}dt=\int_{-\infty}^{\infty}f(u)e^{-i\omega(u+a)}du$$
$$=e^{-i\omega a}\int_{-\infty}^{\infty}f(u)e^{-i\omega u}du=e^{-i\omega a}\mathcal{F}[f(t)]. \tag{16.32}$$

The convolution property can also be proved by direct integration. From

$$\mathcal{F}[f*g]=\int_{-\infty}^{\infty}\Big[\int_{-\infty}^{\infty}f(\tau)g(t-\tau)d\tau\Big]e^{-i\omega t}dt, \tag{16.33}$$

we have (with $u=t-\tau$ or $t=u+\tau$)

$$\mathcal{F}[f*g]=\int_{-\infty}^{\infty}e^{-i\omega t}dt\int_{-\infty}^{\infty}[f(\tau)g(t-\tau)]d\tau=\int_{-\infty}^{\infty}f(\tau)d\tau\int_{-\infty}^{\infty}g(u)e^{-i\omega(u+\tau)}du$$
$$=\int_{-\infty}^{\infty}f(\tau)e^{-i\omega\tau}d\tau\int_{-\infty}^{\infty}g(u)e^{-i\omega u}du=\mathcal{F}[f(t)]\mathcal{F}[g(t)]. \tag{16.34}$$

Example 16.8

For a Gaussian function $g(x)=e^{-x^2/2}$, its Fourier transform is given by

$$G(\omega)=\mathcal{F}[g(x)]=\int_{-\infty}^{\infty}e^{-x^2/2}e^{-i\omega x}dx,$$

which is a complex integral that is not easy to calculate. However, if we use Euler's formula, we have

$$G(\omega)=\int_{-\infty}^{\infty}e^{-x^2/2}[\cos(\omega x)-i\sin(\omega x)]dx=\int_{-\infty}^{\infty}e^{-x^2/2}\cos(\omega x)dx-i\int_{-\infty}^{\infty}e^{-x^2/2}\sin(\omega x)dx.$$

Here, the first integral is still not easy to calculate, though the second integral is zero because $e^{-x^2/2}\sin(\omega x)$ is an odd function.

TABLE 16.1 Fourier transforms.

$f(t)$	$F(\omega) = \mathcal{F}[f(t)]$
$f(t - t_0)$	$F(\omega)e^{-i\omega t_0}$
$f(t)e^{-i\omega_0 t}$	$F(\omega - \omega_0)$
$\delta(t)$	1
$\cos(\omega_0 t)$	$\pi[\delta(\omega - \omega_0) + \delta(\omega + \omega_0)]$
$\sin(\omega_0 t)$	$\frac{\pi}{i}[\delta(\omega - \omega_0) - \delta(\omega + \omega_0)]$
$e^{i\omega_0 t}$	$2\pi\delta(\omega - \omega_0)$
$e^{-a\|t\|}$	$\frac{2a}{a^2+\omega^2}$ $(a > 0)$
$u(t)e^{-at}$	$\frac{1}{a+i\omega}$ $(a > 0)$
$f'(t)$	$i\omega F(\omega)$
$f''(t)$	$(i\omega)^2 F(\omega)$
$tf(t)$	$iF'(\omega)$
$f(at)$	$\frac{1}{\|a\|}F(\frac{\omega}{a})$

If we use more advanced mathematical handbooks, we may find that

$$\int_{-\infty}^{\infty} e^{-x^2/2}\cos(\omega x)dx = \sqrt{2\pi}e^{-\omega^2/2}.$$

Then, we have

$$G(\omega) = \sqrt{2\pi}e^{-\omega^2/2}.$$

But this provides little insight about the procedure of obtaining the final results.

There are some variations of the transforms such as the Fourier sine transform and the Fourier cosine transform. The Fourier transforms of some common functions, also called Fourier transform pairs, are listed in Table 16.1.

The above example shows that Fourier transforms may involve highly nonlinear and complex integrals that can be very hard to calculate analytically. Even with some detailed Fourier transform tables, difficulty in computing such integrals may often occur. The good news is that we can use computer software and discrete Fourier transforms (to be introduced later in this chapter) to carry out sophisticated calculations.

16.3 SOLVING DIFFERENTIAL EQUATIONS USING FOURIER TRANSFORMS

Fourier transforms can also be used to solve differential equations and partial differential equations. The main idea is to transform the original linear differential equation about $y(t)$ (in the time domain) into their counterpart of Fourier transforms $Y(\omega)$ (in the frequency domain).

Under such a transform, the differential equation can usually become an algebraic equation, and we then solve it for $Y(\omega)$. Because the solution is in terms of $y(t)$, we have to use the inverse Fourier transform to convert $Y(\omega)$ in the frequency domain to $y(t)$ in the time domain.

Here, we only demonstrate this through an example.

Example 16.9

Let us solve a simple linear first-order ordinary differential equation

$$y'(t) + ay(t) = u(t), \quad a > 0,$$

where $u(t)$ is a unit step function or a Heaviside function

$$u(t) = \begin{cases} 1, & t \geq 0, \\ 0, & t < 0. \end{cases}$$

We can take the Fourier transforms of both sides and using the derivative properties

$$\mathcal{F}[y'(t)] = i\omega\mathcal{F}[y(t)] = i\omega Y(\omega),$$

we have

$$\mathcal{F}[y'(t) + ay(t)] = \mathcal{F}[u(t)],$$

or

$$\mathcal{F}[y'(t)] + a\mathcal{F}[y(t)] = \mathcal{F}[u(t)],$$

which becomes the following algebraic equation

$$[i\omega + a]Y(\omega) = 1,$$

where

$$Y(\omega) = \mathcal{F}[y(t)], \quad 1 = \int_{-\infty}^{\infty} u(t)e^{-i\omega t}\,dt.$$

Some straightforward algebraic operations give

$$Y(\omega) = \frac{1}{a + i\omega}.$$

From Table 16.1, we know that

$$\mathcal{F}^{-1}[Y(\omega)] = \mathcal{F}^{-1}[\frac{1}{(a+i\omega)}] \Longrightarrow u(t)e^{-at}.$$

So the final solution is

$$y(t) = u(t)e^{-at} = \begin{cases} e^{-at}, & t \geq 0, \\ 0, & 0. \end{cases}$$

You may wonder why such a simple example requires quite a few steps, compared with the simple ODE techniques such as direction integration or integration factors. The fact is that Fourier transforms can be used to solve ODEs; however, the full advantages may become more significant when applying to solve partial differential equations to be introduce later in this book.

On the other hand, discrete Fourier transforms, especially the Fast Fourier Transform (FFT), can be evaluated very efficiently in practice. In the rest of this chapter, we will introduce such transforms.

16.4 DISCRETE AND FAST FOURIER TRANSFORMS

Now we try to write the Fourier series (16.1) in a complex form using

$$\cos\theta = (e^{i\theta} + e^{-i\theta})/2$$

and

$$\sin\theta = (e^{i\theta} - e^{-i\theta})/2i,$$

we have the nth term

$$f_n(t) = a_n\cos(\frac{n\pi t}{T}) + b_n\sin(\frac{n\pi t}{T}) = \frac{a_n[e^{\frac{in\pi t}{T}} + e^{\frac{-in\pi t}{T}}]}{2} + \frac{b_n[e^{\frac{in\pi t}{T}} - e^{\frac{-in\pi t}{T}}]}{2i}$$
$$= \frac{(a_n - ib_n)}{2}e^{in\pi t/T} + \frac{(a_n + ib_n)}{2}e^{-in\pi t/T} \tag{16.35}$$

If we define $\beta_n = \frac{(a_n - ib_n)}{2}$, and $\beta_{-n} = \frac{(a_n + ib_n)}{2}$ where $(n = 0, 1, 2, ...)$, and set $\beta_0 = a_0/2$, we get

$$f_n(t) = \beta_n e^{-in\pi t/T} + \beta_{-n} e^{-in\pi t/T}.$$

Therefore, the Fourier series can be written in the complex form

$$f(t) = \sum_{n=-\infty}^{\infty} \beta_n e^{i\pi nt/T}. \tag{16.36}$$

In signal processing, we are often interested in $f(t)$ in $[0, 2T]$ (rather than $[-T, T]$). Without loss of generality, we can set $T = \pi$. In this case, the Fourier coefficients become

$$\beta_n = \frac{1}{2\pi} \int_0^{2\pi} f(t) e^{-int} dt. \tag{16.37}$$

As it is not easy to compute these coefficients accurately, we often use the numerical integration to approximate the above integral with a step size $h = 2\pi/N$. This is equivalent to sampling t with N sample points $t_k = 2\pi k/N$ where $k = 0, 1, ..., N-1$. Therefore, the coefficient can be estimated by

$$\beta_n = \frac{1}{2\pi} \int_0^{2\pi} f(t) e^{-int} dt \approx \frac{1}{N} \sum_{k=0}^{N-1} f(\frac{2\pi k}{N}) e^{-2\pi i n k/N}.$$

Once we know β_n, we know the whole spectrum. Let f_k denote $f(2\pi k/N)$, we can define the discrete Fourier transform (DFT) as

$$F_n = \sum_{k=0}^{N-1} f_k e^{-\frac{2\pi nk}{N}}, \tag{16.38}$$

which is for periodic discrete signals $f(k)$ with a period of N. A periodic signal $f(k+N) = f(k)$ has a periodic spectrum $F(n+N) = F(n)$. The discrete Fourier transform consists of N multiplications and $N-1$ additions for each F_n, thus the computational complexity is of $O(N^2)$.

In fact, by rearranging the formulas, we can get a class of the Fast Fourier Transform (FFT) whose computational complexity is about $O(N \log(N))$. Using the notation $\omega = e^{-2\pi i/N}$ and $\omega^N = e^{2\pi i} = 1$, we can rewrite (16.38) as

$$F_n = \sum_{k=0}^{N-1} f_k \omega^{kn}, \qquad -\infty < n < \infty. \tag{16.39}$$

It is worth pointing out that

$$\omega = e^{-2\pi i/N}$$

is the Nth root of unity in the complex plane, thus the powers of ω always lie on a unit circle in the complex plane. Here, the computations only involve the summation and the power of ω.

If we are more interested in the range of $n = 0, 1, ..., N-1$, then the above equation can be rewritten as a matrix form

$$\begin{pmatrix} F_0 \\ F_0 \\ \vdots \\ F_{N-1} \end{pmatrix} = \begin{pmatrix} 1 & 1 & 1 & \dots & 1 \\ 1 & \omega & \omega^2 & \dots & w^{N-1} \\ 1 & \omega^2 & \omega^4 & \dots & \omega^{2(N-1)} \\ 1 & \omega^{N-1} & \omega^{2(N-1)} & \dots & \omega^{(N-1)^2} \end{pmatrix} \begin{pmatrix} f_0 \\ f_0 \\ \vdots \\ f_{N-1} \end{pmatrix}.$$

As we know from matrix algebra, the original signal f_k can conveniently be reconstructed by the inverse of the matrix.

This usually requires a lot of computations; however, in the case when N can be factorized, some of the calculations can be decomposed into different steps and many of calculations become unnecessary. In this case, we often use $N = 2^m$ where m is a positive integer; it becomes the so-called FFT, and the computational complexity is now reduced to $2N \log_2(N)$.

For example, when $N = 2^{20}$, FFT will reduce the computational time from three weeks for full Fourier transforms to less than a minute for FFT on modern desktop computers. There is a huge amount of literature about FFT, filter design, signal reconstruction and their applications in signal and image processing.

EXERCISES

16.1. Another way of representing the square wave is $g(t) = 1$ for $0 \le t < \pi$ and $g(t) = 0$ for $\pi \le t < 0$, show that its Fourier series is

$$g(t) = \frac{1}{2} + \frac{2}{\pi}\Big[\sin t + \frac{\sin 3t}{3} + \frac{\sin 5t}{5} + \dots + \frac{\sin(2n-1)t}{(2n-1)} + \dots\Big].$$

16.2. Find a way (e.g., Fourier series) to show that

$$1 + \frac{1}{2^2} + \frac{1}{3^2} + \frac{1}{4^2} + \dots + \frac{1}{n^2} + \dots = \frac{\pi^2}{6}.$$

16.3. Show that the Fourier series of $f(t) = |\sin t|$ is

$$f(t) = \frac{2}{\pi}\Big[1 - 2\sum_{n=1}^{\infty} \frac{\cos(2nt)}{4n^2 - 1}\Big].$$

Discuss the Fourier series for $g(t) = Af(t)$ where A is a constant.

Chapter 17

Laplace Transforms

Chapter Points

- Laplace transforms are introduced, together with their scaling properties. Examples of some commonly used Laplace transform pairs are also given.
- The concept of transfer functions and their relevance to linear systems are presented with illustrative examples.
- The main procedure for solving differential equations using Laplace transforms is explained with worked examples.
- Other transforms such as Z-transforms are briefly introduced and their relationships to Laplace transforms are also discussed briefly.

17.1 LAPLACE TRANSFORM

The Laplace transform $\mathcal{L}[f(t)]$ of a function $f(t)$ is defined as

$$F(s) = \mathcal{L}[f(t)] = \int_0^{\infty} f(t)e^{-st}dt, \tag{17.1}$$

where $s > 0$ (or strictly $\Re(s) > 0$). The inverse Laplace transform of $F(s)$ is $f(t)$, and is denoted as $\mathcal{L}^{-1}[F(s)]$, or $f(t) = \mathcal{L}^{-1}[F(s)]$.

In theory, the inverse Laplace transform should be written as

$$f(t) = \frac{1}{2\pi i} \int_{\sigma - i\infty}^{\sigma + i\infty} F(s)e^{st}ds,$$

where σ should be large enough so that $F(s)$ is defined for $\Re(s) \geq \sigma$. However, this is not part of the current syllabus and thus is not the focus here.

In comparison with Fourier transforms, we can consider that the Laplace transform is an extension of Fourier transforms by setting $s = i\omega$, or more generally $s = \sigma + i\omega$, for a signal $f(t)$ with $t \geq 0$.

The Laplace transforms of most simple functions can be obtained by direct integration.

Example 17.1

For simple functions t and $e^{\alpha t}$, we have

$$\mathcal{L}[t] = \int_0^{\infty} te^{-st}dt = \int_0^{\infty} \frac{1}{s}e^{-st}dt + \Big[-\frac{t}{s}e^{-st}\Big]_0^{\infty} = \frac{1}{s^2},$$

and

$$\mathcal{L}[e^{\alpha t}] = \int_0^{\infty} e^{\alpha t}e^{-st}dt = \int_0^{\infty} e^{-(s-\alpha)t}dt = \Big[-\frac{1}{s-\alpha}e^{-(s-\alpha)t}\Big]_0^{\infty} = \frac{1}{s-\alpha}.$$

Conversely, we have

$$\mathcal{L}^{-1}[\frac{1}{s^2}] = t, \quad \mathcal{L}^{-1}[\frac{1}{s-\alpha}] = e^{\alpha t}. \tag{17.2}$$

For the Dirac δ-function, we have its Laplace transform

$$\mathcal{L}[\delta(t)] = \int_0^\infty \delta(t) e^{-st} dt = e^{-st}\Big|_{t=0}^{\infty} = 1. \tag{17.3}$$

Similarly, for a (unit) step function $f(t) = 1$ for $t \geq 0$, we have

$$\mathcal{L}(s) = \int_0^\infty e^{-st} dt = -\frac{1}{s} e^{-st}\Big|_0^\infty = \frac{1}{s}. \tag{17.4}$$

Here, we have implicitly assumed that $\exp(-st) \to 0$ as $t \to \infty$ and also implicitly assumed that $\Re(s) > 0$. That is,

$$\mathcal{L}(1) = \int_0^\infty 1 e^{-st} dt = \frac{1}{s}. \tag{17.5}$$

Example 17.2

For $f(t) = t^n$ where $n \geq 1$, we can use integration by parts by setting $u = t^n$ and $v' = e^{-st}$, and we have

$$\begin{aligned} \mathcal{L}(f(t)) &= \int_0^\infty t^n e^{-st} dt \\ &= -\frac{t^n e^{-st}}{s}\Big|_0^\infty + \frac{n}{s} \int_0^\infty t^{n-1} e^{-st} dt \\ &= \frac{n}{s} \mathcal{L}(t^{n-1}) = \ldots = \frac{n!}{s^{n+1}}, \end{aligned} \tag{17.6}$$

where we have assumed that $\Re(s) > 0$.

Sometime, it is necessary to use other formulas and their combinations to obtain the Laplace transform of some functions.

Example 17.3

In order to obtain the Laplace transform of $f(t) = \cos\omega t$, we shall first write

$$f(t) = \cos\omega t = \frac{1}{2}(e^{i\omega t} + e^{-i\omega t}).$$

Then, we have

$$\begin{aligned} \mathcal{L}[f(t)] = F(s) &= \int_0^\infty [\frac{1}{2}(e^{i\omega t} + e^{-i\omega t})] e^{-st} dt \\ &= \frac{1}{2}[\int_0^\infty e^{-(s-i\omega)t} dt + \int_0^\infty e^{-(s+i\omega)t} dt] \\ &= \frac{1}{2}\int_0^\infty e^{-u} \frac{du}{s-i\omega} + \frac{1}{2}\int_0^\infty e^{-u} \frac{du}{s+i\omega} \\ &= \frac{1}{2}\frac{1}{(s-i\omega)} \int_0^\infty e^{-u} du + \frac{1}{2}\frac{1}{(s+i\omega)} \int_0^\infty e^{-u} du \\ &= \frac{1}{2}[\frac{1}{s-i\omega} + \frac{1}{s+i\omega}] = \frac{s}{s^2+\omega^2}. \end{aligned}$$

Here we have used $u = (s - i\omega)t$ assuming s and ω are constants or parameters. We have also used the fact that $\int_0^\infty e^{-u} du = 1$.

Similarly, for the sine function, we have

$$\mathcal{L}[\sin(\omega t)] = \frac{\omega}{s^2+\omega^2}. \tag{17.7}$$

TABLE 17.1 Laplace transform pairs.

Function $f(t)$	Laplace transform $F(s)$
1	$\frac{1}{s}$
$\delta(t)$	1
$t^n, n > 0$	$\frac{n!}{s^{n+1}}$
$\cos(\alpha t)$	$\frac{s}{s^2+\alpha^2}$
$\sin(\alpha t)$	$\frac{\alpha}{s^2+\alpha^2}$
$e^{\alpha t}$	$\frac{1}{s-\alpha}$
$t^{1/2}$	$\frac{1}{2}(\frac{\pi}{s^3})^{1/2}$
$t^{-1/2}$	$\sqrt{\frac{\pi}{s}}$
$t^n f(t)$	$(-1)^n \frac{d^n F(s)}{ds^n}$
$\cos(\alpha t + \beta)$	$\frac{s\cos(\beta)-\alpha\sin(\beta)}{s^2+\alpha^2}$
$\sinh(\alpha t)$	$\frac{\alpha}{s^2-\alpha^2}$
$\cosh(\alpha t)$	$\frac{s}{s^2-\alpha^2}$
$\mathrm{erfc}(\frac{\alpha}{2\sqrt{t}})$	$\frac{1}{s}e^{-\alpha\sqrt{s}}$
$\frac{1}{\sqrt{\pi t}}e^{-\frac{\alpha^2}{4t}}$	$\frac{1}{\sqrt{s}}e^{-\alpha\sqrt{s}}$
$\sin\alpha\sqrt{t}$	$\frac{\alpha}{2}\sqrt{\frac{\pi}{s^3}}e^{-\frac{\alpha^2}{4s}}$
$\frac{1-e^{-\alpha t}}{t}$ $(\alpha > 0)$	$\ln(1+\frac{\alpha}{s})$
$\frac{1}{\alpha-\beta}(e^{\alpha t}-e^{\beta t})$ $(\alpha \neq \beta)$	$\frac{1}{(s-\alpha)(s-\beta)}$

17.1.1 Laplace Transform Pairs

The Laplace transform of a function can often be obtained by direct integration. However, the inverse Laplace transform is usually more complicated. It often involves the partial fractions of polynomials and usage of different rules of Laplace transforms.

The original function $f(t)$ and its Laplace transform $F(s)$ form a Laplace pair. The Laplace transform pairs of common functions are listed below in Table 17.1.

Both Fourier and Laplace transforms follow the convolution theorem. For two functions f and g, their convolution $f * g$ is defined by

$$f * g = \int_0^t f(t-\alpha)g(\alpha)d\alpha. \tag{17.8}$$

and their Laplace transforms follow

$$\mathcal{L}[f(t) * g(t)] = F(s)G(s), \tag{17.9}$$

$$\mathcal{L}^{-1}[F(s)G(s)] = \int_0^t f(t-\alpha)g(\alpha)d\alpha. \tag{17.10}$$

The Fourier transform has the similar properties

$$f(t) * g(t) = \int_{-\infty}^{\infty} f(t)g(t-u)du, \tag{17.11}$$

$$\mathcal{F}[f(t) * g(t)] = F(\omega)G(\omega). \tag{17.12}$$

17.1.2 Scalings and Properties

From the basic definition, it is straightforward to prove that the Laplace transform has the following properties:

$$\mathcal{L}[\alpha f(t) + \beta g(t)] = \alpha\mathcal{L}[f(t)] + \beta\mathcal{L}[g(t)], \tag{17.13}$$

$$\mathcal{L}[e^{\alpha t} f(t)] = F(s-\alpha), \;\; s > \alpha, \quad F(s) = \mathcal{L}[f(t)], \tag{17.14}$$

$$\mathcal{L}[f(t-\alpha)] = e^{-\alpha s}\mathcal{L}[f(t)], \tag{17.15}$$

$$\mathcal{L}[f'(t)] = s\mathcal{L}[f(t)] - f(0), \tag{17.16}$$

$$\mathcal{L}[\int_0^t f(\tau)d\tau] = \frac{1}{s}\mathcal{L}[f]. \tag{17.17}$$

These properties are easy to verify, and we will show how to derive some of these properties below. For example, it is straightforward to verify that

$$\mathcal{L}[e^{at} f(t)] = \int_0^\infty f(t)e^{at}e^{-st}dt = \int_0^\infty f(t)e^{-(s-a)t}dt = F(s-a). \tag{17.18}$$

Example 17.4

For a function $g(t) = f(at)$ where $a > 0$, we can set $\tau = at$, and thus we have $d\tau = adt$ (or $dt = d\tau/a$). We finally get

$$G(s) = \int_0^\infty f(at)e^{-st}dt = \frac{1}{a}\int_0^\infty f(\tau)e^{s\tau/a}d\tau = \frac{1}{a}F(\frac{s}{a}). \tag{17.19}$$

The properties of Laplace transforms can be used to simplify the calculations of Laplace transforms for certain complex expressions. Let us see an example.

Example 17.5

From $\mathcal{L}(\cos t) = \frac{s}{s^2+1}$ and using the above equation, we have

$$\mathcal{L}[e^{-t}\cos(t)] = \frac{s+1}{(s+1)^2+1} = \frac{s+1}{s^2+2s+2}. \tag{17.20}$$

For a more complicated expression such as

$$f(t) = e^{-t} + e^{-t}\cos t + \sinh(at) + 2\sqrt{t},$$

we can use the above linear properties and we have

$$\begin{aligned}\mathcal{L}[f(t)] &= \mathcal{L}[e^{-t}] + \mathcal{L}[e^{-t}\cos t] + \mathcal{L}[\sinh(at)] + 2\mathcal{L}[\sqrt{t}] \\ &= \frac{1}{s+1} + \frac{s+1}{(s+1)^2+1} + \frac{a}{s^2-a^2} + 2\frac{\sqrt{\pi}}{2s^{3/2}} \\ &= \frac{1}{s+1} + \frac{s+1}{s^2+2s+2} + \frac{a}{s^2-a^2} + \frac{\sqrt{\pi}}{s^{3/2}},\end{aligned}$$

where we have used the result of $\mathcal{L}[e^{-t}\cos t]$.

For a given time delay, $w(t) = f(t-T)$ where $t \geq T$, we have

$$\begin{aligned}W(s) &= \int_0^\infty w(t)e^{-st}dt = \int_T^\infty f(t-T)e^{-st}dt \\ &= \int_0^\infty f(\tau)e^{-s(\tau+T)}d\tau = e^{-sT}F(s).\end{aligned} \tag{17.21}$$

Example 17.6

Using the above properties, it is easy to obtain the following:

$$\mathcal{L}[(e^{3t}(t^2+1)] = \frac{2}{(s-3)^2} + \frac{1}{(s-3)}. \tag{17.22}$$

because

$$\mathcal{L}[(t^2+1)] = \frac{2}{s^3} + \frac{1}{s}, \quad \mathcal{L}[e^{3t}(t^2+1)] = \frac{2}{(s-3)^2} + \frac{1}{s-3}. \tag{17.23}$$

Similarly, we have

$$\mathcal{L}[e^{at}\cos\omega t] = \frac{(s-a)}{(s-a)^2+\omega^2}, \tag{17.24}$$

and

$$\mathcal{L}[t^3+2t+7] = \mathcal{L}(t^3) + 5\mathcal{L}(t) + 7\mathcal{L}(1) = \frac{3!}{s^4} + \frac{2}{s^2} + \frac{7}{s}. \tag{17.25}$$

17.1.3 Derivatives and Integrals

It is straightforward to check that

$$\begin{aligned}\mathcal{L}(f') &= \int_0^\infty f'(t)e^{-st}dt = e^{-st}f(t)\Big|_0^\infty - \int_0^\infty f(t)(-se^{-st})dt \\ &= \lim_{t\to\infty}[f(t)e^{-st}] - f(0) + s\int_0^\infty f(t)e^{-st}dt = sF(s) - f(0),\end{aligned} \tag{17.26}$$

where we have used two facts: s is independent of t and $e^{-st}f \to 0$ as $t \to \infty$.

Following the similar procedure, we can prove that

$$\begin{aligned}\mathcal{L}(f'') &= s\mathcal{L}(f') - f'(0) = s[sF(s) - f(0)] - f'(0) \\ &= s^2F(s) - sf(0) - f'(0).\end{aligned} \tag{17.27}$$

Example 17.7

Since $\sin(\omega t) = -(\cos\omega t)'/\omega$, we have

$$\begin{aligned}\mathcal{L}(\sin\omega t) &= -\frac{1}{\omega}\mathcal{L}(\cos\omega t) \\ &= -\frac{1}{\omega}\Big[s(\frac{s}{s^2+\omega^2}) - 1\Big] = \frac{\omega}{s^2+\omega^2}.\end{aligned} \tag{17.28}$$

Furthermore, for $g(t) = tf(t)$, we have

$$\mathcal{L}[g(t)] = -F'(s). \tag{17.29}$$

We can prove this formula by differentiating (with respect to s)

$$\begin{aligned}F'(s) &= \Big[\int_0^\infty f(t)e^{-st}\Big]' = \int_0^\infty \Big(\frac{d[f(t)e^{-st}]}{ds}\Big)dt \\ &= \int_0^\infty (-t)e^{-st}f(t)dt = -\mathcal{L}[tf(t)].\end{aligned} \tag{17.30}$$

That is

$$\mathcal{L}[tf(t)] = -F'(s). \tag{17.31}$$

On the other hand, for integration $h(t) = \int_0^t f(\tau)d\tau$, we have

$$H(s) = \int_0^\infty \Big(\int_0^t f(\tau)d\tau\Big)e^{-st}dt = \int_0^\infty \Big[\int_0^t f(\tau)e^{-st}dt\Big]d\tau. \tag{17.32}$$

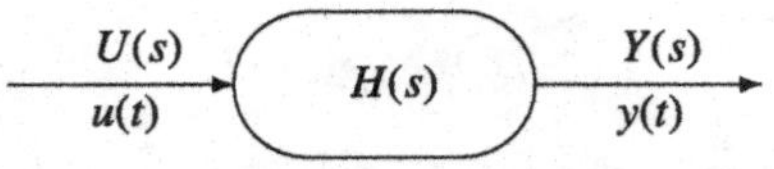

$$U(s),\ u(t) \longrightarrow \boxed{H(s)} \longrightarrow Y(s),\ y(t)$$

FIGURE 17.1 System response $Y(s) = H(s)U(s)$ for a given input $u(t)$.

Because we are integrating horizontally first over a triangular area over $0 \le \tau < t$, it is possible to switch the order to integrating vertically first. This means that we have changed the integral limits and we have

$$\int_{\tau=0}^{\infty}\int_{t=\tau}^{\infty} f(\tau)e^{-st}dtd\tau = \int_{\tau=0}^{\infty} f(\tau)\Big(\int_{t=\tau}^{\infty} e^{-st}dt\Big)d\tau$$
$$= \int_{0}^{\infty} f(\tau)(\frac{1}{s})e^{-s\tau}d\tau = \frac{F(s)}{s}. \tag{17.33}$$

That is

$$\mathcal{L}\Big(\int_0^t f(\tau)d\tau\Big) = \frac{F(s)}{s}. \tag{17.34}$$

So with Laplace transforms, differentiation becomes multiplication, while integration becomes division, which makes the system analysis in control theory much easier.

Transfer functions are used widely in control systems when describing the characteristics and response behavior of linear systems. Laplace transforms typically transform the differential equation that describes the system into a polynomial transfer function in the state space or in the complex frequency domain.

17.2 TRANSFER FUNCTION

A transfer function $H(s)$ is a representation of a system in the frequency domain, often in terms of Laplace transforms. From any input $u(t)$ to a system (see Fig. 17.1), the output or response of the system $y(t)$ can be obtained by

$$Y(s) = H(s)U(s), \tag{17.35}$$

where $U(s) = \mathcal{L}(u(t))$ and $Y(s) = \mathcal{L}(y(t))$ are the Laplace transforms of $u(t)$ and $y(t)$, respectively. Therefore, $H(s)$ is a system function. The behavior and characteristics of a system can be fully described by its transfer function $H(s)$.

At the same time, a system (such as a damped spring system or a robotic arm) can be modeled as a second-order ordinary differential equation

$$my''(t) + by'(t) + ky(t) = u(t), \tag{17.36}$$

where $u(t)$ is the driving force of a unit-step type. This can be converted into a transfer function

$$G(s) = \frac{Y(s)}{U(s)} = \frac{1}{ms^2 + bs + k}. \tag{17.37}$$

Example 17.8

An RLC circuit can be modeled as

$$L\frac{\partial i}{\partial t} + Ri + \frac{1}{C}\int idt = V, \tag{17.38}$$

or

$$Lq''(t) + Rq'(t) + \frac{q(t)}{C} = V, \quad i = dq/dt, \tag{17.39}$$

where R, L and C are constants. This differential equation is equivalent to the following transfer function

$$G(s) = \frac{Q(s)}{V(s)} = \frac{1}{Ls^2 + Rs + 1/C}. \tag{17.40}$$

It is worth pointing out that transfer functions are usually about zero initial states and thus all initial conditions are zero.

A transfer function $H(s)$ can be written as

$$H(s) = \frac{N(s)}{D(s)}, \tag{17.41}$$

where the degree k of the numerator $N(s)$ should be lower than (or equal to) the degree n of the denominator $D(s)$. In essence, the transfer function is a function of $s = \sigma + i\omega$. The above form can be factorized and written as

$$\begin{aligned} H(s) &= \frac{N(s)}{D(s)} = \frac{b_k s^k + b_{k-1}s^{k-1} + ... + b_1 s + b_0}{a_n s^n + a_{n-1}s^{n-1} + ... + a_1 s + a_0} \\ &= \frac{b_k}{a_n} \cdot \frac{(s - z_1)(s - z_2)...(s - z_k)}{(s - p_1)(s - p_2)...(s - p_n)}, \end{aligned} \tag{17.42}$$

where $k \le n$. Here, all the coefficients a_j and b_j are real. Here the roots are z_j and the poles are p_j.

Let us look at a concrete example.

Example 17.9

For the transfer function

$$H(s) = \frac{N(s)}{D(s)} = \frac{s^2 - s}{s^3 + 2s^2 + s + 2},$$

we have

$$N(s) = s^2 - s = s(s - 1),$$

and

$$D(s) = s^3 + 2s^2 + s + 2 = (s + 2)(s^2 + 1).$$

There are two zeros: 0 and 1, obtained by solving $N(s) = s(s - 1) = 0$, and there are three poles: -2 and $\pm i$, obtained by solving $D(s) = (s + 2)(s^2 + 1) = 0$. It is easy to see that $H(s) \to 0$ when s approaches zeros at 0 and 1 for this transfer function. At the same time, if s approaches any of the three poles (-2 or $\pm i$), $H(s)$ can become very large ($H(s) \to \infty$).

In general, the asymptotic behavior becomes

$$\lim_{s \to z_j} H(s) \to 0, \quad \lim_{s \to p_j} H(s) \to \infty. \tag{17.43}$$

The solutions or roots of $N(s) = 0$ are called zeros, while the roots of $D(s) = 0$ are called poles. In fact, the poles (locations in the complex plane) determine the main characteristics of the system.

For the unforced response of a system, the system poles determine the system response. That is, the unforced response can be written as

$$y_u(t) = \sum_{j=1}^{n} A_j e^{p_j t}, \tag{17.44}$$

where p_j's are the poles, and A_j's are the coefficients to be determined. Therefore, a system will be stable only if

$$\lim_{t \to \infty} \sum_{j=1}^{n} A_j e^{p_j t} \to 0. \tag{17.45}$$

We can expect to see why poles and zeros are important to analyze system responses in many applications such as control systems.

The above introduction provides a brief background for control theory and state-space models for control systems and signal processing.

17.3 SOLVING ODE VIA LAPLACE TRANSFORM

The beauty of the Laplace transform method is to transform an ordinary differential equation (ODE) into an algebraic equation. The solution in the s-domain can be obtained by solving the algebraic equation, and the final solution in the original domain can be obtained by the inverse Laplace transform after some algebraic manipulations. Let us look at two examples.

Example 17.10

For a first-order ODE

$$y'(t) + y(t) = u(t), \quad y(0) = 0,$$

where $u(t)$ is a unit step function ($u = 1$ for $t \geq 0$ and $u = 0$ when $t < 0$). Taking Laplace transform, we have

$$sY(s) - y(0) + Y(s) = \frac{1}{s}, \tag{17.46}$$

where we have used $\mathcal{L}(u(t)) = 1/s$. The solution $Y(s)$ is

$$Y(s) = \frac{1/s}{s+1} = \frac{1}{s(s+1)} = \frac{1}{s} - \frac{1}{s+1}. \tag{17.47}$$

The true solution $y(t)$ can be obtained by transforming $Y(s)$ back

$$y(t) = \mathcal{L}^{-1}[\frac{1}{s} - \frac{1}{s+1}] = \mathcal{L}^{-1}(\frac{1}{s}) - \mathcal{L}^{-1}(\frac{1}{s+1}) = 1 - e^{-t}. \tag{17.48}$$

This above ODE is relatively simple. To solve the higher-order differential equations, the procedure is the same, though the calculations can be more complicated.

Example 17.11

Let us now solve a second-order ODE:

$$y'' - 3y' + 2y = 0, \quad y(0) = 0, \quad y'(0) = 1. \tag{17.49}$$

Taking Laplace transform of both sides, we have

$$\mathcal{L}[y'' - 3y' + 2y] = \mathcal{L}(0), \tag{17.50}$$

which becomes

$$[s^2Y(s) - sy(0) - y'(0)] - 3[sY(s) - y(0)] + 2Y(s) = 0, \tag{17.51}$$

where we have use the fact that $\mathcal{L}(0) = 0$. Rearranging the above equation, we have

$$(s^2 - 3s + 2)Y(s) - 1 = 0. \tag{17.52}$$

Solving for $Y(s)$, we get

$$Y(s) = \frac{1}{s^2 - 3s + 2}. \tag{17.53}$$

In order to make inverse Laplace transform easier, we have to use the method of partial fractions. That is, we set the right-hand side as

$$Y(s) = \frac{1}{s^2 - 3s + 2} = \frac{A}{s-1} + \frac{B}{s-2}. \tag{17.54}$$

By direct algebraic manipulations, we have

$$\frac{1}{s^2 - 3s + 2} = \frac{A(s-2) + B(s-1)}{(s-1)(s-2)} = \frac{(A+B)s - 2A - B}{s^2 - 3s + 2}. \tag{17.55}$$

This requires that

$$A + B = 0, \quad -2A - B = 1. \tag{17.56}$$

So we have

$$A = -1, \quad B = 1. \tag{17.57}$$

Now we have

$$Y(s) = \frac{-1}{s-1} + \frac{1}{s-2}. \tag{17.58}$$

Finally, the solution becomes

$$y(t) = \mathcal{L}^{-1}(\frac{-1}{s-1}) + \mathcal{L}^{-1}(\frac{1}{s-2}) = -1e^t + 1e^{2t} = e^{2t} - e^t. \tag{17.59}$$

For the damped harmonic motion in Eq. (17.36), we can write it as

$$y''(t) + \frac{b}{m} y'(t) + \frac{k}{m} y(t) = \frac{u(t)}{m}, \tag{17.60}$$

or

$$y''(t) + 2\beta y'(t) + \omega_0^2 y(t) = f(t), \quad \omega_0 = \sqrt{\frac{k}{m}}, \quad \beta = \frac{b}{2m}, \tag{17.61}$$

where ω_0 is the natural angular frequency of the system and β is the damping coefficient. Taking the Laplace transform of the both sides, we have

$$\mathcal{L}y''(t) + 2\beta \mathcal{L}y'(t) + \omega_0^2 \mathcal{L}y(t) = \mathcal{L}f(t), \tag{17.62}$$

or

$$[s^2 Y(s) - sy(0) - y'(0)] + 2\beta[sY(s) - y(0)] + \omega_0^2 Y(s) = F(s), \tag{17.63}$$

which can be written as

$$[s^2 + 2\beta s + \omega_0^2]Y(s) = (s + 2\beta) + y'(0) + F(s). \tag{17.64}$$

So the solution in the s-domain (or Laplace domain) of this damped harmonic system is

$$Y(s) = \frac{(s + 2\beta)y(0) + y'(0) + F(s)}{s^2 + 2\beta s + \omega_0^2}. \tag{17.65}$$

For the denominator, we can write it as

$$s^2 + 2\beta s + \omega_0^2 = (s + \beta)^2 + (\omega_0^2 - \beta^2) = (s + \beta)^2 + \omega^2, \tag{17.66}$$

where $\omega = \sqrt{\omega_0^2 - \beta^2}$ is the damped natural (angular) frequency. we have

$$Y(s) = \frac{(s+2\beta)y(0) + y'(0)}{[(s+\beta)^2 + \omega^2]} + \frac{F(s)}{[(s+\beta)^2 + \omega^2]}. \tag{17.67}$$

The final solution $y(t)$ is obtained by the inverse Laplace transform, and we have

$$y(t) = \underbrace{\mathcal{L}^{-1}\Big\{\frac{(s+\beta)y(0) + \beta y(0) + y'(0)}{[(s+\beta)^2 + \omega^2]}\Big\}}_{\text{transient}} + \underbrace{\mathcal{L}^{-1}\Big\{\frac{F(s)}{[(s+\beta)^2 + \omega^2]}\Big\}}_{\text{steady state}}. \tag{17.68}$$

The exact inverse and analytical form will depend on the conditions and forcing term $f(t)$. Let us look at an example.

Example 17.12

In the special case when there is no forcing term (i.e., $f(t) = 0$ and thus $F(s) = 0$), we only have the transient term. In addition, if we assume $y'(0) = 0$ and $y(0) = 1$, we have the solution

$$y(t) = \mathcal{L}^{-1}\Big\{\frac{(s+\beta)+\beta}{[(s+\beta)^2+\omega^2]}\Big\} = \mathcal{L}^{-1}\Big\{\frac{(s+\beta)}{(s+\beta)^2+\omega^2}\Big\} + \mathcal{L}^{-1}\Big\{\frac{\beta}{(s+\beta)^2+\omega^2}\Big\}.$$

From the Laplace transform table, we know that

$$\mathcal{L}^{-1}\Big\{\frac{(s+\beta)}{(s+\beta)^2+\omega^2}\Big\} = e^{-\beta t}\cos(\omega t), \quad \mathcal{L}^{-1}\Big\{\frac{\beta}{(s+\beta)^2+\omega^2}\Big\} = \frac{\beta}{\omega}e^{-\beta t}\sin(\omega t).$$

So the final transient solution is

$$\begin{aligned} y(t) &= e^{-\beta t}\cos(\omega t) + \frac{\beta}{\omega}e^{-\beta t}\sin(\omega t) \\ &= e^{-\beta t}\Big[\cos(\sqrt{\omega_0^2-\beta^2}\ t) + \frac{\beta}{\sqrt{\omega_0^2-\beta^2}}\sin(\sqrt{\omega_0^2-\beta^2}\ t)\Big]. \end{aligned}$$

Since $\beta > 0$, the transient term will die out exponentially as t increases.

17.4 Z-TRANSFORM

The Laplace transform $X(s)$ works for a continuous-time signal $x(t)$. That is,

$$X(s) = \int_0^\infty x(t)e^{-st}dt. \tag{17.69}$$

When signal $x(t)$ is sampled, at $t = nT_s$ where T_s is a fixed constant and $n = 0, 1, 2, ...$, we can denote the signal as $x(n)$. If we let $T_s = 1$, the above equation (17.69) becomes

$$x(e^s) = \sum_{n=0}^{\infty} x(n)e^{-sn}. \tag{17.70}$$

Setting $z = e^s$, we have

$$X(z) = \sum_{n=0}^{\infty} x(n)z^{-n}, \tag{17.71}$$

which is a one-sided z-transform. It is straightforward to extend to a two-sided z-transform

$$X(z) = \sum_{n=-\infty}^{\infty} x(n)z^{-n}. \tag{17.72}$$

Obviously, if we let the signal $x(n)=0$ for $n<0$, then we have the original one-sided z-transform, as usual. Therefore, the z-transform can be considered as a discrete form of the Laplace transform.

The z-transform have many similar properties to those for the Laplace transform, including linearity.

For two signals $x_1(n)$ and $x_2(n)$, their transforms $X_1(z)=\mathcal{Z}(x_1)$ and $X_2(z)=\mathcal{Z}(x_2)$ obey

$$\mathcal{Z}(\alpha x_1+\beta x_2)=\alpha X_1(z)+\beta X_2(z). \tag{17.73}$$

For a time shift k, we can use $m=n-k$ and we have

$$\begin{aligned}\mathcal{Z}[x(n-k)] &= \sum_{n=-\infty}^{\infty} x(n-k)z^{-n} \\ &= \sum_{m=-\infty}^{\infty} x(m)z^{-(m+k)} \\ &= z^{-k}\sum_{m=-\infty}^{\infty} x(m)z^{-m} = z^{-k}X(z).\end{aligned} \tag{17.74}$$

Other properties can be derived, in a similar way, from their corresponding Laplace transform counterparts.

17.5 RELATIONSHIPS BETWEEN FOURIER, LAPLACE AND Z-TRANSFORMS

It is worth pointing out that there are intrinsic links between Fourier transforms, Laplace transforms and z-transforms. Using $z=e^s$ and $s=\sigma+i\omega$, we have

$$z=e^s=e^{\sigma}e^{i\omega}=re^{i2\pi f},$$

and

$$r=e^{\sigma}, \quad \omega=2\pi f. \tag{17.75}$$

When $r=e^{\sigma}=1$ (or $\sigma=0$, corresponding to a unit circle in the z-plane or complex plane), the z-transform becomes the (discrete) Fourier transform.

$$X(z=e^{-i2\pi f})=\sum_{n=-\infty}^{\infty} x(n)e^{-i2\pi fn}. \tag{17.76}$$

As a result, we can consider the z-transform as a generalized form of the Fourier transform (of a sampled signal). In addition, the z-transform is the Laplace transform of a discrete signal.

We also know before that the Laplace transform can be considered as a one-sided transform of the Fourier transform when $\sigma=0$ so that $s=\sigma+i\omega=i\omega$.

The relationships between the Fourier transform, the Laplace transform and the z-transform can be summarized as follows:

- The Fourier transform is the Laplace transform when evaluating along the $s=i\omega$ axis.
- The z-transform is the Laplace transform for a sampled signal, and becomes the latter when $T_s\to 0$.
- The discrete-time Fourier transform (DTFT) can be obtained by substituting $z=e^{i\omega}$ in the z-transform.

EXERCISES

17.1. Show that $\mathcal{L}[\sinh(at)]=a/(s^2-a^2)$.

17.2. Show that

$$\mathcal{L}[\sin(at+b)]=\frac{s\sin b+a\cos b}{(s^2+a^2)},$$

where a and b are constants.

17.3. Find the zeros and poles of the transfer function

$$H(s) = (s^2 + 1)/(s^2 + 3s + 2).$$

17.4. Solve $y'(t) - y(t) = e^{-t}$ with $y(0) = 0$ using the Laplace transform technique.

Part VI

Statistics and Curve Fitting

Chapter 18

Probability and Statistics

Chapter Points

- Random variables and their probability distributions are introduced, including binomial, Poisson and Gaussian distributions and others.
- Key concepts such as mean, expectation, variance, moment and moment-generating functions are also briefly presented with examples.
- The central limit theorem is explained, followed by the introduction of other distributions such as the Weibull distribution.

18.1 RANDOM VARIABLES

Randomness such as roulette-rolling and noise arises from the lack of information, or incomplete knowledge of reality. It can also come from the intrinsic complexity, diversity and perturbations of the system. The theory of probability is mainly the studies of random phenomena so as to find non-random regularity. Probability P is a number or an expected frequency assigned to an event A that indicates how likely it is that the event will occur when a random experiment is performed. This probability is often written as $P(A)$ to show that the probability P is associated with event A.

For a discrete random variable X with distinct values such as the number of cars passing through a junction, each value x_i may occur with a certain probability $p(x_i)$. In other words, the probability varies and is associated with its corresponding random variable. A probability function $p(x_i)$ is a function that defines probabilities to all the discrete values x_i of the random variable X. As an event must occur inside a sample space, the requirement that all the probabilities must be summed to one leads to

$$\sum_{i=1}^{n} p(x_i) = 1. \tag{18.1}$$

Example 18.1

The outcomes of tossing a fair coin form a sample space. The outcome of a head (H) is an event with a probability of $P(H) = 1/2$, and the outcome of a tail (T) is also an event with a probability of $P(T) = 1/2$. The sum of both probabilities should be one; that is

$$P(H) + P(T) = \frac{1}{2} + \frac{1}{2} = 1.$$

The cumulative probability function of X is defined by

$$P(X \le x) = \sum_{x_i < x} p(x_i). \tag{18.2}$$

For a continuous random variable X that takes a continuous range of values (such as the level of noise), its distribution is continuous and the probability density function $p(x)$ is defined for a range of values $x \in [a, b]$ for given limits a and b [or even over the whole real axis $x \in (-\infty, \infty)$]. In this case, we always use the interval $(x, x + dx]$ so that $p(x)$ is the probability that the random variable X takes the value $x < X \le x + dx$ is

$$\Phi(x) = P(x < X \le x + dx) = p(x)dx. \tag{18.3}$$

Engineering Mathematics with Examples and Applications

As all the probabilities of the distribution shall be added to unity, we have

$$\int_a^b p(x)dx = 1. \tag{18.4}$$

The cumulative probability function becomes

$$\Phi(x) = P(X \le x) = \int_a^x p(x)dx, \tag{18.5}$$

which is the definite integral of the probability density function between the lower limit a up to the present value $X = x$.

18.2 MEAN AND VARIANCE

Two main measures for a random variable X with a given probability distribution $p(x)$ are its mean and variance. The mean μ or the expectation value of $E[X]$ is defined by

$$\mu \equiv E[X] \equiv \langle X \rangle = \int xp(x)dx, \tag{18.6}$$

for a continuous distribution and the integration is within the integration limits. If the random variable is discrete, then the integration becomes the weighted sum

$$E[X] = \sum_i x_i p(x_i). \tag{18.7}$$

The variance var$[X] = \sigma^2$ is the expectation value of the deviation squared $(X - \mu)^2$. That is

$$\sigma^2 \equiv \text{var}[X] = E[(X-\mu)^2] = \int (x-\mu)^2 p(x)dx. \tag{18.8}$$

The square root of the variance $\sigma = \sqrt{\text{var}[X]}$ is called the standard deviation, which is simply σ.

Example 18.2

A simple distribution is the so-called uniform distribution, which has a probability density

$$p(x) = \frac{1}{b-a}, \quad a \le x \le b, \ b > a > 0,$$

which means that $p(x)$ is simply a constant.

The mean can be calculated by

$$\mu = \int_a^b xp(x)dx = \int_a^b \frac{x}{(b-a)}dx = \frac{1}{(b-a)}\int_a^b xdx$$
$$= \frac{1}{(b-a)}\Big[\frac{x^2}{2}\Big]_a^b = \frac{1}{(b-a)} \cdot [\frac{b^2}{2} - \frac{a^2}{2}] = \frac{a+b}{2}.$$

In addition, the variance is

$$\sigma^2 = E[(X-\mu)^2] = \int_a^b (x-\mu)^2 p(x)dx = \int_a^b [x - \frac{(a+b)}{2}]^2 \cdot \frac{1}{(b-a)}dx$$
$$= \frac{1}{(b-a)}\int_a^b [x^2 - (a+b)x + \frac{(a+b)^2}{4}]dx = \frac{1}{(b-a)}\Big[\frac{x^3}{3}\Big|_a^b - \frac{(a+b)}{2}x^2\Big|_a^b + \frac{(a+b)^2}{4}x\Big|_a^b\Big]$$
$$= \frac{1}{12(b-a)}(b^3 + 3a^2b - 3ab^3 - a^3) = \frac{1}{12(b-a)}(b-a)^3 = \frac{(b-a)^2}{12}.$$

For a discrete distribution, the variance simply becomes the following sum

$$\sigma^2 = \sum_i (x-\mu)^2 p(x_i). \tag{18.9}$$

In addition, any other formulas for a continuous distribution can be converted to their counterpart for a discrete distribution if the integration is replaced by the sum. Therefore, we will mainly focus on the continuous distribution in the rest of the section.

From these definitions, it is straightforward to prove

$$E[\alpha x + \beta] = \alpha E[X] + \beta, \qquad E[X^2] = \mu^2 + \sigma^2, \tag{18.10}$$

and

$$\mathrm{var}[\alpha x + \beta] = \alpha^2 \mathrm{var}[X], \tag{18.11}$$

where α and β are constants.

Other frequently used measures are the mode and median. The mode of a distribution is defined by the value at which the probability density function $p(x)$ is the maximum. For an even number of data sets, the mode may have two values. The median m of a distribution corresponds to the value at which the cumulative probability function $\Phi(m) = 1/2$. The upper and lower quartiles Q_U and Q_L are defined by $\Phi(Q_U) = 3/4$ and $\Phi(Q_L) = 1/4$.

18.3 BINOMIAL AND POISSON DISTRIBUTIONS

A discrete random variable is said to follow the binomial distribution $B(n, p)$ if its probability distribution is given by

$$B(n, p) = {}^nC_x p^x (1-p)^{n-x}, \qquad {}^nC_x = \binom{n}{x} = \frac{n!}{x!(n-x)!}, \tag{18.12}$$

where $x = 0, 1, 2, ..., n$ are the values that the random variable X may take, and n is the number of trials. There are only two possible outcomes: success or failure. p is the probability of a so-called 'success' of the outcome. Subsequently, the probability of the failure of a trial is $q = 1 - p$. Therefore, $B(n, p)$ represents the probability of x successes and $n - x$ failures in n trials. The coefficients come from the coefficients of the binomial expansions

$$(p+q)^n = \sum_{x=0}^{n} {}^nC_x p^x q^{n-x} = 1, \tag{18.13}$$

which is exactly the requirement that all the probabilities should be summed to unity.

Example 18.3

Tossing a fair coin 20 times, the probability of getting 15 heads is $B(n, 1/2)$. Since $p = 1/2$ and $x = 15$, then we have

$$^{20}C_{15}(\frac{1}{2})^{15}(\frac{1}{2})^5 = \frac{15504}{1048576} \approx 0.01479.$$

It is straightforward to prove that $\mu = E[X] = np$ and $\sigma^2 = npq = np(1-p)$ for a binomial distribution, and thus their proofs are left as an exercise.

Another related distribution is the geometric distribution whose probability function is defined by

$$P(X = n) = pq^{n-1} = p(1-p)^{n-1}, \tag{18.14}$$

where $n \geq 1$. This distribution is used to calculate the first success, thus the first $n - 1$ trials must be a failure if n trials are needed to observe the first success. The mean and variance of this distribution are $\mu = 1/p$ and $\sigma^2 = (1-p)/p^2$.

The Poisson distribution can be thought of as the limit of the binomial distribution when the number of trial is very large $n \to \infty$ and the probability $p \to 0$ (small probability) with the constraint that $\lambda = np$ is finite. For this reason, it is often called the distribution for small-probability events. Typically, it is concerned with the number of events that occur in a certain time interval (e.g., number of telephone calls in an hour) or spatial area. The Poisson distribution is

$$P(X = x) = \frac{\lambda^x e^{-\lambda}}{x!}, \qquad \lambda > 0, \tag{18.15}$$

where $x = 0, 1, 2, ..., n$ and λ is the mean of the distribution.

Obviously, the sum of all the probabilities must be equal to one. That is

$$\sum_{x=0}^{\infty} \frac{\lambda^x e^{-\lambda}}{x!} = \frac{\lambda^0 e^{-\lambda}}{0!} + \frac{\lambda^1 e^{-\lambda}}{1!} + \frac{\lambda^2 e^{-\lambda}}{2!} + \frac{\lambda^3 e^{-\lambda}}{3!} + ...$$

$$= e^{-\lambda}[1 + \lambda + \frac{\lambda^2}{2!} + \frac{\lambda^3}{3!} + ...] = e^{-\lambda} e^{\lambda} = e^{-\lambda+\lambda} = e^0 = 1. \tag{18.16}$$

Many stochastic processes such as the number of phone calls in a call center and the number of cars passing through a junction obey the Poisson distribution. Let us look at an example.

Example 18.4

Suppose you receive one email per hour on average. If you attend a 1-hour lesson, what is the probability of receiving exactly two emails after the lesson? What is the probability of no email at all?

Since the distribution is Poisson with $\lambda = 1$, the probability of receiving two emails is

$$P(X=2) = \frac{\lambda^2 e^{-\lambda}}{2!} = \frac{1^2 e^{-1}}{2!} \approx 0.184.$$

The probability of no email is

$$P(X=0) = \frac{\lambda^0 e^{-\lambda}}{0!} = \frac{1^0 e^{-1}}{0!} \approx 0.368.$$

The probability of receiving exactly 1 email is

$$P(X=1) = \frac{\lambda^1 e^{-\lambda}}{1!} = \frac{1^1 e^{-1}}{1} \approx 0.368.$$

On the other hand, what is the probability of receiving two or more emails? This means $X = 2, 3, 4, ...$, which have an infinite number terms, that is

$$P(X \geq 2) = P(X=2) + P(X=3) + P(X=4) + ...,$$

but how do we calculate this probability? Since the total probability is one or

$$1 = P(X=0) + P(X=1) + P(X=2) + P(X=3) + P(X=4) + ...$$
$$= P(X=0) + P(X=1) + P(X \geq 2),$$

we have

$$P(X \geq 2) = 1 - P(X=0) - P(X=1) = 1 - 0.184 - 0.368 \approx 0.264.$$

That is, the probability of receiving 2 or more emails is about 0.264.

Furthermore, what is the probability of receiving exactly one email in a 15 min interval?

We know that $\lambda = 1$ for one hour, so $\lambda_* = 1/4 = 0.25$ for a 15-minute period. The probability of receiving exactly one email in a 15-min period is

$$P(X=1) = \frac{\lambda_*^1 e^{-\lambda_*}}{1!} = \frac{0.25^1 e^{-0.25}}{1} \approx 0.195.$$

In this example, we may have realized that we use $\lambda_* = \lambda t$ where t is the time interval in the same time unit when defining λ. In general, we should use λt to replace t when dealing with the arrivals in a fixed period t. Thus, the Poisson distribution becomes

$$P(X=n) = \frac{(\lambda t)^n e^{-\lambda t}}{n!}. \tag{18.17}$$

Therefore, in the above example, if the lesson is a 2-hour session, the probability of getting exactly 1 email after the 2-hour session is

$$P(X=1)=\frac{(1\times 2)^1 e^{-1\times 2}}{1!}\approx 0.271.$$

As the Poisson distribution is so widely used for describing discrete events, let us look another example in a different context.

Example 18.5

Suppose your experiment requires some resistors and the resistors (by a manufacturer) have a 1.5% probability of defects. If you buy 200 resistors, what is the probability of this batch containing exactly 2 defect resistors?

The parameter λ can be calculated as

$$\lambda = 200\times 1.5\% = 200\times 0.015 = 3.$$

So the probability of getting exactly 2 defective resistors is

$$P(X=2)=\frac{3^2 e^{-3}}{2!}\approx 0.224.$$

The probability of at least one defective resistor is

$$P(X\geq 1)=1-P(X=0)=1-\frac{3^0 e^{-3}}{0!}\approx 0.95.$$

Using the definitions of mean and variance, it is straightforward to prove that $\mu=\lambda$ and $\sigma^2=\lambda$ for the Poisson distribution.

Example 18.6

The mean or expectation $E(X)$ can be calculated by

$$\begin{aligned}E(X)&=\sum_{x=0}^{\infty} xP(X=x)=\sum_{x=0}^{\infty} x\frac{\lambda^x e^{-\lambda}}{x!}\\&=0\times e^{-\lambda}+1\times(\lambda e^{-\lambda})+2\times(\frac{\lambda^2 e^{-\lambda}}{2!}+3\times(\frac{\lambda^3 e^{-\lambda}}{3!}+...)\\&=\lambda e^{-\lambda}[1+\lambda+\frac{\lambda^2}{2!}+\frac{\lambda^3}{3!}+...]=\lambda e^{-\lambda}e^{+\lambda}=\lambda.\end{aligned}$$

The parameter λ controls the location of the peak as shown in Fig. 18.1, and λ essentially describes a Poisson distribution uniquely. Therefore, some textbooks use Poisson(λ) to denote a Poisson distribution with parameter λ. Poisson distributions have an interesting property. For two independent random variables U and V that obey Poisson distributions Poisson(λ_1) and Poisson(λ_2), respectively, $S=U+V$ obeys Poisson($\lambda_1+\lambda_2$). From the basic Poisson distribution, we know that

$$\begin{aligned}P(S=n)&=\sum_{k=0}^{n} P(U=k,V=n-k)=\sum_{k=1}^{n} P(U=k)P(V=n-k)\\&=\sum_{k=0}^{n}\frac{\lambda_1^k e^{-\lambda_1}}{k!}\cdot\frac{\lambda_2^{n-k}e^{-\lambda_2}}{(n-k)!}\\&=e^{-(\lambda_1+\lambda_2)}\frac{1}{n!}\sum_{k=0}^{n}\frac{n!}{k!(n-k)!}\lambda_1^k\lambda_2^{n-k}\\&=e^{-(\lambda_1+\lambda_2)}\frac{(\lambda_1+\lambda_2)^n}{n!}=e^{-\lambda}\frac{\lambda^n}{n!},\end{aligned}\tag{18.18}$$

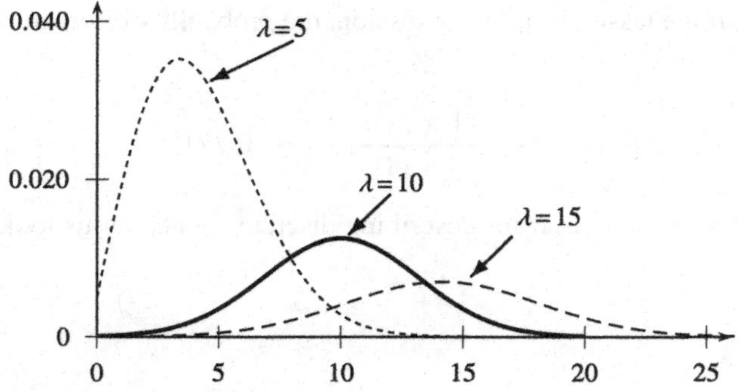

FIGURE 18.1 Poisson distributions for different values of $\lambda = 5, 10, 15$.

where $\lambda = \lambda_1 + \lambda_2$. In the above calculations, we have used the fact that U and V are independent (and thus the joint probability is the product of their probabilities). We have also used the binomial expansions

$$(\lambda_1 + \lambda_2)^n = \sum_{k=0}^{n} \frac{n!}{k!(n-k)!} \lambda_1^k \lambda_2^{n-k}. \tag{18.19}$$

Let us look at an example.

Example 18.7

For two students A and B, Student A receives on average one email per hour and Student B receives on average 1.5 emails per hour. What is the probability of receiving a total of exactly 4 emails in one hour?

Since $\lambda_1 = 1$ and $\lambda_2 = 1.5$, we have A obeys Poisson(1) and B obeys Poisson(1.5), so A+B obeys Poisson($\lambda_1 + \lambda_2$) = Poisson(2.5). The probability of receiving exactly 4 emails is

$$P(A+B=4) = \frac{(\lambda_1 + \lambda_2)^4 e^{-(\lambda_1+\lambda_2)}}{4!} = \frac{2.5^4 e^{-2.5}}{4!} \approx 0.134.$$

In addition, the probability of none of them receiving any email is

$$P(A+B=0) = \frac{2.5^0 e^{-2.5}}{0!} \approx 0.082.$$

In probability, an important concept is the moment-generating function which is defined as

$$G_X(v) = E[e^{vX}], \tag{18.20}$$

where X is the random variable and $G_X(0) = 1$.

The moment-generating function for the Poisson distribution is given by

$$G_X(v) = \sum_{x=0}^{\infty} \frac{e^{vx} \lambda^x e^{-\lambda}}{x!} = \exp[\lambda(e^v - 1)]. \tag{18.21}$$

The nth moment can be calculated by

$$m_n = E[X^n] = \frac{d^n G_X}{dv^n}\Big|_{v=0}, \tag{18.22}$$

where $n = 1, 2, ...$ is an integer. Obviously, the mean of a random variable is the first moment, while the variance is the second central moment. That is

$$\mu = m_1 = E[X] = \frac{dG_X(v)}{dv}\Big|_{v=0}, \tag{18.23}$$

and

$$\sigma^2 = m_2 - m_1^2 = E[X^2] - (E[X])^2 = m_2 - \mu^2 = \frac{d^2 G_X(v)}{dv^2}\Big|_{v=1} - \mu^2. \quad (18.24)$$

Example 18.8

For the Poisson distribution, the moment-generating function is

$$G(v) = e^{\lambda(e^v - 1)},$$

which means that the mean μ can be obtained by

$$\mu = m_1 = \frac{dG(v)}{dv}\Big|_{v=0} = e^{\lambda(e^v-1)} \cdot \lambda e^v\Big|_{v=0} = e^{\lambda(e^0-1)} \cdot \lambda e^0 = \lambda,$$

and the second moment m_2 is

$$m_2 = \frac{d^2 G(v)}{dv^2}\Big|_{v=0} = e^{\lambda(e^v-1)}[\lambda e^v + (\lambda e^v)^2]\Big|_{v=0} = e^{\lambda(e^0-1)}[\lambda e^0 + \lambda^2 e^0] = \lambda + \lambda^2.$$

Thus, the variance becomes

$$\sigma^2 = m_2 - m_1^2 = (\lambda + \lambda^2) - \lambda^2 = \lambda.$$

These are indeed what we have obtained earlier.

For the binomial distribution $B(n, p)$, its moment-generating function is

$$G_b(v) = (1 - p + pe^v)^n,$$

its mean can be computed by

$$\mu = \frac{dG_b(v)}{dv} = n(1 - p + pe^v)^{n-1} \cdot pe^v\Big|_{v=0} = n(1 - p + pe^0)^{n-1} \cdot pe^0 = np,$$

and the second moment m_2 is

$$m_2 = \frac{d^2 G_b(v)}{dv^2}\Big|_{v=0} = \Big[np(1 - p + pe^v)^{n-1}e^v + np^2(n-1)(1 - p + pe^v)^{n-2}\Big]\Big|_{v=0}$$
$$= np(1 - p + pe^0)^{n-1} + n(n-1)p^2(1 - p + pe^0)^{n-2} = np + n(n-1)p^2.$$

Thus, the variance of the binomial distribution is

$$\sigma^2 = m_2 - m_1^2 = np + n(n-1)p^2 - (np)^2 = np(1-p),$$

which is what we got in Section 3 of this chapter.

18.4 GAUSSIAN DISTRIBUTION

The Gaussian distribution or normal distribution is the most important continuous distribution in probability and it has a wide range of applications. For a continuous random variable X, the probability density function (PDF) of a Gaussian distribution is given by

$$p(x) = \frac{1}{\sigma\sqrt{2\pi}} e^{-\frac{(x-\mu)^2}{2\sigma^2}}, \quad (18.25)$$

where $\sigma^2 = \text{var}[X]$ is the variance and $\mu = E[X]$ is the mean of the Gaussian distribution. From the Gaussian integral, it is straightforward to verify that

$$\int_{-\infty}^{\infty} p(x)dx = 1, \quad (18.26)$$

and this is exactly the reason why the factor $1/\sqrt{2\pi}$ is required in the normalization of all the probabilities.

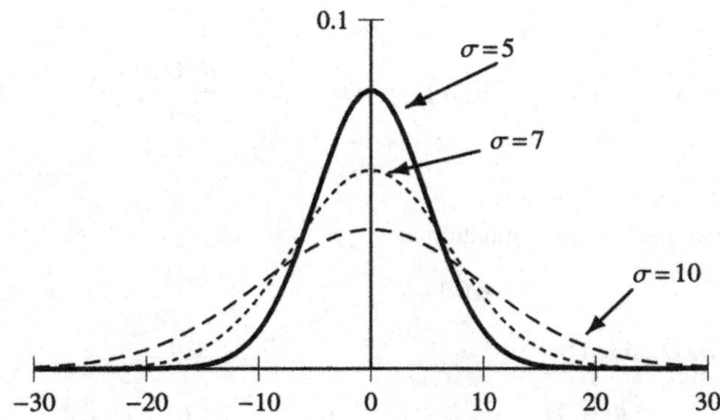

FIGURE 18.2 Gaussian distributions for $\sigma = 5, 7, 10$.

Example 18.9

Let us try to evaluate the integral

$$I = \int_{-\infty}^{\infty} e^{-\frac{(x-\mu)^2}{2\sigma^2}} dx.$$

Let $u = x - \mu$, we have $du = dx$ and

$$I = \int_{-\infty}^{\infty} e^{-\frac{(x-\mu)^2}{2\sigma^2}} dx = \int_{-(\infty-\mu)}^{\infty-\mu} e^{-\frac{u^2}{2\sigma^2}} du = \int_{-\infty}^{\infty} e^{-\frac{u^2}{2\sigma^2}} du.$$

Using the result of the Gaussian integral (see Section 14.4.2)

$$\int_{-\infty}^{\infty} e^{-\alpha u^2} du = \sqrt{\frac{\pi}{\alpha}},$$

and $\alpha = 1/(2\sigma^2)$, we get

$$I = \sqrt{\frac{\pi}{1/(2\sigma^2)}} = \sqrt{2\pi\sigma^2} = \sigma\sqrt{2\pi},$$

which leads to

$$\int_{-\infty}^{\infty} \frac{1}{\sigma\sqrt{2\pi}} e^{-\frac{(x-\mu)^2}{2\sigma^2}} dx = 1.$$

The probability function reaches a peak at $x = \mu$ and the variance σ^2 controls the width of the peak (see Fig. 18.2). The cumulative probability function (CPF) for a normal distribution is the integral of $p(x)$, which is defined by

$$\Phi(x) = P(X < x) = \frac{1}{\sqrt{2\pi\sigma^2}} \int_{-\infty}^{x} e^{-\frac{(\zeta-\mu)^2}{2\sigma^2}} d\zeta. \tag{18.27}$$

Using the error function defined in early chapters, we can write the above equation as

$$\Phi(x) = \frac{1}{\sqrt{2}} \Big[1 + \text{erf}\Big(\frac{x-\mu}{\sqrt{2}\sigma} \Big) \Big], \tag{18.28}$$

where

$$\text{erf}(x) = \frac{2}{\sqrt{\pi}} \int_0^x e^{-\zeta^2} d\zeta. \tag{18.29}$$

TABLE 18.1 Function ϕ defined by (18.34).

ξ	$\phi(\xi)$	ξ	$\phi(\xi)$
0.0	0.500	1.0	0.841
0.1	0.540	1.1	0.864
0.2	0.579	1.2	0.885
0.3	0.618	1.3	0.903
0.4	0.655	1.4	0.919
0.5	0.692	1.5	0.933
0.6	0.726	1.6	0.945
0.7	0.758	1.7	0.955
0.8	0.788	1.8	0.964
0.9	0.816	1.9	0.971

The moment generating function for the Gaussian distribution is given by

$$G_X(\nu) = e^{\mu\nu + \frac{1}{2}(\sigma\nu)^2}. \tag{18.30}$$

The Gaussian distribution can be considered as the limit of the Poisson distribution when $\lambda \gg 1$. Using the Sterling's approximation $x! \sim \sqrt{2\pi x}(x/e)^x$ for $x \gg 1$, and setting $\mu = \lambda$ and $\sigma^2 = \lambda$, it can be verified that the Poisson distribution can be written as a Gaussian distribution

$$P(x) \approx \frac{1}{\sqrt{2\pi\lambda}} e^{-\frac{(x-\mu)^2}{2\lambda}}, \tag{18.31}$$

where $\mu = \lambda$. In statistical applications, the normal distribution is often written as $N(\mu, \sigma^2)$ or $N(\mu, \sigma)$ to emphasize that the probability density function depends on two parameters μ and σ.

The standard normal distribution is a normal distribution $N(\mu, \sigma^2)$ with a mean of $\mu = 0$ and standard deviation $\sigma = 1$, that is $N(0, 1)$. This is useful to normalize or standardize data for statistical analysis. If we define a normalized variable

$$\xi = \frac{x - \mu}{\sigma}, \tag{18.32}$$

it is equivalent to giving a score so as to place the data above or below the mean in the unit of standard deviation. In terms of the area under the probability density function, ξ sorts where the data falls. It is worth pointing out that some books define $z = \xi = (x - \mu)/\sigma$ in this case, and call the standard normal distribution the Z distribution.

Now the probability density function of the standard normal distribution becomes

$$p(x) = \frac{1}{\sqrt{2\pi}} e^{-\frac{\xi^2}{2}}. \tag{18.33}$$

Its cumulative probability function is

$$\begin{aligned} \phi(\xi) &= \frac{1}{\sqrt{2\pi}} \int_{-\infty}^{\xi} e^{-\frac{\xi^2}{2}} d\xi \\ &= \frac{1}{2}\Big[1 + \text{erf}(\frac{\xi}{\sqrt{2}})\Big]. \end{aligned} \tag{18.34}$$

As the calculations of ϕ and the error function involve numerical integration, it is a traditional practice to tabulate ϕ in a table (see Table 18.1) so that we do not have to calculate their values each time we use this distribution.

18.5 OTHER DISTRIBUTIONS

There are a number of other important distributions such as the exponential distribution, log-normal distribution, uniform distribution and the χ^2-distribution. The uniform distribution has a probability density function

$$p = \frac{1}{\beta - \alpha}, \qquad x = [\alpha, \beta], \tag{18.35}$$

whose mean is $E[X] = (\alpha + \beta)/2$ and variance is $\sigma^2 = (\beta - \alpha)^2/12$.

The exponential distribution has the following probability density function

$$f(x) = \lambda e^{-\lambda x}, \quad \lambda > 0, \quad (x > 0), \tag{18.36}$$

and $f(x) = 0$ for $x \leq 0$. Its mean and variance are

$$\mu = 1/\lambda, \qquad \sigma^2 = 1/\lambda^2. \tag{18.37}$$

Example 18.10

The expectation $E(X)$ of an exponential distribution is

$$\mu = E(X) = \int_{-\infty}^{\infty} x\lambda e^{-\lambda x} dx = \int_0^{\infty} x\lambda e^{-\lambda x} dx$$

$$= \Big[-xe^{-\lambda x} - \frac{1}{\lambda} e^{-\lambda x} \Big]_0^{\infty} = \frac{1}{\lambda}.$$

For $E(X^2)$, we have

$$E(X^2) = \int_0^{\infty} x^2 \lambda e^{-\lambda x} dx = \Big[-x^2 e^{-\lambda x} \Big]_0^{\infty} + 2 \int_0^{\infty} x e^{-\lambda x} dx$$

$$= \Big[-x^2 e^{-\lambda x} \Big]_0^{\infty} + \Big[-\frac{2x}{\lambda} e^{-\lambda x} - \frac{2}{\lambda^2} e^{-\lambda x} \Big]_0^{\infty} = \frac{2}{\lambda^2}.$$

Here, we have used the fact that x and x^2 grow slower than $\exp(-\lambda x)$ decreases. That is, $x\exp(-\lambda x) \to 0$ and $x^2 \exp(-\lambda x) \to 0$ when $x \to \infty$.

Since $E(X^2) = \mu^2 + \sigma^2 = \mu^2 + \mathrm{Var}(X)$ from (18.10), we have

$$\mathrm{Var}(X) = \frac{2}{\lambda^2} - (\frac{1}{\lambda})^2 = \frac{1}{\lambda^2}.$$

Exponential distributions are widely used in queuing theory and simulating discrete events. For example, the arrival process of customers in a bank is a Poisson process and the time interval between arrivals (or inter-arrival time) obeys an exponential distribution.

The service time of a queue typically obeys an exponential distribution

$$P(t) = \begin{cases} \lambda e^{-\lambda t}, & t \geq 0, \\ 0, & t < 0, \end{cases} \tag{18.38}$$

where λ is the average number of customer served per unit time, and thus $\mu = 1/\lambda$ is the mean service time. Thus, the service time X less than some time (t) is the cumulative distribution

$$P(X \geq t) = \int_{-\infty}^{t} \lambda e^{-\lambda \tau} d\tau = \int_0^t \lambda e^{-\lambda \tau} d\tau = -e^{-\lambda \tau}\Big|_0^t = 1 - e^{-\lambda t}. \tag{18.39}$$

Example 18.11

If you are in a queue in a bank, you observed that it takes 5 minutes on average to service a customer. The service time obeys a cumulative distribution function

$$P(X \leq t) = 1 - e^{-\lambda t},$$

what is the probability of taking less than 2 minutes to the next customer?

We know that $\lambda = 1/5 = 0.2$ (5 minutes per customer or 0.2 customer per minute), so the probability is thus

$$P(t \leq 2) = 1 - e^{-0.2 \times 2} \approx 0.39.$$

The log-normal distribution has a probability density function

$$f(x) = \frac{1}{x\sqrt{2\pi\sigma^2}} \exp\Big[-\frac{(\ln x - \mu)^2}{2\sigma^2}\Big], \tag{18.40}$$

whose mean and variance are

$$E[X] = e^{\mu+\sigma^2/2}, \qquad \text{var}[X] = e^{\sigma^2+2\mu}(e^{\sigma^2} - 1). \tag{18.41}$$

The χ^2-distribution, called chi-square or chi-squared distribution, is very useful in statistical inference and the method of least squares. This distribution is for the quantity

$$\chi_n^2 = \sum_{i=1}^{n} \Big(\frac{X_i - \mu_i}{\sigma_i}\Big)^2, \tag{18.42}$$

where the n independent variables X_i are normally distributed with means μ_i and variances σ_i^2. The probability density function for χ^2-distribution is given by

$$p(x) = \frac{1}{2^{n/2}\Gamma(n/2)} x^{\frac{n}{2}-1} e^{-x/2}, \tag{18.43}$$

where $x \geq 0$, and n is called the degree of freedom. Here the gamma function $\Gamma(n)$ is defined by

$$\Gamma(n) = \int_0^\infty x^{n-1} e^{-x} dx, \tag{18.44}$$

for $n > 0$. In the simplest case, $n = 1$, we have

$$\Gamma(1) = \int_0^\infty x^0 e^{-x} dx = -e^{-x}\Big|_0^\infty = 1. \tag{18.45}$$

Using integration by parts, we have

$$\Gamma(n+1) = \int_0^\infty x^n e^{-x} dx = -xe^{-x}\Big|_0^\infty + \int_0^\infty nx^{n-1} e^{-x} dx$$
$$= n\int_0^\infty x^{n-1} e^{-x} dx = n\Gamma(n), \tag{18.46}$$

which leads to

$$\Gamma(n+1) = n!, \tag{18.47}$$

for any integer $n > 0$.

It can be verified that the mean of the χ^2-distribution is n and its variance is $2n$. For other distributions, readers can refer to any books that are devoted to advanced topics in probability theory and statistical analysis.

18.6 THE CENTRAL LIMIT THEOREM

The most important theorem in probability is the central limit theorem which concerns a large number of trials and explains why the normal distribution occurs so widely.

Let X_i $(i = 1, 2, ..., n)$ be n independent random variables, each of which is defined by a probability density function $p_i(x)$ with a corresponding mean μ_i and a variance σ_i^2. The sum of all these random variables

$$\Theta = \sum_{i=1}^{n} X_i = X_1 + X_2 + ... + X_n, \tag{18.48}$$

is also a random variable whose distribution approaches the Gaussian distribution as $n \to \infty$. Its mean $E[\Theta]$ and variance var$[\Theta]$ are given by

$$E[\Theta] = \sum_{i=1}^{n} E[X_i] = \sum_{i=1}^{n} \mu_i, \tag{18.49}$$

and

$$\mathrm{var}[\Theta] = \sum_{i=1}^{n} \mathrm{var}[X_i] = \sum_{i=1}^{n} \sigma_i^2. \tag{18.50}$$

The proof of this theorem is beyond the scope of this book as it involves the moment generating functions, characteristics functions and other techniques. In engineering mathematics, we simply use these important results for statistical analysis.

In the special case when all the variables X_i are described by the same probability density function with the same mean μ and variance σ^2, these results become

$$E[\Theta] = n\mu, \qquad \mathrm{var}[\Theta] = n\sigma^2. \tag{18.51}$$

By defining a new variable

$$\xi_n = \frac{\Theta - n\mu}{\sigma\sqrt{n}}, \tag{18.52}$$

then the distribution of ξ_n converges towards the standard normal distribution $N(0, 1)$ as $n \to \infty$.

18.7 WEIBULL DISTRIBUTION

Although the distribution functions of the real-world random processes are dominated by the Gaussian or normal distribution, however, there are some cases where other distributions describe the related phenomena more accurately. Weibull's distribution is such a distribution with many applications in areas such as reliability analysis, engineering design and quality assessment. Therefore, it deserves a special introduction in detail. This distribution was originally developed by Swedish physicist, A. Weibull in 1939, to try to explain the fact, well known but unexplained at that time, that the relative strength of a specimen decreases with its increasing dimension. Since then, it has been applied to study many real-world stochastic processes even including the distributions of wind speeds, rainfalls, energy resources and earthquakes.

Weibull's distribution is a three-parameter distribution given by

$$p(x, \lambda, \beta, n) = \begin{cases} \frac{n}{\lambda}(\frac{x-\beta}{\lambda})^{n-1} \exp[-(\frac{x-\beta}{\lambda})^n] & (x \geq \beta), \\ 0 & (x < \beta), \end{cases} \tag{18.53}$$

where λ is a scaling parameter, and n is the shape parameter, often referred to as the Weibull modulus. The parameter β is the threshold of the distribution. For example, $n = 1$ gives an exponential distribution. When $n > 3.5$, the distribution can be used to approximate a Gaussian distribution.

By straightforward integration, we have the cumulative probability density distribution

$$\Phi(x, \lambda, \beta, n) = 1 - e^{-(\frac{x-\beta}{\lambda})^n}. \tag{18.54}$$

For the fixed values $\lambda = 1$ and $\beta = 0$, the variation of n will give a broad range of shapes and can be used to approximate various distributions as shown in Fig. 18.3.

In reliability analysis, especially for a large infrastructure such as a dam or a tall building under stress, the survival probability is more conveniently represented as

$$P_s(V) = \exp\Big[\int_V -(\frac{\sigma}{\sigma_0})^n \frac{dV}{V_0}\Big], \tag{18.55}$$

where V is the volume of the system. σ_0 is the failure stress (either tensile or shear) for the reference volume V_0. The failure probability is

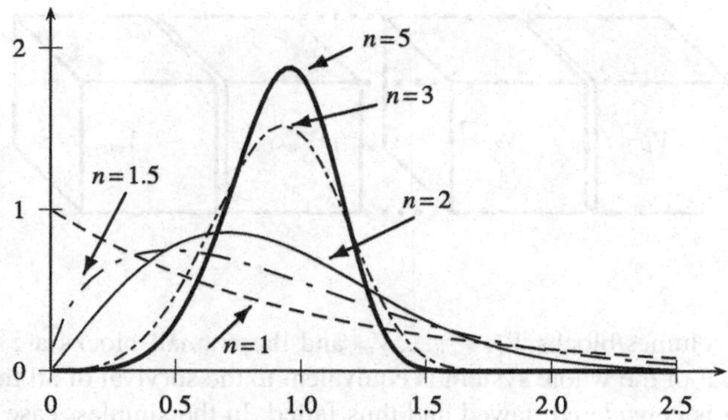

FIGURE 18.3 Weibull density function for different values of λ.

$$P_f(V) = 1 - P_s(V). \tag{18.56}$$

For constant stress σ over the whole volume V, we simply have

$$P_s(V) = \exp\Big[- (\frac{\sigma}{\sigma_0})^n \frac{V}{V_0} \Big]. \tag{18.57}$$

At the reference point $\sigma = \sigma_0$ and $V = V_0$ often obtained using laboratory tests, we have

$$P_s(V_0) = e^{-1} \approx 0.3679. \tag{18.58}$$

As the stress becomes extreme, $\sigma \to \infty$, then $P_s \to 0$ and $P_f \to 1$.

Example 18.12

A batch of some critical components (such as turbine blades) were manufactured. Assuming that the distribution is a Weibull distribution with a characteristic stress $\sigma_0 = 240$ MPa, a shape parameter $n = 6$ and $V = V_0$, what is the service stress that can be applied so as to ensure that no less than 99.99% of all components will survive?

The survival probability is 99.99%, so the probability of failure is

$$1 - \exp[-(\frac{\sigma}{240})^6] < 0.01\% = 10^{-4}.$$

Thus, we have

$$1 - 10^{-4} < \exp[-(\frac{\sigma}{240})^6], \quad \text{or} \quad (\frac{\sigma}{240})^6 \le 10^{-4},$$

where we have used $\ln(1 + x) \approx x$ when $x \to 0$. The above inequality requires that

$$\sigma \le \sqrt[6]{10^{-4} \times 240^6} \approx 51.7 \text{ MPa}.$$

Suppose during the manufacturing process, the materials properties can vary about 10%, and thus the minimum characteristics becomes $\sigma_0 = 215$ MPa instead of 240 MPa. How much will the above results change? The new allowable service stress becomes

$$\sigma \le \sqrt[6]{10^{-4} \times 215^6} \approx 46.3 \text{ MPa}.$$

However, if the previous stress level 51.7 MPa is still applied, then the failure probability or reliability will become

$$P = 1 - \exp[-(51.7/215)^6] \approx 1.9 \times 10^{-4},$$

which is almost twice the previous probability.

The fundamental idea of this volume-related probability function is the weakest link theory. The larger the volume of a system, the more likely it is to have critical flaws that could cause potential failure. We can consider that the whole large

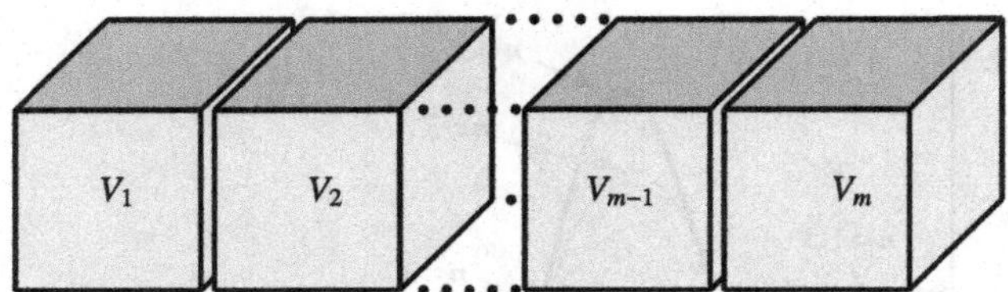

FIGURE 18.4 The weakest link analogy.

volume V consists of m small volumes/blocks $V_1, V_2, ..., V_m$ and these small blocks are glued together (see Fig. 18.4), thus the probability of the survival of the whole system is equivalent to the survival of all the subsystem blocks. If any one of these blocks fails, the system is considered flawed and thus failed. In the simplest case, $V_1 = V_2 = ... = V_m = V_0$ and $m = V/V_0$, the survival probability of the whole system under constant stress σ is

$$P_s(V) = P_s(mV_0) = \overbrace{P_s(V_0) \times P_s(V_0) \times ... \times P_s(V_0)}^{m}$$
$$= [P_s(V_0)]^m = [e^{-(\frac{\sigma}{\sigma_0})^m}]^{\frac{V}{V_0}} = \exp\Big[-\frac{V}{V_0}(\frac{\sigma}{\sigma_0})^m\Big]. \tag{18.59}$$

There are a whole range of statistical methods based on solid probability theory. However, we will not discuss any theory further, as our focus in this book is engineering mathematics and applicable numerical methods. Interested readers can refer to more advanced literature.

EXERCISES

18.1. If the number scored by a football team in a match in a World Cup can be modeled by a Poisson distribution with a mean number of goals of 2.5. What is the probability of no goal at all in a match? What is the probability of scoring two and more goals in a match?

18.2. The so-called clean tile problem concerns the probability of dropping a coin on a square grid floor. Let L be the edge length of the tile and d be the diameter of the coin. What is the probability of a randomly dropped coin landing on any of the edges of the tails?

18.3. Laplace's distribution can be described by

$$p(x) = \frac{1}{2a}\exp\Big(-\frac{|x-\mu|}{a}\Big), \quad a > 0.$$

Show that its cumulative probability distribution for $x > \mu$ is

$$F(x) = 1 - \frac{1}{2}\exp[-\frac{x-\mu}{a}].$$

18.4. Find the cumulative probability distribution and the mean for the Rayleigh distribution

$$r(x) = \frac{x}{\sigma^2}e^{-x^2/(2\sigma^2)}, \quad x \geq 0.$$

18.5. The moment-generating function for a normal distribution is $G(t) = e^{\mu t + \frac{1}{2}\sigma^2 t^2}$. Find the mean and the variance of its corresponding normal distribution.

Chapter 19

Regression and Curve Fitting

Chapter Points

- Linear regression is explained in detail as it is a very useful technique to find trends in data, and it belongs to the general class of least-squares methods for curve-fitting.
- Correlation coefficient is defined and linearization for curve-fitting is also explained with some examples to illustrate how the technique works.
- Hypothesis testing such as Student's t-test is presented with some examples, together with the introduction of confidence intervals.

19.1 SAMPLE MEAN AND VARIANCE

If a sample consists of n independent observations $x_1, x_2, ..., x_n$ on a random variable x such as the noise level on a road or the price of a cup of coffee, two important and commonly used parameters are sample mean and sample variance, which can easily be estimated from the sample. The sample mean is calculated by

$$\bar{x} \equiv \langle x \rangle = \frac{1}{n}(x_1 + x_2 + ... + x_n) = \frac{1}{n}\sum_{i=1}^{n} x_i, \tag{19.1}$$

which is essentially the arithmetic average of the values x_i.

The sample variance S^2 is defined by

$$S^2 = \frac{1}{n-1}\sum_{i=1}^{n}(x_i - \bar{x}\,)^2. \tag{19.2}$$

Let us look at an example.

Example 19.1

A small class has 10 students and their final marks in a maths exam are:

$$86,\ 93,\ 93,\ 84,\ 83,\ 90,\ 99,\ 87,\ 92,\ 85.$$

From the data, we know that $n = 10$ and the mode is 93 as 93 appears twice (all the rest only appears once).

The sample mean of the marks is

$$\begin{aligned}\bar{x} &= \frac{1}{10}(x_1 + x_2 + ... + x_{10}) \\ &= \frac{1}{10}(86 + 93 + 93 + 84 + 83 + 90 + 99 + 87 + 92 + 85) = \frac{892}{10} = 89.2.\end{aligned}$$

The corresponding sample variance can be calculated by

$$\begin{aligned}S^2 &= \frac{1}{n-1}\sum_{i=1}^{n}(x_i - \bar{x})^2 = \frac{1}{10-1}\sum_{i=1}^{10}(x_i - 89.2)^2 \\ &= \frac{1}{9}[(86 - 89.2)^2 + (93 - 89.2)^2 + ... + (85 - 89.2)^2] = \frac{1}{9}[(-3.2)^2 + (-3.8)^2 + ... + (2.8)^2 + (-4.2)^2] \\ &= \frac{1}{9}[10.24 + 14.44 + ... + 17.64] = \frac{231.6}{9} \approx 25.73.\end{aligned}$$

Generally speaking, if u is a linear combination of n independent random variables $y_1, y_2, ..., y_n$ and each random variable y_i has an individual mean μ_i and a corresponding variance σ_i^2, we have the linear combination

$$u = \sum_{i=1}^{n} \alpha_i y_i = \alpha_1 y_1 + \alpha_2 y_2 + ... + \alpha_n y_n, \tag{19.3}$$

where the parameters α_i $(i = 1, 2, ..., n)$ are the weighting coefficients. From the central limit theorem, we have the mean μ_u of the linear combination

$$\mu_u = E(u) = E(\sum_{i=1}^{n} \alpha_i y_i) = \sum_{i=1}^{n} \alpha E(y_i) = \sum \alpha_i \mu_i. \tag{19.4}$$

Then, the variance σ_u^2 of the combination is

$$\sigma_u^2 = E[(u - \mu_u)^2] = E\Big[\sum_{i=1}^{n} \alpha_i (y_i - \mu_i)^2\Big], \tag{19.5}$$

which can be expanded as

$$\sigma_u^2 = \sum_{i=1}^{n} \alpha_i^2 E[(y_i - \mu_i)^2] + \sum_{i,j=1;i\neq j}^{n} \alpha_i \alpha_j E[(y_i - \mu_i)(y_j - \mu_j)], \tag{19.6}$$

where $E[(y_i - \mu_i)^2] = \sigma_i^2$. Since y_i and y_j are independent, we have

$$E[(y_i - \mu_i)(y_j - \mu_j)] = E[(y_i - \mu_i)]E[(y_j - \mu_j)] = 0. \tag{19.7}$$

Therefore, we get

$$\sigma_u^2 = \sum_{i=1}^{n} \alpha_i^2 \sigma_i^2. \tag{19.8}$$

The sample mean defined in Eq. (19.1) can also be viewed as a linear combination of all the x_i assuming that each of which has the same mean $\mu_i = \mu$ and variance $\sigma_i^2 = \sigma^2$, and the same weighting coefficient $\alpha_i = 1/n$. Hence, the sample mean is an unbiased estimate of the sample due to the fact $\mu_{\bar{x}} = \sum_{i=1}^{n} \mu/n = \mu$. In this case, however, we have the variance

$$\sigma_{\bar{x}}^2 = \sum_{i=1}^{n} \frac{1}{n^2} \sigma^2 = \frac{\sigma^2}{n}, \tag{19.9}$$

which means the variance becomes smaller as the size n of the sample increases by a factor of $1/n$.

For the sample variance S^2 defined earlier by

$$S^2 = \frac{1}{n-1} \sum_{i=1}^{n} (x_i - \bar{x})^2, \tag{19.10}$$

we can see that the factor is $1/(n-1)$ not $1/n$ because only $1/(n-1)$ will give the correct and unbiased estimate of the variance. From the probability theory in the previous chapter, we know that $E[x^2] = \mu^2 + \sigma^2$. The mean of the sample variance is

$$\mu_{S^2} = E[\frac{1}{n-1} \sum_{i=1}^{n} (x_i - \bar{x})^2] = \frac{1}{n-1} \sum_{i=1}^{n} E[(x_i^2 - n\bar{x}^2)]. \tag{19.11}$$

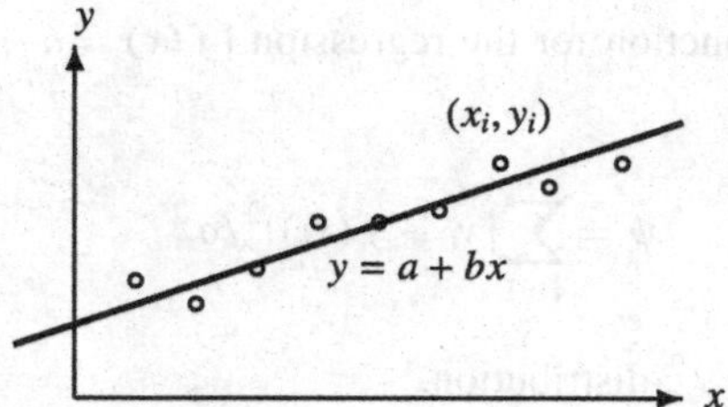

FIGURE 19.1 Least square and the best fit line.

Using $E[\bar{x}^2] = \mu^2 + \sigma^2/n$, we get

$$\mu_{S^2} = \frac{1}{n-1} \sum_{i=1}^{n} \{E[x_i^2] - nE[\bar{x}^2]\}$$

$$= \frac{1}{n-1} \Big[n(\mu^2 + \sigma^2) - n(\mu^2 + \frac{\sigma^2}{n}) \Big] = \sigma^2. \tag{19.12}$$

Obviously, if we used the factor $1/n$ instead of $1/(n-1)$, we would get $\mu_{S^2} = \frac{n-1}{n}\sigma^2 < \sigma^2$, which would underestimate the sample variance. The other way to think about the factor $1/(n-1)$ is that we need at least one value to estimate the mean, we need at least 2 values to estimate the variance. Thus, for n observations, only $n-1$ different values of variance can be obtained to estimate the total sample variance.

19.2 METHOD OF LEAST SQUARES

19.2.1 Maximum Likelihood

For a sample of n values $x_1, x_2, ..., x_n$ of a random variable X whose probability density function $p(x)$ depends on a set of k parameters $\beta_1, ..., \beta_k$, the joint probability is the product of all the probabilities, that is

$$\Phi(\beta_1, ..., \beta_k) = \prod_{i=1}^{n} p(x_i, \beta_1, ..., \beta_k)$$

$$= p(x_1, \beta_1, ..., \beta_k) \cdot p(x_2, \beta_1, ..., \beta_k) \cdots p(x_n, \beta_1, ..., \beta_k), \tag{19.13}$$

where Π means the product of all its components. For example, $\Pi_{n=1}^{3} n = 1 \times 2 \times 3$. The essence of the maximum likelihood is to maximize Φ by choosing the parameters β_j. As the sample can be considered as given values, the maximum likelihood requires the following stationary conditions:

$$\frac{\partial \Phi}{\partial \beta_j} = 0, \quad (j = 1, 2, ..., k), \tag{19.14}$$

whose solutions for β_j are the maximum likelihood estimates.

19.2.2 Linear Regression

For experiments and observations, we usually plot one variable such as pressure or price y against another variable x such as time or spatial coordinates. We try to present the data in such a way that we can see some trend in the data.

For a set of n data points (x_i, y_i), the usual practice is to try to draw a straight line $y = a + bx$ so that it represents the major trend. Such a line is often called the regression line or the best fit line as shown in Fig. 19.1.

The method of linear least squares is to try to determine the two parameters, a (intercept) and b (slope), for the regression line from n data points, assuming that x_i are known more precisely, and the values of y_i obey a normal distribution around the potentially best fit line with a variance σ^2. So we have the probability

$$P = \prod_{i=1}^{n} p(y_i) = A \exp\Big\{ -\frac{1}{2\sigma^2} \sum_{i=1}^{n} [y_i - f(x_i)]^2 \Big\}, \tag{19.15}$$

where A is a constant, and $f(x)$ is the function for the regression $[f(x) = a + bx$ for the linear regression]. It is worth pointing out that the exponent

$$\psi = \sum_{i=1}^{n} [y_i - f(x_i)]^2/\sigma^2$$

is similar to the quantity χ_n^2 defined in the χ^2-distribution.

The maximization of Φ is equivalent to the minimization of ψ. In order to minimize ψ as a function of a and b (via $f(x) = a + bx$), its derivatives should be zero. That is

$$\frac{\partial \psi}{\partial a} = -2\sum_{i=1}^{n} [y - (a + bx_i)] = 0, \tag{19.16}$$

and

$$\frac{\partial \psi}{\partial b} = -2\sum_{i=1}^{n} x_i[y_i - (a + bx_i)] = 0, \tag{19.17}$$

where we have used $\sigma^2 \neq 0$ and thus omitted this factor.

By expanding these equations, we have

$$na + b\sum_{i=1}^{n} x_i = \sum_{i=1}^{n} y_i, \tag{19.18}$$

and

$$a\sum_{i=1}^{n} x_i + b\sum_{i=1}^{n} x_i^2 = \sum_{i=1}^{n} x_i y_i, \tag{19.19}$$

which is a system of linear equations for a and b, and it is straightforward to obtain the solutions as

$$a = \frac{1}{n}[\sum_{i=1}^{n} y_i - b\sum_{i=1}^{n} x_i] = \bar{y} - b\bar{x}, \tag{19.20}$$

$$b = \frac{n\sum_{i=1}^{n} x_i y_i - (\sum_{i=1}^{n} x_i)(\sum_{i=1}^{n} y_i)}{n\sum_{i=1}^{n} x_i^2 - (\sum_{i=1}^{n} x_i)^2}, \tag{19.21}$$

where

$$\bar{x} = \frac{1}{n}\sum_{i=1}^{n} x_i, \qquad \bar{y} = \frac{1}{n}\sum_{i=1}^{n} y_i. \tag{19.22}$$

If we use the following notations

$$K_x = \sum_{i=1}^{n} x_i, \qquad K_y = \sum_{i=1}^{n} y_i, \tag{19.23}$$

and

$$K_{xx} = \sum_{i=1}^{n} x_i^2, \qquad K_{xy} = \sum_{i=1}^{n} x_i y_i, \tag{19.24}$$

then the above equations for a and b become

$$a = \frac{K_{xx}K_y - K_x K_{xy}}{nK_{xx} - (K_x)^2}, \tag{19.25}$$

and

$$b = \frac{nK_{xy} - K_x K_y}{nK_{xx} - (K_x)^2}. \tag{19.26}$$

The residual error is defined by

$$\epsilon_i = y_i - (a + bx_i), \tag{19.27}$$

whose sample mean is given by

$$\mu_\epsilon = \frac{1}{n}\sum_{i=1}^{n} \epsilon_i = \frac{1}{n} y_i - a - b\frac{1}{n}\sum_{i=1}^{n} x_i = \bar{y} - a - b\bar{x} = 0. \tag{19.28}$$

The sample variance S^2 is

$$S^2 = \frac{1}{n-2}\sum_{i=1}^{n} [y_i - (a + bx_i)]^2, \tag{19.29}$$

where the factor $1/(n-2)$ comes from the fact that two constraints are needed for the best fit, and therefore the residuals have $n-2$ degrees of freedom.

19.3 CORRELATION COEFFICIENT

The correlation coefficient $r_{x,y}$ is a very useful parameter for finding any potential relationship between two sets of data x_i and y_i for two random variables x and y, respectively. If x has a mean μ_x and a sample variance S_x^2, and y has a mean μ_y and a sample variance S_y^2, the correlation coefficient is defined by

$$r_{x,y} = \frac{\text{cov}(x, y)}{S_x S_y} = \frac{E[xy] - \mu_x \mu_y}{S_x S_y}, \tag{19.30}$$

where

$$\text{cov}(x, y) = E[(x - \mu_x)(y - \mu_y)], \tag{19.31}$$

is the covariance. If the two variables are independent or $\text{cov}(x, y) = 0$, there is no correlation between them ($r_{x,y} = 0$). If $r_{x,y}^2 = 1$, then there is a linear relationship between these two variables. $r_{x,y} = 1$ is an increasing linear relationship where the increase of one variable will lead to the increase of another. On the other hand, $r_{x,y} = -1$ is a decreasing relationship when one increases while the other decreases.

For a set of n data points (x_i, y_i), the correlation coefficient can be calculated by

$$r_{x,y} = \frac{n\sum_{i=1}^{n} x_i y_i - \sum_{i=1}^{n} x_i \sum_{i=1}^{n} y_i}{\sqrt{[n\sum x_i^2 - (\sum_{i=1}^{n} x_i)^2][n\sum_{i=1}^{n} y_i^2 - (\sum_{i=1}^{n} y_i)^2]}},$$

or

$$r_{x,y} = \frac{nK_{xy} - K_x K_y}{\sqrt{(nK_{xx} - K_x^2)(nK_{yy} - K_y^2)}}, \tag{19.32}$$

where $K_{yy} = \sum_{i=1}^{n} y_i^2$.

Example 19.2

Is there any relationship between shoe size and height among the general population? By collecting data randomly, say, we have the following data:

Height (h): 162, 167, 168, 171, 174, 176, 183, 179 (cm);

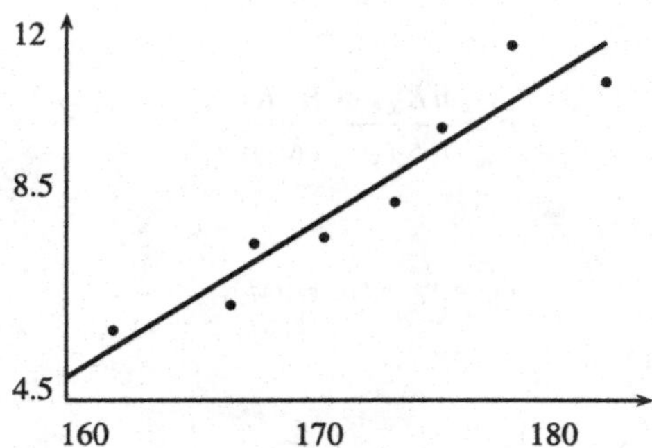

FIGURE 19.2 A best fit line by linear least square.

Shoe size (s): 5.5, 6, 7.5, 7.5, 8.5, 10, 11, 12.
Let us try the linear regression. We have

$$s = a + bh.$$

Since

$$K_h = \sum_{i=1}^{8} h_i = 1380, \quad K_s = \sum_{i=1}^{8} s_i = 68,$$

$$K_{hs} = \sum_{i=1}^{8} h_i s_i = 11835.5, \quad K_{hh} = \sum_{i=1}^{8} h_i^2 = 238380,$$

we get

$$a = \frac{K_{hh}K_s - K_h K_{xy}}{nK_{hh} - K_h^2} = \frac{238380 \times 68 - 1380 \times 11835.5}{8 \times 238380 - 1380^2} \approx -46.6477,$$

and

$$b = \frac{nK_{hs} - K_h K_s}{nK_{hh} - K_h^2} = \frac{8 \times 11835.5 - 1380 \times 68}{8 \times 238380 - 1380^2} \approx 0.3197.$$

So the regression line becomes

$$s = -46.6477 + 0.3197h,$$

which is shown in Fig. 19.2.

From the data set, we know that the sample means are $\mu_h = 172.5$, and $\mu_s = 8.5$. The covariance $\text{cov}(h, s) = E[(h - \mu_h)(s - \mu_s)] = 13.2$. We also have the standard deviation of height $S_h = 6.422$ and the standard deviation of shoe size $S_s = 2.179$. Therefore, their correlation coefficient r is given by

$$r = \frac{\text{cov}(h, s)}{S_h S_s} \approx \frac{13.2}{6.422 * 2.179} \approx 0.94.$$

This is a relatively strong correlation indeed. It is worth pointing out this conclusion is based on a small set of samples, which may not extend to the general population.

It is worth pointing out that the above formulations are based on the fact that the curve-fitting function $y = f(x) = ax + b$ is linear in terms of the independent variable x and the parameters (a and b). Here, the key linearity is about parameters, but about the basis function x. Thus, the above technique can still be applicable to both $f(x) = ax^2 + b$ and $g(x) = a + b\sin(x)$ functions with some minor adjustments to be discussed later in this chapter. However, if we have a function in the form

$$y = \ln(ax + b),$$

then the above technique cannot be applied directly, and some linearization approximations should be used.

19.4 LINEARIZATION

Sometimes, some obviously nonlinear functions can be transformed into linear forms so as to carry out linear regression, instead of more complicated nonlinear regression. However, there is no general formula for such linearization and thus it has often to deal each case individually. This can be illustrated by some examples.

Example 19.3

For example, the following nonlinear function

$$f(x) = \alpha e^{-\beta x}, \tag{19.33}$$

can be transformed into a linear form by taking logarithms of both sides. We have

$$\ln f(x) = \ln(\alpha) - \beta x, \tag{19.34}$$

which is equivalent to $y = a + bx$ if we let $y = \ln f(x)$, $a = \ln(\alpha)$ and $b = -\beta$.

In addition, the following function

$$f(x) = \alpha e^{-\beta x + \gamma} = A e^{-\beta x},$$

where $A = \alpha e^{\gamma}$ is essentially the same as the above function.

Similarly, function

$$f(x) = \alpha x^{\beta}, \tag{19.35}$$

can also be transformed into

$$\ln[f(x)] = \ln(\alpha) + \beta \ln(x), \tag{19.36}$$

which is a linear regression $y = a + b\zeta$ between $y = \ln[f(x)]$ and $\zeta = \ln(x)$ where $a = \ln(\alpha)$ and $b = \beta$.

Furthermore, function

$$f(x) = \alpha \beta^{x}, \tag{19.37}$$

can also be converted into the standard linear form

$$\ln f(x) = \ln \alpha + x \ln \beta, \tag{19.38}$$

by letting $y = \ln[f(x)]$, $a = \ln \alpha$, and $b = \ln \beta$.

It is worth pointing out that the data points involving zeros should be taken out due to the potential singularity of the logarithm. Fortunately, these points are rarely in the regression for the functions in the above form.

Example 19.4

If a set of data can fit to a nonlinear function

$$y = ax \exp(-x/b),$$

in the range of $(0, \infty)$, it is then possible to convert it to a linear regression.

As $x = 0$ is just a single point, so we can leave this out. For $x \neq 0$, we can divide both sides by x, we have

$$\frac{y}{x} = a\exp(-x/b).$$

Taking the logarithm of both sides, we have

$$\ln\frac{y}{x} = \ln a - \frac{1}{b}x,$$

which is a linear regression of y/x versus x.

19.5 GENERALIZED LINEAR REGRESSION

The most widely used linear regression is the so-called generalized least square as a linear combination of basic functions. Fitting to a polynomial of degree p

$$y(x) = \alpha_0 + \alpha_1 x + \alpha_2 x^2 + .. + \alpha_p x^p, \tag{19.39}$$

is probably the most widely used. This is equivalent to the regression to the linear combination of the basis functions 1, x, x, ..., and x^p. However, there is no particular reason why we have to use these basis functions. In fact, the basis functions can be any arbitrary known functions such as $\sin(x)$, $\cos(x)$ and even $\exp(x)$, and the main requirement is that they can be explicitly expressed as basic functions. In this sense, the generalized least square can be written as

$$y(x) = \sum_{j=0}^{p} \alpha_j f_j(x), \tag{19.40}$$

where the basis functions f_j are known functions of x. Now the sum of least squares is defined as

$$\psi = \sum_{i=1}^{n} \frac{[y_i - \sum_{j=0}^{p} \alpha_j f_j(x_i)]^2}{\sigma_i^2}, \tag{19.41}$$

where σ_i $(i = 1, 2, ..., n)$ are the standard deviations of the ith data point at (x_i, y_i). There are n data points in total. In order to determine the coefficients uniquely, it requires that

$$n \geq p + 1. \tag{19.42}$$

In the case of unknown standard deviations σ_i, we can always set all the values σ_i as the same constant $\sigma_i = \sigma = 1$.

Let $\mathbf{D}$ be the design matrix which is given by

$$D_{ij} = \frac{f_j(x_i)}{\sigma_i}. \tag{19.43}$$

The minimum of ψ is determined by

$$\frac{\partial \psi}{\partial \alpha_j} = 0, \qquad (j = 0, 1, ..., p). \tag{19.44}$$

That is

$$\sum_{i=1}^{n} \frac{f_k(x_i)}{\sigma_i^2}\Big[y_i - \sum_{j=0}^{p} \alpha_j f_j(x_i)\Big] = 0, \qquad k = 0, ..., p. \tag{19.45}$$

Rearranging the terms and interchanging the order of summations, we have

$$\sum_{j=0}^{p}\sum_{i=1}^{n} \frac{\alpha_j f_j(x_i) f_k(x_i)}{\sigma_i^2} = \sum_{i=1}^{n} \frac{y_i f_k(x_i)}{\sigma_i^2}, \tag{19.46}$$

which can be written compactly as the following matrix equation

$$\sum_{j=0}^{p} A_{kj}\alpha_j = b_k, \tag{19.47}$$

or

$$\mathbf{A}\alpha = \mathbf{b}, \tag{19.48}$$

where

$$\mathbf{A} = \mathbf{D}^{\mathbf{T}} \cdot \mathbf{D},$$

is a $(p+1) \times (p+1)$ matrix. That is

$$A_{kj} = \sum_{i=1}^{n} \frac{f_k(x_i) f_j(x_i)}{\sigma_i^2}. \tag{19.49}$$

Here b_k is a column vector given by

$$b_k = \sum_{i=1}^{n} \frac{y_i f_k(x_i)}{\sigma_i^2}, \tag{19.50}$$

where $(k = 0, ..., p)$. Equation (19.47) is a linear system of the so-called normal equations which can be solved using the standard methods for solving linear systems. The solution of the coefficients is $\alpha = \mathbf{A}^{-1}\mathbf{b}$ or

$$\alpha_k = \sum_{j=0}^{p} [A]_{kj}^{-1} b_j, \qquad (k = 0, ..., p), \tag{19.51}$$

where $\mathbf{A}^{-1} = [A]_{ij}^{-1}$.

Example 19.5

If we have a data set which can fit to a sine function, we can assume that

$$y = a + b \sin x.$$

For simplicity, we also assume that $\sigma_i = 1$. Thus, we have

$$\psi = \sum_{i=1}^{n} [y_i - (a + b \sin x_i)]^2.$$

The requirement of least squares leads to

$$\frac{\partial \psi}{\partial a} = 2na + \sum_{i=1}^{n} (-2y_i + 2b \sin x_i) = 0,$$

and

$$\frac{\partial \psi}{\partial b} = \sum_{i=1}^{n} (-2y_i \sin x_i + 2a \sin x_i + 2b \sin^2 x_i) = 0.$$

Solving the first equation to get a in terms of b, we have

$$a = \frac{1}{n}\sum_{i=1}^{n}(y_i - b\sin x_i).$$

Substituting it into the other equation, we can obtain b. Therefore, the solutions for a and b are

$$a = \frac{1}{\Delta}[(\sum_{i=1}^{n} y_i)(\sum_{i=1}^{n}\sin^2 x_i) - (\sum_{i=1}^{n}\sin x_i)(\sum_{i=1}^{n} y_i \sin x_i)],$$

$$b = \frac{1}{\Delta}[n(\sum_{i=1}^{n} y_i \sin x_i) - (\sum_{i=1}^{n}\sin x_i)(\sum_{i=1}^{n} y_i)],$$

where

$$\Delta = n\sum_{i=1}^{n}\sin^2 x_i - (\sum_{i=1}^{n}\sin x_i)^2.$$

A special case of the generalized linear least squares is the so-called polynomial least squares when the basis functions are simple power functions $f_i(x) = x^i$, $(i = 0, 1, ..., p)$. That is

$$f_i(x) = 1,\ x,\ x^2, ...,\ x^p. \tag{19.52}$$

For simplicity, we assume that $\sigma_i = \sigma = 1$. The matrix equation (19.47) simply becomes

$$\begin{pmatrix} \sum_{i=1}^n 1 & \sum_{i=1}^n x_i & \dots & \sum_{i=1}^n x_i^p \\ \sum_{i=1}^n x_i & \sum_{i=1}^n x_i^2 & \dots & \sum_{i=1}^n x_i^{p+1} \\ \vdots & & \ddots & \\ \sum_{i=1}^n x_i^p & \sum_{i=1}^n x_i^{p+1} & \dots & \sum_{i=1}^n x_i^{2p} \end{pmatrix}\begin{pmatrix} \alpha_0 \\ \alpha_1 \\ \vdots \\ \alpha_p \end{pmatrix} = \begin{pmatrix} \sum_{i=1}^n y_i \\ \sum_{i=1}^n x_i y_i \\ \vdots \\ \sum_{i=1}^n x_i^p y_i \end{pmatrix}.$$

In the simplest case when $p = 1$, it becomes the standard linear regression

$$y = \alpha_0 + \alpha_1 x = a + bx.$$

Now we have

$$\begin{pmatrix} n & \sum_{i=1}^n x_i \\ \sum_{i=1}^n x_i & \sum_{i=1}^n x_i^2 \end{pmatrix}\begin{pmatrix} \alpha_0 \\ \alpha_1 \end{pmatrix} = \begin{pmatrix} \sum_{i=1}^n y_i \\ \sum_{i=1}^n x_i y_i \end{pmatrix}. \tag{19.53}$$

Its solution is

$$\begin{aligned}\begin{pmatrix} \alpha_0 \\ \alpha_1 \end{pmatrix} &= \frac{1}{\Delta}\begin{pmatrix} \sum_{i=1}^n x_i^2 & -\sum_{i=1}^n x_i \\ -\sum_{i=1}^n x_i & n \end{pmatrix}\begin{pmatrix} \sum_{i=1}^n y_i \\ \sum_{i=1}^n x_i y_i \end{pmatrix} \\ &= \frac{1}{\Delta}\begin{pmatrix} (\sum_{i=1}^n x_i^2)(\sum_{i=1}^n y_i) - (\sum_{i=1}^n x_i)(\sum_{i=1}^n x_i y_i) \\ n\sum_{i=1}^n x_i y_i - (\sum_{i=1}^n x_i)(\sum_{i=1}^n y_i) \end{pmatrix},\end{aligned} \tag{19.54}$$

where

$$\Delta = n\sum_{i=1}^{n} x_i^2 - (\sum_{i=1}^{n} x_i)^2. \tag{19.55}$$

These are exactly the same coefficients as those in Eq. (19.26).

Example 19.6

We now use a quadratic function to best fit the following data:

$$x: -0.98,\ 1.00,\ 2.02,\ 3.03,\ 4.00$$
$$y: 2.44,\ -1.51,\ -0.47,\ 2.54,\ 7.52$$

For the formula $y = \alpha_0 + \alpha_1 x + \alpha_2 x^2$, we have

$$\begin{pmatrix} n & \sum_{i=1}^n x_i & \sum_{i=1}^n x_i^2 \\ \sum_{i=1}^n x_i & \sum_{i=1}^n x_i^2 & \sum_{i=1}^n x_i^3 \\ \sum_{i=1}^n x_i^2 & \sum_{i=1}^n x_i^3 & \sum_{i=1}^n x_i^4 \end{pmatrix} \begin{pmatrix} \alpha_0 \\ \alpha_1 \\ \alpha_2 \end{pmatrix} \begin{pmatrix} \sum_{i=1}^n y_i \\ \sum_{i=1}^n x_i y_i \\ \sum_{i=1}^n x_i^2 y_i \end{pmatrix}.$$

Using the data set, we have $n = 5$, $\sum_{i=1}^n x_i = 9.07$ and $\sum_{i=1}^n y_i = 10.52$. Other quantities can be calculated in a similar way. Therefore, we have

$$\begin{pmatrix} 5.0000 & 9.0700 & 31.2217 \\ 9.0700 & 31.2217 & 100.119 \\ 31.2217 & 100.119 & 358.861 \end{pmatrix} \begin{pmatrix} \alpha_0 \\ \alpha_0 \\ \alpha_2 \end{pmatrix} = \begin{pmatrix} 10.52 \\ 32.9256 \\ 142.5551 \end{pmatrix}.$$

By direct inversion, we have

$$\begin{pmatrix} \alpha_0 \\ \alpha_1 \\ \alpha_2 \end{pmatrix} = \begin{pmatrix} -0.5055 \\ -2.0262 \\ 1.0065 \end{pmatrix}.$$

Finally, the best fit equation is

$$y(x) = -0.5055 - 2.0262x + 1.0065x^2,$$

which is quite close to the formula $y = x^2 - 2x - 1/2$ used to generate the original data with a random component of about 2.5%.

19.6 HYPOTHESIS TESTING

Hypothesis testing forms a class of important statistical techniques to carry out some inference about the whole population based on a sample. Such tests are used to distinguish the hull hypothesis from an alternative hypothesis to see if there is sufficient evidence at a predefined level of confidence.

19.6.1 Confidence Interval

The confidence interval is defined as the interval $\theta_1 \le X \le \theta_2$ so that the probabilities at these two limits θ_1 and θ_2 are equal to a given probability $\gamma = 1 - \alpha$ (say, 95% or 99%). That is

$$P(\theta_1 \le X \le \theta_2) = \gamma = 1 - \alpha. \tag{19.56}$$

The predetermined parameter γ is always near 1 so that it can be expressed as a small deviation $\alpha \ll 1$ from 1 (see Fig. 19.3). If we choose $\gamma = 95\%$, it means that we can expect that about 95% of the sample will fall within the confidence interval while 5% of the data will not.

For the standard normal distribution, this means $P(-\theta \le \xi \le \theta) = 1 - \alpha$, so that

$$\phi(\xi \le \theta) = 1 - \frac{\alpha}{2}. \tag{19.57}$$

If $\alpha = 0.05$, we have $\phi(\xi \le \theta) = 0.975$ or $\theta = 1.960$. That is to say, $-\theta \le \xi \le \theta$ or $\mu - \theta\sigma \le x \le \mu + \theta\sigma$. We also know that if you repeat an experiment n times, the variance will decrease from σ^2 to σ^2/n, which is equivalent to saying that the

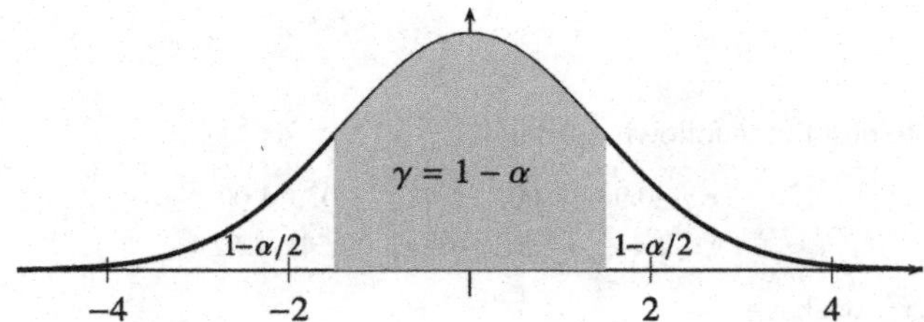

FIGURE 19.3 Confidence interval $\gamma = 1 - \alpha$.

standard deviation becomes $\sigma/\sqrt{n}$ for a sample size n. If $\alpha = 0.01$, then $\theta = 2.579$, we have

$$\mu - 2.579\frac{\sigma}{\sqrt{n}} \le x \le \mu + 2.579\frac{\sigma}{\sqrt{n}}. \tag{19.58}$$

On the other hand, for $\theta = 1$, we get $\mu - \sigma \le x \le \mu + \sigma$ and $\gamma = 0.682$. In other words, only 68.2% of the sample data will fall within the interval $[\mu - \sigma, \mu + \sigma]$ or

$$x = \mu \pm \sigma, \tag{19.59}$$

with a 68.2% confidence level.

It is conventional to use $\gamma = 0.95$ for probably significant, 0.99 for significant, and 0.999 for highly significant.

19.6.2 Student's t-Distribution

The Student's t-test is a very powerful method for testing the null hypothesis to see if the means of two normally distributed samples are equal. This method was designed by W. S. Gosset in 1908 and he had to use a pen name 'Student' because of his employer's policy in publishing research results at that time. This is a powerful method for hypothesis testing using small-size samples. This test can also be used to test if the slope of the regression line is significantly different from 0. It has become one of the most popular methods for hypothesis testing. The theoretical basis of the t-test is the Student's t-distribution for a sample population with the unknown standard deviation σ, which of course can be estimated in terms of the sample variance S^2 from the sample data.

For n independent measurements/data $x_1, x_2, ..., x_n$ with an estimated sample mean $\bar{x}$ and a sample variance S^2 as defined by equation (19.10), the t-variable is defined by

$$t = \frac{\bar{x} - \mu}{(S/\sqrt{n})}. \tag{19.60}$$

The Student's t-distribution with $k = n - 1$ degrees of freedom is the distribution for the random variable t, and the probability density function is

$$p(t) = \frac{\Gamma(\frac{k+1}{2})}{\sqrt{k\pi}\,\Gamma(k/2)}\Big[1 + \frac{t^2}{k}\Big]^{-\frac{k+1}{2}}. \tag{19.61}$$

It can be verified that the mean is $E[t] = 0$. The variance is $\sigma^2 = k/(k-2)$ for $k > 2$ and infinite for $0 < k \le 2$.

The corresponding cumulative probability function is

$$F(t) = \frac{\Gamma(\frac{k+1}{2})}{\sqrt{k\pi}\,\Gamma(k/2)}\int_{\infty}^{t}[1 + \frac{\zeta^2}{k}]^{-\frac{k+1}{2}}\,d\zeta. \tag{19.62}$$

This integral leads to a hypergeometric function that is not straightforward to calculate, which is why they are tabulated in many statistical tables. For a confidence level of $\gamma = 1 - \alpha$, the confidence interval is given by

$$F(\theta) = 1 - \frac{\alpha}{2}. \tag{19.63}$$

For $\alpha = 0.05$ and 0.01 (or $1 - \alpha/2 = 0.975$ and 0.995), the values are tabulated in Table 19.1.

TABLE 19.1 Limits defined by $F(\theta) = 1 - \alpha/2$ in (19.63).

k	$F(\theta)_{0.975}$	$F(\theta)_{0.995}$
1	12.7	63.7
2	4.30	9.93
3	3.18	5.84
4	2.78	4.60
5	2.57	4.03
6	2.45	3.71
7	2.37	3.50
8	2.31	3.36
9	2.26	3.25
10	2.23	3.17
20	2.09	2.85
50	2.01	2.68
100	1.98	2.63
∞	1.96	2.58

Suppose we are dealing with the 95% confidence interval, we have $p(-\theta \le t \le \theta) = 1 - \alpha = 0.95$ or $p(t \le \theta) = 1 - \alpha/2 = 0.975$, so we have $\theta = t_{\alpha,k} = 12.70$ $(k = 1)$, 4.30 $(k = 2)$, 3.18 $(k = 3)$, ..., 2.228 $(k = 10)$, ..., 1.959 for $k \to \infty$. Hence,

$$\mu - \theta \frac{S}{\sqrt{n}} \le t \le \mu + \theta \frac{S}{\sqrt{n}}. \tag{19.64}$$

This is much more complicated than its counterpart, the standard normal distribution.

19.6.3 Student's *t*-Test

There are quite a few variations of the Student's t-test, and the most common t-tests are the one-sample t-test and the two-sample t-test. The one sample t-test is used for measurements that are randomly drawn from a population to compare the sample mean with a known number.

In order to do statistical testing, we first have to pose precise questions or form a hypothesis, which is conventionally called the null hypothesis. The basic steps of a t-test are as follows:

1. The null hypothesis: H_0: $\mu = \mu_0$ (often known value) for one sample, or H_0: $\mu_1 = \mu_2$ for two samples;
2. Calculate the t-test statistic t and find the critical value θ for a given confidence level $\gamma = 1 - \alpha$ by using $F(t \le \theta) = 1 - \alpha/2$;
3. If $|t| > \theta$, reject the hypothesis. Otherwise, accept the hypothesis.

Another important t-test is the two-sample paired test. Assuming that two pairs of n sample data sets U_i and V_i are independent and drawn from the same normal distribution, the paired t-test is used to determine whether they are significantly different from each other. The t-variable is defined by

$$t = \frac{(\bar{U} - \bar{V})}{S_d/\sqrt{n}} = (\bar{U} - \bar{V})\sqrt{\frac{n(n-1)}{\sum_{i=1}^{n}(\tilde{U}_i - \tilde{V}_i)^2}}, \tag{19.65}$$

where $\tilde{U}_i = U_i - \bar{U}$ and $\tilde{V}_i = V_i - \bar{V}$. In addition,

$$S_d^2 = \frac{1}{n-1}\sum_{i=1}^{n}(\tilde{U}_i - \tilde{V}_i)^2. \tag{19.66}$$

This is equivalent to applying the one-sample test to the difference $U_i - V_i$ data sequence. Let us look at a short example.

Example 19.7

A novel teaching method of teaching children sciences was tried in a class (say class B), while a standard method was used in another class (say class A). At the end of the assessment, 8 students are randomly drawn from each class, and their science scores are as follows:
Class A: $U_i = 76, 77, 76, 81, 77, 76, 75, 82$;
Class B: $V_i = 79, 81, 77, 86, 82, 81, 82, 80$.
At a 95% confidence level, can you say the new method is really better than the standard method?

If we suppose that the two methods do not produce any difference in results, that is to say, their means are the same. Thus the null hypothesis is:

$$H_0 : \mu_A = \mu_B.$$

We know that $\bar{U} = 77.5$, $\bar{V} = 81$. The combined sample variance $S_d = 2.828$. We now have

$$t = \frac{\bar{U} - \bar{V}}{S_d/\sqrt{n}} = \frac{77.5 - 81}{2.828/\sqrt{8}} = -3.5.$$

We know from the statistical table that the critical value $\theta = 2.37$ for $F(\theta) = 1 - \alpha/2$ and $k = n - 1 = 7$. As $t < -\theta$ or $t > \theta$, we can reject the null hypothesis. That is to say, the new method does produce better results in teaching sciences.

The variance analysis and hypothesis testing are important topics in applied statistics, and there are many excellent books on these topics. Readers can refer to the relevant books listed at the end of this book. It is worth pointing out that other important methods for hypothesis testing are Fisher's F-test, χ^2-test, and non-parametric tests. What we have discussed in this chapter is just the tip of the iceberg; however, it forms the solid basis for further studies.

EXERCISES

19.1. For a simple data set of $1, 2, 3, 4, 5, 6, 7, 8$, what are its sample mean and sample variance?

19.2. Use linearization to derive formulas for fitting $y = \exp[ax^2 + b \sin x]$.

19.3. Two types of robots to carry out a task with the following times (in seconds):
Robot A: 23, 21, 19, 40, 20, 27, 25.
Robot B: 22, 18, 18, 25, 17, 21, 23.
If the manufacturers claim that Type B is better (quicker) than Type A, do you believe their claim?

Part VII

Numerical Methods

Chapter 20

Numerical Methods

Chapter Points

- Root-finding algorithms are used to solve polynomial equations and such methods include bisection methods, Newton's method and others.
- Many integrals cannot be expressed using explicit analytical formulas and it is thus necessary to use numerical integration techniques such as the trapezium rule and Simpson's rule.
- Numerical solutions of ODEs are introduced with detailed discussions of Euler's method and the Runge-Kutta method.

There are many different numerical methods for solving a wide range of problems with different orders of accuracy and various levels of complexity. For example, for numerical solutions of ODEs, we can have a simple Euler integration scheme or higher-order Runge-Kutta scheme. For PDEs, we can use finite difference methods, finite element methods, finite volume methods and others. As this book is an introductory textbook, we will only introduce the most basic methods that are most useful to a wide range of problems in engineering.

To demonstrate how these numerical methods work, we will use step-by-step examples to find the roots of nonlinear equations, to estimate integrals by numerical integration, and to solve ODEs by direct integration and higher-order Runge-Kutta methods.

20.1 FINDING ROOTS

Let us start by trying to find the square root of a number a. It is essentially equivalent to finding the solution of

$$x^2 - a = 0. \tag{20.1}$$

We can rearrange it as

$$x = \frac{1}{2}(x + \frac{a}{x}), \tag{20.2}$$

which makes it possible to estimate the root x iteratively. If we start from a guess, say $x_0 = 1$, we can calculate the new estimate x_{n+1} from any previous value x_n using

$$x_{n+1} = \frac{1}{2}(x_n + \frac{a}{x_n}). \tag{20.3}$$

Let us look at an example.

Example 20.1

In order to find $\sqrt{23}$, we have $a = 23$ with an initial guess $x_0 = 1$. The first five iterations are as follows:

$$x_1 = \frac{1}{2}(x_0 + \frac{23}{x_0}) = \frac{1}{2}(1 + \frac{23}{1}) = 12,$$

$$x_2 = \frac{1}{2}(x_1 + \frac{23}{x_1}) = \frac{1}{2}(12 + \frac{23}{12}) \approx 6.9583333,$$

$$x_3 = \frac{1}{2}(x_2 + \frac{23}{x_2}) \approx 5.1318613, \quad x_4 \approx 4.8068329, \quad x_5 \approx 4.79584411.$$

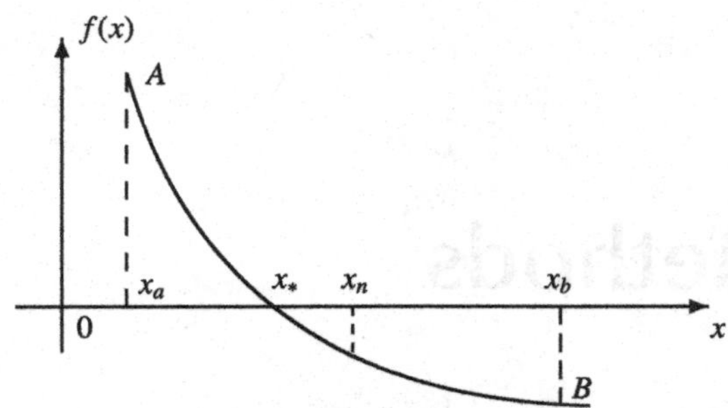

FIGURE 20.1 Bisection method for finding the root x_* of $f(x_*) = 0$ between two bounds x_a and x_b in the domain $x \in [a, b]$.

We know that the exact solution is $\sqrt{23} = 4.7958315233$. We can see that after only five iterations, x_5 is accurate to the 4th decimal place.

Iteration is in general very efficient; however, we have to be careful about the proper design of the iteration formula and the selection of an appropriate initial guess. For example, we cannot use $x_0 = 0$ as the initial value. Similarly, if we start $x_0 = 1000$, we have to do more iterations to get the same accuracy.

There are many methods for finding the roots such as the bisection method, though the Newton-Raphson method is by far the most successful and widely used.

20.2 BISECTION METHOD

The above-mentioned iteration method to find $x = \sqrt{k}$ is in fact equivalent to finding the solution or the root of the function $f(x) = x^2 - k = 0$. For any function $f(x)$ in the interval $[a, b]$, the root-finding bisection method works in the following way as shown in Fig. 20.1.

The iteration procedure starts with two initial guessed bounds x_a (lower bound), and x_b (upper bound) so that the true root $x = x_*$ lies between these two bounds. This requires that $f(x_a)$ and $f(x_b)$ have different signs. In our case shown in Fig. 20.1, $f(x_a) > 0$ and $f(x_b) < 0$, but $f(x_a)f(x_b) < 0$. The obvious choice is $x_a = a$ and $x_b = b$. The next estimate is just the midpoint of A and B, and we have

$$x_n = \frac{1}{2}(x_a + x_b). \tag{20.4}$$

We then have to test the sign of $f(x_n)$. If $f(x_n) < 0$ (having the same sign as $f(x_b)$), we then update the new upper bound as $x_b = x_n$. If $f(x_n) > 0$ (having the same sign as $f(x_a)$), we update the new lower bound as $x_a = x_n$. In a special case when $f(x_n) = 0$, you have found the true root. The iterations continue in the same manner until a given accuracy is achieved or the prescribed number of iterations is reached.

Example 20.2

If we want to find $\sqrt{\pi}$, we have

$$f(x) = x^2 - \pi = 0.$$

We can use $x_a = 1$ and $x_b = 2$ since $\pi < 4$ (thus $\sqrt{\pi} < 2$). The first bisection point is

$$x_1 = \frac{1}{2}(x_a + x_b) = \frac{1}{2}(1 + 2) = 1.5.$$

Since $f(x_a) < 0$, $f(x_b) > 0$ and $f(x_1) = -0.8916 < 0$, we update the new lower bound $x_a = x_1 = 1.5$. The second bisection point is

$$x_2 = \frac{1}{2}(1.5 + 2) = 1.75,$$

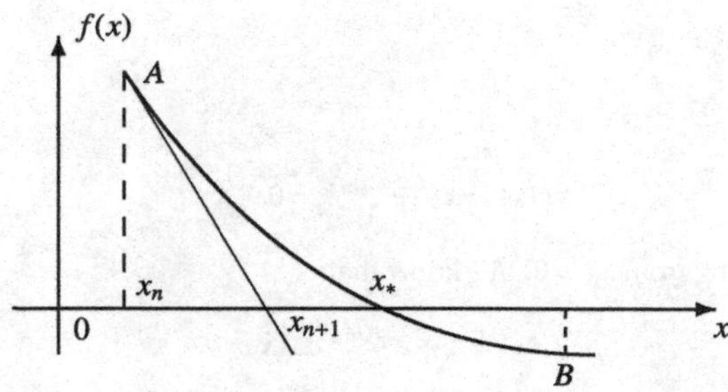

FIGURE 20.2 Newton's method of approximating the root x_* by x_{n+1} from x_n.

and $f(x_2) = -0.0791 < 0$, so we update lower bound again $x_a = 1.75$. The third bisection point is

$$x_3 = \frac{1}{2}(1.75 + 2) = 1.875.$$

Since $f(x_3) = 0.374 > 0$, we now update the new upper bound $x_b = 1.875$. The fourth bisection point is

$$x_4 = \frac{1}{2}(1.75 + 1.875) = 1.8125.$$

It is within 2.5% of the true value of $\sqrt{\pi} \approx 1.7724538509$.

In general, the convergence of the bisection method is very slow, and Newton's method is a much better choice in most cases.

20.3 NEWTON-RAPHSON METHOD

Newton's method is a widely-used classical method for finding the solution to a nonlinear univariate function of $f(x)$ on the interval $[a, b]$. It is also referred to as the Newton-Raphson method. At any given point x_n shown in Fig. 20.2, we can approximate the function by a Taylor series

$$f(x_{n+1}) = f(x_n + \Delta x) \approx f(x_n) + f'(x_n)\Delta x, \tag{20.5}$$

where

$$\Delta x = x_{n+1} - x_n, \tag{20.6}$$

which leads to

$$x_{n+1} - x_n = \Delta x \approx \frac{f(x_{n+1}) - f(x_n)}{f'(x_n)}, \tag{20.7}$$

or

$$x_{n+1} \approx x_n + \frac{f(x_{n+1}) - f(x_n)}{f'(x_n)}. \tag{20.8}$$

Since we try to find an approximation to $f(x) = 0$ with $f(x_{n+1})$, we can use the approximation $f(x_{n+1}) \approx 0$ in the above expression. Thus we have the standard Newton iterative formula

$$x_{n+1} = x_n - \frac{f(x_n)}{f'(x_n)}. \tag{20.9}$$

The iteration procedure starts from an initial guess value x_0 and continues until a predefined criterion is met. A good initial guess will use fewer steps; however, if there is no obvious initial good starting point, you can start at any point on the interval $[a, b]$. But if the initial value is too far from the true zero, the iteration process may fail. So it is a good idea to limit the number of iterations.

Example 20.3

To find the root of

$$f(x) = x - e^{-\sin x} = 0,$$

we use Newton-Raphson method starting from $x_0 = 0$. We know that

$$f'(x) = 1 + e^{-\sin x} \cos x,$$

and thus the iteration formula becomes

$$x_{n+1} = x_n - \frac{x_n - e^{-\sin x_n}}{1 + e^{-\sin x_n} \cos x_n}.$$

Since $x_0 = 1$, we have

$$x_1 = 0 - \frac{0 - e^{-\sin 0}}{1 + e^{-\sin 0} \cos 0} = 0.5000000000,$$
$$x_2 = 0.5 - \frac{0.5 - e^{-\sin 0.5}}{1 + e^{-\sin 0.5} \cos 0.5} \approx 0.5771952598,$$
$$x_3 \approx 0.5787130827, \qquad x_4 \approx 0.5787136435.$$

We can see that x_3 (only three iterations) is very close (to the 6th decimal place) to the true root which is $x_* \approx 0.57871364351972$, while x_4 is accurate to the 10th decimal place.

We have seen that Newton-Raphson's method is very efficient and that is why it is so widely used. Using this method, we can virtually solve almost all root-finding problems, though care should be taken when dealing with multiple roots. Obviously, this method is not applicable to carrying out integration.

20.4 NUMERICAL INTEGRATION

For any smooth function, we can always calculate its derivatives by direct differentiation; however, integration is often difficult even for seemingly simple integrals such as the error function

$$\mathrm{erf}(x) = \frac{2}{\sqrt{\pi}} \int_0^x e^{-u^2} du. \tag{20.10}$$

The integration of this simple integrand $\exp(-u^2)$ does not lead to any simple explicit expression, which is why it is often written as erf(), referred to as the error function. If we pick up a mathematical handbook, we know that $\mathrm{erf}(0) = 0$, and $\mathrm{erf}(\infty) = 1$, while

$$\mathrm{erf}(0.5) \approx 0.52049, \qquad \mathrm{erf}(1) \approx 0.84270. \tag{20.11}$$

If we want to calculate such integrals, numerical integration is the best alternative.

Now if we want to numerically evaluate the following integral

$$\mathcal{I} = \int_a^b f(x)dx, \tag{20.12}$$

where a and b are fixed and finite; we know that the value of the integral is exactly the total area under the curve $y = f(x)$ between a and b. As both the integral and the area can be considered as the sum of the values over many small intervals, the simplest way of evaluating such numerical integration is to divide up the integral interval into n equal small sections and split the area into n thin strips of width h so that $h \equiv \Delta x = (b - a)/n$, $x_0 = a$ and $x_i = ih + a$ $(i = 1, 2, ..., n)$. The values of the functions at the dividing points x_i are denoted as $y_i = f(x_i)$, and the value at the midpoint between x_i and x_{i+1} is labeled as $y_{i+1/2} = f_{i+1/2}$

$$y_{i+1/2} = f(x_{i+1/2}) = f_{i+1/2}, \qquad x_{i+1/2} = \frac{x_i + x_{i+1}}{2}. \tag{20.13}$$

The accuracy of such approximations depends on the number n and the way to approximate the curve in each interval.

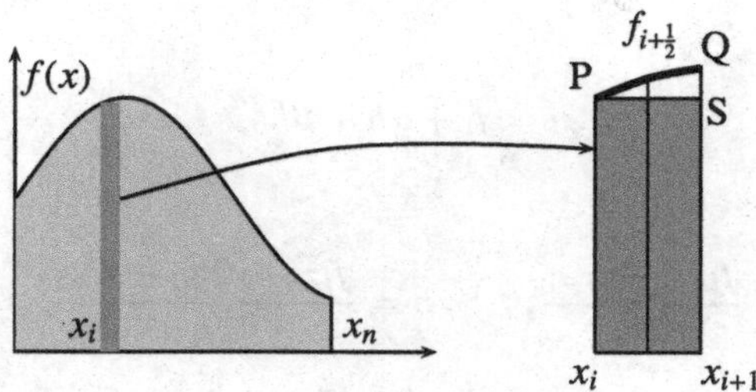

FIGURE 20.3 Integral as a sum of multiple thin stripes.

Fig. 20.3 shows such an interval $[x_i, x_{i+1}]$ which is exaggerated in the figure for clarity. The curve segment between P and Q is approximated by a straight line with a slope

$$\frac{\Delta y}{\Delta x} = \frac{f(x_{i+1}) - f(x_i)}{h}, \tag{20.14}$$

which approaches $f'(x_{i+1/2})$ at the midpoint point when $h \to 0$.

The trapezium (formed by P, Q, x_{i+1}, and x_i) is a better approximation than the rectangle (P, S, x_{i+1} and x_i) because the former has an area

$$A_i = \frac{f(x_i) + f(x_{i+1})}{2} h, \tag{20.15}$$

which is closer to the area

$$\mathcal{I}_i = \int_{x_i}^{x_{i+1}} f(x)dx, \tag{20.16}$$

under the curve in the small interval x_i and x_{i+1}. If we use the area A_i to approximate $\mathcal{I}_i$, we have the trapezium rule of numerical integration. Thus, the integral is simply the sum of all these small trapeziums, and we have

$$\begin{aligned} \mathcal{I} &\approx \frac{h}{2}[f_0 + 2(f_1 + f_2 + ... + f_{n-1}) + f_n] \\ &= h[f_1 + f_2 + ... + f_{n-1} + \frac{(f_0 + f_n)}{2}]. \end{aligned} \tag{20.17}$$

From the Taylor series, we know that

$$\begin{aligned} \frac{f(x_i) + f(x_{i+1})}{2} &\approx \frac{1}{2}\Big\{[f(x_{i+1/2}) - \frac{h}{2} f'(x_{i+1/2}) + \frac{1}{2!}(\frac{h}{2})^2 f''(x_{i+1/2})] \\ &\quad + [f(x_{i+1/2}) + \frac{h}{2} f'(x_{i+1/2}) + \frac{1}{2!}(\frac{h}{2})^2 f''(x_{i+1/2})]\Big\} \\ &= f(x_{i+1/2}) + \frac{h^2}{8} f''(x_{i+1/2}), \end{aligned} \tag{20.18}$$

where $O(h^2 f'')$ means that the value is about the order of $h^2 f''$, or $O(h^2) = Kh^2 f''$ where K is a constant.

The trapezium rule is just one of the simple and popular schemes for numerical integration with the error of $O(h^3 f'')$. If we want higher accuracy, we can either reduce h or use a better approximation for $f(x)$. A small h means a large n, which implies that we have to do the sum of many small sections, and it may increase the computational time.

On the other hand, we can use higher-order approximations for the curve. Instead of using straight lines or linear approximations for curve segments, we can use parabolas or quadratic approximations. For any consecutive three points x_{i-1}, x_i and x_{i+1}, we can construct a parabola in the form

$$f(x_i + t) = f_i + \alpha t + \beta t^2, \qquad t \in [-h, h]. \tag{20.19}$$

As this parabola must go through the three known points (x_{i-1}, f_{i-1}) at $t = -h$, (x_i, f_i) at $t = 0$ and x_{i+1}, f_{i+1} at $t = h$, we have the following equations for α and β

$$f_{i-1} = f_i - \alpha h + \beta h^2, \tag{20.20}$$

and

$$f_{i+1} = f_i + \alpha h + \beta h^2, \tag{20.21}$$

which lead to

$$\alpha = \frac{f_{i+1} - f_{i-1}}{2h}, \qquad \beta = \frac{f_{i-1} - 2f_i + f_{i+1}}{2h^2}. \tag{20.22}$$

In fact, α is the centered approximation for the first derivative f_i' and β is related to the central difference scheme for the second derivative f_i''. Therefore, the integral from x_{i-1} to x_{i+1} can be approximated by

$$\mathcal{I}_i = \int_{x_{i-1}}^{x_{i+1}} f(x)dx \approx \int_{-h}^{h} [f_i + \alpha t + \beta t^2]dt = \frac{h}{3}[f_{i-1} + 4f_i + f_{i+1}],$$

where we have substituted the expressions for α and β. To ensure the whole interval $[a, b]$ can be divided up to form three-point approximations without any point left out, n must be even. Therefore, the estimate of the integral becomes

$$\mathcal{I} \approx \frac{h}{3}[f_0 + 4(f_1 + f_3 + ... + f_{n-1}) + 2(f_2 + f_4 + ... + f_{n-2}) + f_n], \tag{20.23}$$

which is the standard Simpson's rule.

As the approximation for the function $f(x)$ is quadratic, an order higher than the linear form, the error estimate of Simpson's rule is thus $O(h^4)$ or $O(h^4 f''')$ to be more specific. There are many variations of Simpson's rule with higher-order accuracies such as $O(h^5 f^{(4)})$.

Example 20.4

We know the value of the integral

$$I = \text{erf}(1) = \frac{2}{\sqrt{\pi}} \int_0^1 e^{-x^2} dx = 0.8427007929.$$

Let us now estimate

$$I = \frac{2}{\sqrt{\pi}} J, \qquad J = \int_0^1 e^{-x^2} dx,$$

using the Simpson rule with $n = 8$ and $h = (1 - 0)/8 = 0.125$. We have

$$J \approx \frac{h}{3}[f_0 + 4(f_1 + f_3 + f_5 + f_7) + 2(f_2 + f_4 + f_6) + f_8].$$

Since $f_i = e^{-x_i^2} = e^{-(i*h)^2}$, we have $f_0 = 1$, $f_1 = 0.984496$, $f_2 = 0.939413$, $f_3 = 0.868815$, $f_4 = 0.778801$, $f_5 = 0.676634$, $f_6 = 0.569783$, $f_7 = 0.465043$, and $f_8 = 0.367879$. Now the integral estimate of J is

$$J \approx \frac{0.125}{3}[1 + 4 \times 2.9949885 + 2 \times 2.2879967 + 0.367879]$$
$$\approx \frac{0.125}{3} \times 17.923827 \approx 0.746826.$$

Finally, the integral estimate of I is

$$I = \frac{2}{\sqrt{\pi}} J = \frac{2}{\sqrt{3.1415926}} \times 0.746826 \approx 0.842703.$$

We can see that this estimate is accurate to the 5th decimal place.

There are other even better ways for evaluating the integral more accurately using fewer points of evaluation. Such numerical integration is called the Gaussian integration or Gaussian quadrature. Interested readers can refer to more advanced books on numerical methods or numerical analysis.

20.5 NUMERICAL SOLUTIONS OF ODES

The simplest first-order differential equation can be written as

$$\frac{dy}{dx} = f(x, y), \tag{20.24}$$

where $f(x, y)$ is a known function of x and y. In principle, the solution can be obtained by direct integration

$$y(x) = y(x_0) + \int_{x_0}^{x} f(x, y(x))dx, \tag{20.25}$$

but in practice it is usually impossible to do the integration analytically, as it requires the solution $y(x)$ to evaluate the right-hand side. Numerical integration in this case is the most common technique for obtaining approximate solutions.

A naive approach is to use the standard techniques such as Simpson's rule for numerical integration to evaluate the integral numerically. However, since we have to use some approximation for $y(x)$, such techniques hardly work here without modification. A better approach is to start from the known initial value $y(x_0)$ at x_0, and try to march to the next point at $x = x_0 + h$ where h is a small increment. In this manner, the solution at all other values of x can be estimated numerically. This is essentially the Euler scheme.

20.5.1 Euler Scheme

The basic idea of the Euler method is to approximate the integral (20.25) using an iterative procedure

$$y_{n+1} = y_n + \int_{x_n}^{x_{n+1}} f(x, y)dx = y_n + hf(x_n, y_n), \tag{20.26}$$

where $h = \Delta x = x_{n+1} - x_n$ is a small increment. Here we use the notations $x_n = x_0 + nh$ for $(n = 0, 1, 2, ..., N)$ and $y_n = y(x_n)$. This is essentially to divide the interval from x_0 to $x = Nh$ into N small interval of width h, so the value of $f(x, y)$ is approximated by its value in the interval as $f(x_n, y_n)$.

The above equation can also be viewed as the approximation of the first derivative

$$\frac{dy_n}{dx} = y_n' = \frac{y_{n+1} - y_n}{h}. \tag{20.27}$$

This is a forward difference scheme for approximating the first derivative. From the differentiation and integration discussed earlier in this book, we know that such an approximation is very crude, and the accuracy is at most $O(h^2)$.

A potentially better way is to use the so-called central difference to approximate the first derivative by

$$y_n' = \frac{y_{n+1} - y_{n-1}}{2h}, \tag{20.28}$$

which uses two steps. This does not always solve the problem. In order to get a reasonable accuracy, we have to use a very small value of h. There is also an issue called numerical stability. If the step size h is too large, there is a possibility that any error will be amplified during each iteration, and after some iterations, the value y_n might be unstable and become meaningless. A far better and more stable numerical scheme is the well-known Runge-Kutta method.

20.5.2 Runge-Kutta Method

The Runge-Kutta method uses a trial step to march the solution to the midpoint of the interval by the central difference

$$y_{n+1/2} = y_n + \frac{h}{2} f(x_n, y_n), \tag{20.29}$$

then it combines with the standard forward difference as used in Euler scheme

$$y_{n+1} = y_n + hf(x_{n+1/2}, y_{n+1/2}). \tag{20.30}$$

This scheme can be viewed as a predictor-corrector method as it first predicts a value in the midpoint, and then corrects it to an even better estimate at the next step.

If we follow the same idea, we can devise higher-order methods with multiple steps, which will give higher accuracy. The popular classical Runge-Kutta method is a fourth-order method, which involves four steps to advance from n to $n+1$. We have

$$a = hf(x_n, y_n), \qquad b = hf(x_n + h/2, y_n + a/2),$$
$$c = hf(x_n + h/2, y_n + b/2), \qquad d = hf(x_n + h, y_n + c),$$
$$y_{n+1} = y_n + \frac{a + 2(b+c) + d}{6}, \tag{20.31}$$

which is fourth-order accurate.

Example 20.5

Let us solve the nonlinear equation numerically

$$\frac{dy}{dx} + y^2 = -1, \qquad x \in [0, 2],$$

with the initial condition $y(0) = 1$. We know that $\frac{dy}{dx} = -(1 + y^2)$, or

$$-\int \frac{1}{1+y^2} dy = -\tan^{-1} y = \int dx = x + K,$$

where K is the constant of the integration. This gives

$$y = -\tan(x + K).$$

Using the initial condition $y = 1$ and $x = 0$, we have $1 = -\tan(K)$, or

$$K = -\tan^{-1}(1) = -\frac{\pi}{4}.$$

The analytical solution becomes

$$y(x) = -\tan(x - \frac{\pi}{4}).$$

On the interval $[0, 2]$, let us first solve the equation using the Euler scheme for $h = 0.5$. There are five points $x_i = ih$ (where $i = 0, 1, 2, 3, 4$). As $dy/dx = f(y) = -1 - y^2$, we have the Euler scheme

$$y_{n+1} = y_n + hf(y_n) = y_n - h - hy_n^2.$$

From the initial condition $y_0 = 1$, we now have

$$y_1 = y_0 - h - hy_0^2 = 1 - 0.5 - 0.5 \times 1^2 = 0,$$
$$y_2 \approx -0.5, \qquad y_3 \approx -1.125, \qquad y_4 = -2.2578.$$

These are significantly different (about 30%) from the exact solution

$$y_0* = 1, \quad y_1^* = 0.2934079, \quad y_2^* = -0.21795809,$$
$$y_3^* = -0.86756212, \quad y_4^* = -2.68770693.$$

Now let us use the Runge-Kutta method to solve the same equation to see if it is better. Since $f(x_n, y_n) = -1 - y_n^2$, we have

$$a = hf(x_n, y_n) = -h(1 + y_n^2), \qquad b = -h[1 + (y_n + \frac{a}{2})^2],$$
$$c = -h[1 + (y_n + \frac{b}{2})^2], \qquad d = -h[1 + (y_n + c)^2],$$

and

$$y_{n+1} = y_n + \frac{a+2(b+c)+d}{6}.$$

From $y_0 = 1$ and $h = 0.5$, we have

$$y_1 \approx 0.29043, \ y_2 \approx -0.22062, \ y_3 \approx -0.87185, \ y_4 \approx -2.67667.$$

These values are within about 1% of the analytical solution y_n^*.

We can see that even with the same step size h, the Runge-Kutta method is more efficient and accurate than the Euler scheme. Generally speaking, higher-order schemes are better than lower-order schemes, but not always. Let us look at another example.

Example 20.6

We know from the example in an earlier chapter on ODEs that the solution to the differential equation

$$\frac{dy}{dx} = (x^4 + e^x)e^{-y},$$

with the initial condition $y(0) = 0$ at $x_0 = 0$ is

$$y = \ln(e^x + \frac{1}{5}x^5).$$

Now let us try to solve it numerically to estimate $y(1)$.

First, we divide the interval [0, 1] into two steps so that $h = (1-0)/2 = 0.5$. We have $x_0 = 0$ and $y_0 = 0$. Then, we use the Euler scheme, and we have y_1 at $x_1 = 0.5$

$$y_1 = y_0 + hf(x_0, y_0) = y_0 + h(x_0^4 + e^{x_0})e^{-y_0} = 0 + 0.5 \times (0^4 + e^0)e^{-0} = 0.5.$$

The next step is to evaluate y_2 at $x_2 = 2h = 1$, and we have

$$y_2 = y_1 + h(x_1^4 + e^{x_1})e^{-y_1} = 0.5 + 0.5 \times (0.5^4 + e^{0.5})e^{-0.5} \approx 1.0189.$$

We know the exact value at $x = 1$ is

$$y_* = y(1) = \ln(e^1 + \frac{1^5}{5}) = 1.07099503.$$

We can see that the value y_2 obtained by the Euler scheme is about 5% from the true solution. Now let us see whether the Runge-Kutta method can do better.

For the first step to get y_1, we have

$$a = hf(x_0, y_0) = h(x_0^4 + e^{x_0})\epsilon^{-y_0} = 0.5 \times (0^4 + e^0) \times e^{-0} = 0.5,$$
$$b = hf(x_0 + h/2, y_0 + a/2) = 0.5 \times [(0 + h/2)^4 + e^{0+h/2}] \times e^{-(0+a/2)} = 0.50152,$$

and

$$c = hf(x_0 + h/2, y_0 + b/2) \approx 0.50114, \ d = hf(x_0 + h, y_0 + c) \approx 0.518363.$$

Now we get

$$y_1 = y_0 + \frac{a+2(b+c)+d}{6} = 0.503947.$$

The next step to get y_2 from $x_1 = 0.5$ and $y_1 = 0.503947$, we have

$$a = hf(x_1, y_1) = 0.5 \times (0.5^4 + e^{0.5}) \times e^{-0.503947} \approx 0.51691,$$

$$b = hf(x_1 + \frac{h}{2}, y_1 + \frac{a}{2}) \approx 0.56765, c = hf(x_1 + \frac{h}{2}, y_1 + \frac{b}{2}) \approx 0.55343,$$

and $d = hf(x_1 + h, y_1 + c) \approx 0.64580$. Finally, we have

$$y_2 = y_1 + \frac{a + 2(b+c) + d}{6} \approx 1.07142.$$

We can see that Runge-Kutta is more accurate than the Euler method, and the solution is about 0.04% from the true value because this Runge-Kutta method is of $O(h^4)$ accurate.

This example demonstrates that higher-order methods are usually the best choice, and they generally work better for most problems. That is why they are more widely used and have been implemented in many software packages. The general guidance is that you should try to use these methods first until you find a better method for a particular problem.

So far, we have only introduced the numerical method to the first-order equations. Higher-order ordinary differential equations can always be converted to a first-order system. For example, the following second-order equation

$$y''(x) + p(x)y'(x) + q(x)y(x) = r(x), \tag{20.32}$$

can be rewritten as a system if we let $u = y'(x)$

$$y' = u, \qquad u' = r(x) - q(x)y(x) - p(x)u. \tag{20.33}$$

The above system of two equations can be written as

$$\frac{d}{dx}\begin{pmatrix} y \\ u \end{pmatrix} = \begin{pmatrix} u \\ r(x) - q(x)y - p(x)u \end{pmatrix}. \tag{20.34}$$

This is essentially the same as (20.24) in the context of vectors and matrices, and it can essentially be solved using similar methods with little modifications.

Example 20.7

Let us try to solve the following second-order ordinary differential equation:

$$y''(x) + xy'(x) + x^2 y(x) = e^{-x}.$$

By setting $u(x) = y'(x)$, we have a system of two equations

$$y'(x) = u(x), \quad u' = e^{-x} - x^2 y(x) - xu(x),$$

which can be rewritten as

$$\begin{cases} y' = f(x, y, u) = u, \\ u' = g(x, y, u) = e^{-x} - x^2 y - xu. \end{cases}$$

This can be solved any method discussed earlier. For example, with the simple Euler method, we have

$$\begin{cases} y_{n+1} = y_n + hy'_n = y_n + hf(x_n, y_n, u_n), \\ u_{n+1} = u_n + hu'_n = u_n + hg(x_n, y_n, u_n), \end{cases}$$

where $h > 0$ is the step size. To actually solve this system, it requires two initial conditions:

$$y(x_0) = y_0, \quad u(x_0) = y'(x_0) = u_0.$$

Numerical solutions of partial differential equations are even more difficult, especially when the domain is multidimensional and irregular. For partial differential equations to be introduced later, numerical methods can be even more complicated, including finite different methods, finite volume methods, finite element methods, boundary element methods, spectral methods, and meshless methods. Interested readers can refer to more advanced literature.

EXERCISES

20.1. Write a program in any programming language to compute $\sqrt{\pi}$ to the 7th decimal place.
20.2. Solve $e^{-x} - x^2 = 0$ using Newton's method and write a compute program.
20.3. Find the integral $\int_0^1 \sqrt{1 - x^{3/2}}dx$ numerically.
20.4. Solve $y'(x) - xy(x) = x$ with the initial condition $y(0) = 0$.

Chapter 21

Computational Linear Algebra

Chapter Points

- Systems of linear equations are introduced, together with the fundamentals of Gauss elimination and LU factorization.
- Solutions of linear systems, especially large, sparse systems, are rarely done by inverse methods. In most cases, iteration methods are used. Thus, we introduce a few iterative methods such as Jacobi and Gauss-Seidel iterations, relaxation method, and conjugate gradient method.

21.1 SYSTEM OF LINEAR EQUATIONS

A linear system of m equations for n unknowns $\mathbf{u} = (u_1, ..., u_n)^T$

$$\sum_{j=1}^{n} a_{ij} u_j = b_i, \qquad (i = 1, 2, ..., m), \tag{21.1}$$

can be written in the compact form as

$$\begin{pmatrix} a_{11} & a_{12} & \dots & a_{1n} \\ a_{21} & a_{22} & \dots & a_{2n} \\ \vdots & & \ddots & \vdots \\ a_{m1} & a_{m2} & \dots & a_{mn} \end{pmatrix} \begin{pmatrix} u_1 \\ u_2 \\ \vdots \\ u_n \end{pmatrix} = \begin{pmatrix} b_1 \\ b_2 \\ \vdots \\ b_m \end{pmatrix}, \tag{21.2}$$

or simply

$$\mathbf{Au} = \mathbf{b}. \tag{21.3}$$

Here both the matrix $\mathbf{A} = [a_{ij}]$ and the column vector $\mathbf{b} = (b_1, ..., b_m)^T$ are known. If $m < n$, the system is under-determined as the conditions are not sufficient to guarantee a unique solution set. On the other hand, the system is over-determined if $m > n$ because there are too many conditions and a solution may not exist at all. The unique solution is only possible when $m = n$ or when $\mathbf{A}$ is a non-singular square matrix.

The solution of this matrix equation is important to many numerical problems, ranging from the solution of a large system of linear equations to linear mathematical programming, and from data interpolation to finding solutions to finite element problems.

The inverse of $\mathbf{A}$ is possible only if $m = n$. If the inverse $\mathbf{A}^{-1}$ does not exist, then the linear system is under-determined or there are no unique solutions (or even no solution at all). In order to find the solutions, we multiply both sides by $\mathbf{A}^{-1}$,

$$\mathbf{A}^{-1}\mathbf{Au} = \mathbf{A}^{-1}\mathbf{b}, \tag{21.4}$$

and we obtain the solution

$$\mathbf{u} = \mathbf{A}^{-1}\mathbf{b}. \tag{21.5}$$

A special case of this equation is when $\mathbf{b} = \lambda\mathbf{u}$, and this becomes an eigenvalue problem as discussed earlier in this book.

It is worth pointing out that matrix $\mathbf{A}$ can also be complex with its entries or elements being complex numbers. In addition, not every element are non-zero. In fact, in many real-world applications, matrix $\mathbf{A}$ can usually be sparse but large;

that is, most of the entries of $\mathbf{A}$ are zeros; however, the size of the matrix tends to be very large with millions of entries. Thus, this can cause problems for storing the matrices and thus some special approaches of memory-saving techniques can be used to store such large, sparse matrices.

Mathematically speaking, a linear system can in principle be solved analytically using Cramer's rule,

$$u_i = \frac{\det \mathbf{A}_i}{\det \mathbf{A}}, \qquad i = 1, 2, ..., n, \tag{21.6}$$

where the matrix $\mathbf{A}_i$ is obtained by replacing the ith column by the column vector $\mathbf{b}$.

Example 21.1

For example, for three linear equations with three unknowns u_1, u_2 and u_3,

$$a_{11}u_1 + a_{12}u_2 + a_{13}u_3 = b_1,$$
$$a_{21}u_1 + a_{22}u_2 + a_{23}u_3 = b_2,$$
$$a_{31}u_1 + a_{32}u_2 + a_{33}u_3 = b_3,$$

its solution vector is given by the following Cramer's rule

$$u_1 = \frac{1}{\Delta}\begin{vmatrix} b_1 & a_{12} & a_{13} \\ b_2 & a_{22} & a_{23} \\ b_3 & a_{32} & a_{33} \end{vmatrix}, \quad u_2 = \frac{1}{\Delta}\begin{vmatrix} a_{11} & b_1 & a_{13} \\ a_{21} & b_2 & a_{23} \\ a_{31} & b_3 & a_{33} \end{vmatrix}, \quad u_3 = \frac{1}{\Delta}\begin{vmatrix} a_{11} & a_{12} & b_1 \\ a_{21} & a_{22} & b_2 \\ a_{31} & a_{32} & b_3 \end{vmatrix},$$

where

$$\Delta = \begin{vmatrix} a_{11} & a_{12} & a_{13} \\ a_{21} & a_{22} & a_{23} \\ a_{31} & a_{32} & a_{33} \end{vmatrix}.$$

Though it is straightforward to extend the rule to any dimensions in theory, this is not an easy task in practice because the calculation of the determinant of a large matrix is not easy. Though Cramer's rule is good for proving theorems, it is not good for numerical implementation. Better methods are to use the inverse matrix and/or iterations.

Finding the inverse $\mathbf{A}^{-1}$ of a square $n \times n$ matrix $\mathbf{A}$ is not an easy task either, especially when the size of the matrix is large, and it usually requires the algorithm complexity of $O(n^3)$. For example, in a typical engineering simulation using the finite element method, we can easily have $n = 10^7$. In order to calculate the inverse of $\mathbf{A}$, it may require $O(n^3) = O(10^{21})$, which is a huge number of calculations. Even with the supercomputers such as IBM's Blue Gene that can nowadays do 10 petaflops of calculations (that is about 10^{16} floating-point operations per second), it can still take more than a day. In fact, many solution methods are designed to avoid the necessity of calculating the inverse $\mathbf{A}^{-1}$ if possible.

There are many ways of solving linear equations, but they fall into two categories: direct algebraic methods and iteration methods. The purpose of the former is to find the solution by elimination, decomposition of matrix and substitutions, while the latter involves certain iterations to find the approximate solutions. The choice of these methods depends on the characteristics of the matrix $\mathbf{A}$, size of the problem, computational time, the type of problem and the required solution quality.

21.2 GAUSS ELIMINATION

The basic idea of Gauss elimination is to transform a square matrix into a triangular matrix by elementary row operations, so that the simplified triangular system can be solved by direct back substitution. For the linear system

$$\begin{pmatrix} a_{11} & a_{12} & a_{13} & \dots & a_{1n} \\ a_{21} & a_{22} & a_{23} & \dots & a_{2n} \\ & & \vdots & & \\ a_{n1} & a_{n2} & a_{n3} & \dots & a_{nn} \end{pmatrix} \begin{pmatrix} u_1 \\ u_2 \\ \vdots \\ u_n \end{pmatrix} = \begin{pmatrix} b_1 \\ b_2 \\ \vdots \\ b_n \end{pmatrix}, \tag{21.7}$$

the aim in the first step is to try to make all the coefficients in the first column $(a_{21}, ..., a_{n1})$ become zeros by elementary row operations, except for the first element. This is based on the principle that a linear system will remain the same if its rows are multiplied by some non-zero coefficients, or any two rows are interchanged, or any two (or more) rows are combined through addition and subtraction.

To do this, we first divide the first equation by a_{11} (we can always assume $a_{11} \neq 0$; if not, we rearrange the order of the equations to achieve this). We now have

$$\begin{pmatrix} 1 & \frac{a_{12}}{a_{11}} & \frac{a_{13}}{a_{11}} & \dots & \frac{a_{1n}}{a_{11}} \\ a_{21} & a_{22} & a_{23} & \dots & a_{2n} \\ \vdots & & & & \\ a_{n1} & a_{n2} & a_{n3} & \dots & a_{nn} \end{pmatrix} \begin{pmatrix} u_1 \\ u_2 \\ \vdots \\ u_n \end{pmatrix} = \begin{pmatrix} \frac{b_1}{a_{11}} \\ b_2 \\ \vdots \\ b_n \end{pmatrix}. \tag{21.8}$$

Then multiplying the first row by $-a_{21}$ and adding it to the second row, multiplying the first row by $-a_{i1}$ and adding it to the i-th row, we finally have

$$\begin{pmatrix} 1 & \frac{a_{12}}{a_{11}} & \frac{a_{13}}{a_{11}} & \dots & \frac{a_{1n}}{a_{11}} \\ 0 & a_{22} - \frac{a_{21}a_{12}}{a_{11}} & & \dots & a_{2n} - \frac{a_{21}a_{1n}}{a_{11}} \\ \vdots & & & & \\ 0 & a_{n2} - \frac{a_{n1}a_{12}}{a_{11}} & & \dots & a_{nn} - \frac{a_{n1}a_{1n}}{a_{11}} \end{pmatrix} \begin{pmatrix} u_1 \\ u_2 \\ \vdots \\ u_n \end{pmatrix} = \begin{pmatrix} \frac{b_1}{a_{11}} \\ b_2 - \frac{a_{21}b_1}{a_{11}} \\ \vdots \\ b_n - \frac{a_{n1}b_n}{a_{11}} \end{pmatrix}.$$

We then repeat the same procedure for the third row to the n-th row, and the final form of the linear system should be in the following generic form

$$\begin{pmatrix} \alpha_{11} & \alpha_{12} & \alpha_{13} & \dots & \alpha_{1n} \\ 0 & \alpha_{22} & \alpha_{23} & \dots & \alpha_{2n} \\ \vdots & & & \ddots & \\ 0 & 0 & 0 & \dots & \alpha_{nn} \end{pmatrix} \begin{pmatrix} u_1 \\ u_2 \\ \vdots \\ u_n \end{pmatrix} = \begin{pmatrix} \beta_1 \\ \beta_2 \\ \vdots \\ \beta_n \end{pmatrix}, \tag{21.9}$$

where $\alpha_{1j} = a_{1j}/a_{11}$, $\alpha_{2j} = a_{2j} - a_{1j}a_{21}/a_{11} (j = 1, 2, ..., n)$, ..., $\beta_1 = b_1/a_{11}$, $\beta_2 = b_2 - a_{21}b_1/a_{11}$ and others. From the above form, we can see that $u_n = \beta_n/\alpha_{nn}$ because there is only one unknown u_n in the n-th row. We can then use the back substitutions to obtain u_{n-1} and up to u_1. Therefore, we have

$$u_n = \frac{\beta_n}{\alpha_{nn}},$$

$$u_i = \frac{1}{\alpha_{ii}}(\beta_i - \sum_{j=i+1}^{n} \alpha_{ij}x_j), \tag{21.10}$$

where $i = n-1, n-2, ..., 1$. Obviously, in our present case, $\alpha_{11} = ... = \alpha_{nn} = 1$. Let us look at an example.

Example 21.2

For the linear system

$$\begin{pmatrix} 2 & -1 & 3 & 4 \\ 3 & 2 & -5 & 6 \\ -2 & 1 & 0 & 5 \\ 4 & -5 & -6 & 0 \end{pmatrix} \begin{pmatrix} u_1 \\ u_2 \\ u_3 \\ u_4 \end{pmatrix} = \begin{pmatrix} 21 \\ 9 \\ 12 \\ -3 \end{pmatrix},$$

we first divide the first row by $a_{11} = 2$, we have

$$\begin{pmatrix} 1 & -1/2 & 3/2 & 2 \\ 3 & 2 & -5 & 6 \\ -2 & 1 & 0 & 5 \\ 4 & -5 & -6 & 0 \end{pmatrix} \begin{pmatrix} u_1 \\ u_2 \\ u_3 \\ u_4 \end{pmatrix} = \begin{pmatrix} 21/2 \\ 9 \\ 12 \\ -3 \end{pmatrix}.$$

Multiplying the first row by 3 and subtracting it from the second row, and carrying out similar row manipulations for the other rows, we have

$$\begin{pmatrix} 1 & -1/2 & 3/2 & 2 \\ 0 & 7/2 & -19/2 & 0 \\ 0 & 0 & 3 & 9 \\ 0 & -3 & -12 & -8 \end{pmatrix} \begin{pmatrix} u_1 \\ u_2 \\ u_3 \\ u_4 \end{pmatrix} = \begin{pmatrix} 21/2 \\ -45/7 \\ 33 \\ -45 \end{pmatrix}.$$

For the second row, we repeat this procedure again, we have

$$\begin{pmatrix} 1 & -1/2 & 3/2 & 2 \\ 0 & 1 & -19/7 & 0 \\ 0 & 0 & 3 & 9 \\ 0 & 0 & -141/7 & -8 \end{pmatrix} \begin{pmatrix} u_1 \\ u_2 \\ u_3 \\ u_4 \end{pmatrix} = \begin{pmatrix} 21/2 \\ -45/7 \\ 33 \\ -450/7 \end{pmatrix}.$$

Similarly, for the third row, we have

$$\begin{pmatrix} 1 & -1/2 & 3/2 & 2 \\ 0 & 1 & -19/7 & 0 \\ 0 & 0 & 1 & 3 \\ 0 & 0 & 0 & 367/7 \end{pmatrix} \begin{pmatrix} u_1 \\ u_2 \\ u_3 \\ u_4 \end{pmatrix} = \begin{pmatrix} 21/2 \\ -45/7 \\ 11 \\ 1101/7 \end{pmatrix}.$$

The fourth row gives that $u_4 = 3$. Using the back substitution, we have $u_3 = 2$ from the third row. Similarly, we have $u_2 = -1$ and $u_1 = 1$. So the solution vector is

$$\begin{pmatrix} u_1 \\ u_2 \\ u_3 \\ u_4 \end{pmatrix} = \begin{pmatrix} 1 \\ -1 \\ 2 \\ 3 \end{pmatrix}.$$

We have seen from the example that there are many floating-point calculations even for the simple system of four linear equations. In fact, the full Gauss elimination is computationally extensive with an algorithmic complexity of $O(2n^3/3)$.

Gauss-Jordan elimination, a variant of Gauss elimination, solves a linear system and, at the same time, can also compute the inverse of a square matrix. The first step is to formulate an augmented matrix from **A**, **b** and the unit matrix **I** (with the same size of **A**). That is

$$\mathbf{B} = [\mathbf{A}|\mathbf{b}|\mathbf{I}] = \begin{pmatrix} a_{11} & \dots & a_{1n} & |b_1| & 1 & 0 & \dots & 0 \\ a_{21} & \dots & a_{2n} & |b_2| & 0 & 1 & \dots & 0 \\ & \vdots & & \vdots & & & \vdots & \\ a_{n1} & \dots & a_{nn} & |b_n| & 0 & 0 & \dots & 1 \end{pmatrix}, \tag{21.11}$$

where the notation $\mathbf{A}|\mathbf{b}$ denotes the augmented form of two matrix **A** and **b**. The aim is to reduce **B** to the following form by elementary row reductions in a way similar to those carried out in Gauss elimination

$$\begin{pmatrix} 1 & 0 & \dots & 0 & |u_1| & a'_{11} & \dots & a'_{1n} \\ 0 & 1 & \dots & 0 & |u_2| & a'_{21} & \dots & a'_{2n} \\ & \vdots & & & \vdots & & \vdots & \\ 0 & 0 & \dots & 1 & |u_n| & a'_{n1} & \dots & a'_{nn} \end{pmatrix} = [\mathbf{I}|\mathbf{u}|\mathbf{A}^{-1}], \tag{21.12}$$

where $\mathbf{A}^{-1} = [a'_{ij}]$ is the inverse. This is better demonstrated by an example.

Example 21.3

In order to solve the following system

$$\mathbf{Au} = \begin{pmatrix} 1 & 2 & 3 \\ -2 & 2 & 5 \\ 4 & 0 & -5 \end{pmatrix} \begin{pmatrix} u_1 \\ u_2 \\ u_3 \end{pmatrix} = \begin{pmatrix} 5 \\ -2 \\ 14 \end{pmatrix} = \mathbf{b},$$

we first write it in an augmented form

$$\mathbf{B} = \begin{pmatrix} 1 & 2 & 3 & | & 5 & | & 1 & 0 & 0 \\ -2 & 2 & 5 & | & -2 & | & 0 & 1 & 0 \\ 4 & 0 & -5 & | & 14 & | & 0 & 0 & 1 \end{pmatrix}.$$

By elementary row operations, this could be changed into

$$\mathbf{B}' = [\mathbf{I}|\mathbf{u}|\mathbf{A}^{-1}] = \begin{pmatrix} 1 & 0 & 0 & | & 1 & | & \frac{5}{7} & -\frac{5}{7} & -\frac{2}{7} \\ 0 & 1 & 0 & | & 5 & | & -\frac{5}{7} & \frac{17}{14} & \frac{11}{14} \\ 0 & 0 & 1 & | & -2 & | & \frac{4}{7} & -\frac{4}{7} & -\frac{3}{7} \end{pmatrix},$$

which gives

$$\mathbf{u} = \begin{pmatrix} 1 \\ 5 \\ -2 \end{pmatrix}, \qquad \mathbf{A}^{-1} = \frac{1}{14} \begin{pmatrix} 10 & -10 & -4 \\ -10 & 17 & 11 \\ 8 & -8 & -6 \end{pmatrix}.$$

We can see that both the solution $\mathbf{u}$ and the inverse $\mathbf{A}^{-1}$ are simultaneously obtained in the Gauss-Jordan elimination.

The Gauss-Jordan elimination is not quite stable numerically. In order to get better and more stable schemes, a common practice is to use pivoting. Basically, pivoting is a scaling procedure by dividing all the elements in a row by the element with the largest magnitude or norm. If necessary, rows can be interchanged so the largest element is moved so that it becomes the leading coefficient, especially on the diagonal position. This means that all the scaled elements are in the range of $[-1, 1]$. Thus, exceptionally large numbers are removed, which makes the scheme more numerically stable.

An important issue in both Gauss elimination and Gauss-Jordan elimination methods is the non-zero requirement of leading coefficients such as $a_{11} \neq 0$. For a_{11}, it is possible to rearrange the equations to achieve this requirement. However, there is no guarantee that other coefficients such as $a_{22} - a_{21}a_{12}/a_{11}$ should be nonzero. If it is zero, there is a potential difficulty due to the division by zero. In order to avoid this problem, we can use other methods such as the pivoting method and LU factorization or decomposition.

21.3 LU FACTORIZATION

Any square matrix $\mathbf{A}$ can be written as the product of two triangular matrices in the form

$$\mathbf{A} = \mathbf{LU}, \tag{21.13}$$

where $\mathbf{L}$ and $\mathbf{U}$ are the lower and upper triangular matrices, respectively. A lower (upper) triangular matrix has elements only on the diagonal and below (above). That is

$$\mathbf{L} = \begin{pmatrix} \beta_{11} & 0 & \dots & 0 \\ \beta_{21} & \beta_{22} & \dots & 0 \\ \vdots & & \ddots & \\ \beta_{n1} & \beta_{n2} & \dots & \beta_{nn} \end{pmatrix}, \tag{21.14}$$

and

$$\mathbf{U} = \begin{pmatrix} \alpha_{11} & \dots & \alpha_{1,n-1} & \alpha_{1,n} \\ & \ddots & & \vdots \\ 0 & \dots & \alpha_{n-1,n-1} & \alpha_{n-1,n} \\ 0 & \dots & 0 & \alpha_{nn} \end{pmatrix}. \tag{21.15}$$

The linear system $\mathbf{Au} = \mathbf{b}$ can be written as the following two steps

$$\mathbf{Au} = (\mathbf{LU})\mathbf{u} = \mathbf{L}(\mathbf{Uu}) = \mathbf{b}, \tag{21.16}$$

or

$$\mathbf{Uu} = \mathbf{v}, \qquad \mathbf{Lv} = \mathbf{b}, \tag{21.17}$$

which are two linear systems with triangular matrices only, and these systems can be solved by forward and back substitutions. The solutions of v_i are given by

$$v_1 = \frac{b_1}{\beta_{11}}, \qquad v_i = \frac{1}{\beta_{ii}}(b_i - \sum_{j=1}^{i-1} \beta_{ij} v_j), \tag{21.18}$$

where $i = 2, 3, ..., n$. The final solutions u_i are then given by

$$u_n = \frac{v_n}{\alpha_{nn}}, \quad u_i = \frac{1}{\alpha_{ii}}(v_i - \sum_{j=i+1}^{n} \alpha_{ij} u_j), \tag{21.19}$$

where $i = n - 1, ..., 1$.

For triangular matrices such as $\mathbf{L}$, there are some interesting properties. The inverse of a lower (upper) triangular matrix is also a lower (upper) triangular matrix. The determinant of a triangular matrix is simply the product of its diagonal entries. That is

$$\det(\mathbf{L}) = |\mathbf{L}| = \prod_{i=1}^{n} \beta_{ii} = \beta_{11}\beta_{22}...\beta_{nn}. \tag{21.20}$$

More interestingly, the eigenvalues of a triangular matrix are the diagonal entries: $\beta_{11}, \beta_{22}, ..., \beta_{nn}$. These properties are convenient in determining the stability of an iteration scheme.

But there is another issue here and that is how to decompose a square matrix $\mathbf{A} = [a_{ij}]$ into $\mathbf{L}$ and $\mathbf{U}$. As there are $n(n+1)/2$ coefficients α_{ij} and $n(n+1)/2$ coefficients β_{ij}, so there are $n^2 + n$ unknowns. For the equation (21.13), we know that it could provide only n^2 equations (as there are only n^2 coefficients a_{ij}). They are

$$\sum_{k=1}^{i} \beta_{ik}\alpha_{kj} = a_{ij}, \qquad (i < j), \tag{21.21}$$

$$\sum_{k=1}^{j=i} \beta_{ik}\alpha_{kj} = a_{ij}, \qquad (i = j), \tag{21.22}$$

and

$$\sum_{k=1}^{j} \beta_{ik}\alpha_{kj} = a_{ij}, \qquad (i > j), \tag{21.23}$$

which again forms another system of n equations.

As $n^2 + n > n^2$, there are n free coefficients. Therefore, such factorization or decomposition is not uniquely determined. We have to impose some extra conditions. Fortunately, we can always set either $\alpha_{ii} = 1$ or $\beta_{ii} = 1$ where $i = 1, 2, ..., n$.

If we set $\beta_{ii} = 1$, we can use Crout's algorithm to determine α_{ij} and β_{ij}. We have the coefficients for the upper triangular matrix

$$\alpha_{ij} = a_{ij} - \sum_{k=1}^{i-1} \beta_{ik}\alpha_{kj}, \tag{21.24}$$

for $(i = 1, 2, ..., j)$ and $j = 1, 2, ..., n$. For the lower triangular matrix, we have

$$\beta_{ij} = \frac{1}{\alpha_{jj}}(a_{ij} - \sum_{k=1}^{j-1} \beta_{ik}\alpha_{kj}), \tag{21.25}$$

for $i = j+1, j+2, ..., n$.

Let us look at a very simple example.

Example 21.4

For a simple matrix $\mathbf{A} = \begin{pmatrix} 1 & 2 \\ 3 & 4 \end{pmatrix}$, we can decompose it into

$$\mathbf{A} = \mathbf{LU} = \begin{pmatrix} 1 & 0 \\ l_{21} & 1 \end{pmatrix} \begin{pmatrix} u_{11} & u_{12} \\ 0 & u_{22} \end{pmatrix}.$$

We can use the above procedure to get the unknown coefficients. Alternatively, we can simply expand to give

$$\mathbf{A} = \begin{pmatrix} 1 & 2 \\ 3 & 4 \end{pmatrix} = \begin{pmatrix} u_{11} & u_{12} \\ l_{21}u_{11} & l_{21}u_{12} + u_{22} \end{pmatrix},$$

which gives

$$u_{11} = 1, \quad u_{12} = 2,$$

and

$$l_{21} \times 1 = 3, \quad l_{21} \times 2 + u_{22} = 4.$$

Thus, we have

$$l_{21} = 3, \quad u_{22} = -2.$$

Finally, we get

$$\mathbf{L} = \begin{pmatrix} 1 & 0 \\ 3 & 1 \end{pmatrix}, \quad \mathbf{U} = \begin{pmatrix} 1 & 2 \\ 0 & -2 \end{pmatrix}.$$

It is easy to verify that

$$\mathbf{LU} = \begin{pmatrix} 1 & 0 \\ 3 & 1 \end{pmatrix} \begin{pmatrix} 1 & 2 \\ 0 & -2 \end{pmatrix} = \begin{pmatrix} 1 & 2 \\ 3 & 4 \end{pmatrix} = \mathbf{A}.$$

The same issue appears again, that is, all the leading coefficients α_{ii} must be non-zero. For sparse matrices with many zero entries, this could often cause some significant problems numerically. Better methods such as iteration methods should be used in this case.

21.4 ITERATION METHODS

For a linear system $\mathbf{Au} = \mathbf{b}$, the solution $\mathbf{u} = \mathbf{A}^{-1}\mathbf{b}$ generally involves the inversion of a large matrix. The direct inversion becomes impractical if the matrix is very large (say, if $n > 100{,}000$). Many efficient algorithms have been developed for solving such systems. Jacobi and Gauss-Seidel iteration methods are just two examples.

21.4.1 Jacobi Iteration Method

The basic idea of the Jacobi-type iteration method is to decompose an $n \times n$ square matrix $\mathbf{A}$ into three simpler matrices

$$\mathbf{A} = \mathbf{D} + \mathbf{L} + \mathbf{U}, \tag{21.26}$$

where $\mathbf{D}$ is a diagonal matrix. $\mathbf{L}$ and $\mathbf{U}$ are the strictly lower and upper triangular matrices, respectively. Here the 'strict' means that the lower (or upper) triangular matrices do not include the diagonal elements. That is say, all the diagonal elements of the triangular matrices are zeros.

It is worth pointing out that here the triangular matrices $\mathbf{L}$ and $\mathbf{U}$ are different from those in the LU decomposition where it requires matrix products. In comparison with the LU decomposition where $\mathbf{LU} = \mathbf{A}$, we have used the simple additions here and this makes the decomposition an easier task. Using such decomposition, the linear system $\mathbf{Au} = \mathbf{b}$ becomes

$$\mathbf{Au} = (\mathbf{D} + \mathbf{L} + \mathbf{U})\mathbf{u} = \mathbf{b}, \tag{21.27}$$

which can be written as an iteration procedure

$$\mathbf{Du}^{(n+1)} = \mathbf{b} - (\mathbf{L} + \mathbf{U})\mathbf{u}^{(n)}. \tag{21.28}$$

This can be used to calculate the next approximate solution $\mathbf{u}^{(n+1)}$ from the current estimate $\mathbf{u}^{(n)}$. As the inverse of any diagonal matrix $\mathbf{D} = \text{diag}[d_{ii}]$ is easy, we have

$$\mathbf{u}^{(n+1)} = \mathbf{D}^{-1}[\mathbf{b} - (\mathbf{L} + \mathbf{U})\mathbf{u}^{(n)}]. \tag{21.29}$$

Writing in terms of the elements, we have

$$u_i^{(n+1)} = \frac{1}{d_{ii}}[b_i - \sum_{j \neq i} a_{ij} u_j^{(n)}], \tag{21.30}$$

where $d_{ii} = a_{ii}$ are the diagonal elements only.

This iteration usually starts from an initial guess $\mathbf{u}^{(0)}$ (say, $\mathbf{u}^{(0)} = \mathbf{0}$). However, this iteration scheme is only stable under the condition that the square matrix is strictly diagonally dominant. That is to require that

$$|a_{ii}| > \sum_{j=1, j \neq i}^{n} |a_{ij}|, \tag{21.31}$$

for all $i = 1, 2, ..., n$.

In order to show how the iteration works, let us look at an example by solving the following linear system

$$5u_1 + u_2 - 2u_3 = 5,$$
$$u_1 + 4u_2 = -10,$$
$$2u_1 + 2u_2 - 7u_3 = -9.$$

We know its exact solution is

$$\mathbf{u} = \begin{pmatrix} u_1 \\ u_2 \\ u_3 \end{pmatrix} = \begin{pmatrix} 2 \\ -3 \\ 1 \end{pmatrix}. \tag{21.32}$$

Now let us solve the system by the Jacobi-type iteration method.

Example 21.5

We first write the above simple system as the compact matrix form $\mathbf{Au} = \mathbf{b}$ or

$$\begin{pmatrix} 5 & 1 & -2 \\ 1 & 4 & 0 \\ 2 & 2 & -7 \end{pmatrix} \begin{pmatrix} u_1 \\ u_2 \\ u_3 \end{pmatrix} = \begin{pmatrix} 5 \\ -10 \\ -9 \end{pmatrix}.$$

Then we decompose the matrix $\mathbf{A}$ as

$$\mathbf{A} = \begin{pmatrix} 5 & 1 & -2 \\ 1 & 4 & 0 \\ 2 & 2 & -7 \end{pmatrix} = \mathbf{D} + (\mathbf{L} + \mathbf{U})$$
$$= \begin{pmatrix} 5 & 0 & 0 \\ 0 & 4 & 0 \\ 0 & 0 & -7 \end{pmatrix} + \begin{pmatrix} 0 & 0 & 0 \\ 1 & 0 & 0 \\ 2 & 2 & 0 \end{pmatrix} + \begin{pmatrix} 0 & 1 & -2 \\ 0 & 0 & 0 \\ 0 & 0 & 0 \end{pmatrix}.$$

The inverse of $\mathbf{D}$ is simply

$$\mathbf{D}^{-1} = \begin{pmatrix} 1/5 & 0 & 0 \\ 0 & 1/4 & 0 \\ 0 & 0 & -1/7 \end{pmatrix}.$$

The Jacobi-type iteration formula becomes

$$\mathbf{u}^{(n+1)} = \mathbf{D}^{-1}[\mathbf{b} - (\mathbf{L} + \mathbf{U})\mathbf{u}^{(n)}],$$

or

$$\begin{pmatrix} u_1 \\ u_2 \\ u_3 \end{pmatrix}_{n+1} = \begin{pmatrix} 1/5 & 0 & 0 \\ 0 & 1/4 & 0 \\ 0 & 0 & -1/7 \end{pmatrix} \left[\begin{pmatrix} 5 \\ -10 \\ -9 \end{pmatrix} - \begin{pmatrix} 0 & 1 & -2 \\ 1 & 0 & 0 \\ 2 & 2 & 0 \end{pmatrix} \begin{pmatrix} u_1 \\ u_2 \\ u_3 \end{pmatrix}_n \right].$$

If we start from the initial guess $\mathbf{u}^{(0)} = \begin{pmatrix} 0 & 0 & 0 \end{pmatrix}^T$, we have

$$\mathbf{u}^{(1)} \approx \begin{pmatrix} 1 \\ -2.5 \\ 1.2857 \end{pmatrix}, \quad \mathbf{u}^{(2)} \approx \begin{pmatrix} 2.0143 \\ -2.7500 \\ 0.8571 \end{pmatrix}, \quad \mathbf{u}^{(3)} \approx \begin{pmatrix} 1.8929 \\ 3.0036 \\ 1.0755 \end{pmatrix},$$

$$\mathbf{u}^{(4)} \approx \begin{pmatrix} 2.0309 \\ -2.9732 \\ 0.9684 \end{pmatrix}, \quad \mathbf{u}^{(5)} \approx \begin{pmatrix} 1.9820 \\ -3.0077 \\ 1.0165 \end{pmatrix}.$$

We can see that after 5 iterations, the approximate solution is quite close to the true solution $\mathbf{u} = \begin{pmatrix} 2 & -3 & 1 \end{pmatrix}^T$.

However, there is an important issue here. If we interchange the second row (equation) and the third row (equation), the new diagonal matrix becomes

$$\begin{pmatrix} 5 & 0 & 0 \\ 0 & 2 & 0 \\ 0 & 0 & 0 \end{pmatrix},$$

which has no inverse as it is singular. This means the order of the equations is important to ensure that the matrix is diagonally dominant.

Example 21.6

Furthermore, if we interchange the first equation (row) and second equation (row), we have an equivalent system

$$\begin{pmatrix} 1 & 4 & 0 \\ 5 & 1 & -2 \\ 2 & 2 & -7 \end{pmatrix}\begin{pmatrix} u_1 \\ u_2 \\ u_3 \end{pmatrix} = \begin{pmatrix} -10 \\ 5 \\ -9 \end{pmatrix}.$$

Now the new decomposition becomes

$$\mathbf{A} = \begin{pmatrix} 1 & 4 & 0 \\ 5 & 1 & -2 \\ 2 & 2 & -7 \end{pmatrix} = \begin{pmatrix} 1 & 0 & 0 \\ 0 & 1 & 0 \\ 0 & 0 & -7 \end{pmatrix} + \begin{pmatrix} 0 & 0 & 0 \\ 5 & 0 & 0 \\ 2 & 2 & 0 \end{pmatrix} + \begin{pmatrix} 0 & 4 & 0 \\ 0 & 0 & -2 \\ 0 & 0 & 0 \end{pmatrix},$$

which gives the following iteration formula

$$\begin{pmatrix} u_1 \\ u_2 \\ u_3 \end{pmatrix}_{n+1} = \begin{pmatrix} 1 & 0 & 0 \\ 0 & 1 & 0 \\ 0 & 0 & -\frac{1}{7} \end{pmatrix}\left[\begin{pmatrix} -10 \\ 5 \\ -9 \end{pmatrix} - \begin{pmatrix} 0 & 4 & 0 \\ 5 & 0 & -2 \\ 2 & 2 & 0 \end{pmatrix}\begin{pmatrix} u_1 \\ u_2 \\ u_3 \end{pmatrix}_n\right].$$

Starting from $\mathbf{u}^{(0)} = (\ 0 \quad 0 \quad 0\ )^T$ again, we have

$$\mathbf{u}^{(1)} = \begin{pmatrix} -10 \\ 5 \\ 1.2857 \end{pmatrix}, \ \mathbf{u}^{(2)} = \begin{pmatrix} -30 \\ 57.5714 \\ -0.1429 \end{pmatrix}, \ \mathbf{u}^{(3)} = \begin{pmatrix} -240.28 \\ 154.71 \\ 9.16 \end{pmatrix}, \ \mathbf{u}^{(4)} = \begin{pmatrix} -628.9 \\ 1224.8 \\ -23.2 \end{pmatrix}, \ \ldots.$$

We can see that it diverges. So what is the problem? How can the order of the equations affect the results so significantly?

There are two important criteria for the iterations to converge correctly, and they are: the inverse of $\mathbf{D}^{-1}$ must exist, and the spectral radius of the right matrix must be less than 1. The first condition is obvious; if $\mathbf{D}^{-1}$ does not exist (say, when any of the diagonal elements is zero), then we cannot carry out the iteration process at all. The second condition requires

$$\rho(\mathbf{D}^{-1}) \le 1, \qquad \rho[\mathbf{D}^{-1}(\mathbf{L}+\mathbf{U})] \le 1, \tag{21.33}$$

where $\rho(\mathbf{A})$ is the spectral radius of the matrix $\mathbf{A}$. From the diagonal matrix $\mathbf{D}$, its largest absolute eigenvalue is 1. So $\rho(\mathbf{D}^{-1}) = \max(|\lambda_i|) = 1$ seems no problem. How about the following matrix?

$$\mathbf{N} = \mathbf{D}^{-1}(\mathbf{L}+\mathbf{U}) = \begin{pmatrix} 0 & 4 & 0 \\ 5 & 0 & -2 \\ -2/7 & -2/7 & 0 \end{pmatrix}. \tag{21.34}$$

The three eigenvalues of $\mathbf{N}$ are $\lambda_i = 4.590, -4.479, -0.111$. So its spectral radius is $\rho(\mathbf{N}) = \max(|\lambda_i|) = 4.59 > 1$, which means that the iteration scheme will diverge.

Example 21.7

If we revisit our earlier example, we have

$$\mathbf{D}^{-1} = \begin{pmatrix} 1/5 & 0 & 0 \\ 0 & 1/4 & 0 \\ 0 & 0 & -1/7 \end{pmatrix}, \ \text{eig}(\mathbf{D}^{-1}) = \frac{1}{5}, \frac{1}{4}, \frac{-1}{7}, \tag{21.35}$$

and

$$\mathbf{N} = \mathbf{D}^{-1}(\mathbf{L}+\mathbf{U}) = \begin{pmatrix} 0 & 1/5 & -2/5 \\ 1/4 & 0 & 0 \\ -2/7 & -2/7 & 0 \end{pmatrix}, \tag{21.36}$$

whose eigenvalues are

$$\lambda_1 = 0.4739, \qquad \lambda_{2,3} = -0.2369 \pm 0.0644i. \tag{21.37}$$

So we have

$$\rho(\mathbf{D}^{-1}) = 1/4 < 1, \qquad \rho(\mathbf{N}) = 0.4739 < 1. \tag{21.38}$$

That is why the earlier iteration procedure is convergent.

21.4.2 Gauss-Seidel Iteration

In the Jacobi-type iterations, we have to store both $\mathbf{u}^{(n+1)}$ and $\mathbf{u}^{(n)}$ as we will use all the $\mathbf{u}_j^{(n)}$ values to compute the values at the next level $t = n + 1$, this means that we cannot use the running update when the new approximate has just been computed

$$\mathbf{u}_j^{(n+1)} \rightarrow \mathbf{u}_j^{(n)}, \qquad (j = 1, 2, ...).$$

If the vector size $\mathbf{u}$ is large (it usually is), then we can devise other iteration schemes to save memory by using the running update. So only one-vector storage is needed.

The Gauss-Seidel iteration procedure is such a procedure to provide an efficient way of solving the linear matrix equation $\mathbf{Au} = \mathbf{b}$. It uses the same decomposition as the Jacobi-type iteration by splitting $\mathbf{A}$ into

$$\mathbf{A} = \mathbf{L} + \mathbf{D} + \mathbf{U}, \tag{21.39}$$

but the difference from the Jacobi method is that we use $\mathbf{L} + \mathbf{D}$ instead of $\mathbf{D}$ for the inverse so that the running update is possible. The nth step iteration is updated by

$$(\mathbf{L} + \mathbf{D})\mathbf{u}^{(n+1)} = \mathbf{b} - \mathbf{U}\mathbf{u}^{(n)}, \tag{21.40}$$

or

$$\mathbf{u}^{(n+1)} = (\mathbf{L} + \mathbf{D})^{-1}[\mathbf{b} - \mathbf{U}\mathbf{u}^{(n)}]. \tag{21.41}$$

This procedure, starting from an initial vector $\mathbf{u}^{(0)}$, stops if a prescribed criterion is reached.

It is worth pointing out that Gauss-Seidel iteration requires the same criteria of convergence as for the Jacobi-type iteration method. That is to say, the inverse of the matrix must exist, and the largest spectral radius must be less than 1.

21.4.3 Relaxation Method

The above Gauss-Seidel iteration method is still slow, and the relaxation method provides a more efficient iteration procedure. A popular method is the successive over-relaxation method which consists of two steps

$$\mathbf{v}^{(n)} = (\mathbf{L} + \mathbf{D} + \mathbf{U})\mathbf{u}^{(n)} - \mathbf{b}, \tag{21.42}$$

and

$$\mathbf{u}^{(n+1)} = \mathbf{u}^{(n)} - \omega(\mathbf{L} + \mathbf{D})^{-1}\mathbf{v}^{(n)}, \tag{21.43}$$

where $0 < \omega < 2$ is the over-relaxation parameter. If we combine the above equations and rearrange them, we have

$$\mathbf{u}^{(n+1)} = (1 - \omega)\mathbf{u}^{(n)} + \omega\tilde{\mathbf{u}}^{(n)}, \tag{21.44}$$

where $\tilde{\mathbf{u}}^{(n)} = (\mathbf{L} + \mathbf{D})^{-1}(\mathbf{b} - \mathbf{U}\mathbf{u}^{(n)})$ is the standard Gauss-Seidel procedure. Therefore, this method is essentially the weighted average between the previous iteration and the successive Gauss-Seidel iteration. Clearly, if $\omega = 1$, then it reduces to the standard Gauss-Seidel iteration method.

Broadly speaking, a small value of $0 < \omega < 1$ corresponds to under-relaxation with slower convergence, while $1 < \omega < 2$ leads to over-relaxation and faster convergence. It has been proved theoretically that the scheme will not converge if $\omega < 0$ or $\omega > 2$.

21.5 NEWTON-RAPHSON METHOD

Sometimes, the algebraic equations under consideration are nonlinear, and direct inversion is not the best technique. In this case, more elaborate techniques should be used.

The nonlinear algebraic equation

$$\mathbf{A}(\mathbf{u})\mathbf{u} = \mathbf{b}(\mathbf{u}), \quad \text{or} \quad \mathbf{F}(\mathbf{u}) = \mathbf{A}(\mathbf{u})\mathbf{u} - \mathbf{b}(\mathbf{u}) = \mathbf{0}, \tag{21.45}$$

can be solved using a simple iteration technique

$$\mathbf{A}(\mathbf{u}^{(n)})\mathbf{u}^{(n+1)} = \mathbf{b}(\mathbf{u}^{(n)}), \qquad n = 0, 1, 2, ... \tag{21.46}$$

until $\|\mathbf{u}^{(n+1)} - \mathbf{u}^{(n)}\|$ is sufficiently small. Iterations require a starting vector $\mathbf{u}^{(0)}$. This method is also referred to as the successive substitution method.

If this simple method does not work, the relaxation method can be used. The relaxation technique first gives a tentative new approximation $\mathbf{u}^*$ from $\mathbf{A}(\mathbf{u}^{(n)})\mathbf{u}^* = \mathbf{b}(\mathbf{u}^{(n)})$, then we use

$$\mathbf{u}^{(n+1)} = \omega \mathbf{u}^* + (1-\omega)\mathbf{u}^{(n)}, \qquad \omega \in (0, 1], \tag{21.47}$$

where ω is a prescribed relaxation parameter.

The nonlinear equation (21.45) can also be solved using the Newton-Raphson procedure. We approximate $\mathbf{F}(\mathbf{u})$ by a linear function $\mathbf{R}(\mathbf{u}; \mathbf{u}^{(n)})$ in the vicinity of an existing approximation $\mathbf{u}^{(n)}$ to $\mathbf{u}$:

$$\mathbf{R}(\mathbf{u}; \mathbf{u}^{(n)}) = \mathbf{F}(\mathbf{u}^{(n)}) + \mathbf{J}(\mathbf{u}^{(n)})(\mathbf{u} - \mathbf{u}^{(n)}), \quad \mathbf{J}(\mathbf{u}) = \nabla \mathbf{F}, \tag{21.48}$$

where $\mathbf{J}$ is the Jacobian of $\mathbf{F}(\mathbf{u}) = (F_1, F_2, ..., F_M)^T$.

For $\mathbf{u} = (u_1, u_2, ..., u_M)^T$, we have

$$\mathbf{J}_{ij} = \frac{\partial F_i}{\partial u_j}. \tag{21.49}$$

To find the next approximation $\mathbf{u}^{(n+1)}$ from $\mathbf{R}(\mathbf{u}^{(n+1)}; \mathbf{u}^{(n)}) = 0$, we can solve the linear system with $\mathbf{J}$ as the coefficient matrix

$$\mathbf{u}^{(n+1)} = \mathbf{u}^{(n)} - \mathbf{J}^{-1}\mathbf{F}(\mathbf{u}^{(n)}), \tag{21.50}$$

under a given termination criterion $\|\mathbf{u}^{(n+1)} - \mathbf{u}^{(n)}\| \le \epsilon$. Here, $0 < \epsilon \ll 1$ is a very small tolerance.

There are some extensive literature concerning nonlinear iteration methods. Interested readers can refer to more advanced literature.

21.6 CONJUGATE GRADIENT METHOD

The method of conjugate gradient can be used to solve the following linear system

$$\mathbf{A}\mathbf{u} = \mathbf{b}, \tag{21.51}$$

where $\mathbf{A}$ is often a symmetric positive definite matrix. The above system is equivalent to minimizing the following function $f(\mathbf{u})$

$$f(\mathbf{u}) = \frac{1}{2}\mathbf{u}^T\mathbf{A}\mathbf{u} - \mathbf{b}^T\mathbf{u} + \mathbf{v}, \tag{21.52}$$

where $\mathbf{v}$ is a constant vector and can be taken to be zero. We can easily see that $\nabla f(\mathbf{u}) = 0$ leads to $\mathbf{A}\mathbf{u} = \mathbf{b}$. The theory behind these iterative methods is closely related to the Krylov subspace $\mathcal{K}$ spanned by $\mathbf{A}$ and $\mathbf{b}$ as defined by

$$\mathcal{K}_n(\mathbf{A}, \mathbf{b}) = \{\mathbf{I}\mathbf{b}, \mathbf{A}\mathbf{b}, \mathbf{A}^2\mathbf{b}, ..., \mathbf{A}^{n-1}\mathbf{b}\}, \tag{21.53}$$

where $\mathbf{A}^0 = \mathbf{I}$.

If we use an iterative procedure to obtain the estimate $\mathbf{u}_n$ at nth iteration, the residual is given by

$$\mathbf{r}_n = \mathbf{b} - \mathbf{A}\mathbf{u}_n, \tag{21.54}$$

which is essentially the negative gradient $\nabla f(\mathbf{u}_n)$. The search direction vector in the conjugate gradient method is subsequently determined by

$$\mathbf{d}_{n+1} = \mathbf{r}_n - \frac{\mathbf{d}_n^T \mathbf{A}\mathbf{r}_n}{\mathbf{d}_n^T \mathbf{A}\mathbf{d}_n}\mathbf{d}_n. \tag{21.55}$$

The solution often starts with an initial guess $\mathbf{u}_0$ at $n = 0$, and proceeds iteratively. The above steps can compactly be written as

$$\mathbf{u}_{n+1} = \mathbf{u}_n + \alpha_n \mathbf{d}_n, \quad \mathbf{r}_{n+1} = \mathbf{r}_n - \alpha_n \mathbf{A}\mathbf{d}_n, \tag{21.56}$$

and

$$\mathbf{d}_{n+1} = \mathbf{r}_{n+1} + \beta_n \mathbf{d}_n, \tag{21.57}$$

where

$$\alpha_n = \frac{\mathbf{r}_n^T \mathbf{r}_n}{\mathbf{d}_n^T \mathbf{A}\mathbf{d}_n}, \qquad \beta_n = \frac{\mathbf{r}_{n+1}^T \mathbf{r}_{n+1}}{\mathbf{r}_n^T \mathbf{r}_n}. \tag{21.58}$$

Iterations stop when a prescribed accuracy is reached. This can easily be programmed in any programming language, especially Matlab.

The conjugate gradient method is a powerful method with a diverse range of applications. In fact, it was voted as one of the top 10 most popular methods in scientific computing.

EXERCISES

21.1. Compute the determinant and inverse of

$$\mathbf{L} = \begin{pmatrix} 1 & 0 & 0 \\ 1 & 1 & 0 \\ 3 & 2 & 1 \end{pmatrix}.$$

21.2. Find the LU decomposition of

$$\mathbf{A} = \begin{pmatrix} 2 & 3 & 2 \\ 2 & 2 & 4 \\ 6 & 7 & 17 \end{pmatrix},$$

by letting that the diagonal elements of the lower triangular matrix are ones. Find the inverse of **A** and the determinant of **A**.

21.3. Use any analytical method introduced in this chapter to solve the following system:

$$\begin{cases} 2x + 3y + 2z = 10, \\ 2x + 2y + 4z = 10, \\ 6x + 7y + 17z = 37. \end{cases}$$

Hint. Write in a matrix form and try to use results from the previous exercise.

21.4. Solve the same system as given in the previous question using any of the iteration methods.

Part VIII

Optimization

Chapter 22

Linear Programming

Chapter Points

- Linear programming concerns the maximization or minimization of a linear cost function subject to various linear constraints. Examples will be provided to show how linear programming problems can be formulated in the right form.
- Though linear, such linear programs tend to have many variables with multiple constraints and thus necessitate specialized methods such as the Simplex method to be introduced with a worked example.

22.1 LINEAR PROGRAMMING

Linear programming is a powerful mathematical programming technique which is widely used in business planning, transport routing and many other optimization applications. It is worth pointing out that the term *programming* here means *planning*, it has nothing to do with computer programming, and the similarity in wording is purely coincidental.

The basic idea in linear programming is to find the maximum or minimum of a linear objective under linear constraints. As constraints are linear, the feasible region formed or enclosed by all the constraints is convex. In addition, the linearity of the cost function indicates that the optimal solutions can occur only at the corners of the convex region in the search domain.

Example 22.1

For example, an Internet service provider (ISP) can provide two different services x_1 and x_2. The first service is, say, the fixed monthly rate with limited download limits and bandwidth, while the second service is the higher rate with no download limit. The profit of the first service is αx_1 while the second is βx_2, though the profit of the second product is higher $\beta > \alpha > 0$, so the total profit is

$$P(\mathbf{x}) = \alpha x_1 + \beta x_2, \qquad \beta/\alpha > 1, \tag{22.1}$$

which is the objective function because the aim of the ISP company is to increase the profit as much as possible. Suppose the provided service is limited by the total bandwidth of the ISP company, thus at most $n_1 = 16$ (in 1,000,000 units) of the first and at most $n_2 = 10$ (in 1,000,000 units) of the second can be provided per unit of time, say, each day. Therefore, we have

$$x_1 \le n_1, \qquad x_2 \le n_2. \tag{22.2}$$

If the management of each of the two service packages takes the same staff time, so that a maximum of $n = 20$ (in 1,000,000 units) can be maintained, which means

$$x_1 + x_2 \le n. \tag{22.3}$$

The additional constraints are that both x_1 and x_2 must be non-negative since negative numbers are unrealistic. We now have the following constraints

$$0 \le x_1 \le n_1, \qquad 0 \le x_2 \le n_2. \tag{22.4}$$

The problem now is to find the best x_1 and x_2 so that the profit P is maximized. Mathematically, we have

$$\begin{aligned} &\underset{(x_1,x_2)\in\mathcal{N}^2}{\text{maximize}} \quad P(x_1, x_2) = \alpha x_1 + \beta x_2, \\ &\text{subject to} \quad x_1 + x_2 \le n, \\ &\qquad\qquad\quad 0 \le x_1 \le n_1, \; 0 \le x_2 \le n_2. \end{aligned} \tag{22.5}$$

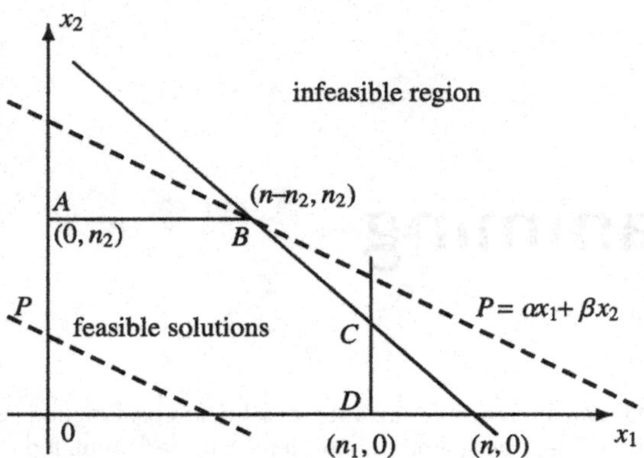

FIGURE 22.1 Schematic representation of linear programming. If $\alpha = 2$, $\beta = 3$, $n_1 = 16$, $n_2 = 10$ and $n = 20$, then the optimal solution is at $B(10, 10)$.

Now we have a simple linear programming problem where $P(x_1, x_2)$ is the objective or cost function. x_1 and x_2 are the design variables, while $0 \le x_1 \le n_1$ and $0 \le x_2 \le n_2$ are simple bounds or limits. The space spanned by the design variable is called the search space and in this simple case it is simple a two-dimensional (2D) plane.

In addition, the budget constraint is an inequality constraint: $x_1 + x_2 \le n$. In general, there are multiple inequality constraints for a linear program. Both the inequality constraints and the simple bounds form the constraints of the problem. The domain formed by the constraints in the search space is called the feasible domain. Any solution (or a point in 2D in the present example) is called a feasible solution. In other words, a feasible solution is a solution vector that satisfies all the constraints.

The solution that maximizes (or minimizes) the objective is called an optimal solution. There may be more than one optimal solution. Among all the optimal solution in the feasible region, there is one global optimal solution that gives the maximum (or minimum) objective for a maximization (or minimization) problem. To find the global optimal solution is the main aim of linear programming.

Example 22.2

The feasible solutions to the Internet provider's problem can graphically be represented as the inside region of the polygon $OABCD$ as shown in Fig. 22.1. As the aim is to maximize the profit P, thus the optimal solution is at the extreme point B with $(n - n_2, n_2)$ and $P = \alpha(n - n_2) + \beta n_2$.

For example, if $\alpha = 2$, $\beta = 3$, $n_1 = 16$, $n_2 = 10$, and $n = 20$, then the optimal solution occurs at $x_1 = n - n_2 = 10$ and $x_2 = n_2 = 10$ with the total profit $P = 2 \times (20 - 10) + 3 \times 10 = 50$ (million pounds).

Since the solution (x_1 and x_2) must be integers, an interesting thing is that the solution is independent of β/α if and only if $\beta/\alpha > 1$. However, the profit P does depend on the parameters α and β.

The number of feasible solutions is infinite if x_1 and x_2 are real numbers. Even for $x_1, x_2 \in \mathbb{N}$ are integers, the number of feasible solutions is quite large. Therefore, there is a need to use a systematic method to find the optimal solution. In order to find the best solution, we first plot out all the constraints as straight lines, and all the feasible solutions satisfying all the constraints form the inside region of the polygon $OABCD$. The vertices of the polygon form the set of the extreme points. Then, we plot the objective function P as a family of parallel lines (shown as dashed lines) so as to find the maximum value of P. Obviously, the highest value of P corresponds to the case when the objective line goes through the extreme point B. Therefore, $x_1 = n - n_2$ and $x_2 = n_2$ at the point B are the best solutions.

The current example is relatively simple because it has only two decision variables and three constraints, which can be solved easily using a graphic approach. For more complicated problems, we need a formal approach. One of the most widely used methods is the simplex method.

22.2 SIMPLEX METHOD

The simplex method was introduced by George Dantzig in 1947. The simplex method essentially works in the following way: for a given linear optimization problem such as the example of the ISP service we discussed earlier, it assumes that all the extreme points are known.

If the extreme points are not known, the first step is to determine these extreme points or to check whether there are any feasible solutions. With known extreme points, it is easy to test whether or not an extreme point is optimal using the algebraic relationship and the objective function. If the test for optimality is not passed, then move to an adjacent extreme point to do the same test. This process stops until an optimal extreme point is found or the unbounded case occurs.

22.2.1 Basic Procedure

Mathematically, the simplex method first transforms the constraint inequalities into equalities by using slack variables.

To convert an inequality such as

$$5x_1 + 6x_2 \le 20, \tag{22.6}$$

we can use a new variable x_3 or

$$s_1 = 20 - 5x_1 - 6x_2$$

so that the original inequality becomes an equality

$$5x_1 + 6x_2 + s_1 = 20, \tag{22.7}$$

with an auxiliary non-negativeness condition

$$s_1 \ge 0. \tag{22.8}$$

Such a variable is referred to as a slack variable.

Example 22.3

The inequalities in our example

$$x_1 + x_2 \le n, \qquad 0 \le x_1 \le n_1, \qquad 0 \le x_2 \le n_2, \tag{22.9}$$

can be written, using three slack variables s_1, s_2, s_3, as the following equalities:

$$x_1 + x_2 + s_1 = n, \tag{22.10}$$

$$x_1 + s_2 = n_1, \qquad x_2 + s_3 = n_2, \tag{22.11}$$

and

$$x_i \ge 0 \ (i = 1, 2), \quad s_j \ge 0 \ (j = 1, 2, 3). \tag{22.12}$$

The original problem (22.5) becomes

$$\underset{\mathbf{x}\in\Re^5}{\text{maximize}} \; P(\mathbf{x}) = \alpha x_1 + \beta x_2 + 0s_1 + 0s_2 + 0s_3,$$

$$\text{subject to} \begin{pmatrix} 1 & 1 & 1 & 0 & 0 \\ 1 & 0 & 0 & 1 & 0 \\ 0 & 1 & 0 & 0 & 1 \end{pmatrix} \begin{pmatrix} x_1 \\ x_2 \\ s_1 \\ s_2 \\ s_3 \end{pmatrix} = \begin{pmatrix} n \\ n_1 \\ n_2 \end{pmatrix},$$

$$x_i \ge 0, \qquad (i = 1, 2, ..., 5), \tag{22.13}$$

which has two control variables (x_1, x_2) and three slack variables $x_3 = s_1$, $x_4 = s_2$, $x_5 = s_3$.

In general, a linear programming problem can be written in the following standard form

$$\underset{\mathbf{x}\in\Re^n}{\text{maximize}}\ f(\mathbf{x}) = Z = \sum_{i=1}^{p} \alpha_i x_i = \alpha^T \mathbf{x},$$

$$\text{subject to } \mathbf{A}\mathbf{x} = \mathbf{b},\ x_i \geq 0\ (i = 1, ..., p), \tag{22.14}$$

where $\mathbf{A}$ is a $q \times p$ matrix, $\mathbf{b} = (b_1, ..., b_q)^T$, and

$$\mathbf{x} = [\mathbf{x}_p\ \mathbf{x}_s]^T = (x_1, ..., x_m, s_1, ..., s_{p-m})^T. \tag{22.15}$$

This problem has p variables, and q equalities and all p variables are non-negative. In the standard form, all constraints are expressed as equalities and all variables including slack variables are non-negative.

A basic solution to the linear system $\mathbf{A}\mathbf{x} = \mathbf{b}$ of q linear equations in p variables in the standard form is usually obtained by setting $p - q$ variables equal to zero, and subsequently solving the resulting $q \times q$ linear system to get a unique solution of the remaining q variables. The q variables (that are not bound to zero) are called the basic variables of the basic solution. The $p - q$ variables at zero are called non-basic variables. Any basic solution to this linear system is referred to as a basic feasible solution (BFS) if all its variables are non-negative.

The important property of the basic feasible solutions is that there is a unique corner point (extreme point) for each basic feasible solution, and there is at least one basic feasible solution for each corner or extreme point. These corners or extreme points are points on the intersection of two adjacent boundary lines such as A and B in Fig. 22.1. Two basic feasible solutions are said to be adjacent if they have $q - 1$ basic variables in common in the standard form.

Suppose $q = 500$, even the simplest integer equalities $x_i + x_j = 1$ where x_i and x_j are just 0 and 1. For $i, j = 1, 2, ..., 500$, it would give a huge number of combinations 2^{500}. Thus the number of basic feasible solutions will be the order of $2^{500} \approx 3 \times 10^{150}$, which is larger than the number of particles in the whole universe. This huge number of basic feasible solutions and extreme points necessitates a systematic and efficient search method. Simplex method is a powerful method to carry out such a mathematical programming task.

22.2.2 Augmented Form

The linear optimization problem is usually converted into the following standard augmented form or the canonical form

$$\begin{pmatrix} 1 & -\alpha^T \\ 0 & \mathbf{A} \end{pmatrix} \begin{pmatrix} Z \\ \mathbf{x} \end{pmatrix} = \begin{pmatrix} 0 \\ \mathbf{b} \end{pmatrix}, \tag{22.16}$$

with the objective to maximize Z. In this canonical form, all the constraints are expressed as equalities for all non-negative variables. All the right-hand sides for all constraints are also non-negative, and each constraint equation has a single basic variable. The intention of writing in this canonical form is to identify basic feasible solutions, and move from one basic feasible solution to another via a so-called pivot operation. Geometrically speaking, this means to find all the corners or extreme points first, then evaluate the objective function by going through the extreme points so as to determine if the current basic feasible solution can be improved or not.

In the framework of the canonical form, the basic steps of the simplex method are: 1) to find a basic feasible solution to start the algorithm. Sometimes, it might be difficult to start, which may either imply there is no feasible solution or that it is necessary to reformulate the problem in a slightly different way by changing the canonical form so that a basic feasible solution can be found; 2) to see if the current basic feasible solution can be improved (even marginally) by increasing the non-basic variables from zero to non-negative values; 3) stop the process if the current feasible solution cannot be improved, which means that it is optimal. If the current feasible solution is not optimal, then move to an adjacent basic feasible solution. This adjacent basic feasible solution can be obtained by changing the canonical form via elementary row operations.

The pivot manipulations are based on the fact that a linear system will remain an equivalent system by multiplying a non-zero constant on a row and adding it to the other row. This procedure continues by going to the second step and repeating the evaluation of the objective function. The optimality of the problem will be reached, or we can stop the iteration if the solution becomes unbounded in the event that you can improve the objective indefinitely.

22.3 A WORKED EXAMPLE

Now we come back to our example, if we use $\alpha = 2$, $\beta = 3$, $n_1 = 16$, $n_2 = 10$ and $n = 20$, we then have

$$\begin{pmatrix} 1 & -2 & -3 & 0 & 0 & 0 \\ 0 & 1 & 1 & 1 & 0 & 0 \\ 0 & 1 & 0 & 0 & 1 & 0 \\ 0 & 0 & 1 & 0 & 0 & 1 \end{pmatrix} \begin{pmatrix} Z \\ x_1 \\ x_2 \\ s_1 \\ s_2 \\ s_3 \end{pmatrix} = \begin{pmatrix} 0 \\ 20 \\ 16 \\ 10 \end{pmatrix}, \tag{22.17}$$

where $x_1, x_2, s_1, ..., s_3 \geq 0$. Now the first step is to identify a corner point or basic feasible solution by setting non-isolated variables $x_1 = 0$ and $x_2 = 0$ (thus the basic variables are s_1, s_2, s_3). We now have

$$s_1 = 20, \; s_2 = 16, \; s_3 = 10. \tag{22.18}$$

The objective function $Z = 0$, which corresponds to the corner point O in Fig. 22.1. In the present canonical form, the corresponding column associated with each basic variable has only one non-zero entry (marked by a box) for each constraint equality, and all other entries in the same column are zero. The non-zero value is usually converted into 1 if it is not unity. This is shown as follows:

$$\begin{array}{cccccc} Z & x_1 & x_2 & s_1 & s_2 & s_3 \end{array} \tag{22.19}$$
$$\begin{pmatrix} 1 & -2 & -3 & 0 & 0 & 0 \\ 0 & 1 & 1 & \boxed{1} & 0 & 0 \\ 0 & 1 & 0 & 0 & \boxed{1} & 0 \\ 0 & 0 & 1 & 0 & 0 & \boxed{1} \end{pmatrix}$$

When we change the set or the bases of basic variables from one set to another, we will aim to convert to a similar form using pivot row operations. There are two ways of numbering this matrix. One way is to call the first row $[1 - 2 - 3 0 0 0]$ as the zero-th row, so that all other rows correspond to their corresponding constraint equation. The other way is simply to use its order in the matrix, so $[1 - 2 - 3 0 0 0]$ is simply the first row. We will use this standard notation.

Now the question is whether we can improve the objective by increasing one of the non-basic variables x_1 and x_2? Obviously, if we increase x_1 by a unit, then Z will also increase by 2 units. However, if we increase x_2 by a unit, then Z will increase by 3 units. Since our objective is to increase Z as much as possible, we choose to increase x_2. As the requirement of the non-negativeness of all variables, we cannot increase x_2 without limits. So we increase x_2 while holding $x_1 = 0$, and we have

$$s_1 = 20 - x_2, \; s_2 = 16, \; s_3 = 10 - x_2. \tag{22.20}$$

Thus, the highest possible value of x_2 is $x = 10$ when $s_1 = s_3 = 0$. If x_2 increases further, both s_1 and s_3 will become negative, thus it is no longer a basic feasible solution.

The next step is either to set $x_1 = 0$ and $s_1 = 0$ as non-basic variables or to set $x_1 = 0$ and $s_3 = 0$. Both cases correspond to the point A in our example, so we simply choose $x_1 = 0$ and $s_3 = 0$ as non-basic variables, and the basic variables are thus x_2, s_1 and s_2. Now we have to do some pivot operations so that s_3 will be replaced by x_2 as a new basic variable. Each constraint equation has only a single basic variable in the new canonical form. This means that each column corresponding to each basic variable should have only a single non-zero entry (usually 1). In addition, the right-hand sides of all the constraints are non-negative and increase the value of the objective function at the same time. In order to convert the third column for x_2 to the form with only a single non-zero entry 1 (all other coefficients in the column should be zero), we first multiply the fourth row by 3 and add it to the first row, and the first row becomes

$$Z - 2x_1 + 0x_2 + 0s_1 + 0s_2 + 3s_3 = 30. \tag{22.21}$$

Then, we multiply the fourth row by -1 and add it to the second row, we have

$$0Z + x_1 + 0x_2 + s_1 + 0s_2 - s_3 = 10. \tag{22.22}$$

So the new canonical form becomes

$$\begin{pmatrix} 1 & -2 & 0 & 0 & 0 & 3 \\ 0 & 1 & 0 & 1 & 0 & -1 \\ 0 & 1 & 0 & 0 & 1 & 0 \\ 0 & 0 & 1 & 0 & 0 & 1 \end{pmatrix} \begin{pmatrix} Z \\ x_1 \\ x_2 \\ s_1 \\ s_2 \\ s_3 \end{pmatrix} = \begin{pmatrix} 30 \\ 10 \\ 16 \\ 10 \end{pmatrix}, \tag{22.23}$$

where the third, fourth, and fifth columns (for x_2, s_1 and s_2, respectively) have only one non-zero coefficient. All the values on the right-hand side are non-negative. From this canonical form, we can find the basic feasible solution by setting non-basic variables equal to zero. This is to set $x_1 = 0$ and $s_3 = 0$. We now have the basic feasible solution

$$x_2 = 10, \; s_1 = 10, \; s_2 = 16, \tag{22.24}$$

which corresponds to the corner point A. The objective $Z = 30$.

Now again the question is whether we can improve the objective by increasing the non-basic variables. As the objective function is

$$Z = 30 + 2x_1 - 3s_3, \tag{22.25}$$

Z will increase 2 units if we increase x_1 by 1, but Z will decrease -3 if we increase s_3. Thus, the best way to improve the objective is to increase x_1. The question is what the limit of x_1 is. To answer this question, we hold s_3 at 0, we have

$$s_1 = 10 - x_1, \; s_2 = 16 - x_1, x_2 = 10. \tag{22.26}$$

We can see if x_1 can increase up to $x_1 = 10$, after that s_1 becomes negative, and this occurs when $x_1 = 10$ and $s_1 = 0$. This also suggests that the new adjacent basic feasible solution can be obtained by choosing s_1 and s_3 as the non-basic variables. Therefore, we have to replace s_1 with x_1 so that the new basic variables are x_1, x_2 and s_2.

Using these basic variables, we have to make sure that the second column (for x_1) has only a single non-zero entry. Thus, we multiply the second row by 2 and add it to the first row, and the first row becomes

$$Z + 0x_1 + 0x_2 + 2s_1 + 0s_2 + s_3 = 50. \tag{22.27}$$

We then multiply the second row by -1 and add it to the third row, and we have

$$0Z + 0x_1 + 0x_2 - s_1 + s_2 + s_3 = 6. \tag{22.28}$$

Thus we have the following canonical form

$$\begin{pmatrix} 1 & 0 & 0 & 2 & 0 & 1 \\ 0 & 1 & 0 & 1 & 0 & -1 \\ 0 & 0 & 0 & -1 & 1 & 1 \\ 0 & 0 & 1 & 0 & 0 & 1 \end{pmatrix} \begin{pmatrix} Z \\ x_1 \\ x_2 \\ s_1 \\ s_2 \\ s_3 \end{pmatrix} = \begin{pmatrix} 50 \\ 10 \\ 6 \\ 10 \end{pmatrix}, \tag{22.29}$$

whose basic feasible solution can be obtained by setting non-basic variables $s_1 = s_3 = 0$. We have

$$x_1 = 10, \; x_2 = 10, \; s_2 = 6, \tag{22.30}$$

which corresponds to the extreme point B in Fig. 22.1. The objective value is $Z = 50$ for this basic feasible solution. Let us see if we can improve the objective further. Since the objective becomes

$$Z = 50 - 2s_1 - s_3, \tag{22.31}$$

any increase of s_1 or s_3 from zero will decrease the objective value. Therefore, this basic feasible solution is optimal. Indeed, this is the same solution as that obtained from the graph method. We can see that a major advantage is that we have reached the optimal solution after searching a certain number of extreme points, and there is no need to evaluate other extreme points. This is exactly why the simplex method is so efficient.

The case study we used here is relatively simple, but it is useful to show how the basic procedure works in linear programming. For more practical applications, there are well-established software packages which will do the work for you once you have set up the objective and constraints properly.

EXERCISES

22.1. For the worked example given in this chapter, will the optimal solution remain the same if an extra constraint $2x_1 + 3x_2 \leq 70$? Will the feasible region change?

22.2. In the same worked example, discuss what happens when an extra constraint $x_1 \leq x_2/2$ is imposed? Will the final solution change?

22.3. Solve the following linear programming problem:

$$\text{maximize } 2x + 3y + 4z,$$

subject to

$$\begin{cases} x - y + z \leq 20, \\ x + 2y + 4z \leq 32, \\ 3x + 2y \leq 40, \\ x, y, z \geq 0. \end{cases}$$

22.4. A final year project in engineering has a budget of 100 dollars to design two small products for a show by your team. Imagine that the profit is 3 dollars for each product A and 5 dollars for each product B. However, the production of products A and B takes 2 hours and 4 hours, respectively, but the team has only 50 hours for the project. In addition, the production costs of A and B are 5 and 10 per unit, respectively. The product on the show will also require some packaging: product A takes twice the space of product B, and the total allowed packages are 25. How do you manage the project and how many of each product your team can produce to maximize the overall profit.

Chapter 23

Optimization

Chapter Points

- The general formulation of optimization is introduced, together with the concepts of objective, constraint, constrained and unstrained optimization.
- Optimality is explained in detail with the introduction of KKT conditions for nonlinear optimization.
- Basic solving methods are explained, including the gradient-based methods, penalty method and method of Lagrange multipliers.
- Quadratic programming and sequential quadratic programming are also explained briefly.

23.1 OPTIMIZATION

Whatever the real-world applications may be, it is usually possible to formulate an optimization problem in a generic form. All optimization problems with explicit objectives can in general be expressed as a nonlinearly constrained optimization problem

$$\begin{aligned} &\underset{\mathbf{x}\in\mathbb{R}^d}{\text{maximize/minimize}}\ f(\mathbf{x}),\ \mathbf{x}=(x_1,x_2,...,x_d)^T\in\mathbb{R}^d,\\ &\text{subject to } \phi_j(\mathbf{x})=0,\ \ (j=1,2,...,M),\\ &\psi_k(\mathbf{x})\geq 0,\ \ (k=1,...,N), \end{aligned} \tag{23.1}$$

where $f(\mathbf{x})$, $\phi_i(\mathbf{x})$ and $\psi_j(\mathbf{x})$ are scalar functions of the design vector $\mathbf{x}$. Here the components x_i of $\mathbf{x}=(x_1,...,x_d)^T$ are called design or decision variables, and they can be either continuous, discrete or mixed of these two. The vector $\mathbf{x}$ is often called the decision vector which varies in a d-dimensional space $\mathbb{R}^d$. It is worth pointing out that we use a column vector here for $\mathbf{x}$ (thus with a transpose T). We can also use a row vector $\mathbf{x}=(x_1,...,x_d)$ and the results will be the same. Different textbooks may use slightly different formulations. Once we are aware of such minor variations, it causes no difficulty or confusion.

In addition, the function $f(\mathbf{x})$ is called the objective function or cost function. In addition, $\phi_i(\mathbf{x})$ are constraints in terms of M equalities, and $\psi_j(\mathbf{x})$ are constraints written as N inequalities. So there are $M+N$ constraints in total. The optimization problem formulated here is a nonlinear constrained problem.

The space spanned by the decision variables is called the search space $\mathbb{R}^d$, while the space formed by the values of the objective function is called the solution space. The optimization problem essentially maps the $\mathbb{R}^d$ domain or space of decision variables into a solution space $\mathbb{R}$ (or the real axis in general).

The objective function $f(\mathbf{x})$ can be either linear or nonlinear. If the constraints ϕ_i and ψ_j are all linear, it becomes a linearly constrained problem. Furthermore, when ϕ_i, ψ_j and the objective function $f(\mathbf{x})$ are all linear, then it becomes a linear programming problem that we have discussed in the previous chapter. If the objective is at most quadratic with linear constraints, then it is called quadratic programming. If all the values of the decision variables can be integers, then this type of linear programming is called integer programming or integer linear programming. Linear programming is very important in applications as we have explained in the previous chapter.

On the other hand, if no constraints are specified so that x_i can take any values in the real axis (or any integers), the optimization problem is referred to as an unconstrained optimization problem.

As a very simple example of optimization problems without any constraints, we discuss the search of the maxima or minima of a univariate function.

Example 23.1

For example, to find the maximum of a univariate function $f(x)$

$$f(x) = x^2 e^{-x^2}, \quad -\infty < x < \infty, \tag{23.2}$$

is a simple unconstrained problem, while the following problem is a simple constrained minimization problem

$$f(x_1, x_2) = x_1^2 + x_1 x_2 + x_2^2, \quad (x_1, x_2) \in \mathbb{R}^2, \tag{23.3}$$

subject to

$$x_1 \geq 1, \quad x_2 - 2 = 0. \tag{23.4}$$

It is worth pointing out that the objectives are explicitly known in all the optimization problems to be discussed in this book. However, in reality, it is often difficult to quantify what we want to achieve, but we still try to optimize certain things such as the degree of enjoyment or a quality service on holiday. In other cases, it might be impossible to write the objective function in any explicit form mathematically.

From basic calculus, we know that, for a given curve described by $f(x)$, its gradient $f'(x)$ describes the rate of change. When $f'(x) = 0$, the curve has a horizontal tangent at that particular point. This means that it becomes a point of special interest. In fact, the maximum or minimum of a curve can only occur at

$$f'(x_*) = 0, \tag{23.5}$$

which is a critical condition or stationary condition. The solution x_* to this equation corresponds to a stationary point and there may be multiple stationary points for a given curve.

In order to see if it is a maximum or minimum at $x = x_*$, we have to use the information of its second derivative $f''(x)$. In fact, $f''(x_*) > 0$ corresponds to a minimum, while $f''(x_*) < 0$ corresponds to a maximum. Let us see a concrete example.

Example 23.2

To find the minimum of $f(x) = x^2 e^{-x^2}$, we have the stationary condition $f'(x) = 0$ or

$$f'(x) = 2x \times e^{-x^2} + x^2 \times (-2x)e^{-x^2} = 2(x - x^3)e^{-x^2} = 0.$$

As $e^{-x^2} > 0$, we have

$$x(1 - x^2) = 0,$$

or

$$x = 0, \quad x = \pm 1.$$

The second derivative is given by

$$f''(x) = 2e^{-x^2}(1 - 5x^2 + 2x^4),$$

which is an even function with respective to x.

So at $x = \pm 1$, $f''(\pm 1) = 2[1 - 5(\pm 1)^2 + 2(\pm 1)^4]e^{-(\pm 1)^2} = -4e^{-1} < 0$. Thus, there are two maxima that occur at $x_* = \pm 1$ with $f_{\max} = e^{-1}$. At $x = 0$, we have $f''(0) = 2 > 0$, thus the minimum of $f(x)$ occurs at $x_* = 0$ with $f_{\min}(0) = 0$.

Whatever the objective is, we have to evaluate it many times. In most cases, the evaluations of the objective functions consume a substantial amount of computational power (which costs money) and design time. Any efficient algorithm that can reduce the number of objective evaluations will save both time and money.

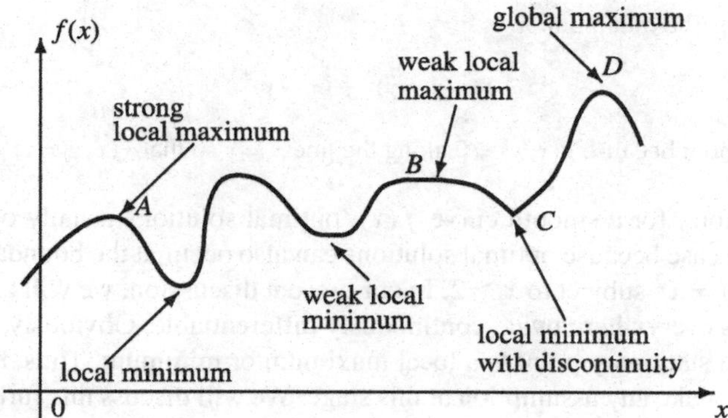

FIGURE 23.1 Strong and weak maxima and minima.

23.2 OPTIMALITY CRITERIA

In mathematical programming, there are many important concepts, and we will first introduce three related concepts: feasible solutions, optimality criteria, the strong local optimum and weak local optimum.

23.2.1 Feasible Solution

A point **x** which satisfies all the constraints is called a feasible point and thus is a feasible solution to the problem. The set of all feasible points is called the feasible region.

Example 23.3

We know that the domain $f(x) = x^2$ consists of all the real numbers. If we want to minimize $f(x)$ without any constraint, all solutions such as $x = -1$, $x = 1$ and $x = 0$ are feasible. In fact, the feasible region is the whole real axis. Obviously, $x = 0$ corresponds to $f(0) = 0$ as the true minimum.

However, if we want to find the minimum of $f(x) = x^2$ subject to $x \geq 2$, it becomes a constrained optimization problem. The points such as $x = 1$ and $x = 0$ are no long feasible because they do not satisfy $x \geq 2$. In this case, the feasible solutions are all the points that satisfy $x \geq 2$. So $x = 2$, $x = 100$ and $x = 10^8$ are all feasible. It is obvious that the minimum occurs at $x = 2$ with $f(2) = 2^2 = 4$. That is, the optimal solution for this problem occurs at the boundary point $x = 2$.

23.2.2 Optimality Criteria

A point $\mathbf{x}_*$ is called a strong local maximum of the nonlinearly constrained optimization problem if $f(\mathbf{x})$ is defined in a δ-neighborhood $N(\mathbf{x}_*, \delta)$ and satisfies $f(\mathbf{x}_*) > f(\mathbf{u})$ for $\forall \mathbf{u} \in N(\mathbf{x}_*, \delta)$ where $\delta > 0$ and $\mathbf{u} \neq \mathbf{x}_*$. If $\mathbf{x}_*$ is not a strong local maximum, the inclusion of equality in the condition $f(\mathbf{x}_*) \geq f(\mathbf{u})$ for $\forall \mathbf{u} \in N(\mathbf{x}_*, \delta)$ defines the point $\mathbf{x}_*$ as a weak local maximum (see Fig. 23.1). The local minima can be defined in the similar manner when $>$ and $\geq$ are replaced by $<$ and $\leq$, respectively.

Fig. 23.1 shows various local maxima and minima. Point A is a strong local maximum, while point B is a weak local maximum because there are many (in fact infinite) different values of **x** which will lead to the same value of $f(\mathbf{x}_*)$. Point D is the global maximum. However, point C is a strong local minimum, but it has a discontinuity in $f'(\mathbf{x}_*)$. So the stationary condition for this point $f'(\mathbf{x}_*) = 0$ is not valid. We will not deal with these types of minima or maxima in detail.

Example 23.4

For example, the minimum of

$$f(x) = x^2$$

at $x = 0$ is a strong local minimum. The minimum of

$$g(x, y) = (x - y)^2 + (x - y)^4$$

at $x = y = 0$ is a weak local minimum because $g(x, y) = 0$ along the line $x = y$ so that $g(x, y = x) = 0 = g(0, 0)$.

As we briefly mentioned before, for a smooth curve $f(x)$, optimal solutions usually occur at stationary points where $f'(x) = 0$. This is not always the case because optimal solutions can also occur at the boundary as we have seen in the previous example of minimizing $f(x) = x^2$ subject to $x \geq 2$. In our present discussion, we will assume that both $f(\mathbf{x})$ and $f'(\mathbf{x})$ are always continuous or $f(\mathbf{x})$ is everywhere twice-continuously differentiable. Obviously, the information of $f'(x)$ is not sufficient to determine whether a stationary point is a local maximum or minimum. Thus, higher-order derivatives such as $f''(x)$ are needed, but we do not make any assumption at this stage. We will discuss this further in detail in the next section.

23.3 UNCONSTRAINED OPTIMIZATION

Optimization problems can be classified as either unconstrained or constrained. Unconstrained optimization problems can in turn be subdivided into univariate and multivariate problems.

23.3.1 Univariate Functions

The simplest optimization problem without any constraints is probably the search for the maxima or minima of a univariate function $f(x)$. For unconstrained optimization problems, the optimality occurs at either boundary or more often at the critical points given by the stationary condition $f'(x) = 0$.

However, this stationary condition is just a necessary condition, but it is not a sufficient condition. If $f'(x_*) = 0$ and $f''(x_*) > 0$, it is a local minimum. Conversely, if $f'(x_*) = 0$ and $f''(x_*) < 0$, then it is a local maximum. However, if $f'(x_*) = 0$ and $f''(x_*) = 0$, care should be taken because $f''(x)$ may be indefinite (both positive and negative) when $x \to x_*$, then x_* corresponds to a saddle point. For example, for $f(x) = x^3$, we have

$$f'(x) = 3x^2, \quad f''(x) = 6x. \tag{23.6}$$

The stationary condition $f'(x) = 3x^2 = 0$ gives $x_* = 0$. However, we also have

$$f''(x_*) = f''(0) = 0.$$

In fact, $f(x) = x^3$ has a saddle point $x_* = 0$ because $f'(0) = 0$ but f'' changes sign from $f''(0+) > 0$ to $f''(0-) < 0$.

Example 23.5

For example, in order to find the maximum or minimum of a univariate function $f(x)$

$$f(x) = xe^{-x^2}, \qquad -\infty < x < \infty, \tag{23.7}$$

we have to find first the stationary point x_* when the first derivative $f'(x)$ is zero. That is

$$\frac{df(x_*)}{dx_*} = e^{-x_*^2} - 2x_*^2 e^{-x_*^2} = 0. \tag{23.8}$$

Since $\exp(-x_*^2) \neq 0$, we have

$$x_* = \pm\frac{\sqrt{2}}{2}. \tag{23.9}$$

From the basic calculus we know that the maximum requires $f''(x_*) \leq 0$ while the minimum requires $f''(x_*) \geq 0$. At $x_* = \sqrt{2}/2$, we have

$$f''(x_*) = (4x_*^2 - 6)x_* e^{-x_*^2} = -2\sqrt{2}e^{-1/2} < 0, \tag{23.10}$$

so this point corresponds to the maximum $f(x_*) = \frac{1}{2}e^{-1/2}$.

Similarly, at $x_* = -\sqrt{2}/2$, $f''(x_*) = 2\sqrt{2}e^{-1/2} > 0$, we have the minimum

$$f(x_*) = -\frac{1}{2}e^{-1/2}.$$

The maximization of a function $f(x)$ can be converted into the minimization of $A - f(x)$ where A is usually a large positive number (though $A = 0$ will do). For example, we know the maximum of $f(x) = e^{-x^2}, x \in (-\infty, \infty)$ is 1 at $x_* = 0$. This problem can be converted to a minimization problem $-f(x)$. For this reason, the optimization problems can be expressed as either minimization or maximization depending on the context and convenience of finding the solutions.

In fact, in the optimization literature, some books formulate all the optimization problems in terms of maximization, while others write these problems in terms of minimization, though they are in essence dealing with the same topics.

23.3.2 Multivariate Functions

We can extend the optimization procedure for univariate functions to multivariate functions using partial derivatives and relevant conditions. Let us start with an example

$$\text{minimize} \quad f(x, y) = x^2 + y^2, \quad x, y \in \Re. \tag{23.11}$$

It is obvious that $x = 0$ and $y = 0$ is a minimum solution because $f(0, 0) = 0$. The question is how to solve this problem formally. We can extend the stationary condition to partial derivatives, and we have

$$\frac{\partial f}{\partial x} = 0, \quad \frac{\partial f}{\partial y} = 0. \tag{23.12}$$

In this case, we have

$$\frac{\partial f}{\partial x} = 2x + 0 = 0, \tag{23.13}$$

and

$$\frac{\partial f}{\partial y} = 0 + 2y = 0. \tag{23.14}$$

The solution is obviously $x_* = 0$ and $y_* = 0$.

Now how do we know it corresponds to a maximum or minimum? If we try to use the second derivatives, we have four different partial derivatives such as f_{xx} and f_{yy} and which one should we use? In fact, we need to define a Hessian matrix from these second partial derivatives and we have

$$\mathbf{H} = \begin{pmatrix} f_{xx} & f_{xy} \\ f_{yx} & f_{yy} \end{pmatrix} = \begin{pmatrix} \frac{\partial^2 f}{\partial x^2} & \frac{\partial^2 f}{\partial x \partial y} \\ \frac{\partial^2 f}{\partial y \partial x} & \frac{\partial^2 f}{\partial y^2} \end{pmatrix}. \tag{23.15}$$

Since $\partial x \partial y = \partial y \partial x$ or

$$\frac{\partial^2 f}{\partial x \partial y} = \frac{\partial^2 f}{\partial y \partial x}, \tag{23.16}$$

we can conclude that the Hessian matrix is always symmetric. In the case of $f = x^2 + y^2$, it is easy to check that the Hessian matrix is

$$\mathbf{H} = \begin{pmatrix} 2 & 0 \\ 0 & 2 \end{pmatrix}. \tag{23.17}$$

Mathematically speaking, if $\mathbf{H}$ is positive definite, then the stationary point (x_*, y_*) corresponds a local minimum. Similarly, if $\mathbf{H}$ is negative definite, the stationary point corresponds to a maximum. Since the Hessian matrix here does not involve any x or y, it is always positive definite in the whole search domain $(x, y) \in \mathbb{R}^2$, we can conclude that the solution at point $(0, 0)$ is the global minimum.

Obviously, this is a special case. In general, the Hessian matrix will depend on the independent variables, but the definiteness test conditions still apply. That is, positive definiteness of a stationary point means a local minimum. Alternatively, for bivariate functions, we can define the determinant of the Hessian matrix in Eq. (23.15) as

$$\Delta = \det(\mathbf{H}) = f_{xx} f_{yy} - (f_{xy})^2. \tag{23.18}$$

At the stationary point (x_*, y_*), if $\Delta > 0$ and $f_{xx} > 0$, then (x_*, y_*) is a local minimum. If $\Delta > 0$ but $f_{xx} < 0$, it is a local maximum. If $\Delta = 0$, it is inconclusive and we have use other information such as higher-order derivatives. However, if $\Delta < 0$, it is a saddle point. A saddle point is a special point where a local minimum occurs along one direction while the maximum occurs along another (orthogonal) direction.

Example 23.6

Let us solve the minimization of $f(x, y) = (x-1)^2 + x^2 y^2$. In this case, we have

$$\frac{\partial f}{\partial x} = 2(x-1) + 2xy^2 = 0, \tag{23.19}$$

and

$$\frac{\partial f}{\partial y} = 0 + 2x^2 y = 0. \tag{23.20}$$

The second condition gives $y = 0$ or $x = 0$. Substituting $y = 0$ to the first condition, we have $x = 1$. However, $x = 0$ does not satisfy the first condition. Therefore, we have a solution $x_* = 1$ and $y_* = 0$.

For our example with $f = (x-1)^2 + x^2 y^2$, we have

$$\frac{\partial^2 f}{\partial x^2} = 2y^2 + 2, \quad \frac{\partial^2 f}{\partial x \partial y} = 4xy, \quad \frac{\partial^2 f}{\partial y \partial x} = 4xy, \quad \frac{\partial^2 f}{\partial y^2} = 2x^2, \tag{23.21}$$

and thus we have

$$\mathbf{H} = \begin{pmatrix} 2y^2 + 2 & 4xy \\ 4xy & 2x^2 \end{pmatrix}. \tag{23.22}$$

At the stationary point $(x_*, y_*) = (1, 0)$, the Hessian matrix becomes

$$\mathbf{H} = \begin{pmatrix} 2 & 0 \\ 0 & 2 \end{pmatrix},$$

which is positive definite because its double eigenvalues 2 are positive. Alternatively, we have $\Delta = 4 > 0$ and $f_{xx} = 2 > 0$. Therefore, $(1, 0)$ is a local minimum.

One disadvantage of using this determinant test is that it is only easier for functions with two independent variables. Even so, care should be take when using these conditions. Let us look at an example of the well-known Monkey Saddle surface.

Example 23.7

For the Monkey Saddle surface

$$f(x, y) = x^3 - 3xy^2,$$

we have

$$\frac{\partial f}{\partial x} = 3x^2 - 3y^2 = 0,$$

$$\frac{\partial f}{\partial y} = 0 - 6xy = 0.$$

Their solution is simple $x = y = 0$.

The Hessian matrix is

$$\mathbf{H} = \begin{pmatrix} 6x & -6y \\ -6y & -6x \end{pmatrix},$$

which is zero at $x = y = 0$. That is

$$\mathbf{H} = \begin{pmatrix} 0 & 0 \\ 0 & 0 \end{pmatrix}.$$

This means that we cannot determine the properties of this point based on the test of Δ. In fact, $\mathbf{H}$ is indefinite, and thus $(0, 0)$ is a saddle point. When you plot out the surface and look at the shape of the surface, it becomes obvious why it is called Monkey's Saddle.

In fact, for a multivariate function $f(x_1, x_2, ..., x_d)$ in a d-dimensional space, the stationary condition can be extended to

$$\mathbf{G} = \nabla f = (\frac{\partial f}{\partial x_1}, \frac{\partial f}{\partial x_2}, ..., \frac{\partial f}{\partial x_d})^T = 0, \tag{23.23}$$

where $\mathbf{G}$ is called the gradient vector. The second derivative test become the definiteness of the Hessian matrix

$$\mathbf{H} = \begin{pmatrix} \frac{\partial^2 f}{\partial x_1{}^2} & \frac{\partial^2 f}{\partial x_1 \partial x_2} & \cdots & \frac{\partial^2 f}{\partial x_1 \partial x_d} \\ \frac{\partial^2 f}{\partial x_2 \partial x_1} & \frac{\partial^2 f}{\partial x_2{}^2} & \cdots & \frac{\partial^2 f}{\partial x_2 \partial x_d} \\ \vdots & \vdots & \ddots & \vdots \\ \frac{\partial^2 f}{\partial x_d \partial x_1} & \frac{\partial^2 f}{\partial x_d \partial x_2} & \cdots & \frac{\partial^2 f}{\partial x_d{}^2} \end{pmatrix}. \tag{23.24}$$

At the stationary point defined by $\mathbf{G} = \nabla f = 0$, the positive definiteness of $\mathbf{H}$ gives a local minimum, while the negative definiteness corresponds a local maximum.

Example 23.8

It is easy to check that the gradient of $f(x, y, z) = xy - ye^{-z} + z\cos(x)$ is simply

$$\mathbf{G} = \begin{pmatrix} y - z\sin x & x - e^{-z} & ye^{-z} + \cos x \end{pmatrix}^T,$$

while the Hessian matrix is

$$\mathbf{H} = \begin{pmatrix} -z\cos x & 1 & -\sin x \\ 1 & 0 & e^{-z} \\ -\sin x & e^{-z} & -ye^{-z} \end{pmatrix}.$$

Now let us discuss the general formulation of multivariate optimization. To find the maximum or minimum of a multivariate function $f(\mathbf{x})$ where $\mathbf{x} = (x_1, ..., x_d)^T$, we can express it as an univariate optimization problem concerning a design vector $\mathbf{x}$

$$\text{minimize/maximize} \quad f(\mathbf{x}), \quad \mathbf{x} \in \mathbb{R}^d. \tag{23.25}$$

For a general function $f(\mathbf{x})$, we can expand it using Taylor series about a point $\mathbf{x} = \mathbf{x}_*$ so that $\mathbf{x} = \mathbf{x}_* + \epsilon\mathbf{u}$

$$f(\mathbf{x} + \epsilon\mathbf{u}) = f(\mathbf{x}_*) + \epsilon\mathbf{u}\mathbf{G}(\mathbf{x}_*) + \frac{1}{2}\epsilon^2\mathbf{u}^T\mathbf{H}(\mathbf{x}_* + \epsilon\mathbf{u})\mathbf{u} + ..., \tag{23.26}$$

where **G** and **H** are its gradient vector and Hessian matrix, respectively. ϵ is a small parameter, and **u** is a vector. For example, for a generic quadratic function

$$f(\mathbf{x}) = \frac{1}{2}\mathbf{x}^T\mathbf{A}\mathbf{x} + \mathbf{k}^T\mathbf{x} + \mathbf{b},$$

where **A** is a constant square matrix, **k** is the gradient vector, and **b** is a vector constant, we have

$$f(\mathbf{x}_* + \epsilon\mathbf{u}) = f(\mathbf{x}_*) + \epsilon\mathbf{u}^T\mathbf{k} + \frac{1}{2}\epsilon^2\mathbf{u}^T\mathbf{A}\mathbf{u} + ..., \tag{23.27}$$

where

$$f(\mathbf{x}_*) = \frac{1}{2}\mathbf{x}_*^T\mathbf{A}\mathbf{x}_* + \mathbf{k}^T\mathbf{x}_* + \mathbf{b}. \tag{23.28}$$

Thus, in order to study the local behavior of a quadratic function, we only need to study **G** and **H**. In addition, for simplicity, we can take $\mathbf{b} = \mathbf{0}$ as it is a constant vector anyway.

At a stationary point $\mathbf{x}_*$, the first derivatives are zero or $\mathbf{G}(\mathbf{x}_*) = 0$. Therefore, equation (23.26) becomes

$$f(\mathbf{x}_* + \epsilon\mathbf{u}) \approx f(\mathbf{x}_*) + \frac{1}{2}\epsilon^2\mathbf{u}^T\mathbf{H}\mathbf{u}. \tag{23.29}$$

If $\mathbf{H} = \mathbf{A}$, then

$$\mathbf{A}\mathbf{v} = \lambda\mathbf{v} \tag{23.30}$$

forms an eigenvalue problem. For an $n \times n$ matrix **A**, there will be n eigenvalues $\lambda_j (j = 1, ..., n)$ with n corresponding eigenvectors **v**. As we have seen earlier that **A** is symmetric, these eigenvectors are orthonormal. That is,

$$\mathbf{v}_i^T\mathbf{v}_j = \delta_{ij}. \tag{23.31}$$

Near any stationary point $\mathbf{x}_*$, if we take $\mathbf{u}_j = \mathbf{v}_j$ as the local coordinate systems, we then have

$$f(\mathbf{x}_* + \epsilon\mathbf{v}_j) = f(\mathbf{x}_*) + \frac{1}{2}\epsilon^2\lambda_j, \tag{23.32}$$

which means that the variations of $f(\mathbf{x})$, when **x** moves away from the stationary point $\mathbf{x}_*$ along the direction $\mathbf{v}_j$, are characterized by the eigenvalues.

If $\lambda_j > 0$, $|\epsilon| > 0$ will lead to $|\Delta f| = |f(\mathbf{x}) - f(\mathbf{x}_*)| > 0$. In other words, $f(\mathbf{x})$ will increase as $|\epsilon|$ increases. Conversely, if $\lambda_j < 0$, $f(\mathbf{x})$ will decrease as $|\epsilon| > 0$ increases. Obviously, in the special case $\lambda_j = 0$, the function $f(\mathbf{x})$ will remain constant along the corresponding direction of $\mathbf{v}_j$.

Example 23.9

We know that function

$$f(x, y) = xy,$$

has a saddle point at $(0, 0)$. It increases along the $x = y$ direction and decreases along $x = -y$ direction. From the above analysis, we know that $\mathbf{x}_* = (x_*, y_*)^T = (0, 0)^T$ and $f(x_*, y_*) = 0$. We now have

$$f(\mathbf{x}_* + \epsilon\mathbf{u}) \approx f(\mathbf{x}_*) + \frac{1}{2}\epsilon^2\mathbf{u}^T\mathbf{A}\mathbf{u},$$

where

$$\mathbf{A} = \mathbf{H} = \nabla^2 f(\mathbf{x}_*) = \begin{pmatrix} \frac{\partial^2 f}{\partial x^2} & \frac{\partial^2 f}{\partial x \partial y} \\ \frac{\partial^2 f}{\partial x \partial y} & \frac{\partial^2 f}{\partial y^2} \end{pmatrix} = \begin{pmatrix} 0 & 1 \\ 1 & 0 \end{pmatrix}.$$

The eigenvalue problem is simply

$$\mathbf{A}v = \lambda_j \mathbf{v}_j, \quad (j = 1, 2),$$

or

$$\begin{vmatrix} -\lambda_j & 1 \\ 1 & -\lambda_j \end{vmatrix} = 0,$$

whose solutions are

$$\lambda_j = \pm 1.$$

For $\lambda_1 = 1$, the corresponding eigenvector is

$$\mathbf{v}_1 = \begin{pmatrix} \sqrt{2}/2 \\ \sqrt{2}/2 \end{pmatrix}.$$

Similarly, for $\lambda_2 = -1$, the eigenvector is

$$\mathbf{v}_2 = \begin{pmatrix} \sqrt{2}/2 \\ -\sqrt{2}/2 \end{pmatrix}.$$

Since A is symmetric, $\mathbf{v}_1$ and $\mathbf{v}_2$ are orthonormal. Indeed this is the case because $\|\mathbf{v}_1\| = \|\mathbf{v}_2\| = 1$ and

$$\mathbf{v}_1^T \mathbf{v}_2 = \frac{\sqrt{2}}{2} \times \frac{\sqrt{2}}{2} + \frac{\sqrt{2}}{2} \times (-\frac{\sqrt{2}}{2}) = 0.$$

Thus, we have

$$f(\epsilon \mathbf{v}_j) = \frac{1}{2}\epsilon^2 \lambda_j, \qquad (j = 1, 2). \tag{23.33}$$

As $\lambda_1 = 1$ is positive, f increases along the direction $\mathbf{v}_1 = \frac{\sqrt{2}}{2}(1\ 1)^T$ which is indeed along the line $x = y$.

Similarly, for $\lambda_2 = -1$, f will decrease along $\mathbf{v}_2 = \frac{\sqrt{2}}{2}(1\ -1)^T$ which is exactly along the line $x = -y$. As there is no zero eigenvalue, the function will not remain constant in the region around $(0, 0)$.

In essence, the eigenvalues of the Hessian matrix $\mathbf{H}$ determine the local behavior of the function. As we mentioned before, if $\mathbf{H}$ is positive semi-definite, it corresponds to a local minimum.

23.4 GRADIENT-BASED METHODS

Gradient-based methods are iterative methods that extensively use the gradient information of the objective function during iterations. For the minimization of a function $f(\mathbf{x})$, the essence of this method is

$$\mathbf{x}^{(n+1)} = \mathbf{x}^{(n)} + \alpha g(\nabla f, \mathbf{x}^{(n)}), \tag{23.34}$$

where α is the step size which can vary during iterations. $g(\nabla f, \mathbf{x}^{(n)})$ is a function of the gradient ∇f and the current location $\mathbf{x}^{(n)}$. Different methods use different forms of $g(\nabla f, \mathbf{x}^{(n)})$.

23.4.1 Newton's Method

We know that Newton's method is a popular iterative method for finding the zeros of a nonlinear univariate function of $f(x)$ on the interval $[a, b]$. It can be modified for solving optimization problems because it is equivalent to finding the zeros of the first derivative $f'(\mathbf{x})$ once the objective function $f(\mathbf{x})$ is given.

For a given function $f(\mathbf{x})$ which is continuously differentiable, we have the Taylor expansion about a known point $\mathbf{x} = \mathbf{x}_n$ (with $\Delta\mathbf{x} = \mathbf{x} - \mathbf{x}_n$)

$$f(\mathbf{x}) = f(\mathbf{x}_n) + (\nabla f(\mathbf{x}_n))^T \Delta\mathbf{x} + \frac{1}{2}\Delta\mathbf{x}^T \nabla^2 f(\mathbf{x}_n) \Delta\mathbf{x} + ...,$$

which is minimized near a critical point when $\Delta\mathbf{x}$ is the solution of the following linear equation

$$\nabla f(\mathbf{x}_n) + \nabla^2 f(\mathbf{x}_n)\Delta\mathbf{x} = 0. \tag{23.35}$$

This leads to

$$\mathbf{x} = \mathbf{x}_n - \mathbf{H}^{-1}\nabla f(\mathbf{x}_n), \tag{23.36}$$

where $\mathbf{H} = \nabla^2 f(\mathbf{x}_n)$ is the Hessian matrix. If the iteration procedure starts from the initial vector $\mathbf{x}^{(0)}$ (usually taken to be a guessed point in the domain), then Newton's iteration formula for the nth iteration is

$$\mathbf{x}^{(n+1)} = \mathbf{x}^{(n)} - \mathbf{H}^{-1}(\mathbf{x}^{(n)})\nabla f(\mathbf{x}^{(n)}). \tag{23.37}$$

It is worth pointing out that if $f(\mathbf{x})$ is quadratic, then the solution can be found exactly in a single step. However, this method may become tricky for non-quadratic functions, especially when we have to calculate the large Hessian matrix.

It can usually be time-consuming to calculate the Hessian matrix for second derivatives. A good alternative is to use an identity matrix to approximate the Hessian by using $\mathbf{H}^{-1} = \mathbf{I}$, and we have the quasi-Newton method

$$\mathbf{x}^{(n+1)} = \mathbf{x}^{(n)} - \alpha\mathbf{I}\,\nabla f(\mathbf{x}^{(n)}), \tag{23.38}$$

where $\alpha \in (0, 1)$ is a step size. In this case, the method is essentially the steepest descent method.

23.4.2 Steepest Descent Method

The essence of this method is to find the lowest possible objective function $f(\mathbf{x})$ from the current point $\mathbf{x}^{(n)}$. From the Taylor expansion of $f(\mathbf{x})$ about $\mathbf{x}^{(n)}$, we have

$$f(\mathbf{x}^{(n+1)}) = f(\mathbf{x}^{(n)} + \Delta\mathbf{s}) \approx f(\mathbf{x}^{(n)} + (\nabla f(\mathbf{x}^{(n)}))^T\Delta\mathbf{s}, \tag{23.39}$$

where $\Delta\mathbf{s} = \mathbf{x}^{(n+1)} - \mathbf{x}^{(n)}$ is the increment vector. Since we try to find a lower (better) approximation to the objective function, it requires that the second term on the right hand is negative. That is

$$f(\mathbf{x}^{(n)} + \Delta\mathbf{s}) - f(\mathbf{x}^{(n)}) = (\nabla f)^T\Delta\mathbf{s} < 0. \tag{23.40}$$

From vector analysis, we know the inner product $\mathbf{u}^T\mathbf{v}$ of two vectors $\mathbf{u}$ and $\mathbf{v}$ is largest when they are parallel but in opposite directions.

Therefore, $(\nabla f)^T\Delta\mathbf{s}$ becomes the largest when

$$\Delta\mathbf{s} = -\alpha\nabla f(\mathbf{x}^{(n)}), \tag{23.41}$$

where $\alpha > 0$ is the step size. This is the case when the direction $\Delta\mathbf{s}$ is along the steepest descent in the negative gradient direction. As we have seen earlier, this method is a quasi-Newton method.

The choice of the step size α is very important. A very small step size means slow movement towards the local minimum, while a large step may overshoot and subsequently makes it move far away from the local minimum. The step size $\alpha = \alpha^{(n)}$ should be different at each iteration step and should be chosen so that it minimizes the objective function $f(\mathbf{x}^{(n+1)}) = f(\mathbf{x}^{(n)}, \alpha^{(n)})$. Therefore, the steepest descent method can be written as

$$\mathbf{x}^{(n+1)} = \mathbf{x}^{(n)} - \alpha^{(n)}(\nabla f(\mathbf{x}^{(n)})). \tag{23.42}$$

In each iteration, the gradient and step size will be calculated. Again, a good initial guess of both the starting point and the step size is useful, though it is not an easy task.

Example 23.10

Let us minimize the function

$$f(x_1, x_2) = 10x_1^2 + 5x_1x_2 + 10(x_2 - 3)^2,$$

where

$$(x_1, x_2) = [-10, 10] \times [-15, 15],$$

using the steepest descent method starting with the initial $\mathbf{x}^{(0)} = (10, 15)^T$. We know that the gradient

$$\nabla f = (20x_1 + 5x_2,\ 5x_1 + 20x_2 - 60)^T,$$

therefore

$$\nabla f(\mathbf{x}^{(0)}) = (275,\ 290)^T.$$

In the first iteration, we have

$$\mathbf{x}^{(1)} = \mathbf{x}^{(0)} - \alpha_0 \begin{pmatrix} 275 \\ 290 \end{pmatrix}.$$

The step size α_0 should be chosen such that $f(\mathbf{x}^{(1)})$ is at the minimum, which means that

$$f(\alpha_0) = 10(10 - 275\alpha_0)^2 + 5(10 - 275\alpha_0)(15 - 290\alpha_0) + 10(12 - 290\alpha_0)^2,$$

should be minimized. This becomes an optimization problem for a single independent variable α_0. All the techniques for univariate optimization problems such as Newton's method can be used to find α_0. We can also obtain the solution by setting

$$\frac{df}{d\alpha_0} = -159725 + 3992000\alpha_0 = 0,$$

whose solution is $\alpha_0 \approx 0.04001$.

At the second step, we have

$$\nabla f(\mathbf{x}^{(1)}) = (-3.078, 2.919)^T,\ \mathbf{x}^{(2)} = \mathbf{x}^{(1)} - \alpha_1 \begin{pmatrix} -3.078 \\ 2.919 \end{pmatrix}.$$

The minimization of $f(\alpha_1)$ gives $\alpha_1 \approx 0.066$, and the new location of the steepest descent process is

$$\mathbf{x}^{(2)} \approx (-0.797, 3.202)^T.$$

At the third iteration, we have

$$\nabla f(\mathbf{x}^{(2)}) = (0.060, 0.064)^T,\ \mathbf{x}^{(3)} = \mathbf{x}^{(2)} - \alpha_2 \begin{pmatrix} 0.060 \\ 0.064 \end{pmatrix}.$$

The minimization of $f(\alpha_2)$ leads to $\alpha_2 \approx 0.040$, and we have

$$\mathbf{x}^{(3)} \approx (-0.8000299, 3.20029)^T.$$

Then, the iterations continue until a prescribed tolerance is met.

From the basic calculus, we know that we can set the first partial derivatives equal to zero

$$\frac{\partial f}{\partial x_1} = 20x_1 + 5x_2 = 0,\quad \frac{\partial f}{\partial x_2} = 5x_1 + 20x_2 - 60 = 0,$$

we know that the minimum occurs exactly at

$$\mathbf{x}_* = (-4/5, 16/5)^T = (-0.8, 3.2)^T.$$

The steepest descent method gives almost the exact solution after only 3 iterations.

In finding the step size α_n in the above steepest descent method, we have used the stationary condition $df(\alpha_n)/d\alpha_n = 0$. Well, you may say that if we use this stationary condition for $f(\alpha_0)$, why not use the same method to get the minimum point of $f(\mathbf{x})$ in the first place. There are two reasons here. The first reason is that this is a simple example for demonstrating how the steepest descent method works. The second reason is that even for complicated multiple variables $f(x_1, ..., x_p)$ (say $p = 500$), then $f(\alpha_n)$ at any step n is still a univariate function, and the optimization task of such $f(\alpha_n)$ is much simpler compared with the original multivariate problem. In practice, the step sizes will be determined by combining with other methods such as line search and step sizes are adjusted during the iteration process.

From our example, we know that the convergence from the second iteration to the third iteration is slow. In fact, the steepest descent is typically slow once the local minimization is near. This is because near the local optimality the gradient is nearly zero, and thus the rate of descent is also slow. If high accuracy is needed near the local minimum, other local search methods should be used.

Though gradient-based methods can be very efficient, the final solution tends to be dependent on the starting point. If the starting point is very far away from the optimal solution, the algorithm can either reach to a complete different solution for multimodal problems or simply fail in some cases. Therefore, there is no guarantee that the global optimal solution can be found.

It is worth pointing out that there are many variations of the steepest descent methods. If such optimization aims is to find the maximum, then this method becomes the *hill-climbing* method because the aim is to climb up the hill to the highest peak.

23.5 NONLINEAR OPTIMIZATION

As most real-world problems are nonlinear, nonlinear mathematical programming forms an important part of mathematical optimization methods. A broad class of nonlinear programming problems is about the minimization or maximization of $f(\mathbf{x})$ subject to no constraints, and another important class is the minimization of a quadratic objective function subject to nonlinear constraints. There are many other nonlinear programming problems as well.

Nonlinear programming problems are often classified according to the convexity of the defining functions. An interesting property of a convex function f is that the vanishing of the gradient $\nabla f(\mathbf{x}_*) = 0$ guarantees that the point x_* is a global minimum or maximum of f. If a function is not convex or concave, then it is much more difficult to find global minima or maxima.

23.5.1 Penalty Method

For the simple function optimization with equality and inequality constraints, a common method is the penalty method. For the optimization problem

$$\begin{aligned} &\text{minimize } f(\mathbf{x}), \ \mathbf{x} = (x_1, ..., x_n)^T \in \mathbb{R}^n, \\ &\text{subject to } \phi_i(\mathbf{x}) = 0, \ (i = 1, ..., M), \\ &\qquad\qquad \psi_j(\mathbf{x}) \le 0, \ (j = 1, ..., N), \end{aligned} \tag{23.43}$$

the idea is to define a penalty function so that the constrained problem is transformed into an unconstrained problem. Now we define $\Pi(\mathbf{x}, \mu_i, \nu_j)$

$$\Pi(\mathbf{x}, \mu_i, \nu_j) = f(\mathbf{x}) + \sum_{i=1}^{M} \mu_i \phi_i^2(\mathbf{x}) + \sum_{j=1}^{N} \nu_j \psi_j^2(\mathbf{x}), \tag{23.44}$$

where $\mu_i \gg 1$ and $\nu_j \ge 0$.

Example 23.11

For example, in order to solve the following problem of Gill-Murray-Wright type

$$\begin{aligned} &\underset{x \in \mathbb{R}}{\text{minimize}} \ f(x) = 100(x - b)^2 + 1, \\ &\text{subject to } g(x) = x - a \ge 0, \end{aligned} \tag{23.45}$$

where $a > b$ is a given value, we can define a penalty function $\Pi(x)$ using a penalty parameter $\mu \gg 1$. We have

$$\Pi(x, \mu) = f(x) + \frac{\mu}{2} g(x)^T g(x) = 100(x - b)^2 + 1 + \frac{\mu}{2}(x - a)^2, \tag{23.46}$$

where the typical value for μ is $2000 \sim 10000$.

This essentially transforms the original constrained optimization into an unconstrained problem. From the stationary condition $\Pi'(x) = 0$, we have

$$200(x_* - b) - \mu(x_* - a) = 0, \tag{23.47}$$

which gives

$$x_* = \frac{200b + \mu a}{200 + \mu}. \tag{23.48}$$

For $\mu \to \infty$, we have $x_* \to a$. For $\mu = 2000$, $a = 2$ and $b = 1$, we have $x_* \approx 1.9090$.

This means the solution depends on the value of μ, and it is very difficult to use extremely large values without causing extra computational difficulties.

Ideally, the formulation using the penalty method should be properly designed so that the results will not depend on the penalty coefficient or at least the dependence should be sufficiently weak.

23.5.2 Lagrange Multipliers

Another powerful method without the above limitation of using large μ is the method of Lagrange multipliers. If we want to minimize a function $f(\mathbf{x})$

$$\text{minimize } f(\mathbf{x}), \qquad \mathbf{x} = (x_1, ..., x_n)^T \in \mathbb{R}^n, \tag{23.49}$$

subject to the following nonlinear equality constraint

$$h(\mathbf{x}) = 0, \tag{23.50}$$

then we can combine the objective function $f(\mathbf{x})$ with the equality to form a new function, called the Lagrangian

$$\Pi = f(\mathbf{x}) + \lambda h(\mathbf{x}), \tag{23.51}$$

where λ is the Lagrange multiplier, which is an unknown scalar to be determined.

This again converts the constrained optimization into an unstrained problem for $\Pi(\mathbf{x})$, which is the beauty of this method. If we have M equalities,

$$h_j(\mathbf{x}) = 0, \qquad (j = 1, ..., M), \tag{23.52}$$

then we need M Lagrange multipliers $\lambda_j (j = 1, ..., M)$. We thus have

$$\Pi(x, \lambda_j) = f(\mathbf{x}) + \sum_{j=1}^{M} \lambda_j h_j(\mathbf{x}). \tag{23.53}$$

The requirement of stationary conditions leads to

$$\frac{\partial \Pi}{\partial x_i} = \frac{\partial f}{\partial x_i} + \sum_{j=1}^{M} \lambda_j \frac{\partial h_j}{\partial x_i}, \quad (i = 1, ..., n), \tag{23.54}$$

and

$$\frac{\partial \Pi}{\partial \lambda_j} = h_j = 0, \quad (j = 1, ..., M). \tag{23.55}$$

These $M + n$ equations will determine the n-component of $\mathbf{x}$ and M Lagrange multipliers. As $\frac{\partial \Pi}{\partial g_j} = \lambda_j$, we can consider λ_j as the rate of the change of the quantity Π as a functional of h_j.

Example 23.12

To solve the optimization problem

$$\text{maximize } f(x, y) = xy^2, \quad (x, y) \in \mathbb{R}^2,$$

subject to the equality condition

$$h(x, y) = x^2 + y^2 - 3 = 0,$$

we can now define

$$\Pi = f(x, y) + \lambda h(x, y) = xy^2 + \lambda(x^2 + y^2 - 3).$$

The stationary conditions become

$$\frac{\partial \Pi}{\partial x} = y^2 + 2\lambda x = 0,$$
$$\frac{\partial \Pi}{\partial y} = 2xy + 2\lambda y = 0,$$

and

$$\frac{\partial \Pi}{\partial \lambda} = x^2 + y^2 - 3 = 0.$$

The condition $xy + \lambda y = 0$ implies that $y = 0$ or $\lambda = -x$. The case of $y = 0$ can be eliminated as it leads to $x = 0$ from $y^2 + 2\lambda x = 0$, which does not satisfy the last condition $x^2 + y^2 = 3$. Therefore, the only valid solution is

$$\lambda = -x.$$

From the first stationary condition, we have

$$y^2 - 2x^2 = 0, \quad \text{or} \quad y^2 = 2x^2.$$

Substituting this into the third stationary condition, we have

$$x^2 + 2x^2 - 3 = 0,$$

which gives

$$x = \pm 1, \quad y = \pm\sqrt{2}.$$

So we have four stationary points

$$P_1(1, \sqrt{2}), \ P_2(1, -\sqrt{2}), \ P_3(-1, \sqrt{2}), \ P_4(-1, -\sqrt{(2)}).$$

The values of the function $f(x, y)$ at these four points are

$$f(P_1) = 2, \ f(P_2) = 2, \ f(P_3) = -2, \ f(P_4) = -2.$$

Thus, the function reaches its maxima at $(1, \sqrt{2})$ and $(1, -\sqrt{2})$. The other two points correspond to the minima. The Lagrange multiplier in this case is $\lambda = -1$.

There is a counterpart of the Lagrange multipliers for nonlinear optimization with constraint inequalities. The Karush-Kuhn-Tucker (KKT) conditions concern the requirement for a solution to be optimal in nonlinear programming.

23.6 KARUSH-KUHN-TUCKER CONDITIONS

Let us know focus on the following nonlinear optimization problem

$$\text{minimize } f(\mathbf{x}), \quad \mathbf{x} \in \mathbb{R}^n,$$

$$\text{subject to } \phi_i(\mathbf{x}) = 0, \ (i = 1, ..., M),$$
$$\psi_j(\mathbf{x}) \le 0, \ (j = 1, ..., N). \tag{23.56}$$

If all the functions are continuously differentiable, at a local minimum $\mathbf{x}_*$, there exist constants $\lambda_0, \lambda_1, ..., \lambda_q$ and $\mu_1, ..., \mu_p$ such that

$$\lambda_0 \nabla f(\mathbf{x}_*) + \sum_{i=1}^{M} \mu_i \nabla \phi_i(\mathbf{x}_*) + \sum_{j=1}^{N} \lambda_j \nabla \psi_j(\mathbf{x}_*) = 0, \tag{23.57}$$

$$\psi_j(\mathbf{x}_*) \le 0, \tag{23.58}$$

and

$$\lambda_j \psi_j(\mathbf{x}_*) = 0, \ (j = 1, 2, ..., N), \tag{23.59}$$

where $\lambda_j \ge 0$, $(i = 0, 1, ..., N)$. The constants satisfy the following condition

$$\sum_{j=0}^{N} \lambda_j + \sum_{i=1}^{M} |\mu_i| \ge 0. \tag{23.60}$$

This is essentially a generalized method of the Lagrange multipliers. However, there is a possibility of degeneracy when $\lambda_0 = 0$ under certain conditions.

It is worth pointing out that such KKT conditions can be useful to prove theorems and sometimes useful to gain insightful into certain types of problems. However, they are not really helpful in practice in the sense that they do not give any indication where the optimal solutions can lie in the search domain so as to guide the search process.

23.7 SEQUENTIAL QUADRATIC PROGRAMMING

Quadratic programming is a special class of mathematical programming and it deserves a special discussion due to its popularity and good mathematical properties.

23.7.1 Quadratic Programming

The objective function in quadratic programming (QP) is a quadratic form

$$f(\mathbf{x}) = \frac{1}{2}\mathbf{x}^T \mathbf{Q} \mathbf{x} + \mathbf{b}^T \mathbf{x} + \mathbf{c}, \tag{23.61}$$

subject to

$$\mathbf{A}\mathbf{x} \le \mathbf{d}, \quad \mathbf{x} \ge 0, \tag{23.62}$$

where $\mathbf{x} = (x_1, x_2, ..., x_n)^T$ an n-dimensional vector. In addition, $\mathbf{b}$ and $\mathbf{c}$ are constant vectors. In most textbooks, $\mathbf{c} = 0$ is used. Here, $\mathbf{Q}$ is a symmetric square matrix of size $n \times n$, while $\mathbf{A}$ is an $m \times n$ matrix where m is the number of inequality constraints.

The constraints can be incorporated using Lagrange multipliers and KKT formulations. In case of $\mathbf{c} = 0$, we have the Lagrangian

$$L = \frac{1}{2}\mathbf{x}^T \mathbf{Q} \mathbf{x} + \mathbf{b}^T \mathbf{x} + \lambda^T (\mathbf{A}\mathbf{x} - \mathbf{d}) - \mu^T x, \tag{23.63}$$

where the Lagrange multipliers or vectors must be non-negative; that is, $\lambda \ge 0$ and $\mu \ge 0$. The KKT conditions can be summarized as follows:

- Stationarity (or stationary conditions): $\nabla_x L = 0$

$$\nabla_x L = \mathbf{Q}\mathbf{x} + \mathbf{b} + \mathbf{A}^T \lambda - \mu = 0. \tag{23.64}$$

Sequential Quadratic Programming

begin
Choose a starting point $\mathbf{x}_0$ *and approximation* H_0 *to the Hessian*
repeat $k = 1, 2, ...$
  Solve a QP subproblem: QP_k *to get the search direction* $\mathbf{s}_k$
  Given $\mathbf{s}_k$*, find* α *so as to determine* $\mathbf{x}_{k+1}$
  Update the approximate Hessian $\mathbf{H}_{k+1}$ *using the BFGS scheme*
  $k = k + 1$
until *(stop criterion)*
end

FIGURE 23.2 Procedure of sequential quadratic programming.

- Complementarity: $\lambda^T(\mathbf{A} - \mathbf{d}) = 0$ and $\mu^T \mathbf{x} = 0$.

From the quadratic objective, we have

$$\nabla^2 f = \mathbf{Q}. \tag{23.65}$$

If $\mathbf{Q}$ is positive semi-definite, the solution given by the stationary condition will be the global minimum.

23.7.2 Sequential Quadratic Programming

Sequential (or successive) quadratic programming (SQP) represents one of the state-of-art and most popular methods for solving nonlinear constrained optimization. It is also one of the robust methods. Let us consider a general nonlinear optimization problem

$$\text{minimize } f(\mathbf{x}), \tag{23.66}$$

$$\text{subject to } h_i(\mathbf{x}) = 0, \ (i = 1, ..., p), \tag{23.67}$$

$$g_j(\mathbf{x}) \le 0, \ (j = 1, ..., q). \tag{23.68}$$

The fundamental idea of sequential quadratic programming is to approximate the computationally extensive full Hessian matrix using a quasi-Newton updating method. Subsequently, this generates a subproblem of quadratic programming (called QP subproblem) at each iteration, and the solution to this subproblem can be used to determine the search direction and next trial solution. The main idea is outlined in Fig. 23.2.

Using the Taylor expansions, the above problem can be approximated, at each iteration, as the following problem

$$\text{minimize } \frac{1}{2}\mathbf{s}^T \nabla^2 L(\mathbf{x}_k)\mathbf{s} + \nabla f(\mathbf{x}_k)^T \mathbf{s} + f(\mathbf{x}_k), \tag{23.69}$$

$$\text{subject to } \nabla h_i(\mathbf{x}_k)^T \mathbf{s} + h_i(\mathbf{x}_k) = 0, \ (i = 1, ..., p), \tag{23.70}$$

$$\nabla g_j(\mathbf{x}_k)^T \mathbf{s} + g_j(\mathbf{x}_k) \le 0, \ (j = 1, ..., q), \tag{23.71}$$

where the Lagrange function, also called the merit function, is defined by

$$\begin{aligned} L(\mathbf{x}) &= f(\mathbf{x}) + \sum_{i=1}^{p} \lambda_i h_i(\mathbf{x}) + \sum_{j=1}^{q} \mu_j g_j(\mathbf{x}) \\ &= f(\mathbf{x}) + \lambda^T \mathbf{h}(\mathbf{x}) + \mu^T \mathbf{g}(\mathbf{x}), \end{aligned} \tag{23.72}$$

where $\lambda = (\lambda_1, ..., \lambda_p)^T$ is the vector of Lagrange multipliers, and $\mu = (\mu_1, ..., \mu_q)^T$ is the vector of KKT multipliers. Here we have used the notation $\mathbf{h} = (h_1(\mathbf{x}), ..., h_p(\mathbf{x}))^T$ and $\mathbf{g} = (g_1(\mathbf{x}), ..., g_q(\mathbf{x}))^T$.

To approximate the Hessian $\nabla^2 L(\mathbf{x}_k)$ by a positive definite symmetric matrix $\mathbf{H}_k$, the standard Broydon-Fletcher-Goldfarbo-Shanno (BFGS) approximation of the Hessian can be used, and we have

$$\mathbf{H}_{k+1} = \mathbf{H}_k + \frac{\mathbf{v}_k \mathbf{v}_k^T}{\mathbf{v}_k^T \mathbf{u}_k} - \frac{\mathbf{H}_k \mathbf{u}_k \mathbf{u}_k^T \mathbf{H}_k^T}{\mathbf{u}_k^T \mathbf{H}_k \mathbf{u}_k}, \tag{23.73}$$

where

$$\mathbf{u}_k = \mathbf{x}_{k+1} - \mathbf{x}_k, \tag{23.74}$$

and

$$\mathbf{v}_k = \nabla L(\mathbf{x}_{k+1}) - \nabla L(\mathbf{x}_k). \tag{23.75}$$

The QP subproblem is solved to obtain the search direction

$$\mathbf{x}_{k+1} = \mathbf{x}_k + \alpha \mathbf{s}_k, \tag{23.76}$$

using a line search method by minimizing a penalty function, also commonly called merit function

$$\Phi(\mathbf{x}) = f(\mathbf{x}) + \rho\Big[\sum_{i=1}^{p} |h_i(\mathbf{x})| + \sum_{j=1}^{q} \max\{0, g_j(\mathbf{x})\}\Big], \tag{23.77}$$

where ρ is the penalty parameter.

It is worth pointing out that any SQP method requires a good choice of $\mathbf{H}_k$ as the approximate Hessian of the Lagrangian L. Obviously, if $\mathbf{H}_k$ is exactly calculated as $\nabla^2 L$, SQP essentially becomes Newton's method solving the optimality condition. A popular way to approximate the Lagrangian Hessian is to use a quasi-Newton scheme as we used the BFGS formula described earlier.

There are many software packages that have implemented SQP properly and they can be used to solve a wide range of problems in real-world applications.

EXERCISES

23.1. Find the global minimum of $f(x, y) = (1 - x)^2 + 100(y - x^2)^2$.

23.2. Solve the following constrained problem

$$f = x^2 + y^2 + 2xy,$$

subject to

$$y - x^2 + 2 = 0.$$

23.3. For $f(x) = x^2 + 5y^2 + xy$, calculate first the gradients and Hessian matrix, and then find its minimum.

23.4. Find the minimum of $\ln[x^2 + (y - x)^2 + 10(z - y)^2 + 1]$.

23.5. Use any method to find the global minimum of

$$f(x_1, ..., x_n) = \Big[\sum_{i=1}^{n} \sin^2(x_i) - e^{-\sum_{i=1}^{n} x_i^2}\Big] e^{-\sum_{i=1}^{n} \sin^2 \sqrt{|x_i|}},$$

where $-10 \le x_i \le 10$ for $i = 1, 2, ..., n$, and $n \ge 2$ is an integer.

Part IX

Advanced Topics

Chapter 24

Partial Differential Equations

Chapter Points

- The basic concepts of partial differential equations (PDEs) are explained, including the first-order PDEs and second-order PDEs with some examples.
- Classifications of second-order PDEs are introduced with examples about parabolic equations, hyperbolic equations and elliptical equations.
- Some basic techniques for solving linear PDEs are explained with a few worked examples.

Most physical and chemical processes can be modeled by physical laws such as the conservation of mass, energy and momentum. Such laws are often expressed in terms of partial differential equations that describe the variations of some measurable physical quantities with time. Therefore, PDEs are widely used in almost all disciplines of sciences and engineering as well as economics and finance.

24.1 INTRODUCTION

A partial differential equation (PDE) is a relationship containing one or more partial derivatives. Similar to the ordinary differential equation, the highest nth partial derivative is referred to as the order n of the partial differential equation. The general form of a partial differential equation can be written as

$$\psi(u, x, y, ..., \frac{\partial u}{\partial x}, \frac{\partial u}{\partial y}, \frac{\partial^2 u}{\partial x^2}, \frac{\partial^2 u}{\partial y^2}, \frac{\partial^2 u}{\partial x \partial y}, ...) = 0, \tag{24.1}$$

where u is the dependent variable, and $x, y, ...$ are the independent variables.

Example 24.1

For example, the following equation

$$\frac{\partial u}{\partial x} + \frac{\partial u}{\partial y} = 0,$$

is a simple first-order PDE that is also linear. Here, the unknown function u is essentially a function of two independent variables x and y.

In addition, the well-known inviscid Burgers' equation

$$\frac{\partial u}{\partial t} + u\frac{\partial u}{\partial x} = 0,$$

is a simple nonlinear PDE, which can lead to a shock wave solution.

A relatively general example of partial differential equations is the linear first-order partial differential equation, which can be written as

$$a(x, y)\frac{\partial u}{\partial x} + b(x, y)\frac{\partial u}{\partial y} = f(x, y), \tag{24.2}$$

Engineering Mathematics with Examples and Applications

for two independent variables and one dependent variable u. If the right-hand side is zero or simply $f(x, y) = 0$, then the equation is said to be homogeneous. The equation is said to be linear if a, b and f are functions of x, y only, not u itself.

For simplicity in notation in the studies of PDEs, compact subscript forms are often used in the literature. They are

$$u_x \equiv \frac{\partial u}{\partial x}, \quad u_y \equiv \frac{\partial u}{\partial y}, \quad u_{xx} \equiv \frac{\partial^2 u}{\partial x^2}, \quad u_{xy} \equiv \frac{\partial^2 u}{\partial x \partial y}, \quad \ldots \tag{24.3}$$

and thus we can write (24.2) as

$$au_x + bu_y = f. \tag{24.4}$$

In the rest of the chapters in this book, we will use this notation whenever no confusion occurs.

Example 24.2

The first-order PDE

$$y\frac{\partial u}{\partial x} + x\frac{\partial u}{\partial y} = 0,$$

can be written more compactly as

$$yu_x + xu_y = 0.$$

24.2 FIRST-ORDER PDES

A first-order linear partial differential equation can be written as

$$a(x, y)u_x + b(x, y)u_y = f(x, y), \tag{24.5}$$

which can be solved analytically. In case of a homogeneous equation when $f(x, y) = 0$ and if the functions $a(x, y)$ and $b(x, y)$ are separable, it may be possible to solve it using the so-called separation of variables. Let us demonstrate this by an example.

Example 24.3

Let us solve

$$yu_x + xu_y = 0.$$

The idea of the separation of variables is to assume that the solution $u(x, y)$ is the product of only two functions $X(x)$ and $Y(y)$ where $X(x)$ is a function of x only and $Y(y)$ is a function of y only. That is, we have

$$u(x, y) = X(x)Y(y). \tag{24.6}$$

This means that

$$u_x = \frac{\partial u}{\partial x} = X'(x)Y(y), \quad u_y = \frac{\partial u}{\partial y} = X(x)Y'(x). \tag{24.7}$$

Thus, the original equation becomes

$$yX'(x)Y(y) + xX(x)Y'(y) = 0.$$

Assuming $u \neq 0$ (otherwise, we have a trivial zero solution), we can divide both sides by $u = X(x)Y(y)$ and we have

$$y\frac{X'(x)}{X(x)} + x\frac{Y'(y)}{Y(y)} = 0,$$

which can be written as

$$\frac{1}{x}\frac{X'(x)}{X(x)} = -\frac{1}{y}\frac{Y'(y)}{Y(y)}.$$

Since the left-hand side is a function of only x, while the right-hand side is a function of only y, both sides should equal to some fixed constant (say, k), which does not depend on x or y. That is

$$\frac{1}{x}\frac{X'}{X} = k, \quad -\frac{1}{y}\frac{Y'}{Y} = k.$$

The first part becomes

$$\frac{X'}{X} = \frac{d \ln X}{dx} = kx,$$

which gives

$$\ln X = \frac{1}{2}kx^2 + A,$$

or

$$X = e^{A+kx^2/2},$$

where A is an integration constant. Similarly, from the second part after the separation of variables, we have

$$\frac{Y'}{Y} = \frac{d \ln Y}{dy} = -ky,$$

which gives

$$\ln Y = -\frac{ky^2}{2} + B, \quad \text{or} \quad Y = e^{B-ky^2/2},$$

where B is another integration constant. Thus, the final solution is

$$u = X(x)Y(y) = e^{A+kx^2/2} \cdot e^{B-ky^2/2} = De^{k(x^2-y^2)/2},$$

where $D = e^{A+B}$ is an unknown constant.

This example clearly demonstrates that the procedure for solving a seemingly simple PDE can be lengthy. Thus, the choice of solution techniques can be important, depending on the type of problem.

The linear PDE in the form (24.5) can be solved using the method of characteristics in terms of a parameter t

$$\frac{dx}{dt} = a(x, y), \quad \frac{dy}{dt} = b(x, y), \tag{24.8}$$

and

$$\frac{du}{dt} = f(x, y). \tag{24.9}$$

They essentially form a system of first-order ordinary differential equations.

Example 24.4

One of the simplest examples of first-order linear partial differential equations is the first-order hyperbolic equation

$$u_x + vu_y = 0, \tag{24.10}$$

where $v > 0$ is a constant. Obviously, it is easy to check that a constant $u = K$ (constant) is a trial solution to this equation.

Its characteristics are

$$\frac{dx}{dt} = 1, \quad \frac{dy}{dt} = v.$$

The integration with respect to t gives

$$x = t + A_1, \quad y = vt + A_2,$$

where A_1 and A_2 are two integration constants. If we multiply x by v, we have

$$u = y - vx = (vt + A_2) - v(t + A_1) = A_2 - vA_1 = \text{constant},$$

which is also a solution of the original PDE.

In fact, if ϕ is a function (such as the initial profile), the general form

$$u = \psi(y - vx), \tag{24.11}$$

is also a solution, forming a traveling wave along the x-axis with a constant speed v. If the initial shape is $u(x, 0) = \psi(x)$, then $u(x, y) = \psi(y - vx)$ at x (being a time). Therefore, the shape of the wave does not change with time though its position is constantly changing.

The method of characteristics can give some insight into the solution, even though we can obtain the solution by other techniques such as the separation of variables. Let us revisit the example given earlier in Example 24.3.

Example 24.5

From

$$yu_x + xu_y = 0,$$

we have its characteristics

$$\frac{dx}{dt} = \dot{x} = y, \quad \frac{dy}{dt} = \dot{y} = x,$$

which gives

$$x\dot{x} - y\dot{y} = 0.$$

This is equivalent to

$$\frac{1}{2}\frac{d(x^2 - y^2)}{dt} = 0,$$

or

$$u = x^2 - y^2 = K = \text{constant}.$$

It is straightforward to verify that $u = x^2 - y^2$ is a solution to the original PDE. In fact, a more general function in the form $\phi(x^2 - y^2)$ is also a solution to the original PDE.

24.3 CLASSIFICATION OF SECOND-ORDER PDES

A linear second-order partial differential equation can be written in the generic form in terms of two independent variables x and y,

$$au_{xx} + bu_{xy} + cu_{yy} + gu_x + hu_y + ku = f, \tag{24.12}$$

where a, b, c, g, h, k and f are functions of x and y only. If $f(x, y, u)$ is also a function of u, then we say that this equation is quasi-linear.

- If $\Delta = b^2 - 4ac < 0$, the equation is elliptic. One famous example is the Laplace equation

$$\frac{\partial^2 u}{\partial x^2} + \frac{\partial^2 u}{\partial y^2} = 0, \tag{24.13}$$

or

$$u_{xx} + u_{yy} = 0. \tag{24.14}$$

In this case, we have $a = c = 1$ and $b = 0$, and it is obvious $\Delta = b^2 - 4ac = -4 < 0$.

- If $\Delta > 0$, it is hyperbolic. A good example is the wave equation

$$\frac{\partial^2 u}{\partial t^2} = v^2 \frac{\partial^2 u}{\partial x^2},$$

where $v > 0$ is a speed constant. This PDE can be written as

$$u_{tt} = v^2 u_{xx}.$$

In this case, $a = 1$, $b = 0$ and $c = -v^2$ so that $\Delta = b^2 - 4ac = 4v^2 > 0$.

- If $\Delta = 0$, it is parabolic. A good example of this type is the diffusion and/or heat conduction

$$\frac{\partial u}{\partial t} = \kappa \frac{\partial^2 u}{\partial x^2},$$

where $\kappa > 0$ is the diffusion constant. This equation can also be written as

$$u_t = \kappa u_{xx}.$$

In this case, we have $a = 0$, $b = 0$, $c = -\kappa$, which means that $\Delta = b^2 - 4ac = 0$.

However, the type of a PDE is not always clear and sometimes, a mixed type can occur, depending on the parameters and domain. Let us look at an example.

Example 24.6

The following PDE

$$\frac{\partial^2 u}{\partial x^2} + x \frac{\partial^2 u}{\partial y^2} = 0,$$

is a mixed type. It is elliptic if $x > 0$ and hyperbolic if $x < 0$. However,

$$\frac{\partial^2 u}{\partial x^2} + x^2 \frac{\partial^2 u}{\partial y^2} = 0,$$

is always elliptic in the real domain.

24.4 CLASSIC MATHEMATICAL MODELS: SOME EXAMPLES

Three types of classic partial differential equations are widely used and they occur in a vast range of applications. In fact, almost all books or studies on partial differential equations will have to deal with these three types of basic partial differential equations.

24.4.1 Laplace's and Poisson's Equation

In heat transfer problems, the steady state of heat conduction with a source is governed by the Poisson equation

$$k\nabla^2 u = f(x, y, t), \qquad (x, y) \in \Omega, \tag{24.15}$$

or

$$u_{xx} + u_{yy} = q(x, y, t), \tag{24.16}$$

for two independent variables x and y. Here k is thermal diffusivity and $f(x, y, t)$ is the heat source. Ω is the domain of interest, usually a physical region. If there is no heat source ($q = f/\kappa = 0$), it becomes the Laplace equation.

Example 24.7

Laplace's equation in a two-dimensional domain takes form

$$\frac{\partial^2 u}{\partial x^2} + \frac{\partial^2 u}{\partial y^2} = 0,$$

while its three-dimensional case becomes

$$\frac{\partial^2 u}{\partial x^2} + \frac{\partial^2 u}{\partial y^2} + \frac{\partial^2 u}{\partial z^2} = 0.$$

The solution of an equation is said to be harmonic if it satisfies Laplace's equation. It leaves an exercise to verify that

$$u(x, y, z) = \frac{1}{\sqrt{x^2 + y^2 + z^2}},$$

indeed satisfies the Laplace equation.

In order to determine the temperature u completely, the appropriate boundary conditions are needed. A simple boundary condition is to specify the temperature $u = u_0$ on the boundary $\partial\Omega$. This type of problem is the Dirichlet problem.

On the other hand, if the temperature is not known, but the gradient $\partial u/\partial \mathbf{n}$ is known on the boundary where $\mathbf{n}$ is the outward-pointing unit normal, this forms the Neumann problem. Furthermore, some problems may have a mixed type of boundary conditions in the combination of

$$\alpha u + \beta \frac{\partial u}{\partial \mathbf{n}} = \gamma,$$

which naturally occurs as a radiation or cooling boundary condition.

24.4.2 Parabolic Equation

Time-dependent problems, such as diffusion and transient heat conduction, are governed by the parabolic equation

$$u_t = k u_{xx}. \tag{24.17}$$

Written in the n-dimensional case $x_1 = x, x_2 = y, x_3 = z, \ldots$, it can be extended to the reaction-diffusion equation

$$u_t = k\nabla^2 u + f(u, x_1, .., x_n, t). \tag{24.18}$$

Example 24.8

For example, the diffusion equation in 2D can be written as

$$\frac{\partial u}{\partial t} = D\Big(\frac{\partial^2 u}{\partial x^2} + \frac{\partial^2 u}{\partial y^2}\Big),$$

where D is the diffusion coefficient of the media. The 3D version of the diffusion equation becomes

$$\frac{\partial u}{\partial t} = D\nabla^2 u = D\Big(\frac{\partial^2 u}{\partial x^2} + \frac{\partial^2 u}{\partial y^2} + \frac{\partial^2 u}{\partial z^2}\Big).$$

24.4.3 Hyperbolic Equation

The vibration of strings and traveling waves (such as sound in the air) are governed by the hyperbolic wave equation in a three-dimensional space in the following form

$$\frac{\partial^2 u}{\partial t^2} = c^2 \nabla^2 u = c^2 \Big(\frac{\partial^2 u}{\partial x^2} + \frac{\partial^2 u}{\partial y^2} + \frac{\partial^2 u}{\partial z^2} \Big), \tag{24.19}$$

where $c > 0$ is the so-called wave speed.

Example 24.9

The 1-D wave equation in its simplest form is

$$u_{tt} = c^2 u_{xx}, \tag{24.20}$$

where c is the velocity of the wave. Using a transformation of the pair of independent variables

$$\xi = x + ct, \tag{24.21}$$

and

$$\eta = x - ct, \tag{24.22}$$

for $t > 0$ and $-\infty < x < \infty$, we have (using the chain rule)

$$\frac{\partial u}{\partial x} = \frac{\partial u}{\partial \xi}\frac{\partial \xi}{\partial x} + \frac{\partial u}{\partial \eta}\frac{\partial \eta}{\partial x}, \quad \frac{\partial u}{\partial t} = \frac{\partial u}{\partial \xi}\frac{\partial \xi}{\partial t} + \frac{\partial u}{\partial \eta}\frac{\partial \eta}{\partial t},$$

$$\frac{\partial^2 u}{\partial x^2} = \frac{\partial^2 u}{\partial \xi^2}(\frac{\partial \xi}{\partial x})^2 + \frac{\partial^2 u}{\partial \eta^2}(\frac{\partial \eta}{\partial x})^2 + 2\frac{\partial^2 u}{\partial \xi \partial \eta}(\frac{\partial \xi}{\partial x}\frac{\partial \eta}{\partial x}),$$

$$\frac{\partial^2 u}{\partial t^2} = \frac{\partial^2 u}{\partial t^2}(\frac{\partial \xi}{\partial t})^2 + \frac{\partial^2 u}{\partial \eta^2}(\frac{\partial \eta}{\partial t})^2 + 2\frac{\partial^2 u}{\partial \xi \partial \eta}(\frac{\partial \xi}{\partial t}\frac{\partial \eta}{\partial t}).$$

From the definition of ξ and η, we have

$$\frac{\partial \xi}{\partial x} = 1, \quad \frac{\partial \eta}{\partial x} = 1, \quad \frac{\partial \xi}{\partial t} = +c, \quad \frac{\partial \eta}{\partial t} = -c.$$

By using these results, we have

$$\frac{\partial^2 u}{\partial x^2} = \frac{\partial^2 u}{\partial \xi^2} + \frac{\partial^2 u}{\partial \eta^2} + 2\frac{\partial^2 u}{\partial \xi \partial \eta},$$

and

$$\frac{\partial^2 u}{\partial t^2} = c^2 \Big[\frac{\partial^2 u}{\partial t^2} + \frac{\partial^2 u}{\partial \eta^2} - 2\frac{\partial^2 u}{\partial \xi \partial \eta} \Big].$$

Substituting these results into the original PDE ($u_{tt} = c^2 u_{xx}$), we obtain

$$\frac{\partial^2 u}{\partial \xi \partial \eta} = 0,$$

or simply

$$u_{\xi\eta} = 0.$$

Integrating twice and substituting back in terms of x and t, we have

$$u(x,t) = f(x+ct) + g(x-ct), \tag{24.23}$$

where f and g are functions of $x+ct$ and $x-ct$, respectively. We can see that the solution is composed of two independent waves. One wave moves to the right and one travels to the left at the same constant speed c.

24.5 SOLUTION TECHNIQUES

Each type of equation usually requires different solution techniques. However, there are some methods that work for most of the linearly partial differential equations with appropriate boundary conditions on a regular domain. These methods include separation of variables, method of series expansion and transform methods such as the Laplace and Fourier transforms.

24.5.1 Separation of Variables

The separation of variables attempts a solution of the form

$$u = X(x)Y(y)Z(z)T(t), \tag{24.24}$$

where $X(x)$, $Y(y)$, $Z(z)$, $T(t)$ are functions of x, y, z, t, respectively. By determining these functions that satisfy the partial differential equation and the required boundary conditions in terms of eigenvalue problems, the solution of the original problem is then obtained.

Example 24.10

As a classic example, we now try to solve the 1-D heat conduction equation in the domain $x \in [0, L]$ and $t \geq 0$

$$u_t = k u_{xx}, \tag{24.25}$$

with the initial value and boundary conditions

$$u(0,t)=0, \qquad \frac{\partial u(x,t)}{\partial x}\Big|_{x=L}=0, \qquad u(x,0)=\psi(x). \tag{24.26}$$

Letting $u(x,t)=X(x)T(t)$, we have

$$\frac{X''(x)}{X} = \frac{T'(t)}{kT}. \tag{24.27}$$

As the left-hand side depends only on x and the right hand side only depends on t, therefore, both sides must be equal to the same constant, and the constant can be assumed to be $-\lambda^2$. The negative sign is just for convenience because we will see below that the finiteness of the solution $T(t)$ requires that eigenvalues are real and $\lambda^2 > 0$. Hence, we now get two ordinary differential equations

$$X''(x) + \lambda^2 X(x) = 0, \qquad T'(t) + k\lambda^2 T(t) = 0, \tag{24.28}$$

where λ is the eigenvalue. The solution for $T(t)$ is

$$T = A_n e^{-\lambda^2 kt}. \tag{24.29}$$

The basic solution for $X(x)$ is simply

$$X(x) = \alpha \cos \lambda x + \beta \sin \lambda x. \tag{24.30}$$

So the fundamental solution for u is

$$u(x,t) = (\alpha \cos \lambda x + \beta \sin \lambda x) e^{-\lambda^2 kt}, \tag{24.31}$$

where we have absorbed the coefficient A_n into α and β because they are the undetermined coefficients anyway. As the value of λ varies with the boundary conditions, it forms an eigenvalue problem. The general solution for u should be derived by superimposing solutions of (24.31), and we now have

$$u=\sum_{n=1}^{\infty} X_n T_n=\sum_{n=1}^{\infty}(\alpha_n \cos\lambda_n x+\beta_n \sin\lambda_n x)e^{-\lambda_n^2 kt}. \tag{24.32}$$

Obviously, we have not completed the solution yet because the unknown coefficients need to be determined. From the boundary condition $u(0,t)=0$ at $x=0$, we have

$$0=\sum_{n=1}^{\infty}\alpha_n e^{-\lambda_n^2 kt}, \tag{24.33}$$

which leads to $\alpha_n=0$ since $\exp(-\lambda^2 kt)>0$.

From $\frac{\partial u}{\partial x}\Big|_{x=L}=0$, we have

$$\lambda_n \cos\lambda_n L=0, \tag{24.34}$$

which requires

$$\lambda_n L=\frac{(2n-1)\pi}{2}, \quad (n=1,2,...). \tag{24.35}$$

Therefore, λ cannot be continuous, and it only takes an infinite number of discrete values, called eigenvalues.

Each eigenvalue $\lambda=\lambda_n=\frac{(2n-1)\pi}{2L}$, $(n=1,2,...)$ has a corresponding eigenfunction $X_n=\sin(\lambda_n x)$. Substituting into the solution for $T(t)$, we have

$$T_n(t)=A_n e^{-\frac{[(2n-1)\pi]^2}{4L^2}kt}. \tag{24.36}$$

By expanding the initial condition into a Fourier series so as to determine the coefficients, we have

$$u(x,t)=\sum_{n=1}^{\infty}\beta_n \sin(\frac{(2n-1)\pi x}{2L})e^{-[\frac{(2n-1)\pi}{2L}]^2 kt},$$

$$\beta_n=\frac{2}{L}\int_0^L \psi(x)\sin[\frac{(2n-1)\pi x}{2L}]dx. \tag{24.37}$$

This procedure is quite lengthy, though the calculations most involve integrals.

Example 24.11

In the special case when initial condition $u(x,t=0)=\psi=u_0$ is constant, the requirement for $u=u_0$ at $t=0$ becomes

$$u_0=\sum_{n=1}^{\infty}\beta_n \sin\frac{(2n-1)\pi x}{2L}. \tag{24.38}$$

Using the orthogonal relationships

$$\int_0^L \sin\frac{m\pi x}{L}\sin\frac{n\pi x}{L}dx=0, \quad m\neq n,$$

and

$$\int_0^L (\sin\frac{n\pi x}{L})^2 dx=\frac{L}{2}, \quad (n=1,2,...),$$

and multiplying both sides of Eq. (24.38) by $\sin[(2n-1)\pi x/2L]$, we have the integral

$$\beta_n\frac{L}{2}=\int_0^L \sin\frac{(2n-1)\pi x}{2L}u_0 dx$$

$$= \frac{2u_0 L}{(2n-1)\pi}, \quad (n = 1, 2, ...),$$

which leads to

$$\beta_n = \frac{4u_0}{(2n-1)\pi}, \qquad n = 1, 2, ...,$$

and thus the solution becomes

$$u = \frac{4u_0}{\pi} \sum_{n=1}^{\infty} \frac{1}{(2n-1)} e^{-\frac{(2n-1)^2 \pi^2 kt}{4L^2}} \sin \frac{(2n-1)\pi x}{2L}. \tag{24.39}$$

This solution is essentially the same as the classical heat conduction problem discussed by Carslaw and Jaeger in 1986. This same solution can also be obtained using the Fourier series of u_0 in $0 < x < L$.

24.5.2 Laplace Transform

The integral transform can reduce the number of the independent variables. For the 1D time-dependent case, it transforms a partial differential equation into an ordinary differential equation. By solving the ordinary differential equation and inverting it back, we can obtain the solution for the original partial differential equation.

Example 24.12

As an example, we now solve the heat conduction problem over a semi-infinite interval $x \in [0, \infty)$ and $t \geq 0$,

$$u_t = k u_{xx}, \qquad u(x, 0) = 0, \; u(0, t) = T_0. \tag{24.40}$$

Let $\bar{u}(x, s) = \int_0^\infty u(x, t) e^{-st} dt$ be the Laplace transform of $u(x, t)$, then Eq. (24.40) becomes

$$s\bar{u} = k \frac{d^2 \bar{u}}{dx^2},$$

and

$$\bar{u}_{x=0} = \frac{T_0}{s},$$

which is an ordinary differential equation whose general solution can be written as

$$\bar{u} = A e^{-\sqrt{\frac{s}{k}} x} + B e^{\sqrt{\frac{s}{k}} x}.$$

The finiteness of the solution as $x \to \infty$ requires that $B = 0$, and the boundary condition at $x = 0$ leads to

$$\bar{u} = \frac{T_0}{s} e^{-\sqrt{\frac{s}{k}} x}.$$

By using the inverse Laplace transform, we have

$$u = T_0 \text{erfc}\left(\frac{x}{2\sqrt{kt}}\right),$$

where $\text{erfc}(x)$ is the complementary error function.

24.5.3 Fourier Transform

The Fourier transform works in a similar manner to the Laplace transform. A PDE is first transformed into an ordinary differential equation in the frequency domain, and the solution is obtained by solving its corresponding ODE. Then, the solution to the PDE in the time domain is obtained by inverse Fourier transforms.

Example 24.13

The famous example is the classical wave equation

$$u_{tt} = c^2 u_{xx}, \qquad u(x,0) = \psi(x), \;\; u_t(x,0) = 0. \tag{24.41}$$

Let

$$\bar{u}(\omega,t) = \frac{1}{\sqrt{2\pi}} \int_{-\infty}^{\infty} u(x,t) e^{i\omega x} dx$$

be the Fourier transform of $u(x,t)$. This transforms the PDE problem into an ODE

$$\frac{d^2\bar{u}}{dt^2} = -c^2\omega^2\bar{u}, \;\; \bar{u}(\omega,0) = \bar{\psi}(\omega), \qquad \frac{d\bar{u}(\omega,0)}{dt} = 0.$$

The general solution in terms of the parameter ω is

$$\bar{u}(\omega,t) = \bar{\psi}(\omega)\cos(c\omega t).$$

By using the inverse Fourier transform, we finally have

$$u(x,t) = \frac{1}{\sqrt{2\pi}} \int_{-\infty}^{\infty} \bar{\psi}(\omega)\cos(c\omega t) e^{-i\omega x} d\omega$$
$$= \frac{1}{2}[\psi(x+ct) + \psi(x-ct)], \tag{24.42}$$

which implies two traveling waves: one travels along the x-axis and the other along the negative x-axis direction.

24.5.4 Similarity Solution

Sometimes, the diffusion equation

$$u_t = \kappa u_{xx}, \tag{24.43}$$

can be solved by using the so-called similarity method by defining a similar variable

$$\eta = \frac{x}{\sqrt{\kappa t}} \tag{24.44}$$

or

$$\zeta = \frac{x^2}{\kappa t}. \tag{24.45}$$

One can assume that the solution to the equation has the form

$$u = (\kappa t)^{\alpha} f[\frac{x^2}{(\kappa t)^{\beta}}]. \tag{24.46}$$

By substituting it into the diffusion equation, the coefficients α and β can be determined. For most applications, one can assume $\alpha = 0$ so that $u = f(\zeta)$. In this case, we have

$$4\zeta u'' + 2u' + \zeta\beta(\kappa t)^{\beta-1}u' = 0, \tag{24.47}$$

where $u' = du/d\zeta$. In deriving this equation, one has to use the chain rules of differentiations $\frac{\partial}{\partial x} = \frac{\partial}{\partial \zeta}\frac{\partial \zeta}{\partial x}$ and $\frac{\partial}{\partial t} = \frac{\partial}{\partial \zeta}\frac{\partial \zeta}{\partial t}$.

Since the original equation does not have time-dependent terms explicitly, this means that all the exponents for any t-terms must be zero. Therefore, we have

$$\beta = 1. \tag{24.48}$$

Now, the diffusion equation becomes

$$\zeta f''(\zeta) = -(\frac{1}{2} + \frac{\zeta}{4}) f'. \tag{24.49}$$

Using $(\ln f')' = f''/f'$ and integrating the above equation once, we get

$$f' = \frac{K e^{-\zeta/4}}{\sqrt{\zeta}}. \tag{24.50}$$

Integrating it again and using the substitution $\zeta = 4\xi^2$, we obtain

$$u = A \int_0^{\xi} e^{-\xi^2} d\xi = C \operatorname{erf}(\frac{x}{\sqrt{4\kappa t}}) + D, \tag{24.51}$$

where C and D are constants that can be determined from appropriate boundary conditions.

Example 24.14

For the same problem as (24.40), the boundary condition as $x \to \infty$ implies that $C + D = 0$, while $u(0, t) = T_0$ means that $D = -C = T_0$. Therefore, we finally have

$$u = T_0[1 - \operatorname{erf}(\frac{x}{\sqrt{4\kappa t}})] = T_0 \operatorname{erfc}(\frac{x}{\sqrt{4\kappa t}}).$$

24.5.5 Change of Variables

In some cases, a partial differential equation under consideration cannot be written directly in the standard form; however, it can be converted into a known standard equation by a change of variables. It is better to show this using an example.

Example 24.15

For example, the following reaction-diffusion equation

$$\frac{\partial u}{\partial t} = k \frac{\partial^2 u}{\partial x^2} - \alpha u, \tag{24.52}$$

describes the heat conduction along a wire with a heat loss term $-\alpha u$. Carslaw and Jaeger showed that it can be transformed into a standard equation of heat conduction using the following change of variables

$$u = v e^{-\alpha t}, \tag{24.53}$$

where v is the new variable. By simple differentiations, we have

$$\frac{\partial u}{\partial t} = \frac{\partial v}{\partial t} e^{-\alpha t} - \alpha v e^{-\alpha t} = \frac{\partial v}{\partial t} e^{-\alpha t} - \alpha u, \qquad \frac{\partial^2 u}{\partial x^2} = \frac{\partial^2 v}{\partial x^2} e^{-\alpha t}, \tag{24.54}$$

we have

$$\frac{\partial u}{\partial t} = \underbrace{\frac{\partial v}{\partial t} e^{-\alpha t}} - \alpha u = k \frac{\partial^2 u}{\partial x^2} - \alpha u = k \underbrace{\frac{\partial^2 v}{\partial x^2} e^{-\alpha t}} - \alpha u, \tag{24.55}$$

which becomes

$$\frac{\partial v}{\partial t} e^{-\alpha t} = k \frac{\partial^2 v}{\partial x^2} e^{-\alpha t}. \tag{24.56}$$

After dividing both sides by $e^{-\alpha t} > 0$, we have

$$\frac{\partial v}{\partial t} = k\frac{\partial^2 v}{\partial x^2}, \tag{24.57}$$

which is the standard heat conduction equation for v.

For given initial (usually constant) and boundary conditions (usually zero), we can use all the techniques for solving the standard equation to get solutions. However, for some boundary conditions such as $u = u_0$, a more elaborate form of change of variables is needed. Crank introduced Danckwerts's method by using the following transform

$$u = \alpha \int_0^t v e^{-\alpha\tau} d\tau + v e^{-\alpha t}. \tag{24.58}$$

Noting that

$$\frac{\partial u}{\partial t} = \alpha v e^{-\alpha t} - \alpha v e^{-\alpha t} + \frac{\partial v}{\partial t} e^{-\alpha t},$$

it is straightforward to show

$$\frac{\partial u}{\partial t} + \alpha u = k\frac{\partial^2 u}{\partial x^2}. \tag{24.59}$$

For the boundary condition $u = u_0$, we have $v = v_0 = u_0$, and this is because

$$u = u_0 = \alpha v_0 \int_0^t e^{-\alpha\tau} d\tau + v_0 e^{-\alpha t} = v_0 - v_0 e^{-\alpha t} + v_0 e^{-\alpha t} = v_0, \tag{24.60}$$

which is the same boundary condition for u.

There are other important methods for solving partial differential equations. These include Green's function, series methods, asymptotic methods, approximate methods, perturbation methods and naturally the numerical methods. Interested readers can refer to more advanced literature on these topics.

EXERCISES

24.1. Discuss the type of each of the following PDEs:

- $xu_x + yu_y = xyu$
- $(u_t)^2 = u_x + xt$
- $u_t + u^2 u_x = 0$
- $u_{tt} = tu_{xx}$
- $u_t = D\nabla^2 u + \gamma u(1-u)$ where $D > 0$ and $\gamma > 0$ are constants.

24.2. The traffic flow can be modeled by a first-order nonlinear PDE

$$\frac{\partial \rho}{\partial t} + \frac{\partial Q(\rho)}{\partial x} = 0, \quad Q(\rho) = u_0 \rho(1 - \rho/\rho_{\max}),$$

where u_0 and $\rho_{\max}$ are some positive constants. Show that the characteristic speed of the system is

$$v = u_0(1 - 2\rho/\rho_{\max}).$$

24.3. Solve $u_x + u^n u_y = 0$ where $n \geq 1$.

24.4. Show that $\nabla^2 u = 0$ has a solution

$$u = \frac{1}{\sqrt{x^2 + y^2 + z^2}}.$$

Chapter 25

Tensors

Chapter Points

- Summation notations are introduced with detailed examples.
- Tensors and tensor algebra are explained with some basic examples.
- The concepts of tensors are then applied to describe Hooke's law and the generalized elasticity.

25.1 SUMMATION NOTATIONS

In tensor analysis, the summation convention and notations for subscripts are widely used. Any lowercase subscript that appears exactly twice in any term of an expression means that sum is over all the possible values of the subscript. This convention is also called Einstein's summation convention or the index form.

Example 25.1

For example, in the three-dimensional case, we have

$$\alpha_i x_i \equiv \sum_{i=1}^{3} \alpha_i x_i = \alpha_1 x_1 + \alpha_2 x_2 + \alpha_3 x_3. \tag{25.1}$$

The dot product of two 3D vectors **a** and **b** can be written as

$$\mathbf{a} \cdot \mathbf{b} = a_i b_i = a_1 b_1 + a_2 b_2 + a_3 b_3. \tag{25.2}$$

For a 3D vector, its gradient can be written as

$$\nabla \cdot \mathbf{u} \equiv \frac{\partial u_i}{\partial x_i} = \frac{\partial u_1}{\partial x_1} + \frac{\partial u_2}{\partial x_2} + \frac{\partial u_3}{\partial x_3}. \tag{25.3}$$

In general, the product of two matrices can be written

$$A_{ij} B_{jk} \equiv \sum_{j=1}^{n} A_{ij} B_{jk} = A_{i1} B_{1k} + A_{i2} B_{2k} + ... + A_{in} B_{nk}, \tag{25.4}$$

where n is the number of columns of matrix **A** or the number of rows of matrix **B**. Furthermore, the trace of an $n \times n$ square matrix $\mathbf{A} = [a_{ij}]$ can be written as

$$\mathrm{tr}(\mathbf{A}) = a_{ii} = a_{11} + a_{22} + ... + a_{nn}.$$

The Kronecker delta δ_{ij}, which is a unity tensor (like the unity matrix **I** in matrix analysis), is defined as

$$\delta_{ij} = \begin{cases} 1 \text{ (if } i = j), \\ 0 \text{ (if } i \neq j). \end{cases} \tag{25.5}$$

Engineering Mathematics with Examples and Applications

Similar to δ_{ij}, the three subscripts Levi-Civita symbol (not a tensor) is defined as

$$\epsilon_{ijk} = \begin{cases} +1 & \text{(if } i, j, k \text{ is an even permutation of 1,2,3),} \\ -1 & \text{(if } i, j, k \text{ is an odd permutation of 1, 2, 3),} \\ 0 & \text{(otherwise).} \end{cases} \tag{25.6}$$

Both δ_{ij} and ϵ_{ijk} are related by

$$\epsilon_{ijk}\epsilon_{kpq} = \delta_{ip}\delta_{jq} - \delta_{iq}\delta_{jp}. \tag{25.7}$$

Example 25.2

Using the summation conventions, the matrix equation

$$\mathbf{Ax} = \mathbf{b}$$

can alternatively be written as

$$A_{ij}x_j = b_i, \qquad (i = 1, 2, ..., n). \tag{25.8}$$

25.2 TENSORS

When changing the basis from the standard Cartesian $\mathbf{e}_1 = \mathbf{i}$, $\mathbf{e}_2 = \mathbf{j}$, $\mathbf{e}_3 = \mathbf{k}$ to a new basis $\mathbf{e}'_1, \mathbf{e}'_2, \mathbf{e}'_3$, a position vector $\mathbf{x} = (x_1, x_2, x_3)$ can be written as

$$\mathbf{x} = x_1\mathbf{e}_1 + x_2\mathbf{e}_2 + x_3\mathbf{e}_3 = x_i\mathbf{e}_i. \tag{25.9}$$

Mathematically speaking, a vector is a tensor of rank 1, which will be defined more formally later.

The vector $\mathbf{x}$ in the old bases is related to the new vector $\mathbf{x}' = (x'_1, x'_2, x'_3)$ (here $'$ is not the derivative):

$$\mathbf{x}' = x'_1\mathbf{e}'_1 + x'_2\mathbf{e}'_2 + x'_3\mathbf{e}'_3 = x'_i\mathbf{e}'_i, \tag{25.10}$$

through a special matrix $\mathbf{S} = [S_{ij}]$ defined by entries of dot products

$$S_{ij} = \mathbf{e}'_i \cdot \mathbf{e}_j. \tag{25.11}$$

The new vector $\mathbf{x}'$ can be obtained from the old vector $\mathbf{x}$ through

$$\mathbf{x}' = \mathbf{Sx}, \tag{25.12}$$

or in the present notation convention

$$x'_i = S_{ij}x_j. \tag{25.13}$$

Example 25.3

In a special case when the transformation is a simple rotation with an angle θ about a fixed axis such as the z-axis, we have

$$\mathbf{R} = [R_{ij}] = \begin{pmatrix} \cos\theta & \sin\theta & 0 \\ -\sin\theta & \cos\theta & 0 \\ 0 & 0 & 1 \end{pmatrix}. \tag{25.14}$$

The orthogonality of S_{ij} requires that

$$\mathbf{RR}^T = \mathbf{R}^T\mathbf{R} = \mathbf{I},$$

which means that

$$\mathbf{R}^{-1} = \mathbf{R}^T.$$

The above condition can also be written as

$$R_{ij}R_{jk} = \delta_{ik}, \quad R_{ki}R_{kj} = \delta_{ij}. \tag{25.15}$$

25.2.1 Rank of a Tensor

If the components u_i of any variable **u** are transformed to the components u'_i in a new basis in the same manner as

$$u'_i = S_{ij}u_j, \qquad u_i = S_{ji}u'_j, \tag{25.16}$$

then u_i are said to form a first-order Cartesian tensor (or vector in this case). If components of a variable σ_{ij} are transformed as

$$\sigma'_{ij} = S_{ip}S_{jq}\sigma_{pq}, \qquad \sigma_{ij} = S_{pi}S_{qj}\sigma'_{pq}, \tag{25.17}$$

we say these components form a second-order tensor such as stresses and strains.

The order of a tensor is also called its rank or degree. Loosely speaking, the rank of a tensor is the dimensionality (or number of indices) of an array needed to represent the tensor uniquely. Therefore, a vector is a rank-1 tensor, while a stress tensor is a tensor of rank 2. Interesting, a scalar (or a simple number) can be considered as a tensor of rank 0.

25.2.2 Contraction

For a tensor T_{ij}, its contraction (or summation of repetitive indices) T_{ii} is a scalar

$$T_{ij}\delta_{ij} = T_{ii} = T_{jj}, \tag{25.18}$$

and this scalar value remains the same in all frames. For this reason, the contraction here acts like an inner product.

On the other hand, for two vectors **u** and **v**, their outer product gives a tensor

$$T_{ij} = u_i v_j = \mathbf{u} \otimes \mathbf{v}, \tag{25.19}$$

which is also called a dyad. It is worth pointing out that the inner (dot) product of **u** and **v** is a scalar (or a zeroth order/rank tensor), while their cross product $\mathbf{u} \times \mathbf{v}$ is a vector (a first-order/rank tensor). The third possibility of vector products is the tensor product or a dyadic product

$$T_{ij} = \mathbf{u} \otimes \mathbf{v} \tag{25.20}$$

that is a second-order tensor.

Example 25.4

For two vectors $\mathbf{u} = u_1\mathbf{i} + u_2\mathbf{j} + u_3\mathbf{k}$ and $\mathbf{v} = v_1\mathbf{i} + v_2\mathbf{j} + v_3\mathbf{k}$, their dot product is

$$\mathbf{u} \cdot \mathbf{v} = u_i v_i = u_1 v_1 + u_2 v_2 + u_3 v_3.$$

Their outer or cross product is

$$\mathbf{u} \times \mathbf{v} = \epsilon_{ijk} u_i v_j \mathbf{e}_k,$$

where $\mathbf{e}_k = \mathbf{i}, \mathbf{j}, \mathbf{k}$ for $k = 1, 2, 3$, respectively. With this notation, the above two vectors can be written as

$$\mathbf{u} = u_i \mathbf{e}_i, \quad \mathbf{v} = v_j \mathbf{e}_j.$$

In addition, their tensor product is

$$\mathbf{u}\otimes\mathbf{v}=\mathbf{u}\mathbf{v}^T=\begin{pmatrix}u_1\\u_2\\u_3\end{pmatrix}\begin{pmatrix}v_1 & v_2 & v_3\end{pmatrix}=\begin{pmatrix}u_1v_1 & u_1v_2 & u_1v_3\\u_2v_1 & u_2v_2 & u_3v_3\\u_3v_1 & u_3v_2 & u_3v_3\end{pmatrix}.$$

Sometimes, it may be useful to use the notations that include δ_{ij} explicitly for later contractions.

Example 25.5

The dot product of two vectors can be written as

$$\mathbf{u}\cdot\mathbf{v}=u_iv_i=\delta_{ij}u_iv_j,$$

and $\delta_{ii}=1+1+1=3$ in the 3D space. It is also straightforward to check that

$$\delta_{ij}\delta_{jk}=\delta_{ik}.$$

In addition, the Laplace operator is equivalent to

$$\nabla^2\psi=\frac{\partial^2\psi}{\partial x_i\partial x_i}=\delta_{ij}\frac{\partial\psi}{\partial x_i\partial x_j}.$$

Similarly, the divergence theorem can be rewritten as the following form

$$\int_V\frac{\partial u_i}{\partial x_i}dV=\oint_S u_in_idS.$$

In a similar fashion, higher-order tensors can be defined, and for each order increase, then there is one S_{ij} extra in the product for transforming from different bases or frames, but no subscripts are allowed to appear more than twice

$$\tau'_{ij...k}=S_{ip}S_{jq}...S_{kr}\tau_{pq...r},\qquad \tau_{ij...k}=S_{pi}S_{qj}....S_{rk}\tau'_{pq...r}. \tag{25.21}$$

In a similar way to multi-dimensional matrices, two tensors can be added or subtracted component-by-component if and only if they are the tensors of the same order.

An isotropic tensor is a special class of tensors such that its components are the same in all frames or bases. For example, a scalar as a zero-rank tensor remains the same value in all frames. It is worth pointing out that the Kronecker delta δ_{ij} and ϵ_{ijk} can be considered as isotropic because their components are the same in all frames.

25.2.3 Symmetric and Antisymmetric Tensors

For second-order tensors, a tensor τ_{ij} is said to be symmetric if $\tau_{ij}=\tau_{ji}$, and antisymmetric if $\tau_{ij}=-\tau_{ji}$.

Example 25.6

The Kronecker δ_{ij} is symmetric, while ϵ_{ijk} is anti-symmetric in any pairs of its three indices. The strain tensor s_{ij} to be introduced in the next section can be defined by

$$\varepsilon_{ij}=\frac{1}{2}\Big(\frac{\partial u_i}{\partial x_j}+\frac{\partial u_j}{\partial x_i}\Big),$$

is clearly symmetric because

$$\varepsilon_{ji}=\frac{1}{2}\Big(\frac{\partial u_j}{\partial x_i}+\frac{\partial u_i}{\partial x_j}\Big)=\frac{1}{2}\Big(\frac{\partial u_i}{\partial x_j}+\frac{\partial u_j}{\partial x_i}\Big)=\varepsilon_{ij}.$$

An interesting property of a tensor τ_{ij} is that it can always be written as a sum of a symmetric tensor and an antisymmetric tensor

$$\tau_{ij} = \frac{1}{2}(\tau_{ij} + \tau_{ji})[\text{sym.}] + \frac{1}{2}(\tau_{ij} - \tau_{ji})[\text{antisym.}]. \tag{25.22}$$

25.2.4 Tensor Differentiation

All the formulas in vector analysis can be rewritten in the tensor forms using the summation convention, known as index forms. Sometimes, it is convenient to write equations by defining a special differential operator

$$\partial_i = \frac{\partial}{\partial x_i}. \tag{25.23}$$

Example 25.7

For example, we can write the grad of a scalar field Φ as

$$\nabla\Phi = \frac{\partial\Phi}{\partial x_i} = \partial_i\Phi.$$

The div of a vector $\mathbf{u}$ can be written as

$$\nabla\cdot\mathbf{u} = \frac{\partial u_i}{\partial x_i} = \partial_i u_i,$$

and the curl of a vector can be written as

$$[\nabla\times\mathbf{u}]_i = \epsilon_{ijk}\frac{\partial u_k}{\partial x_j} = \epsilon_{ijk}\partial_j u_k.$$

The Laplace operator of a function Φ can be written as

$$\nabla^2\Phi = \frac{\partial^2\Phi}{\partial x_i\partial x_i} = \partial_i\partial_i\Phi.$$

For more complicated operators, we can write (for example)

$$\nabla\times(\nabla\times\mathbf{u})_i = \epsilon_{ijk}\epsilon_{kpq}\frac{\partial^2 u_q}{\partial x_j\partial x_p}. \tag{25.24}$$

These notations can be convenient for proving theorems. Sometimes, the formulas in vector algebra can be difficult to prove using vectors, but they can become surprisingly simple if tensor notations are used.

Example 25.8

Let us prove

$$\nabla\times(\psi\mathbf{u}) = \psi(\nabla\times\mathbf{u}) + (\nabla\psi)\times\mathbf{u},$$

for any function ψ and vector $\mathbf{u}$.

Using the tensor notations, we can write

$$\nabla\times(\psi\mathbf{u}) = \mathbf{e}_i\epsilon_{ijk}\frac{\partial(\psi u_k)}{\partial x_j}$$
$$= \mathbf{e}_i\epsilon_{ijk}\Big[\psi\frac{\partial u_k}{\partial x_j} + \Big(\frac{\partial\psi}{\partial x_j}\Big)u_k\Big] = \psi(\nabla\times\mathbf{u}) + (\nabla\psi)\times\mathbf{u},$$

where we have used the results and notations from the previous examples in this chapter.

The tensor forms are sometimes useful in proving complex relationships among vectors and tensors. They also become handy for the implementation of numerical algorithms.

25.3 HOOKE'S LAW AND ELASTICITY

The basic Hooke's law of elasticity concerns an elastic body such as a spring, and it states that the extension x is proportional to the load F, that is

$$F = kx, \tag{25.25}$$

where k the spring constant. It is worth pointing out that because the force is a restoring force, a negative sign is usually incorporated in most formulations. That is, $F = -kx$.

However, the above equation only works for 1-D deformations. For a bar of uniform cross-section with a length L and a cross section area A, it is more convenient to use strain ε and stress σ. The stress and strain are defined by

$$\sigma = \frac{F}{A}, \qquad \varepsilon = \frac{\Delta L}{L}, \tag{25.26}$$

where ΔL is the extension. The unit of stress is N/m^2, while the strain is dimensionless, though it is conventionally expressed in m/m or % (percentage) in engineering. For the elastic bar, the stress-strain relationship is

$$\sigma = E\varepsilon, \tag{25.27}$$

where E is the Young's modulus of elasticity. Written in terms F and $x = \Delta L$, we have

$$F = \frac{EA}{L}\Delta L = kx, \qquad k = \frac{EA}{L}, \tag{25.28}$$

where k is the equivalent spring constant for the bar. This equation is valid only for any unidirectional compression or extension. For any 2D and 3D deformation, we need to generalize Hooke's law.

The strain tensor is defined in terms of the displacement $\mathbf{u}^T = (u_1, u_2, u_3)$

$$\varepsilon_{ij} = \frac{1}{2}\left(\frac{\partial u_i}{\partial x_j} + \frac{\partial u_j}{\partial x_i}\right), \tag{25.29}$$

where $x_1 = x$, $x_2 = y$, and $x_3 = z$. Sometimes, it is useful to write

$$\boldsymbol{\varepsilon} = \frac{1}{2}(\nabla \mathbf{u} + \nabla \mathbf{u}^T). \tag{25.30}$$

The general stress tensor (also called Cauchy stress tensor) can be written as

$$\boldsymbol{\sigma} = \begin{pmatrix} \sigma_{xx} & \sigma_{xy} & \sigma_{xz} \\ \sigma_{yx} & \sigma_{yy} & \sigma_{yz} \\ \sigma_{zx} & \sigma_{zy} & \sigma_{zz} \end{pmatrix} = \begin{pmatrix} \sigma_{11} & \sigma_{12} & \sigma_{13} \\ \sigma_{21} & \sigma_{22} & \sigma_{23} \\ \sigma_{31} & \sigma_{32} & \sigma_{33} \end{pmatrix}. \tag{25.31}$$

The generalized Hooke's law can concisely be written as

$$\varepsilon_{ij} = \frac{1+\nu}{E}\sigma_{ij} - \frac{\nu}{E}\sigma_{kk}\delta_{ij}, \tag{25.32}$$

where we have used the Einstein's summation convention $\sigma_{kk} = \sigma_{xx} + \sigma_{yy} + \sigma_{zz}$. Here, ν is the Poisson's ratio, and it measures the tendency of extension in transverse directions (say, x and y) when the elastic body is stretched in one direction (say, z). It can be defined as the ratio of the transverse contract strain (normal to the applied load) to the axial strain in a stretched cylindrical bar in the direction of the applied force. For a perfectly incompressible material, $\nu = 0.5$, and $\nu = 0 \sim 0.5$ for most common materials. For example, steels have $\nu = 0.25 \sim 0.3$. Some auxetic material such as polymer foams or anti-rubbers can have a negative Poisson's ratio $\nu < 0$.

Example 25.9

A very special case in 2D configuration is the plane stress condition where $\sigma_{zz} = \sigma_{xz} = \sigma_{zx} = \sigma_{yz} = \sigma_{zy} = 0$. That is, all stresses are constrained in a plane and all out of plane stress components are zero. In this case, the strain and stress tensors take the following form

$$\boldsymbol{\varepsilon} = \begin{pmatrix} \varepsilon_{xx} & \varepsilon_{xy} & 0 \\ \varepsilon_{yx} & \varepsilon_{yy} & 0 \\ 0 & 0 & \varepsilon_{zz} \end{pmatrix}, \quad \boldsymbol{\sigma} = \begin{pmatrix} \sigma_{xx} & \sigma_{xy} & 0 \\ \sigma_{yx} & \sigma_{yy} & 0 \\ 0 & 0 & 0 \end{pmatrix}.$$

It is worth pointing out that the strain is not planar because $\varepsilon_{zz} \neq 0$ due to the so-called Poisson ratio effect.

Since both the strain tensor and the stress tensor are symmetric, we have $\varepsilon_{xy} = \varepsilon_{yx}$ and $\sigma_{xy} = \sigma_{yx}$. In the context of plane stress, we have

$$\begin{pmatrix} \varepsilon_{xx} \\ \varepsilon_{yy} \\ \varepsilon_{xy} \end{pmatrix} = \frac{1}{E} \begin{pmatrix} 1 & -\nu & 0 \\ -\nu & 1 & 0 \\ 0 & 0 & (1+\nu) \end{pmatrix} \begin{pmatrix} \sigma_{xx} \\ \sigma_{yy} \\ \sigma_{xy} \end{pmatrix}.$$

On the other hand, if we solve the above equations for stress components in terms of strains, we have

$$\begin{pmatrix} \sigma_{xx} \\ \sigma_{yy} \\ \sigma_{xy} \end{pmatrix} = \frac{E}{1-\nu^2} \begin{pmatrix} 1 & \nu & 0 \\ \nu & 1 & 0 \\ 0 & 0 & (1-\nu) \end{pmatrix} \begin{pmatrix} \varepsilon_{xx} \\ \varepsilon_{yy} \\ \varepsilon_{xy} \end{pmatrix}.$$

It is worth pointing out that in some engineering textbooks the shear strain $\gamma_{xy} = \varepsilon_{xy}/2$ is used and thus a factor 2 exists.

The tensor forms of elasticity can be very useful in engineering simulations because they can be easily transformed into matrix forms and thus can be implemented easily using most programming languages. Interested readers can refer to more advanced literature in engineering simulations, especially about structural engineering and finite element methods.

EXERCISES

25.1. The Ohm's law for a conduction media relates the current density $\mathbf{J}$ to the applied electric field $\mathbf{E}$ by

$$\mathbf{J} = \sigma \mathbf{E},$$

where σ is the tensor of conductivity. Write this equation in a tensor form using summation notations.

25.2. For 3D rigid body motion, the rotation energy is expressed as $E_K = \frac{1}{2} I \omega^2$, where $\omega = (\omega_1, \omega_2, \omega_3)$ is the angular velocity and I is the tensor of the moment of inertia

$$I = \begin{pmatrix} I_{11} & I_{12} & I_{13} \\ I_{21} & I_{22} & I_{23} \\ I_{31} & I_{32} & I_{33} \end{pmatrix}.$$

Express this energy using tensor notations.

25.3. Prove that the triple vector product can be written as

$$\mathbf{u} \cdot (\mathbf{v} \times \mathbf{w}) = \begin{vmatrix} u_1 & u_2 & u_3 \\ v_1 & v_2 & v_3 \\ w_1 & w_2 & w_3 \end{vmatrix} = \epsilon_{ijk} u_i v_j w_k.$$

Chapter 26

Calculus of Variations

Chapter Points

- Variations of calculus is briefly introduced, together with the derivation of the Euler-Lagrange equation.
- Variations with constraints are explained with a worked example.

The main aim of the calculus of variations is to find a function that makes an integral stationary, making the value of the integral a local maximum or minimum. For example, in mechanics we may want to find the shape $y(x)$ of a rope or chain when suspended under its own weight from two fixed points. In this case, the calculus of variations provides a method for finding the function $y(x)$ so that the curve $y(x)$ minimizes the gravitational potential energy of the hanging rope system.

26.1 EULER-LAGRANGE EQUATION

Before we proceed to the calculus of variations, let us first discuss an important concept, namely the curvature of a curve.

26.1.1 Curvature

In general, a curve $y(x)$ can be described in a parametric form in terms of a vector $\mathbf{r}(s)$ with a parameter s which is the arc length along the curve measured from a fixed point. The curvature κ of a curve is defined as the rate at which the unit tangent $\mathbf{t}$ changes with respect to s. The change of arc length is

$$\frac{ds}{dx} = \sqrt{1 + (\frac{dy}{dx})^2} = \sqrt{1 + y'^2}. \tag{26.1}$$

We have the curvature

$$\frac{d\mathbf{t}}{ds} = \kappa\,\mathbf{n} = \frac{1}{\rho}\mathbf{n}, \tag{26.2}$$

where ρ is the radius of the curvature, and $\mathbf{n}$ is the principal normal shown in Fig. 26.1.

As the direction of the tangent is defined by the angle θ made with the x-axis by $\mathbf{t}$, we have $\tan\theta = y'$. Hence, the curvature becomes

$$\kappa = \frac{d\theta}{ds} = \frac{d\theta}{dx}\frac{dx}{ds}. \tag{26.3}$$

From $\theta = \tan^{-1} y'(x)$, we have

$$\frac{d\theta}{dx} = [\tan^{-1}(y')]' = \frac{y''}{(1 + y'^2)}. \tag{26.4}$$

Using the expression for ds/dx, the curvature can be written in terms of $y(x)$, and we get

$$\kappa = \left|\frac{d^2\mathbf{r}}{ds^2}\right| = \left|\frac{y''}{[1 + (y')^2]^{3/2}}\right|. \tag{26.5}$$

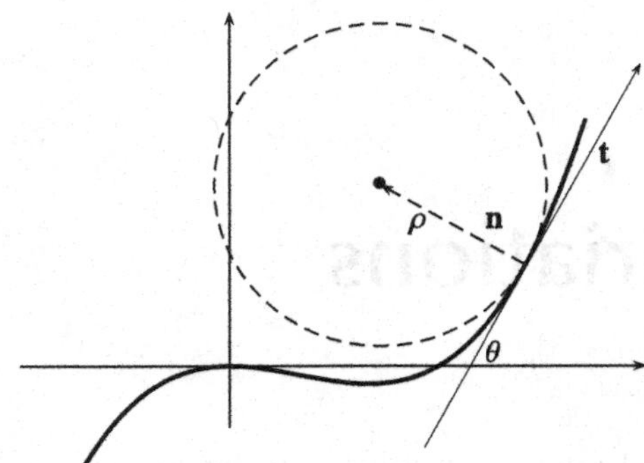

FIGURE 26.1 The basic concept of curvature.

Example 26.1

For a straight line $y = x$, we have $y' = 1$ and $y'' = 0$, thus we have

$$\kappa = \left|\frac{y''}{[1+(y')^2]^{3/2}}\right| = \left|\frac{0}{(1+1^2)^{3/2}}\right| = 0,$$

which means that a straight line has a zero curvature (obviously).

For the simply curve $y = e^x$, we have $y' = e^x$ and $y'' = e^x$, so its curvature is

$$\kappa(x) = \left|\frac{y''}{[1+(y')^2]^{3/2}}\right| = \left|\frac{e^x}{[1+e^{2x}]^{3/2}}\right|.$$

At $x = 0$, we have

$$\kappa(x=0) = \left|\frac{e^0}{(1+e^0)^{3/2}}\right| = \frac{1}{2^{3/2}}.$$

Interestingly, when $x \to -\infty$, we have $e^x \to 0$ and the curvature will approach zero, which means that the asymptote as $x \to 0$ becomes approximately a straight line. On the other hand, for $x \to +\infty$, we have

$$\lim_{x\to\infty} \kappa = \left|\frac{e^x}{(1+e^{2x})^{3/2}}\right|$$
$$\approx \frac{e^x}{(e^{2x})^{3/2}} \approx \frac{e^x}{e^{3x}} = e^{-2x} \to 0,$$

where we have used the fact that e^x is much bigger than 1 when x is sufficiently large (i.e., $e^x \gg 1$).

Once we have the curvature, we can then calculate the radius of curvature of a curve at any point

$$\rho = \frac{1}{\kappa} = \left|\frac{[1+(y')^2]^{3/2}}{y''}\right|. \tag{26.6}$$

In general, the curvature and the radius of curvature vary from point to point.

Obviously, for a straight line, its curvature is zero and thus its radius of curvature becomes infinite. Now let us look another example.

Example 26.2

We know that the equation of a circle centered at (0, 0) with a radius r is (see Fig. 26.2)

$$x^2 + y^2 = r^2.$$

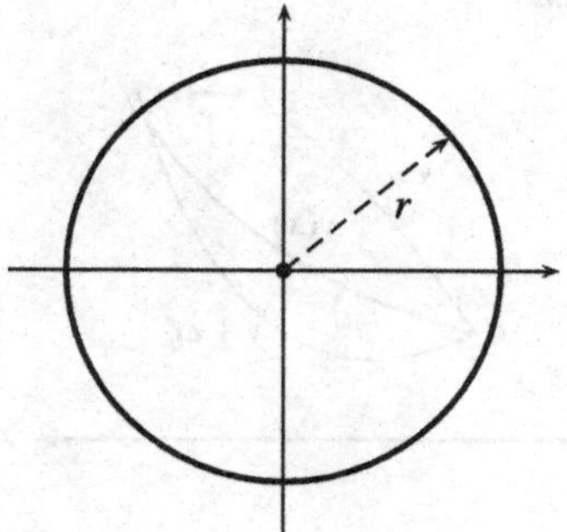

FIGURE 26.2 The radius of curvature of a circle is its radius r.

In order to calculate the curvature at any point (x, y) along the circle, we have to calculate y' and y''. Using implicit differentiation with respect to x, we have

$$2x + 2y'y = 0,$$

which leads to

$$y' = -\frac{x}{y}.$$

The second derivative is

$$y'' = -\frac{x^2 + y^2}{y^3}.$$

Using these expressions and $x^2 + y^2 = r^2$, we have the curvature

$$\kappa = \left|\frac{y''}{[1 + (y')^2]^{3/2}}\right| = \left|-\frac{r^2/y^3}{[1 + (-x/y)^2]^{3/2}}\right|$$

$$= \left|-\frac{1}{y\sqrt{\frac{r^2}{y^2}}}\right| = \frac{1}{r}.$$

Indeed, the curvature of a circle is everywhere $1/r$. Thus, the radius of curvature is the radius of the circle $\rho = 1/\kappa = r$.

26.1.2 Euler-Lagrange Equation

Since the calculus of variations is always related to some minimization or maximization, we can in general assume that the integrand ψ of the integral is a function of the shape or curve $y(x)$ (shown in Fig. 26.3), its derivative $y'(x)$ and the spatial coordinate x (or time t, depending on the context). For the integral

$$I = \int_a^b \psi(x, y, y')dx, \tag{26.7}$$

where a and b are fixed, the aim is to find the solution of the curve $y(x)$ such that it makes the value of I stationary or optimal. In this sense, $I[y(x)]$ is a function of the function $y(x)$, and thus it is referred to as a functional.

Here, stationarity means that a small first-order change in $y(x)$ will only lead to the second-order changes in the values of $I[y(x)]$, and consequently, the change δI of I should be virtually zero due to any small variations in the function $y(x)$. Translating this into mathematical language, we suppose that $y(x)$ has a small change of magnitude of ϵ so that

$$y(x) \to y(x) + \epsilon\zeta(x), \tag{26.8}$$

where $\zeta(x)$ is an arbitrary function. The requirement of I to be stationary means that

$$\delta I = 0, \tag{26.9}$$

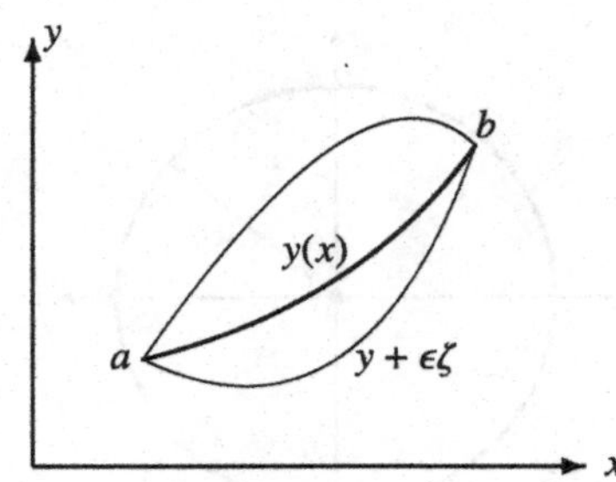

FIGURE 26.3 Variations in the path $y(x)$.

or more accurately,

$$\frac{\delta I}{\delta \epsilon}\Big|_{\epsilon=0} = 0, \qquad \text{for all } \zeta(x). \tag{26.10}$$

Using the multivariate Taylor series, we have

$$\begin{aligned} I(y, \epsilon) &= \int_a^b \psi(x, y + \epsilon\zeta, y' + \epsilon\zeta')dx \\ &= \int_a^b \psi(x, y, y')dx + \int_a^b [\epsilon(\zeta\frac{\partial \psi}{\partial y} + \zeta'\frac{\partial \psi}{\partial y'})]dx + O(\epsilon^2). \end{aligned} \tag{26.11}$$

The first derivative of I should be zero, and we have

$$\frac{\delta I}{\delta \epsilon} = \int_a^b [\frac{\partial \psi}{\partial y}\zeta + \frac{\partial \psi}{\partial y'}\zeta']dx = 0, \tag{26.12}$$

which is exactly what we mean that the change δI (or the first-order variation) in the value of I should be zero. Integrating this equation by parts, we have

$$\int_a^b [\frac{\partial \psi}{\partial y} - \frac{d}{dx}\frac{\partial \psi}{\partial y'}]\zeta dx = -\Big[\zeta\frac{\partial \psi}{\partial y'}\Big]_a^b. \tag{26.13}$$

If we require that $y(a)$ and $y(b)$ are fixed at the points $x = a$ and $x = b$, then these requirements naturally lead to $\zeta(a) = \zeta(b) = 0$. This means that the right-hand side of the equation is zero. That is,

$$\Big[\zeta\frac{\partial \psi}{\partial y'}\Big]_a^b = 0, \tag{26.14}$$

which gives

$$\int_a^b \Big[\frac{\partial \psi}{\partial y} - \frac{d}{dx}\frac{\partial \psi}{\partial y'}\Big]\zeta dx = 0. \tag{26.15}$$

As this equation holds for all $\zeta(x)$, the integrand must be zero. Therefore, we have the well-known Euler-Lagrange equation

$$\frac{\partial \psi}{\partial y} = \frac{d}{dx}(\frac{\partial \psi}{\partial y'}). \tag{26.16}$$

It is worth pointing out that this equation is special in the sense that ψ is known and the unknown is $y(x)$. It has many applications in mathematics, physics and engineering.

Example 26.3

The simplest and classical example is probably to find the shortest path on a plane joining two points, say $(0, 0)$ and $(1, 1)$. We know that the total length along a curve $y(x)$ is

$$L = \int_0^1 \sqrt{1+y'^2}dx. \tag{26.17}$$

Since $\psi = \sqrt{1+y'^2}$ does not contain y, thus $\frac{\partial \psi}{\partial y} = 0$. From the Euler-Lagrange equation, we have

$$\frac{d}{dx}(\frac{\partial \psi}{\partial y'}) = 0, \tag{26.18}$$

its integral is

$$\frac{\partial \psi}{\partial y'} = \frac{y'}{\sqrt{1+y'^2}} = A. \tag{26.19}$$

Rearranging it as

$$y'^2 = \frac{A^2}{1-A^2}, \quad \text{or} \quad y' = \frac{A}{\sqrt{1-A^2}}, \tag{26.20}$$

and integrating again, we have

$$y = kx + c, \qquad k = \frac{A}{\sqrt{1-A^2}}. \tag{26.21}$$

This is a straight line. That is exactly what we expect from the plane geometry.

This example is relatively simple. Let us now study a more complicated case so as to demonstrate the wide range of applications of the Euler-Lagrange equation. In mechanics, Hamilton's principle states that the configuration of a mechanical system is such that the action integral I (the integral of the Lagrangian $\mathcal{L} = T - V$) is stationary with respect to variations in the path. That is to say that the configuration can be uniquely defined by its coordinates q_i and time t, when moving from one configuration at time t_0 to another time $t = t^*$

$$I = \int_{t_0}^{t^*} \mathcal{L}(t, q_i, \dot{q}_i)dt, \qquad i = 1, 2, ..., N, \tag{26.22}$$

where T is the total kinetic energy (usually, a function of $\dot{q}_i$), and V is the potential energy (usually, a function of q_i). Here $\dot{q}_i$ means

$$\dot{q}_i = \frac{\partial q_i}{\partial t}. \tag{26.23}$$

In analytical mechanics, the Lagrangian $\mathcal{L}$ (= kinetic energy − potential energy) is often called the action, thus this principle is also called the principle of least action. The physical configuration or the path of movement follows a path that makes the action integral stationary.

In the special case, $x \to t$, the Euler-Lagrange equation becomes

$$\frac{\partial \mathcal{L}}{\partial q_i} = \frac{d}{dt}(\frac{\partial \mathcal{L}}{\partial \dot{q}_i}), \tag{26.24}$$

which is the well-known Lagrange's equation. Now let us look at a classic example.

Example 26.4

For a simple pendulum shown in Fig. 26.4, we now try to derive its equation of oscillations. We know the kinetic energy T and the potential energy V are

$$T = \frac{1}{2}ml^2(\frac{d\theta}{dt})^2 = \frac{1}{2}ml^2\dot{\theta}^2, \qquad V = mgh = mgl(1-\cos\theta),$$

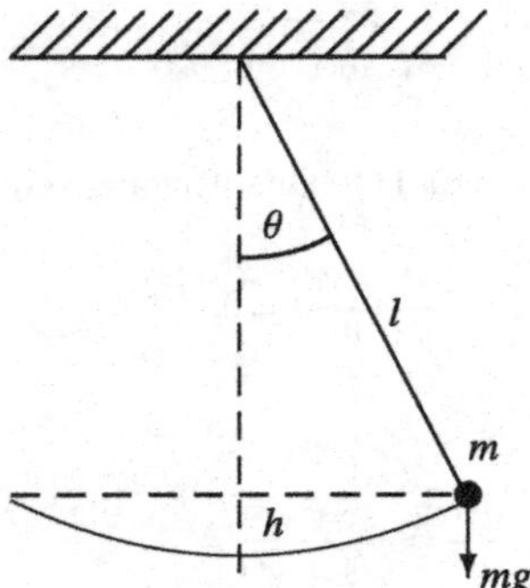

FIGURE 26.4 A simple pendulum.

where g is the acceleration due to gravity. Using $\mathcal{L} = T - V$, $q = \theta$ and $\dot{q} = \dot{\theta}$, we have

$$\frac{\partial \mathcal{L}}{\partial \theta} - \frac{d}{dt}(\frac{\partial \mathcal{L}}{\partial \dot{\theta}}) = 0,$$

which becomes

$$-mgl \sin\theta - \frac{d}{dt}(ml^2\dot{\theta}) = 0.$$

Therefore, we have the pendulum equation

$$\frac{d^2\theta}{dt^2} + \frac{g}{l} \sin\theta = 0.$$

This is a nonlinear equation. If the angle is very small ($\theta \ll 1$), so $\sin\theta \approx \theta$, we then have the standard equation

$$\frac{d^2\theta}{dt^2} + \frac{g}{l}\theta = 0.$$

Using the standard technique for solving linear ordinary differential equations, we have its solution in the form

$$\theta = A\cos(\omega t) + B\sin(\omega t), \quad \omega = \sqrt{\frac{g}{l}},$$

where A and B are two unknown coefficients to be determined by initial conditions. Here, $\omega = 2\omega f$ is the angular frequency and $f = 1/T$ is the frequency where T is the period. The period can be calculated by

$$T = \frac{1}{f} = \frac{2\pi}{\omega} = 2\pi\sqrt{\frac{l}{g}}.$$

Interestingly, the period is independent of the mass of the pendulum. In a special case when $l = 1$ m and $g = 9.8$ m/s^2, we have

$$T = 2\pi\sqrt{\frac{1}{9.8}} \approx 2.007,$$

which is about 2 seconds.

26.2 VARIATIONS WITH CONSTRAINTS

Although the stationary requirement in the calculus of variations leads to the minimization of the integral itself, there is no additional constraint. In this sense, the calculus of variations discussed up to now is unconstrained. However, sometimes these variations have certain additional constraints, for example, the sliding of a bead on a hanging string.

Now we want to make the integral I stationary under another constraint integral Q that is constant. We have

$$I = \int_a^b \psi(x, y, y')dx, \tag{26.25}$$

subjected to the constraint

$$Q = \int_a^b \phi(x, y, y')dx. \tag{26.26}$$

As for most optimization problems under additional constraints, the method of Lagrange multipliers can transform the constrained problem into an unconstrained one by using a combined functional $J = I + \lambda Q$

$$J = \int_a^b [\psi + \lambda\phi]dx, \tag{26.27}$$

where λ is the undetermined Lagrange multiplier. Replacing ψ by $[\psi + \lambda\phi]$ in the Euler-Lagrange equation or following the same derivation, we have

$$[\frac{\partial \psi}{\partial y} - \frac{d}{dx}(\frac{\partial \psi}{\partial y'})] + \lambda[\frac{\partial \phi}{\partial y} - \frac{d}{dx}(\frac{\partial \phi}{\partial y'})] = 0. \tag{26.28}$$

Now we can come back to our example of the hanging rope problem with two fixed points. The total length of the rope is L, and it hangs from two fixed points $(-d, 0)$ and $(d, 0)$. From geometric consideration, it is required that $2d < L$.

Example 26.5

In order to find the shape of the hanging rope under gravity, we now define its total gravitational potential energy E_p as

$$E_p = \int_{x=-d}^{d} [\rho g y(x) ds] = \rho g \int_{-d}^{d} y\sqrt{1 + y'^2}dx, \tag{26.29}$$

where ρ is the mass per unit length of the rope. The additional constraint is that the total length of the rope is a constant (L). Thus,

$$Q = \int_{-d}^{d} \sqrt{1 + y'^2}dx = L. \tag{26.30}$$

By using the Lagrange multiplier λ, we have $J = E_p + \lambda Q$, or

$$J = \int_{-d}^{d} [\rho g y + \lambda]\sqrt{1 + y'^2}dx. \tag{26.31}$$

Since $\Psi = [\rho g y + \lambda]\sqrt{1 + y'^2}$ does not contain x explicitly, or $\frac{\partial \Psi}{\partial x} = 0$, the Euler-Lagrange equation can be reduced into a simpler form in this special case. Using

$$\frac{d\Psi}{dx} = \frac{\partial \Psi}{\partial x} + \frac{\partial \Psi}{\partial y}\frac{dy}{dx} + \frac{\partial \Psi}{\partial y'}\frac{dy'}{dx} = 0 + y'\frac{\partial \Psi}{\partial y} + y''\frac{\partial \Psi}{\partial y'}, \tag{26.32}$$

and the Euler-Lagrange equation $\frac{\partial \Psi}{\partial y} = \frac{d}{dx}(\frac{\partial \Psi}{\partial y'})$, we have

$$\frac{d\Psi}{dx} = y'[\frac{d}{dx}(\frac{\partial \Psi}{\partial y'})] + y''\frac{\partial \Psi}{\partial y'} = \frac{d}{dx}[y'\frac{\partial \Psi}{\partial y'}], \tag{26.33}$$

which can again be written as

$$\frac{d}{dx}[\Psi - y'\frac{\partial \Psi}{\partial y'}] = 0. \tag{26.34}$$

The integration of this equation gives

$$\Psi - y' \frac{\partial \Psi}{\partial y'} = A = \text{const}. \tag{26.35}$$

Substituting the expression for Ψ into the above equation, the stationary value of J requires

$$\sqrt{1+y'^2} - \frac{y'^2}{\sqrt{1+y'^2}} = \frac{A}{\rho g y + \lambda}. \tag{26.36}$$

Multiplying both sides by $\sqrt{1+y'^2}$ and using the substitution $A \cosh \zeta = \rho g y + \lambda$, we have

$$y'^2 = \cosh^2 \zeta - 1, \tag{26.37}$$

whose solution is

$$\cosh^{-1}[\frac{\rho g y + \lambda}{A}] = \pm \frac{x \rho g}{A} + K. \tag{26.38}$$

Using the boundary conditions $y = 0$ at $x = \pm d$ and the constraint $Q = L$, we have $K = 0$ and an implicit equation for A

$$\sinh(\frac{\rho g d}{A}) = \frac{\rho g L}{2A}. \tag{26.39}$$

Finally, the curve for the hanging rope becomes the following catenary

$$y(x) = \frac{A}{\rho g}[\cosh(\frac{\rho g x}{A}) - \cosh(\frac{\rho g d}{A})]. \tag{26.40}$$

Here, the constant A is related to the length of the rope and has to be determined by

$$\int_{-d}^{d} \sqrt{1+y'^2} dx = L.$$

26.3 VARIATIONS FOR MULTIPLE VARIABLES

What we have discussed so far mainly concerns the variations in 2D settings, and the variations are in terms of $y(x)$ or curves only. What happens if we want to study a surface in the full 3D configuration? The principle in the previous sections can be extended to any dimensions with multiple variables; however, we will focus on the minimization of a surface here.

Suppose we want to study the shape of a soap bubble, the principle of least action leads to the minimal surface problem. The surface integral of a soap bubble should be stationary. A similar problem is the shape of the Earth under the influence of gravity. If we assume that the shape of the bubble is $u(x, y)$, then the total surface area is

$$A(u) = \iint_\Omega \Psi dx dy = \iint_\Omega \sqrt{1 + (\frac{\partial u}{\partial x})^2 + (\frac{\partial u}{\partial y})^2} dx dy, \tag{26.41}$$

where

$$\Psi = \sqrt{1 + (\frac{\partial u}{\partial x})^2 + (\frac{\partial u}{\partial y})^2} = \sqrt{1 + u_x^2 + u_y^2}. \tag{26.42}$$

In this case, the extended Euler-Lagrange equation for two variables x and y becomes

$$\frac{\partial \Psi}{\partial u} - \frac{\partial}{\partial x}(\frac{\partial \Psi}{\partial u_x}) - \frac{\partial}{\partial y}(\frac{\partial \Psi}{\partial u_y}) = 0. \tag{26.43}$$

Substituting Ψ into the above equation and using $\frac{\partial \Psi}{\partial u} = \Psi_u = 0$ since Ψ does not contain u explicitly, we get

$$-\frac{\partial}{\partial x}[\frac{1}{\Psi}\frac{\partial u}{\partial x}] - \frac{\partial}{\partial y}[\frac{1}{\Psi}\frac{\partial u}{\partial y}] = 0, \tag{26.44}$$

or

$$(1 + u_y^2)u_{xx} - 2u_x u_y u_{xy} + (1 + u_x^2)u_{yy} = 0. \tag{26.45}$$

This is a nonlinear equation and its solution is beyond the scope of this book. This nonlinear equation has been one of the active research topics for more than a century. It has been proved that the fundamental solution to this equation is a sphere, and in fact we know that all free bubbles are spherical. Similarly, the shape of a planet without rotation should be spherical. However, it becomes a rotational ellipsoid if spin of the planet is included.

For some problems, we can approximately assume that u_x and u_y are small (i.e., $u_x \to 0$ and $u_y \to 0$), thus the above equation becomes Laplace's equation

$$u_{xx} + u_{yy} = 0. \tag{26.46}$$

The calculus of variations has many applications. The other classical examples include Fermat's principle in optics, the Sturm-Liouville problem, the least action principle and finite element analysis.

EXERCISES

26.1. Calculate the curvature of $y = \sin(x)$ at $x = \pi/2$.

26.2. Find the solution $y(x)$ so as to minimize $\int_a^b [\frac{1}{2}y'^2(x) + x^2 y(x)]dx$ where $b > a > 0$.

26.3. Solve $I[u(x)] = \int_0^1 \frac{u'^2}{2} dx$ subject to $\int_0^1 u(x)x = 1$ with the boundary conditions $u(0) = u(1) = 0$.

Chapter 27

Integral Equations

Chapter Points

- Integral equations are introduced, including both Fredholm integral equations and Volterra integral equations.
- Some solution techniques are explained with worked examples.

Differential equations are mathematical equations that relate functions with their derivatives. Ordinary differential equations concern univariate functions, while partial differential equations concern functions in terms of at least two independent variables. All these equations do not explicitly include integrals in their standard forms.

Even for univariate functions, sometimes, integrals can inevitably appear in equations. For example, from the calculus of variations, we know that the unknown $y(x)$ to be optimized is the integrand of I. In a certain sense, this is an integral equation. In fact, many physical processes and laws of conservation are expressed in terms of integral forms rather than their differentiation counterparts. Naturally, one of the ways of constructing an integral equation is to integrate from a differential equation. Integral equations are much more complicated compared with differential equations. There is no universal solution technique for nonlinear equations; even their numerical simulations are usually not straightforward. Thus, we will mainly focus on the simplest types of integral equations.

27.1 INTEGRAL EQUATIONS

An integral equation is an equation that relates a function with its integrals. For example, the following equation

$$y(x) = x + \frac{1}{2}\int_0^x y(u)du, \tag{27.1}$$

is an integral equation, so is the following equation

$$y(x) = \sin(x) + \int_0^1 \cos(xu)y^2(u)du. \tag{27.2}$$

In general, integral equations can be very challenging to solve, and here we only try to solve linear integral equations of two types: Fredholm and Volterra integral equations.

27.1.1 Fredholm Integral Equations

A linear integral equation for $y(x)$ can be written in the following generic form

$$u(x) + \lambda\int_a^b K(x,\eta)y(\eta)d\eta = v(x)y(x), \tag{27.3}$$

where $K(x,\eta)$ is referred to as the kernel of the integral equation. The parameter λ is a known constant. If the function $u(x) = 0$, the equation is then called homogeneous. If $u(x) \neq 0$, the equation is inhomogeneous.

If the function $v(x) = 0$, then the unknown $y(x)$ appears only once in the integral equation, and it is under the integral sign only. This is called the linear integral equation of the first kind

$$u(x) + \lambda\int_a^b K(x,\eta)y(\eta)d\eta = 0. \tag{27.4}$$

On the other hand, if $v(x) = 1$, equation (27.3) becomes the integral equation of the second kind

$$u(x) + \lambda \int_a^b K(x, \eta)y(\eta)d\eta = y(x). \tag{27.5}$$

An integral equation with fixed integration limits a and b, is called a Fredholm equation. If the upper integration limit b is not fixed, then the equation becomes a Volterra equation. The integral equation becomes singular if at least one of its integration limits approaches infinite. In addition, if derivatives also appear in the equation, they are often called integro-differential equations.

Example 27.1

For example,

$$y(x) = x^2 + \lambda \int_0^1 x\eta y(\eta)d\eta$$

is a Fredholm-type integral equation, while

$$y(x) = \sin(x) + \lambda \int_0^x e^{x-\eta} y(\eta)d\eta,$$

is a Voterra-type integral equation. However, the following equation

$$y'(x) = x^2 + \lambda \int_0^1 x\eta y(\eta)d\eta,$$

is an integro-differential equation of Fredholm type and

$$y''(x) = \sin(x) + \lambda \int_0^x e^{x-\eta} y(\eta)d\eta,$$

is an integro-differential equation of Volterra type.

27.1.2 Volterra Integral Equation

In general, the Volterra integral equation can be written as

$$u(x) + \lambda \int_a^x K(x, \eta)y(\eta)d\eta = v(x)y(x). \tag{27.6}$$

The first kind [or $v(x) = 0$] and second kind [or $v(x) = 1$] are defined in a similar manner to the Fredholm equation.

The kernel is said to be separable or degenerate if it can be written in the finite sum form

$$K(x, \eta) = \sum_{i=1}^{N} f_i(x) g_i(\eta), \tag{27.7}$$

where $f_i(x)$ and $g_i(\eta)$ are functions of x and η, respectively.

A kernel is called a convolution kernel (also displacement kernel) if it can be written as a function of the difference $(x - \eta)$ of its two arguments

$$K(x, \eta) = K(x - \eta). \tag{27.8}$$

Example 27.2

For example,

$$y(x) = e^x + \lambda \int_0^x e^{x-\eta} y(\eta) d\eta,$$

is a Volterra-type integral equation with a convolution kernel.

27.2 SOLUTION OF INTEGRAL EQUATIONS

Most integral equations do not have closed-form solutions. For linear integral equations, closed-form solutions are only possible for the special cases of separable and displacement kernels.

27.2.1 Separable Kernels

For a Fredholm integral equation of the second kind with a separable kernel, we can substitute the kernel (27.7) into the equation and we have

$$u(x) + \lambda \int_a^b \sum_{i=1}^N f_i(x) g_i(\eta) y(\eta) d\eta = y(x), \tag{27.9}$$

which becomes

$$u(x) + \lambda \sum_{i=1}^N f_i(x) \int_a^b g_i(\eta) y(\eta) d\eta = y(x). \tag{27.10}$$

Because the integration limits are fixed, the integrals over η are constants that are to be determined. By defining

$$\alpha_i = \int_a^b g_i(\eta) y(\eta) d\eta, \tag{27.11}$$

we now have the solution in the form

$$y(x) = u(x) + \lambda \sum_{i=1}^N \alpha_i f_i(x), \tag{27.12}$$

where the N coefficients α_i are determined by

$$\alpha_i = \int_a^b g_i(\eta) u(\eta) d\eta + \lambda \sum_{i=1}^N \int_a^b [\alpha_i f_i(\eta) g_i(\eta)] d\eta, \tag{27.13}$$

for $i = 1, 2, ..., N$. Only for a few special cases can these coefficients be written as simple explicit expressions.

Example 27.3

Let us try to solve

$$x^3 + \int_0^1 x^2 \eta^2 y(\eta) d\eta = y(x),$$

which means that $u(x) = x^3$ and $\lambda = 1$.

The solution is in the form

$$y(x) = x^3 + x^2\alpha, \quad \alpha = \int_0^1 \eta^2 y(\eta)d\eta.$$

So

$$\alpha = \int_0^1 \eta^2 y(\eta)d\eta = \int_0^1 \eta^2[\eta^3 + \eta^2\alpha]d\eta = \int_0^1 \eta^5 d\eta + \alpha\int_0^1 \eta^4 d\eta$$
$$= \frac{1}{6} + \alpha\frac{1}{5},$$

or

$$\alpha = \frac{1}{6} + \frac{\alpha}{5},$$

which gives $\alpha = 5/24$. Therefore, we finally have

$$y(x) = x^3 + \frac{5}{24}x^2.$$

27.2.2 Volterra Equation

A Volterra equation with separable kernels may be solved by transforming into a differential equation via direct differentiation. Let us demonstrate this by an example.

Example 27.4

To solve an integral equation of Volterra type

$$y(x) = e^x + \int_0^x e^x \sin(\zeta)y(\zeta)d\zeta,$$

we first divide both sides by e^x to get

$$\frac{y(x)}{e^x} = 1 + \int_0^x \sin(\zeta)y(\zeta)d\zeta,$$

whose differentiation with respect to x leads to

$$\Big[\frac{y(x)}{e^x}\Big]' = y(x)\sin(x), \qquad \text{or} \qquad \frac{1}{e^x}y'(x) - y(x)e^{-x} = y(x)\sin(x).$$

Dividing both sides by $y(x)$ and using $[\ln y(x)]' = y'(x)/y(x)$, we have

$$[\ln y(x)]' = e^x \sin x + 1.$$

By direct integration, we have

$$\ln y(x) = x - \frac{1}{2}e^x \cos x + \frac{1}{2}e^x \sin x.$$

Thus, we finally obtain

$$y(x) = \exp[x - \frac{e^x}{2}(\cos x - \sin x)].$$

There are other methods and techniques for solving integral equations such as the operator method, series method and the Fredholm theory. Let us try to solve a Volterra equation by the series method.

Example 27.5

To solve

$$y(x) = x^2 + \lambda \int_0^x y(u)du,$$

we can use the series method. The first or crude approximation is

$$y(x) = x^2.$$

Substituting this first approximation into the original equation, we have a better approximation

$$y(x) = x^2 + \lambda \int_0^x u^2 du = x^2 + \lambda \frac{x^3}{3}.$$

Again, a better approximation can be obtained by using the previous approximation

$$y(x) = x^2 + \lambda \int_0^x [u^2 + \frac{\lambda u^3}{3}]du = x^2 + \lambda \frac{x^3}{3} + \lambda^2 \frac{1}{3}\frac{x^4}{4} = x^2 + \lambda \frac{x^3}{3} + \lambda^2 \frac{x^4}{12}.$$

Then, an even better approximation can be obtained

$$y(x) = x^2 + \lambda \int_0^x [u^2 + \lambda \frac{u^3}{3} + \lambda^2 \frac{u^4}{12}]du = x^2 + \lambda \frac{x^3}{3} + \lambda^2 \frac{x^4}{12} + \lambda^3 \frac{x^5}{60}.$$

If we continue this way, the final solution becomes

$$y(x) = x^2 + 2x^2[\frac{(\lambda x)}{(1+2)!} + \frac{(\lambda x)^2}{(2+2)!} + \frac{(\lambda x)^3}{(3+2)!} + \ldots + \frac{(\lambda x)^n}{(n+2)!}].$$

From the theory of Taylor series, this series will converge for all λ.

As this method does not pose any constraint on the integral limits, the series method can also apply to Fredholm equations. However, the convergence requirement may pose constraint on the value of λ.

Example 27.6

If the equation in the above equation is modified as

$$y(x) = x^2 + \lambda \int_0^1 y(u)du,$$

then the first approximation is still

$$y(x) = x^2,$$

but the second approximation becomes

$$y(x) = x^2 + \lambda \int_0^1 u^2 du = x^2 + \frac{\lambda}{3}.$$

The third approximation becomes

$$y(x) = x^2 + \lambda \int_0^1 [u^2 + \frac{\lambda}{3}]du = x^2 + \lambda \frac{1}{3} + \lambda^2 \frac{1}{3}.$$

The fourth approximation is

$$y(x)=x^2+\lambda\int_0^1 [u^2+\frac{\lambda}{3}+\frac{\lambda^2}{3}]du=x^2+\frac{1}{3}[\lambda+\lambda^2+\lambda^3].$$

Similar calculations lead to

$$y(x)=x^2+\frac{1}{3}[\lambda+\lambda^2+\lambda^3+...+\lambda^n],$$

which only converges if $|\lambda|<1$. As $n\to\infty$, we have

$$y(x)=x^2+\frac{1}{3}\frac{\lambda}{1-\lambda}=x^2+\frac{\lambda}{3(1-\lambda)}.$$

It is straightforward to verify that the above solution indeed satisfies the original equation.

There are other methods for solving integral equations such as Laplace transforms, Fourier transforms and approximation methods. For more advanced techniques, interested readers can refer to more advance literature (Rahman M, *Integral Equations and Their Applications*, WIT Press, 2007). It is worth pointing out that most integral equations do not have closed-form solutions, and thus the analytical methods do not work well. In this case, numerical methods are the best alternatives.

EXERCISES

27.1. Discuss the type of the following equations:

- $y(x)=\cos(x)+\frac{1}{3}\int_0^{\pi} xsy(s)ds.$
- $y(x)=x^2e^{-x}+\int_0^x (x-s)y(s)ds.$
- $y(x)=f(x)+\lambda\int_0^x (x-s)\sin[y(s)]ds.$
- $y'(x)=x+\frac{1}{2}\int_0^1 xsy^2(s)ds.$

27.2. Solve $y(x)=a+b\int_0^1 xy(t)dt$ for $a>0$ and $b=1$.

27.3. Solve $y(x)=a+\int_0^x y(t)dt$ for $a=1$. Can you provide the solution for any $a>0$?

Chapter 28

Mathematical Modeling

Chapter Points

- Mathematical modeling is introduced with the basic modeling procedure, including mathematical model formulation based on physical laws, parameter estimation and normalization.
- Different levels of approximations are explained to discuss the assumptions, abstractions and the balance of accuracy and model complexity.
- Different types of models are explained with some examples relevant to science and engineering applications.
- A worked example is presented in detail to model Brownian motion and diffusion.

28.1 MATHEMATICAL MODELING

Mathematical modeling is the process of formulating an abstract model in terms of mathematical language to describe the complex behavior of a real system. Mathematical models are quantitative models and often expressed in terms of ordinary differential equations and partial differential equations. Mathematical models can also be statistical models, fuzzy logic models and empirical relationships. In fact, any model description using mathematical language can be called a mathematical model. Mathematical modeling is widely used in natural sciences, computing, engineering, meteorology, economics and finance. For example, theoretical physics is essentially all about the modeling of real-world processes using several basic principles (such as the conservation of energy and momentum) and a dozen important equations (such as the wave equation, the Schrödinger equation, the Einstein equation). Most of these equations are partial differential equations.

An important feature of mathematical modeling is its interdisciplinary nature. It involves applied mathematics, computer sciences, physics, chemistry, engineering, biology and other disciplines such as economics, depending on the problem of interest. Mathematical modeling in combination with scientific computing is an emerging interdisciplinary technology. Many international companies use it to model physical processes, to design new products, to find solutions to challenging problems, and to increase their competitiveness in international markets.

Example 28.1

One of the simplest models we learned in school is probably Newton's second law that relates the force F acted on a body with a mass m to its acceleration a. That is

$$F = ma,$$

which is one of the most accurate models in science. This is a linear relationship and thus a linear model, but a very well-tested model.

Apart from a simple mathematical formula, as a mathematical model, all the quantities involved such as force, mass and acceleration must have appropriate units. For example, the unit of F is Newton (N), the unit of mass is kilogram (kg), while the acceleration has a derived unit (a combination of units) of m/s^2. Therefore, a person of 80 kg has a weight (the force acted upon the person by the Earth) is $W = mg$ where $g = 9.8$ m/s^2 is the acceleration due to gravity. That is

$$W = mg = 80 \text{ (kg)} \times 9.8 \text{ (m/s}^2) = 784 \text{ N}.$$

If the units are wrong, even a good model will give wrong values. This highlights the importance of units and the parameters (e.g., g here) in mathematical modeling.

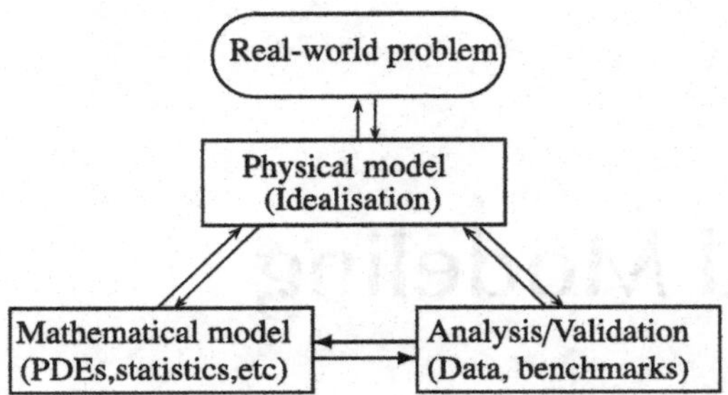

FIGURE 28.1 Mathematical modeling.

Mathematical modeling is an iterative, multidisciplinary process with many steps from the abstraction of the processes in nature to the construction of the full mathematical models. The basic steps of mathematical modeling can be summarized as meta-steps shown in Fig. 28.1. The process typically starts with the analysis of a real world problem so as to extract the fundamental physical processes by idealization and various assumptions. Once an idealized physical model is formulated, it can then be translated into the corresponding mathematical model in terms of partial differential equations (PDEs), integral equations, and statistical models. Then, the mathematical model should be investigated in great detail by mathematical analysis (if possible), numerical simulations and other tools so as to make predictions under appropriate conditions. Then, these simulation results and predictions will be validated against the existing models, well-established benchmarks, and experimental data. If the results are satisfactory (but they rarely are at first), then the mathematical model can be accepted. If not, both the physical model and mathematical model will be modified based on the feedback, then the new simulations and prediction will be validated again.

After a certain number of iterations of the whole process (often many), a good mathematical model can properly be formulated, which will provide great insight into the real world problem and may also predict the behavior of the process under study.

For any physical problem in physics and engineering, for example, there are traditionally two ways to deal with it by either theoretical approaches or field observations and experiments.

The theoretical approach in terms of mathematical modeling is an idealization and simplification of the real problem and the theoretical models often extract the essential or major characteristics of the problem. The mathematical equations obtained even for such over-simplified systems are usually very difficult for mathematical analysis. On the other hand, the field studies and experimental approach can be expensive if not impractical. Apart from financial and practical limitations, other constraining factors include the inaccessibility of the locations, the range of physical parameters, and time for carrying out various experiments. As the computing speed and power of computers have increased dramatically in the last few decades, a practical third way or approach is emerging, which is computational modeling and numerical experimentation based on mathematical models. It is now widely acknowledged that computational modeling and computer simulations serve as a cost-effective alternative, bridging the gap between theory and practice as well as complementing the traditional theoretical and experimental approaches to problem solving.

Mathematical modeling is essentially an abstract art of formulating the mathematical models from their corresponding real-world problems. The mastery of this art requires practice and experience, and it is not easy to teach such skills as the style of mathematical modeling largely depends on each person's own insight, abstraction, type of problems, and experience of dealing with similar problems. Even for the same physical process, different models could be obtained, depending on the emphasis of some part of the process, say, based on your interest in certain quantities in a particular problem, while the same quantities could be viewed as unimportant in other processes and other problems.

28.2 MODEL FORMULATION

Mathematical modeling often starts with the analysis of the physical process and attempts to make an abstract physical model by idealization and approximations. From this idealized physical model, we can use the various first principles such as the conservation of mass, momentum, energy and Newton's laws to translate into mathematical equations. However, such transformation from practice to theory can rarely be achieved in a single step, thus an iterative loop between theory and practice is needed, as pointed out by the famous statistician George Box.

As an example, let us now look at the example of the diffusion process of sugar in a glass of water. We know that the diffusion of sugar will occur if there is any spatial difference in the sugar concentration. The physical process is complicated

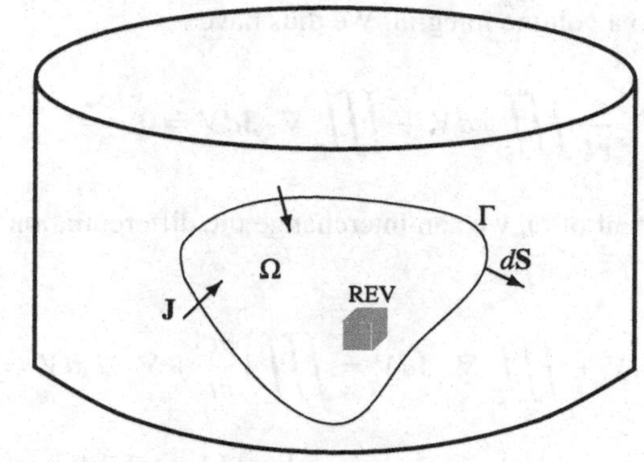

FIGURE 28.2 Representative element volume (REV).

and many factors could affect the distribution of sugar concentration in water, including the temperature, stirring, mass of sugar, type of sugar, how you add the sugar, even geometry of the container and others. We can idealize the process by assuming that the temperature is constant (so as to neglect the effect of heat transfer), and that there is no stirring because stirring will affect the effective diffusion coefficient and introduce the advection of water or even vertices in the (turbulent) water flow.

We then choose a representative element volume (REV) whose size is very small compared with the size of the cup so that we can use a single value of concentration to represent the sugar content inside this REV. If this REV is too large, there is a considerable variation in sugar concentration inside this REV. We also assume that there is no chemical reaction between sugar and water (otherwise, we are dealing with something else). If you drop the sugar into the cup from a considerable height, the water inside the glass will splash and thus fluid volume will change, and this becomes a fluid dynamics problem. So we are only interested in the process after the sugar is added and we are not interested in the initial impurity of the water (to a certain degree).

With these assumptions, the whole process is now idealized as the physical model of the diffusion of sugar in still water at a constant temperature. Now we have to translate this idealized model into a mathematical model, and in the present case, a parabolic partial differential equation or diffusion equation.

Let c be the averaged concentration in a representative element volume with a volume dV inside the cup, and let Ω be an arbitrary, imaginary closed volume Ω (much larger than our REV but smaller than the container, see Fig. 28.2). We know that the rate of change of the mass of sugar per unit time inside Ω is

$$\delta_1 = \frac{\partial}{\partial t}\iiint_{\Omega} c\,dV, \tag{28.1}$$

where t is time. As the mass is conserved, this change of sugar content in Ω must be supplied in or flow out over the surface $\Gamma = \partial\Omega$ enclosing the region Ω. Let $\mathbf{J}$ be the flux through the surface, thus the total mass flux through the whole surface Γ is

$$\delta_2 = \iint_{\Gamma} \mathbf{J} \cdot d\mathbf{S}.$$

Thus the conservation of total mass in Ω requires that

$$\delta_1 + \delta_2 = 0,$$

or

$$\frac{\partial}{\partial t}\iiint_{\Omega} c\,dV + \iint_{\Gamma} \mathbf{J} \cdot d\mathbf{S} = 0. \tag{28.2}$$

This is essentially the integral form of the mathematical model. Using the divergence theorem of Gauss

$$\iint_{\Gamma} \mathbf{J} \cdot d\mathbf{S} = \iiint_{\Omega} \nabla \cdot \mathbf{J}\,dV, \tag{28.3}$$

we can convert the surface integral into a volume integral. We thus have

$$\frac{\partial}{\partial t}\iiint_{\Omega} c dV + \iiint_{\Omega} \nabla \cdot \mathbf{J} dV = 0. \tag{28.4}$$

Since the domain Ω is fixed (independent of t), we can interchange the differentiation and integration in the first term, we now get

$$\iiint_{\Omega} \frac{\partial c}{\partial t} dV + \iiint_{\Omega} \nabla \cdot \mathbf{J} dV = \iiint_{\Omega} [\frac{\partial c}{\partial t} + \nabla \cdot J] dV = 0. \tag{28.5}$$

Since the enclosed domain Ω is arbitrary, the above equation should be valid for any shape or size of Ω; therefore, the integrand must be zero. We finally have

$$\frac{\partial c}{\partial t} + \nabla \cdot \mathbf{J} = 0. \tag{28.6}$$

This is the differential form of the mass conservation. It is a partial differential equation (PDE), and this mathematical model is a PDE.

Example 28.2

As we know that diffusion occurs from the higher concentration to lower concentration, the rate of diffusion is proportional to the gradient ∇c of the concentration. The flux $\mathbf{J}$ over a unit surface area is given by Fick's law

$$\mathbf{J} = -D\nabla c,$$

where D is the diffusion coefficient which depends on the temperature and the type of materials. The negative sign means the diffusion is opposite to the gradient. Substituting this into the mass conservation (28.6), we have

$$\frac{\partial c}{\partial t} - \nabla \cdot (D\nabla c) = 0,$$

or

$$\frac{\partial c}{\partial t} = \nabla \cdot (D\nabla c).$$

In the simplified case when D is constant, we have

$$\frac{\partial c}{\partial t} = D\nabla^2 c, \tag{28.7}$$

which is the well-known diffusion equation.

This equation can be applied to study many phenomena such as heat conduction, pollutant transport, groundwater flow and concentrations if we replace D by their corresponding physical parameters.

28.3 DIFFERENT LEVELS OF APPROXIMATIONS

As we have just seen, we can formulate some mathematical models once we have made some appropriate assumptions such as the process being the same inside the cup as well near the edge of the cup. Otherwise, we may have to deal with the so-called a boundary-layer phenomenon which is not much relevant for the current model.

The assumptions essentially determine the level of consideration and thus the level of approximations. Let us use gravity as an example to explain this issue.

Example 28.3

Newton's law of universal gravitation can be applied to almost any objects. It can often be written as the following formula:

$$F = G\frac{m_1 m_2}{r^2},$$

which essentially states that the force F between two masses m_1 and m_2 is inversely proportional to the square of the distance r. Here, G is the universal gravitational constant $G = 6.674 \times 10^{-11}$ N m²/kg². However, in deriving this formula, some assumptions were made. One of the assumptions is that the two masses are both point masses; that is, their geometrical size does not matter. This is true for a system of planets where the distance between a planet (say, the Earth) to the Sun is sufficiently large compared to the sizes of the celestial bodies.

Imagine that we are trying to calculate the gravitational force between the Earth and its satellites, we have to be careful about the distance r used. Obviously, we can say that r is the distance between the center of the Earth to the center of the satellite. Now the question what about the distribution of mass or the density inside the Earth? Should we consider it as spherically symmetric so that the whole mass of the Earth is essentially concentrated at the center of the Earth? Will the topological variations such as mountainous terrains affect the motion of the satellite? In fact, mountains do affect the motion of a satellite and in this case, we cannot assume that the Earth is a point mass. This means that we cannot directly apply the above formula to the calculations.

It is known that Newton's gravitation can be applied to many scales from small atomic scales to astronomical scales. However, Einstein's general theory of relativity treats this gravity from a completely different perspective, linking energy (rather than mass) to space-time curvature. In this case, Einstein's field equation can be written in the following tensor form:

$$R_{\mu\nu} - \frac{1}{2}Rg_{\mu\nu} + \Lambda g_{\mu\nu} = \frac{8\pi G}{c^4}T_{\mu\nu},$$

where $T_{\mu\nu}$ is the so-called stress-energy tensor, $R_{\mu\nu}$ is the Ricci curvature tensor, R is the curvature, and $g_{\mu\nu}$ is the metric tensor in the four-dimensional (4D) space-time manifold. In addition, c is the speed of light, while G is Newton's gravitational constant as mentioned earlier. The parameter Λ is Einstein's cosmological constant.

Einstein's model can be considered as a generalization of Newton's model, though from a different framework. Their assumptions are different, and the accuracies are also different. Einstein's model works at all scales with a higher accuracy, but it is more challenging to do calculations. Even so, Einstein's model may not work well at Planck's length scale.

This example may be an extreme example where one model seems to be very easy to calculate the force and its calculations can be done using secondary school mathematics by multiplications and division. The other model requires complex tensor theory and sophisticated techniques to compute any observable results. In many applications, different approximations can be linked to different accuracies and the choice of approximations may depend on the accuracy of the solution and ease of calculations. Let us look at another example.

Example 28.4

There is a very simple formula to link the relative humidity (RH) to the dewpoint temperature T_d proposed by M.G. Lawrence (The relationship between relative humidity and the dewpoint temperature in moist air: A simple conversion and applications, Bulletins of the American Meteorological Society, vol. 87, no. 2, 225–233, 2005, http://dx.doi.org/10.1175/BAMS-86-2-225), which states that

$$T_d = T - \Big(\frac{100 - RH}{5}\Big), \tag{28.8}$$

where T is the so-called dry-bulb temperature. Obviously, when RH = 100%, we have $T_d = T$. This is a simple linear relationship and it implies that a 5% decrease in RH would lead to one degree reduction in the dewpoint temperature. It is worth pointing out that the unit here for temperature is degree Celsius (°C); otherwise, the formula will be different.

Obviously, we can simply re-arrange the above equation and get

$$RH = 100 - 5(T - T_d),$$

to calculate RH once we have T_d.

The formula (28.8) is sufficiently accurate only when RH is greater than 50%, and the accuracy is within °C or about 5% for RH in the range of 0 to 30°C. The ease of calculations means that the accuracy is limited, though it is good enough for daily use.

However, for meteorological applications, much higher accuracy may be needed. In this case, we can use

$$T_d = \frac{B[\ln(RH/100) + K]}{A - [\ln(RH/100) + K]}, \quad K = \frac{AT}{B+T},$$

where $A = 17.625$, $B = 243.04°C$. This formula can be accurate within about 0.4% in the range from −40°C to 50°C. However, its calculations involve logarithm and its derivations are based on more realistic but more sophisticated assumptions.

From the above two examples, we can see that we have to balance the ease of calculations and accuracy, depending on the level of approximations. If we insist on getting more accurate results, we may have to consider many factors and minute details. For example, in the previous section, if we wish to know the level of concentration of sugar more accurately at a particular point, we have to know the initial distribution of the water temperature and the sugar cube size, and the way it drops and where it drops and some other details. Such details seem impossible to know. We may wonder if it worths the effort? Will these details provide higher accuracy? As an extreme example, let us imagine that we know the locations of every single atom in the body of a monkey, can we predict the motion and path where a monkey will go at a given moment at a given location? Probably not.

Mathematical modeling is a complicated process, which is essentially an iteration loop between theory and practice. The famous British statistician George Box once said: 'All models are wrong, but some are useful'. Providing more details in a more sophisticated way does not guarantee a higher accuracy. Box's original observations provides us a guiding philosophy for mathematical modeling:

> *'Since all models are wrong the scientist cannot obtain a "correct" one by excessive elaboration. On the contrary following William of Occam he should seek an economical description of natural phenomena. Just as the ability to devise simple but evocative models is the signature of the great scientist so overelaboration and overparameterization is often the mark of mediocrity.' (Box GEP, Science and statistics, J. American Statistical Association, 791–799, 1976.)*
>
> George Box

The philosophy of Occam, often referred to as the Occam's razor, states that "Entities are not to be multiplied without necessity". That is, the simplest answer is usually the correct answer.

Isaac Newton's rules of reasoning in philosophy outlined in his *Principia* also clearly stated: "We are to admit no more causes of natural things than such as are both true and sufficient to explain their appearances." Later, Albert Einstein also famously said: "Make everything as simple as possible, but not simpler."

Such advice suggests that proper mathematical modeling requires to identify the right level of approximations and the right kind of accuracy we wish to achieve, which consequently requires the understanding of the mechanisms of the physical, chemical and/or biological processes at different levels of details, and the abstraction of such processes into the right level of mathematical equations. Therefore, mathematical modeling is an art, requiring practice, practice and more practice.

28.4 PARAMETER ESTIMATION

Another important topic in mathematical modeling is the ability to estimate the orders (not the exact numbers) of certain quantities. If we know the order of a quantity and its range of variations, we can choose the right scales to write the mathematical model in the non-dimensional form so that the right mathematical methods can be used to tackle the problem. It also helps us to choose more suitable numerical methods to find the solution over the correct scales. The estimations will often give us greater insight into the physical process, resulting in more appropriate mathematical models.

Let us look at an example to estimate the rate of heat loss at the Earth's surface, and the temperature gradients in the Earth's crust and the atmosphere. We can also show the importance of the sunlight in the heat energy balance of the atmosphere.

Example 28.5

We know that the average temperature at the Earth's surface is about $T_s = 300$ K, and the thickness of the continental crust varies from $d = 35$ km to 70 km. The temperature at the upper lithosphere is estimated about $T_0 = 900 \sim 1400$ K (very crude estimation). Thus the estimated temperature gradient is about

$$\frac{dT}{dz} = \frac{T_0 - T_s}{d} \approx 9 \sim 31 \text{ K/km}.$$

The observed values of the temperature gradient around the globe are about 10 to 30 K/km. The estimated thermal conductivity k of rocks is about $1.5 \sim 4.5$ W/m K (ignoring the temperature dependence), we can use $k = 3$ W/m K as the estimate for the thermal conductivity of the crust. Thus, the rate of heat loss obeys Fourier's law of conduction

$$q = -k\nabla T = -k\frac{dT}{dz} \approx 0.027 \sim 0.093 \text{ W/m}^2,$$

which is close to the measured average of about 0.07 W/m^2. For oceanic crust with a thickness of $6 \sim 7$ km, the temperature gradient (and thus rate of heat loss) could be five times higher at the bottom of the ocean, and this heat loss provides a major part of the energy to the ocean so as to keep it from being frozen.

If this heat loss goes through the atmosphere, then the energy conservation requires that

$$k\frac{dT}{dz}\Big|_{\text{crust}} + k_a\frac{dT}{dh}\Big|_{\text{air}} = 0,$$

where h is the height above the Earth's surface and $k_a = 0.020 \sim 0.025$ W/m K is the thermal conductivity of the air (again, ignoring the variations with the temperature). Therefore, the temperature gradient in the air is

$$\frac{dT}{dh} = -\frac{k}{k_a}\frac{dT}{dz} \approx -3.6 \sim -4.5 \text{ K/km},$$

if we use $dT/dz = 30$ W/km. The negative sign means the temperature decreases as the height increases. The true temperature gradient in dry air is about 10 K/km in dry air, and $6 \sim 7$ K/km in moist air. As the thermal conductivity increases with the humidity, so the gradient decreases with humidity.

Alternatively, we know the effective thickness of the atmosphere is about 50 km (if we define it as the thickness of layers containing 99.9% of the air mass). We know there is no definite boundary between the atmosphere and outer space, and the atmosphere can extend up to several hundreds of kilometers. In addition, we can also assume that the temperature in space vacuum is about 4 Kelvin (K) and the temperature at the Earth's surface is 300 K, then the temperature gradient in the air is

$$\frac{dT}{dh} \approx \frac{4 - 300}{50} \approx -6 \text{ K/km},$$

which is quite close to the true gradient. The higher rate of heat loss (due to higher temperature gradient) means that the heat supplied from the crust is not enough to balance this higher rate. That is where the energy of sunlight comes into play. We can see that estimates of this kind will provide a good insight in the whole process.

Sometimes, it can be extremely challenging to get a good estimate because the processes involved are too complicated to do any calculations without using sophisticated computer simulations. In this case, an estimation can get the order of the quantities right, but they can be quite different by a factor. As an example, let us estimate the quantity of jet fuel consumption during a takeoff of a typical aircraft such as a jumbo Boeing 747 passenger jet. Obviously, the fuel consumption depends on many factors such as aircraft type, loads, takeoff conditions, weather, altitude and cruise speed and others. Even the same aircraft traveling in the same route can use a different amount of fuel due to the variation of loads, number of passengers and wind direction. If you ask the Boeing company, they may produce a long document with various assumptions without giving you an accurate answer. If you ask airlines or their pilots, they can only give you the average total fuel consumption of a given flight route, though pilots may give the average reading changes of their fuel gauges during takeoffs and their answers may be surprisingly very different.

If we read around and search the Internet, we may have the following data: A typical Boeing 747 (say, from London to New York) can consume about 230,000 liters of jet fuel (and much more for flying from London to Hong Kong). A typical such aircraft can weight 500 tons or $m = 5 \times 10^5$ kg and the fuel consumption during the takeoff (from the sea level to the cruise altitude) can typically consume 2000 liters to 4000 liters of jet fuel. Now the questions are: do all these figures make any sense? How can we provide an estimate without taking too many factors into consideration?

Example 28.6

Let us use the energy conservation to do a very crude estimation. We know that the cruise altitude is about $h = 10$ km (or $h = 10^4$ m) and the cruise speed of a Boeing 747 is about 570 mph or 920 km/h (that is, $v = 255$ m/s). Let us assume the that the aircraft's mass is $m = 5 \times 10^5$ kg (or 500 tons) and the acceleration due to gravity is $g = 10$ m/s^2.

We know its kinetic energy is $\frac{1}{2}mv^2$ at its cruise speed and the potential energy mgh. So the total energy is

$$E = \frac{1}{2}mv^2 + mgh = \frac{1}{2} \times 5 \times 10^5 \times (255)^2 + 5 \times 10^5 \times 10^4 \times 10 \approx 6.6 \times 10^{10} \text{ J}.$$

On the other hand, we know that the energy density of jet fuel is about $e_j = 4.0 \times 10^7$ J/kg and the efficiency of the jet engine is about 98% to 100%. So the fuel consumption W needed during the takeoff to supply the same energy is

$$We_j = E,$$

or

$$W = \frac{E}{e_j} = \frac{6.6 \times 10^{10}}{4.0 \times 10^7} \approx 1600 \text{ kg} \approx 2000 \text{ liters},$$

where we have used 0.8 kg/liter as the approximate value of fuel density. This means that it requires about 2000 liters of jet fuel to get the aircraft to the standard cruise altitude from the ground. Obviously, this estimate is much lower than the real fuel consumption because we have not considered the air resistance and many other factors. The actual fuel needed is typically around 3000 liters to 5000 liters (and more) for a typical Boeing 747, depending on the actual load and other conditions.

Even with this simple estimation, it can provide a lower bound for the fuel consumption or at least some sense of the order of the fuel consumption. That is, the consumption is about thousands of liters, not of a few hundred liters for a jumbo jet.

Of course the choice of typical values is important in order to get a valid estimation. Such choice will depend on the physical process and the scales we are interested in. The right choice will be perfected by expertise and practice.

28.5 TYPES OF MATHEMATICAL MODELS

Mathematical models can take many different forms, and thus it is not an easy task to classify all the mathematical models. However, based on the mathematical forms they can take, we can loosely classify them into eight categories: algebraic models, tensor models, differential equation models, integral models, statistical models, fuzzy models, learned models and data-driven models. We will briefly explain each of these model categories in this section.

28.5.1 Algebraic Equations

Many physical laws can be expressed as algebraic relationships. For example, Newton's second law $F = ma$ is a good example, though strictly speaking we should express it in the vector form:

$$\mathbf{F} = m\mathbf{a}, \tag{28.9}$$

because both force $\mathbf{F}$ and acceleration $\mathbf{a}$ are vectors. Another example is Ohm's law.

Example 28.7

Ohm's law relates the current I through a conductor with a resistance R and the voltage V applied across the conductor. That is

$$I = \frac{V}{R},$$

or

$$V = IR.$$

Here, the main assumption is that the resistance does not vary with V or I. Obviously, in case of R independent of I, we can still write this as an equivalent relationship as an approximation. In fact, R usually depends on temperature T and the flow of current will generate heat (and thus vary the temperature).

Almost all the physical and chemical laws we learned in school are expressed as algebraic equations.

28.5.2 Tensor Relationships

Sometimes, the quantities of interest are tensors such as stress and strain, we have to use tensor relationships to express them. For example, the generalized Hooke's law we discussed earlier in this book can be written as

$$\varepsilon_{ij} = \frac{1+\nu}{E}\sigma_{ij} - \frac{\nu}{E}\sigma_{kk}\delta_{ij}, \tag{28.10}$$

which is a tensor equation.

Example 28.8

The Ohm's law for a simple conductor is very simple; however, it becomes a tensor relations for a complex, anisotropic medium under an electric field **E**:

$$\mathbf{J} = \sigma \mathbf{E},$$

where **J** is the current density that varies with its location. Here, $\sigma = [\sigma_{ij}]$ is the conductivity tensor of anisotropic conducting material such as crystals and nanomaterials. Using the summation convention for tensors, we have

$$J_i = \sigma_{ij} E_j.$$

When the properties of a medium change with certain quantity such as directions, a tensor relationship can occur.

28.5.3 Differential Equations: ODE and PDEs

Though some laws in physics can be expressed in simpler forms, it is sometimes useful to write in a more generalized form for theoretical analysis and this often involves derivatives or integrals.

As we have seen in the chapter about partial differential equations, most mathematical models in physics are expressed in PDEs and the conservation laws such as the conservation of mass and energy are expressed in integral forms.

Example 28.9

Newton's second law $F = ma$ is often written in a differential form

$$\mathbf{F} = \frac{d\mathbf{p}}{dt} = \frac{d(m\mathbf{v})}{dt},$$

where $\mathbf{p} = m\mathbf{v}$ is the linear momentum and $\mathbf{v}$ is the velocity of the object. This form is specially useful in mechanics when discussing variable-mass dynamics such as the trajectory of a rocket where its mass $m(t)$ is not a constant but is varying with time t. In this case, we have

$$\mathbf{F} = \frac{d(m\mathbf{v})}{dt} = \frac{dm}{dt}\mathbf{v} + m\frac{d\mathbf{v}}{dt} = \dot{m}(t)\mathbf{v} + m(t)\dot{\mathbf{v}}.$$

As we have seen earlier in this chapter, mathematical modeling often produces some mathematical equations, often partial differential equations. The next step is to identify the detailed constraints such as the proper boundary conditions and initial conditions so that we can obtain a unique set of solutions.

For the sugar diffusion problem discussed earlier, we cannot obtain the exact solution in the actual domain inside the water-filled glass, because we need to know where the sugar cube or grains were initially added. The geometry of the glass also needs to be specified. In fact, this problem needs numerical methods such as finite element methods or finite volume methods. The only possible solution is the long-time behavior: when $t \to \infty$, we know that the concentration should be uniform $c(z, t \to \infty) \to c_\infty$ (= mass of sugar added/volume of water).

You may say that we know this final state even without mathematical equations, so what is the use of the diffusion equation? The main advantage is that you can calculate the concentration at any time using the mathematical equation with appropriate boundary and initial conditions, either by numerical methods in most cases or by mathematical analysis in some very simple cases. Once you know the initial and boundary conditions, the whole system history will be determined to a certain degree.

The beauty of mathematical models is that many seemingly diverse problems can be reduced to the same mathematical equation. For example, we know that the diffusion problem is governed by the diffusion equation

$$\frac{\partial c}{\partial t} = D\nabla^2 c. \tag{28.11}$$

The heat conduction is governed by the heat conduction equation

$$\frac{\partial T}{\partial t} = \kappa \nabla^2 T, \qquad \kappa = \frac{K}{\rho c_p}, \tag{28.12}$$

where T is temperature and κ is the thermal diffusivity. K is thermal conductivity, ρ is the density and c_p is the specific heat capacity. Mathematically speaking, whether it is the concentration of a pollutant or temperature, it is the same dependent variable u in terms of the same parabolic partial differential equation.

As we have seen earlier in this book, to obtain a solution for a partial differential equation is not an easy task. However, it may be possible to get a good estimate without solving the PDE itself, depending on the purpose of estimation.

In the one-dimensional (1D) case, the above equation becomes

$$\frac{\partial u}{\partial t} = \kappa \frac{\partial^2 u}{\partial x^2}. \tag{28.13}$$

For a typical length L and a typical timescale t_*, we can define a new time $t = t_* \tau =$ and a new (scaled) spatial coordinate $x = L\xi$. Thus, we have

$$dt = t_* d\tau, \quad dx = L d\xi,$$

and

$$\frac{\partial}{\partial t} = \frac{1}{t_*}\frac{\partial}{\partial \tau}, \quad \frac{\partial}{\partial x} = \frac{1}{L}\frac{\partial}{\partial \xi}, \quad \frac{\partial^2}{\partial x^2} = \frac{1}{L^2}\frac{\partial^2}{\partial \xi^2}. \tag{28.14}$$

Now the parabolic PDE (28.13) becomes the following non-dimensional form:

$$\frac{1}{t_*}\frac{\partial u}{\partial \tau} = \frac{\kappa}{L^2}\frac{\partial^2 u}{\partial \xi^2}, \tag{28.15}$$

or

$$\frac{\partial u}{\partial \tau} = \lambda \frac{\partial^2 u}{\partial \xi^2}, \quad \lambda = \frac{\kappa t_*}{L^2}, \tag{28.16}$$

where λ is a non-dimensional parameter. In a special cases $\lambda = 1$, it becomes a parameterless diffusion equation that can describe many phenomena across scales.

Let us look at an example.

Example 28.10

To estimate the cooling time for a hot iron ball or a piece of hot rock, we can use the characteristic time t_* defined by

$$\lambda = 1 = \frac{\kappa t_*}{L^2},$$

or

$$t_* = \frac{L^2}{\kappa}.$$

We know that the thermal conductivity is about $K \approx 35$ W/m K for iron (though the actual values vary greatly with the type of iron and purity of iron), its density is $\rho \approx 7850$ Kg/m^3 and its specific heat capacity $c_p \approx 450$ J/kg K. Thus, the thermal diffusivity of iron is

$$\kappa = \frac{K}{\rho c_p} \approx \frac{35}{7850 \times 450} \approx 9.9 \times 10^{-6} \text{ m}^2/\text{s}. \tag{28.17}$$

For a small iron ball of $d = 0.1$ m, we have

$$t_* = \frac{L^2}{\kappa} = \frac{0.1^2}{9.9 \times 10^{-6}} \approx 1000 \text{ seconds},$$

which is about 16 minutes.

For a larger iron ball of $d = 1$ m, the time scale of cooling is

$$t_* = \frac{L^2}{\kappa} \approx \frac{1}{9.9 \times 10^{-6}} \approx 1.0 \times 10^5 \text{ seconds} \approx 1 \text{ day}. \tag{28.18}$$

For a larger hot body $L = 1000$ m, then that time scale is $t_* = 10^6$ days or 3200 years. This estimate of the cooling time scale is based on the assumption that no more heat generated inside is supplied.

With such simple assumptions, we can do many sensible estimations. It leaves as an exercise to estimate the time for a cup of hot coffee to cool down to the room temperature.

28.5.4 Functional and Integral Equations

Though most mathematical models are written as partial different equations, however, sometimes it might be convenient to write them in terms of integral equations, and these integral forms can be discretized to obtain various numerical methods.

Example 28.11

In the discussion about the conservation of mass earlier in Eq. (28.2), we write it in the integral form

$$\frac{\partial}{\partial t}\Big[\iiint_\Omega c dV\Big] + \iint_\Gamma \mathbf{J} \cdot d\mathbf{S} = 0,$$

which is equivalent to the partial differential equation as follows:

$$\frac{\partial c}{\partial t} + \nabla \cdot \mathbf{J} = 0.$$

One of the advantages of writing in integral forms is that it has a clearer meaning in the representative volume, and such integral form is often used in numerical simulations such as the finite volume methods.

The advantage of writing it in terms of a differential equation is that we can use all the available solution techniques to solve the problem of interest. Obviously, these two forms are closely related and often can be converted from one to another.

However, sometimes, the physical laws or properties can be intrinsically represented by integral equations. Examples include inverse problems, non-local problems and visco-elasticity or creep problems.

Example 28.12

Viscoelasticity is an important phenomenon in engineering. For example, the behavior of chocolate flow, glasses, rubbers, plastic bags and a bicycle pump can be described by viscoelastic behavior. Loosely speaking, linear viscoelastic constitutive relationships or laws can be expressed as

$$\sigma(t) = \sigma_0 + \int_0^t E(t-\eta)\dot{\varepsilon}(\eta) d\eta,$$

where $\sigma(t)$ is the time-dependent stress and $\dot{\varepsilon}$ is the strain rate that can also depend on time t. E is the so-called relaxation modulus, while σ_0 is a constant. This stress-strain relationship, unlike the generalized Hooke's law, is essentially a Volterra integral equation.

28.5.5 Statistical Models

Both differential equations and integral equations are the mathematical models for continuum systems. Other systems are discrete and different mathematical models are needed, though they could reduce to certain forms of differential equations if some averaging is carried out.

On the other hand, many systems have intrinsic randomness, thus the description and proper modeling require statistical models as we have seen in the chapters on probability and statistics. The numbers of telephone calls received in a call center can be described by a Poisson distribution. Statistical models are a very important class of models that are widely used in economics, finance, risk management, data mining and machine intelligence.

Let use discuss two examples: one about the frequency of earthquakes and the other about the Six Sigma design methodology for manufacturing.

Example 28.13

The Gutenberg-Richter law is an empirical, statistical relationship between the magnitude (M) and the total number (N) of earthquakes for a given period in a given region. It can be written as

$$\log_{10} N = a - bM,$$

or

$$N = 10^{a-bM},$$

where a and b are constants. The b-value is of more scientific importance. The typical values are: $a = 0.5$ to 2 and $b \approx 1$. In general, b is relatively stable and does not vary much from region to region.

This model suggests that there will be a 10-fold decrease in seismic activity for a unit increase in magnitude. That is to say, there are about 10 times more magnitude-5 earthquakes than magnitude-6 earthquakes.

Sometimes, a mathematical model can appear as a procedure or methodology, even though it is based on solid mathematics. Let us use Six Sigma as an example.

Six Sigma (6σ) is a quality control methodology, pioneered at Motorola in the 1980s, for eliminating defects in products. Assuming that the normal distribution applies to such scenarios, we have

$$p(x, \mu, \sigma) = \frac{1}{\sigma\sqrt{2\pi}} e^{-\frac{(x-\mu)^2}{2\sigma^2}},$$

where μ is the mean and σ is the standard deviation. Clearly, this function is symmetric on both sides of $x = \mu$. Its cumulative distribution function is

$$\Phi(x) = \frac{1}{2}\Big\{1 + \text{erf}\Big[\frac{(x-\mu)}{\sigma\sqrt{2}}\Big]\Big\},$$

which is the area under the normal distribution curve from $-\infty$ to x. Obviously, when $x \to \infty$, we have $\Phi(\infty) = 1$ because $\text{erf}(\infty) = 1$.

The exact area under the curve from $\mu - k\sigma$ to $\mu + k\sigma$ (where $k > 0$) can be calculated by

$$A_k = \Phi(\mu + k\sigma) - \Phi(\mu - k\sigma) = \frac{1}{2}\Big[\text{erf}(k/\sqrt{2}) - \text{erf}(-k/\sqrt{2})\Big].$$

Since the error function

$$\text{erf}(x) = \frac{2}{\sqrt{\pi}} \int_0^x e^{-\tau^2} d\tau,$$

is an odd function $\text{erf}(-x) = -\text{erf}(x)$, we have

$$A_k = \frac{1}{2}[\text{erf}(k/\sqrt{2}) - \text{erf}(-k/\sqrt{2})] = \text{erf}(\frac{k}{\sqrt{2}}).$$

Thus, for $k = 1$, the area within one standard deviation between $\mu - \sigma$ to $\mu + \sigma$ is

$$A_1 = \text{erf}(1/\sqrt{2}) = 0.68269,$$

which is about 68.3%. In addition, the area within 2σ (between $\mu - 2\sigma$ to $\mu + 2\sigma$) is

$$A_2 = \text{erf}(2/\sqrt{2}) = 0.9545,$$

or about 95%. The area within 3σ is

$$A_3 = \text{erf}(3/\sqrt{2}) = 0.9973,$$

which about 99.7%. Similarly, we have

$$A_4 = 0.99993665, \quad A_5 = 0.9999994267, \quad A_6 = 0.9999999980268.$$

Suppose the product is an electric motor with a no-load speed or the spin speed of $\omega = 3000.0 \pm 0.2$ round per minutes (rpm). If we try to use 3σ, there would 99.7% products without defects, and the defect rate is about 0.27% is still too high for a good product because there about 2700 defective products out of 1 million. If we use 6σ, we have $A_6 = 0.999999998$, which means that there are about 2 defective parts in a billion. This may be a higher quality assurance, but the technologies and costs to ensure this may be extremely expensive.

Example 28.14

The beauty of the Six Sigma methodology is to use $\pm 4.5\sigma$ allowing the mean to shift $\pm 1.5\sigma$ (thus 6σ in essence). Since $k = 4.5$, we have

$$A_{4.5} = 0.999993204653751,$$

which gives 6.8 defective parts per million if the mean is not allowed to shift. Since the mean shift $\pm 1.5\sigma$, the actual defective parts is half of the above value, that is, 3.4 parts per million, which is the essence of the Six Sigma Standards in manufacturing.

Coming back to the motor specification of $\omega = 3000.0 \pm 0.2$ or $3000.0 - 0.2 = 2999.8$ to $3000.0 + 0.2 = 3000.2$, to ensure this standard specification with only about 3.4 defective motors per million, we have to ensure that manufacturing standard deviation σ to be even smaller. Since it requires that $6\sigma = 0.2$, we have

$$\sigma = \frac{0.2}{6} \approx 0.033,$$

which is the standard deviation to be achieved in the actual manufacturing process to ensure the desired quality target.

28.5.6 Fuzzy Models

Sometimes, a model may need to work on situations that are not clear cut as simply yes or no (or 1 or 0) data. In this case, models based on fuzzy logic can be advantageous. The main concepts of fuzzy logic and fuzzy set theory were pioneered by L.A. Zadeh, and they are essentially a form of multi-valued logic with a membership function to measure the closeness.

In the standard binary logic, only there are true (1) and false (0), while fuzzy logic allows any real value from 0 to 1. In fact, fuzzy mathematics is itself a broad subject, and interested readers can refer to more advanced literature.

28.5.7 Learned Models

In many applications such as data mining and machine intelligence, the relationship cannot be directly expressed as a deterministic, explicit function. Instead, the model or relationship is an implicit, dynamically varying model that learned from the data or available information. As the data or information changes, the learned relationship needs to be updated by training or learning to reflect the new information. Such models can be called learned models.

One classical way to construct such model is to use regression analysis as we discussed in the context of statistics. However, there are other, even better, ways to build learned models. Two good examples are artificial neural networks (ANN) and support vector machines. A typical ANN uses a number of neurons arranged into different layers. Each neuron can convert an input signal into a binary output (0 or 1), activated at a threshold, and each neuron in a layer can be connected with other neurons in a layer in front of its layer. Typically, the network contains an input layer, an output layer and one or more hidden layer in the middle. ANNs, especially the so-called convolution deep learning neural networks, have become a powerful tool in machine learning and artificial intelligence.

Support vector machines (SVM) are a class of techniques for classification and regression analysis, they often use the so-called kernel tricks to map data in one space to a higher-dimensional space so that their structures can be identified and different groups or classes can be separated relatively easily by constructing some hyperplanes. Again, SVMs have also become a powerful tool for analyzing data and extracting features in many applications, including engineering, finance and machine intelligence.

28.5.8 Data-Driven Models

As the data volumes have increased dramatically in the two last decades, various techniques are needed to deal with such big data, which have formed a new subject, called the big data science. As these data sets are usually unstructured with high complexity, there is no simple mathematical model that can be constructed to explain the whole data. Even if it may be possible to construct some mathematical models, such models need to be modified as more data flow in. In addition, uncertainties and noise often present in the data, which makes the already challenging data mining tasks even more difficult.

In such applications, models become data-driven, indeterministic and dynamic, and thus it may impossible or meaningless to try to seek a deterministic and simple mathematical model. In this case, we have to live with models that truly data-driven and noisy. This is a relatively new area, and interested readers can refer to more advanced literature on data science.

In this chapter, we have summarized the basic procedures and steps of mathematical modeling. We have also provided a few examples to illustrate the types of models and different levels of approximation. Now let us end this chapter by providing a detailed worked example as a case study.

28.6 BROWNIAN MOTION AND DIFFUSION: A WORKED EXAMPLE

Diffusion is relevant to many phenomena in physics, chemistry and biology. Thus, its importance has attracted much attention in the last two centuries. In essence, diffusion can be modeled by Brown motion as a statistical process, and thus the mathematical models are probabilistic.

Einstein provided in 1905 the first theory of Brownian motion of spherical particles suspended in a liquid, which can be written, after some lengthy calculations, as

$$\overline{\Delta^2} = \frac{k_B}{T} \frac{t}{3\pi \mu a}, \tag{28.19}$$

where a is the radius of the spherical particle, and μ is the viscosity of the liquid. T is the absolute temperature, and t is time. $k_B = R/N_A$ is Boltzmann's constant where R is the universal gas constant and N_A is Avogadro's constant. $\overline{\Delta^2}$ is the mean square of the displacement.

Langevin in 1908 presented a very instructive but much simpler version of the theory. If u is the displacement, and $\xi = \frac{du}{dt}$ is the speed at a given instant, then the kinetic energy of the motion should be equal to the average kinetic energy $\frac{1}{2}k_B T$. (Here Langevin considered only one direction. For three directions, the kinetic energy becomes $\frac{3}{2}k_B T$, thus there is a factor 3. This means that equation (28.28) should be $\overline{u^2} = 6Dt$.) That is

$$\frac{1}{2} m \overline{\xi^2} = \frac{1}{2} k_B T,$$

where m is the mass of the particle. The spherical particle moving at a velocity of ξ will experience a viscous resistance equal to

$$-6\pi\mu a\xi = -6\pi\mu a\frac{du}{dt}$$

according to Stokes' law. If $F(t)$ is the complementary force acting on the particle at the instant so as to maintain the agitation of the particle, we have, according to Newton's second law

$$m\frac{d^2u}{dt^2} = -6\pi\mu a\frac{du}{dt} + F. \tag{28.20}$$

Multiplying both sides by u, we have

$$mu\frac{d^2u}{dt^2} = -6\pi\mu au\frac{d^2u}{dt^2} + Fu. \tag{28.21}$$

Since $(u^2)''/2 = uu'' + u'^2$ or

$$u\frac{d^2u}{dt^2} = \frac{1}{2}\frac{d^2(u^2)}{dt^2} - \left(\frac{du}{dt}\right)^2 = \frac{1}{2}\frac{d}{dt}\left[\frac{d(u^2)}{dt}\right] - \xi^2,$$

we have

$$\frac{m}{2}\frac{d}{dt}\left[\frac{d(u^2)}{dt}\right] - m\xi^2 = -3\pi\mu a\frac{d(u^2)}{dt} + Fu, \tag{28.22}$$

where we have used $uu' = (u^2)'/2$. If we consider a large number of identical particles and take the average, we have

$$\frac{m}{2}\frac{d}{dt}\overline{\left[\frac{d(u^2)}{dt}\right]} - m\overline{\xi^2} = -3\pi\mu a\overline{\frac{d(u^2)}{dt}} + \overline{Fu}, \tag{28.23}$$

where $\overline{()}$ means the average. Since the average of $\overline{Fu} = 0$ due to the fact that the force is random, taking any signs and directions. Let $Z = \overline{d(u^2)/dt}$, we have

$$\frac{m}{2}\frac{dZ}{dt} + 3\pi\mu aZ = k_BT, \tag{28.24}$$

where we have used $m\overline{\xi^2} = k_BT$ discussed earlier. This equation is now known as Langevin's equation. It is a linear first-order differential equation. Its general solution can easily be found using the method discussed in the chapter on differential equations

$$Z_c = Ae^{-t/\tau} + k_BT\frac{1}{3\pi\mu a}, \qquad \tau = \frac{m}{6\pi\mu a}, \tag{28.25}$$

where A is the constant to be determined. For the Brownian motion to be observable, t must be reasonably larger than the characteristic time $\tau \approx 10^{-8}$ seconds, which is often the case. So the first term will decrease exponentially and becomes negligible when $t > \tau$ in the standard timescale of our interest. Now the solution becomes

$$Z = \overline{\frac{d(u^2)}{dt}} = k_BT\frac{1}{3\pi\mu a}. \tag{28.26}$$

Integrating it with respect to t, we have

$$\overline{u^2} = k_BT\frac{t}{3\pi\mu a}, \tag{28.27}$$

which is Einstein's formula for Brownian motion. Furthermore, the Brownian motion is essentially a diffusion process with

$$\overline{u^2} = 2Dt, \quad D = \frac{k_B T}{6\pi \mu a}, \tag{28.28}$$

where D is the equivalent diffusion coefficient. This suggests that large particles diffuse more slowly than smaller particles in the same medium. Mathematically, it is very similar to the heat conduction process discussed in earlier chapters.

We know that the diffusion coefficient of sugar in water at room temperature is $D \approx 0.5 \times 10^{-9}$ m^2/s. In 1905, Einstein was the first to estimate the size of sugar molecules using experimental data in his doctoral dissertation on Brownian motion.

Let us now estimate the size of the sugar molecules. Since $k_B = 1.38 \times 10^{-23}$ J/K, $T = 300$ K, and $\mu = 10^{-3}$ Pa s, we have

$$a = \frac{k_B T}{6\pi \mu D} = \frac{1.38 \times 10^{-23} \times 300}{6\pi \times 10^{-3} \times 0.5 \times 10^{-9}} \approx 4.4 \times 10^{-10} \text{ m}, \tag{28.29}$$

which means that the diameter is about 8.8×10^{-10} m $= 0.88$ nm. In fact, Einstein estimated for the first time the diameter of a sugar molecule was 9.9×10^{-10} m even though the other quantities were not so accurately measured at the time.

From $\overline{u^2} = 2Dt$, we can either estimate the diffusion distance for a given time or estimate the timescale for a given length. For a sugar cube in a cup of water to dissolve completely to form a solution of uniform concentration (without stirring), the time taken will be $t = d^2/2D$, where $d = \sqrt{\overline{u^2}}$ is the size of the cup. Using $d = 5$ cm $= 0.05$ m and $D = 0.5 \times 10^{-9}$ m^2/s, we have $t = d^2/2D = 0.05^2/(2 \times 0.5 \times 10^{-9}) \approx 2.5 \times 10^6$ seconds, which is about one month. This is too slow; that is why we always try to stir a cup of tea or coffee.

EXERCISES

28.1. Discuss how the Lorenz equations for chaos were obtained.

28.2. Build a simple mathematical model to simulate the bungee jump process and discuss any assumptions made.

28.3. In mathematical modeling of epidemic spread of flu, the so-called SIR model is often used. Discuss the assumptions used to derive the SIR model.

28.4. Try to write a simulator to model the pattern-formation equation

$$\frac{\partial u}{\partial t} = D(\frac{\partial^2 u}{\partial x^2} + \frac{\partial^2 u}{\partial y^2}) + \gamma u(1 - u).$$

Discuss the possible boundary and initial conditions.

28.5. If your aim is to automatically identify any hand-written digits and letters such as post codes, discuss the models and methods for completing the task.

Appendix A

Mathematical Formulas

This is a summary of the main mathematical formulas that have appeared in various sections of this book.

A.1 DIFFERENTIATION AND INTEGRATION

Differentiation Rules

$$(uv)' = u'v + uv'$$
$$(\frac{u}{v})' = \frac{u'v - uv'}{v^2}$$
$$\Big\{f[g(x)]\Big\}' = f'[g(x)] \cdot g'(x)$$

Integration by parts

$$\int_a^b u dv = [uv]\Big|_a^b - \int_a^b v du$$

Power Series

$$e^x = 1 + x + \frac{x^2}{2!} + ... + \frac{x^n}{n!} + ..., \quad (x \in \mathbb{R})$$
$$\sin x = x - \frac{x^3}{3!} + \frac{x^5}{5!} - ..., \qquad \cos x = 1 - \frac{x^2}{2!} + \frac{x^4}{4!} - ...$$
$$\sinh x = x + \frac{x^3}{3!} + \frac{x^5}{5!} + ..., \qquad \cosh x = 1 + \frac{x^2}{2!} + \frac{x^4}{4!} + ...$$

A.2 COMPLEX NUMBERS

Euler's formula

$$e^{i\theta} = \cos\theta + i\sin\theta, \quad e^{i\pi} + 1 = 0$$
$$z = x + iy = re^{i\theta} = r(\cos\theta + i\sin\theta)$$

De Moivre's formula

$$[r(\cos\theta + i\sin\theta)]^n = r^n(\cos n\theta + i\sin n\theta)$$

Cauchy's Integral Theorem

$$\oint_\Gamma \frac{f(z)}{(z - z_0)^{n+1}} dz = \frac{2\pi i f^{(n)}(z_0)}{n!}$$

A.3 VECTORS AND MATRICES

Dot Product

$$\mathbf{a} \cdot \mathbf{b} = |\mathbf{a}||\mathbf{b}|\cos\theta = a_i b_j \delta_{ij} = a_1 b_1 + a_2 b_2 + a_3 b_3$$

Cross Product

$$\mathbf{a} \times \mathbf{b} = \mathbf{n}|\mathbf{a}||\mathbf{b}| \sin\theta = \epsilon_{ijk} a_j b_k = \begin{vmatrix} \mathbf{i} & \mathbf{j} & \mathbf{k} \\ a_1 & a_2 & a_3 \\ b_1 & b_2 & b_3 \end{vmatrix}$$

Vector Triple

$$[\mathbf{a}, \mathbf{b}, \mathbf{c}] \equiv \mathbf{a} \cdot (\mathbf{b} \times \mathbf{c}) = \begin{vmatrix} a_1 & a_2 & a_3 \\ b_1 & b_2 & b_3 \\ c_1 & c_2 & c_3 \end{vmatrix}$$

$$\mathbf{a} \cdot (\mathbf{b} \times \mathbf{c}) = \mathbf{b} \cdot (\mathbf{c} \times \mathbf{a}) = \mathbf{c} \cdot (\mathbf{a} \times \mathbf{b}) = -\mathbf{a} \cdot (\mathbf{c} \times \mathbf{b})$$

$$\mathbf{a} \times (\mathbf{b} \times \mathbf{c}) = (\mathbf{a} \cdot \mathbf{c})\mathbf{b} - (\mathbf{a} \cdot \mathbf{b})\mathbf{c}$$

Divergence Theorem of Gauss

$$\iiint_V \nabla \cdot \mathbf{u} dV = \iint_S \mathbf{u} \cdot d\mathbf{S}$$

Green's Theorem

$$\oint (udx + vdy) = \iint \left(\frac{\partial v}{\partial x} - \frac{\partial u}{\partial y}\right) dxdy$$

Identities

$$\nabla \cdot \nabla \times \mathbf{u} = 0, \qquad \nabla \times \nabla\phi = 0$$

$$\nabla \times (\phi\mathbf{u}) = \phi\nabla \times \mathbf{u} + (\nabla\phi) \times \mathbf{u}$$

$$\nabla \cdot (\phi\mathbf{u}) = \phi\nabla \cdot \mathbf{u} + (\nabla\phi) \cdot \mathbf{u}$$

Inverse, Trace and Determinants

$$\mathrm{tr}(\mathbf{A}) = \sum_i \mathbf{A}_{ii} = \sum_i \lambda_i, \qquad \det(\mathbf{A}) = \Pi_i \lambda_i$$

$$\mathrm{tr}(\mathbf{AB}) = \mathrm{tr}(\mathbf{BA}), \qquad \mathrm{tr}(\mathbf{A} + \mathbf{B}) = \mathrm{tr}(\mathbf{A}) + \mathrm{tr}(\mathbf{B})$$

$$\det(\mathbf{A}^{-1}) = \frac{1}{\det(\mathbf{A})}, \qquad \det(\mathbf{AB}) = \det(\mathbf{A})\det(\mathbf{B})$$

A.4 FOURIER SERIES AND TRANSFORM

Fourier Series

$$f(t) = \frac{a_0}{2} + \sum_{n=1}^{\infty} (a_n \cos nt + b_n \sin nt), \quad (-\pi \le t \le \pi),$$

$$a_0 = \frac{1}{\pi} \int_{-\pi}^{\pi} f(t)dt, \quad a_n = \frac{1}{\pi} \int_{-\pi}^{\pi} f(t) \cos nt dt, \quad b_n = \frac{1}{\pi} \int_{-\pi}^{\pi} f(t) \sin nt dt$$

Triangular Wave

$$f(t) = |t| = \frac{1}{2} + \frac{4}{\pi^2} \sum_{n=1,3,5,\ldots}^{\infty} \frac{(-1)^n}{n^2} \cos(n\pi t), \quad t \in [-1, 1]$$

Fourier Transform

$$F(\omega) = \mathcal{F}[f(t)] = \int_{-\infty}^{\infty} f(t) e^{-i\omega t} dt,$$

$$f(t) = \mathcal{F}^{-1}[F(\omega)] = \frac{1}{2\pi} \int_{-\infty}^{\infty} F(\omega) e^{i\omega t} d\omega$$

Laplace Transform

$$F(s) = \mathcal{L}[f(t)] = \int_0^{\infty} f(t)e^{-st}dt$$

$$\mathcal{L}[t^n] = \int_0^{\infty} t^n e^{-st} dt dt = \frac{n!}{s^{n+1}}$$

$$\mathcal{L}[\sin(\omega t)] = \frac{\omega}{s^2+\omega^2}, \quad \mathcal{L}[\cos(\omega t)] = \frac{s}{s^2+\omega^2}$$

A.5 ASYMPTOTICS

Gaussian Distribution

$$p(x; \mu, \sigma) = \frac{1}{\sqrt{2\pi\sigma^2}} \exp[-\frac{(x-\mu)^2}{2\sigma^2}] \to \delta(x), \qquad \sigma \to \infty$$

Binomial Distribution

$$B(k; n, p) = \frac{n!}{(n-k)!k!} p^k (1-p)^{n-k}, \quad (k = 0, 1, 2, ..., n)$$

Poisson Distribution

$$f(x; \lambda) = \frac{e^{-\lambda}\lambda^x}{x!}, \qquad x \in \mathbb{N}$$

A.6 SPECIAL INTEGRALS

Gamma Function

$$\Gamma(x) = \int_0^{\infty} t^{x-1}e^{-t}dt = \int_0^{\infty} e^{-t+(x-1)\ln t}dt$$

$$\Gamma(n+1) = n!, \quad \Gamma(-\frac{1}{2}) = -2\sqrt{\pi}, \quad \Gamma(\frac{1}{2}) = \sqrt{\pi}$$

Stirling's Formula

$$n! \approx (\frac{n}{e})^n \sqrt{2\pi n}, \qquad n \gg 1$$

Gaussian Integral

$$\int_{-\infty}^{\infty} e^{-\alpha x^2} dx = \sqrt{\frac{\pi}{\alpha}}, \quad \alpha > 0$$

$$\int_{-\infty}^{\infty} e^{-x^2} dx = \sqrt{\pi}$$

Error Functions

$$\text{erf}(x) = \frac{2}{\sqrt{\pi}} \int_0^x e^{-\eta^2} d\eta \sim 1 - \frac{e^{-x^2}}{x\sqrt{\pi}}, \quad (x \to \infty)$$

$$\text{erfc}(x) = 1 - \text{erf}(x) = \frac{2}{\sqrt{\pi}} \int_x^{\infty} e^{-t^2} dt$$

$$\text{erf}(x) \sim \frac{2}{\sqrt{\pi}}[x - \frac{x^3}{3} + \frac{x^5}{10} - ...], \quad (x < \infty)$$

Appendix B

Mathematical Software Packages

This is a summary of some common mathematical software packages that are useful to carry out symbolic computation and numerical simulation. It is estimated that there are more than 100 different software packages that can do various degrees of computation, and thus it is not possible to introduce even a good fraction of them. Therefore, the selection here is based on two criteria: their relevance to engineering mathematics and personal experience. Though it may be a biased selection, it does include a good subset of the commonly used software tools.

Here, our list includes Matlab, Mathematica, Maple, R and others. Unlike other introduction to software, we will not focus on their history and syntax, we will instead focus on their mathematical capabilities in our brief introduction.

B.1 MATLAB

Matlab is one of the most widely used in numerical simulation and scientific computation. The name, Matlab, comes from the abbreviation of **mat**rix **lab**oratory, and it is a matrix-based, multi-platform, fourth-generation, script language (http://www.mathworks.com). It a commercial software package with sophisticated visualization tools and many add-on toolboxes such as symbolic computation, image processing and control toolboxes. It has becomes one of the main languages for technical computing.

B.1.1 Matlab

The basic syntax is an interactive and interpretive script with the prompt >>

```
>>2^10+3
ans=
    1027
```

which provides the basic algebraic operations. For a matrix, it can be input as

```
>>A=[1 2 3;4 5 6;7 8 9]
A=
   1 2 3
   4 5 6
   7 8 9
```

Thus, we have

```
>>sum(A)
ans=
    12 15 18
>>t=trace(A)
t=
     15
```

Plotting functions can be easily done by

```
>>x=-10:0.1:10;
>>plot(x,sin(x),x,cos(2*x),'-.')
```

The above commands will produce the graph as shown in Fig. B.1.

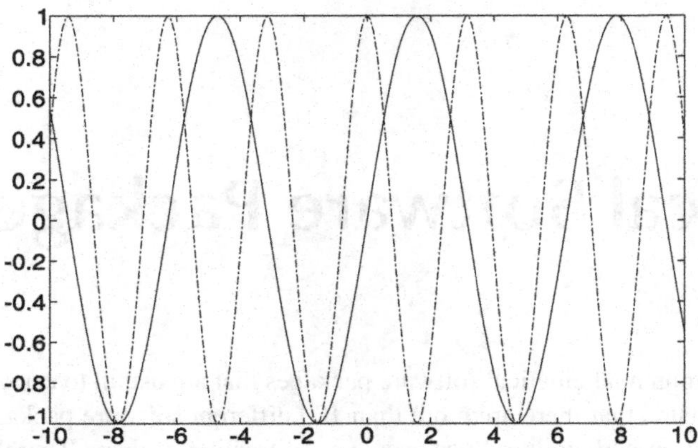

FIGURE B.1 2D plot generated by Matlab.

You can edit and group a set of the commands in a file and call the set of commands by its file name, as a `function`. Mathlab can also carry out symbolic computation such as

```
syms x
int(x*sin(x),x)
```

which gives the answer

```
sin(x)-x*cos(x)
```

while differentiation can be also be done easily, for example, by

```
diff(exp(x)*cos(x),x)
```

which leads to the following answer

```
exp(x)*cos(x)-exp(x)*sin(x)
```

Matlab has a few dozen add-on toolboxes with a diverse range of functionalities and capabilities, including symbolic mathematics (MuPAD), simulink, optimization, statistics, signal processing, control system, image processing, code generation, computational finance, neural networks and many others.

B.1.2 MuPAD

MuPAD is now a symbolic mathematics toolbox, part of the full Matlab distributions. You can invoke any toolbox by typing its name in the prompt. For example, to call the symbolic computing mathematics toolbox MuPAD, we use

```
>>mupad
```

You will get a new window where it prompts for inputs. For example, a simple line within the MuPAD

```
[plot(sin(x),cos(2*x),tan(x))
```

will give the graph as shown in Fig. B.2.

In addition, a simple line such as

```
[plot(sin(x)*sin(y),#3D)
```

can also produce 3D graphs as shown in Fig. B.3.

For calculus, we can do sophisticated manipulations using MuPAD. For example, once type in

```
[diff(sin(x^2)*exp(x),x)
```

we can get (by pressing the enter key)

$$\sin(x^2)e^x + 2x\cos(x^2)e^x$$

which provides a better display of mathematical formulas. Similarly, the input

```
[int(x^2*exp(x),x)
```

gives the answer

$$e^x(x^2 - 2x + 2)$$

In fact, MuPAD is a powerful symbolic mathematical tool that can do almost anything that are relevant to engineering mathematics.

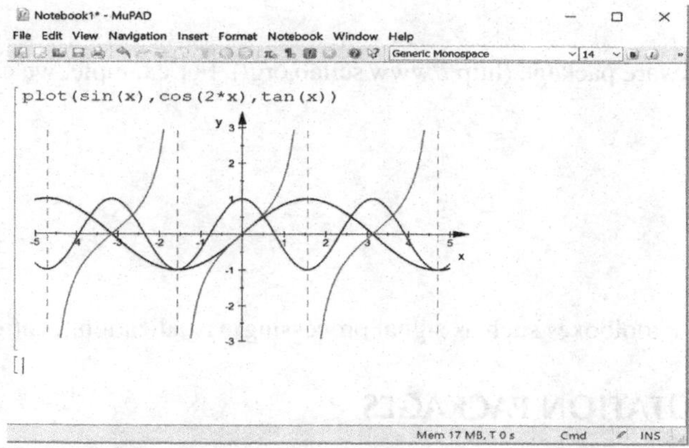

FIGURE B.2 2D plot generated by MuPAD.

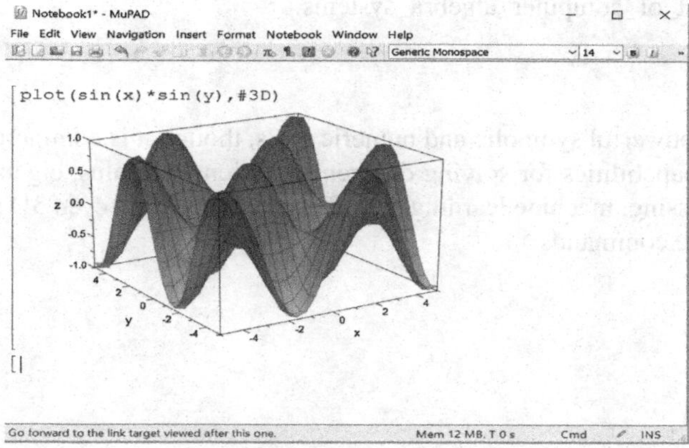

FIGURE B.3 3D plot generated by MuPAD.

B.2 SOFTWARE PACKAGES SIMILAR TO MATLAB

There are other packages that use similar syntax to that used in Matlab. Two good examples are Octave and Scilab.

B.2.1 Octave

Octave is a high-level language for numerical computations (https://www.gnu.org/software/octave/). It is an open source, multi-platform software, distributed under the terms of the GNU general public license (GPL) published by the Free Software Foundation.

It has many similar functions to those in Matlab. For example, in both Matlab and Octave, you can type at the prompt

```
>>A=magic(3)
```

you will get

```
A=
  8  1  6
  3  5  7
  4  9  2
```

In fact, you can type all the examples introduced in the previous section in Matlab (except the part on symbolic computation and MuPAD), and they should all work. The graph and visualization of Octave use the Gnuplot as an engine and Gnuplot is also multi-platform with many amazing visualization functions (http://www.gnuplot.info/).

Similarly, Octave also have many good add-on packages and the full list can be found at http://octave.sourceforge.net/packages.php, including control, finite element analysis, statistics, communications, symbolic mathematics and others.

B.2.2 Scilab

Scilab is also an open source software package (http://www.scilab.org/). For example, we can type in at the prompt -->

```
-->a=cos(%pi)
-->b=sin(%pi)
-->a^2+b^2
```

which gives

```
ans =
    1.
```

Scilab also has some extensive toolboxes such as signal processing, visualization, control systems and optimization.

B.3 SYMBOLIC COMPUTATION PACKAGES

Apart from the MuPAD in Matlab as a symbolic mathematics tool, there are a few dozen tools for symbolic computation and computer algebra. A relatively complete list of such computer algebra systems can be found at the wikipedia page/entry: https://en.wikipedia.org/wiki/List_of_computer_algebra_systems.

B.3.1 Mathematica

Mathematica is one of the most powerful symbolic and numeric tools, though it is commercial (https://www.wolfram.com/mathematica/). It has powerful capabilities for solving different equations, calculus, eigensystems, signal and image processing, text and language processing, machine learning, geometry modeling and even 3D printing.

For example, we can enter the commands

```
In[1]:=2+3
Out[1]= 5
In[2]:=Sin[Pi/4]
```

Out[2]= $\frac{1}{\sqrt{2}}$

```
In[3]:= f[x_]=x^2
```

Out[3]= x^2

```
In[4]:= D[f[x],x]
Out[4]= 2 x
In[5]:= Integrate[f[x],{x,0,1}]
```

Out[5]= $\frac{1}{3}$

Obviously, we can do many things with Mathematica. In fact, Mathematica has also become one of the main engines for the online Wolfram alpha (https://www.wolframalpha.com/), which is a computational knowledge engine and provides interactive, nature-language like queries and answers with an interface as shown in Fig. B.4.

B.3.2 Maple

Maple is another powerful symbolic and numeric computation tool, which is also commercial (http://www.maplesoft.com/products/Maple/students/). Here is an example.

```
>2^34;
17179869184
>cos(Pi/4);
```

$\frac{1}{2}\sqrt{2}$

```
>diff(x^2*cos(x*y),x);
```

$2x\cos(xy) - x^2 y \sin(xy)$

```
> f:=x-> x*cos(x);
>D(f)(x);
```

$\cos(x) - x\sin(x)$

Maple also has many add-on tools such global optimization, grid computing and Maple Net. In addition, MapleSim is a multi-domain modeling and simulation tool that uses Maple as a numerical and symbolic engine.

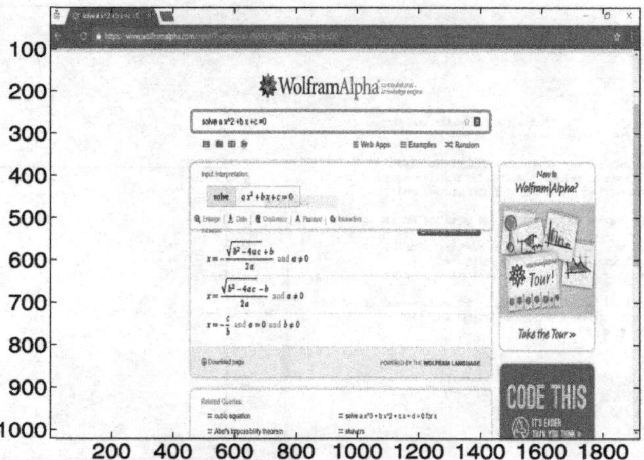

FIGURE B.4 The Wolframe Alpha computational knowledge engine.

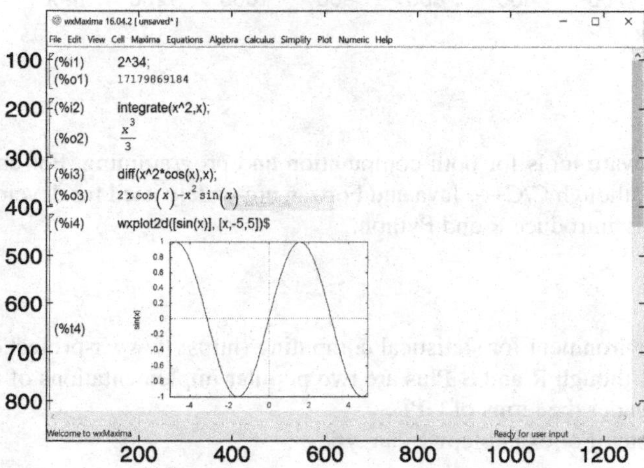

FIGURE B.5 Interface and outputs by Maxima.

B.3.3 Maxima

Maxima is a nice and free software tool that can be very useful to engineering mathematics. Its interface is just a notebook as shown in Fig. B.5.

Here is an example (also shown in Fig. B.5).

(%i1) 2^34;

(%o1) 17179869184

(%i2) integrate(x^2,x);

(%o2) $\frac{x^3}{3}$

(%i3) diff(x^2*cos(x),x);

(%o3) $2x\cos(x) - x^2\sin(x)$

There are other symbolic and numerical tools, and the choice may depend on the availability and personal preference. One has to balance the costs, ease of use, and the main purpose of computing.

Obviously, commercial software packages such as Matlab, Mathematica and Maple can have powerful functions and toolboxes with proper support and documentations. But they may be expensive and not all students can afford.

For the learning of engineering mathematics, it seems that Octave plus Maxima can do a good job for most tasks if your budget is limited.

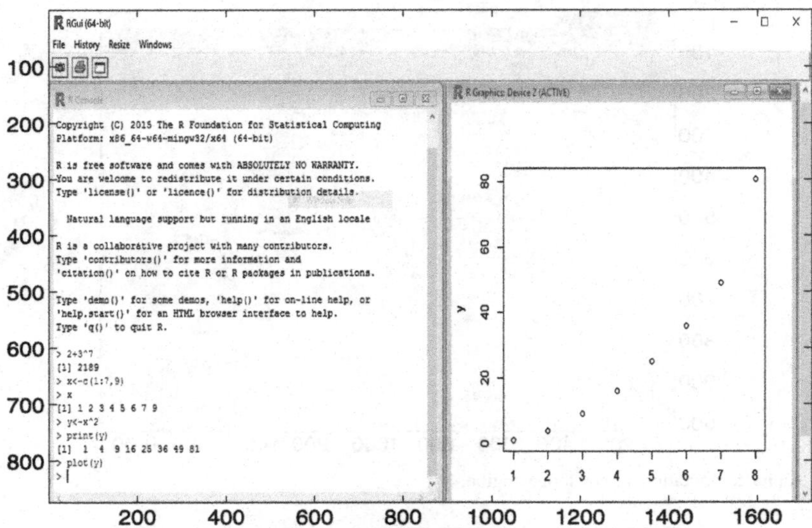

FIGURE B.6 Interface and outputs of R.

B.4 R AND PYTHON

There are other very useful software tools for both computation and programming. For engineering and mathematics, R and Python are also widely used, though C/C++, Java and Fortran are widely used for developing software packages. In the rest of this section, we will briefly introduce R and Python.

B.4.1 R

R is a powerful language and environment for statistical computing (https://www.r-project.org/). In essence, R is a dialect of the programming language S, though R and S-Plus are two popular implementations of S. S-Plus is commercial, while R is an open source distributed under the terms of GPL.

R is a command-line interpreter. For example, we can type

```
>2+3^7
```

and we get

```
[1] 2189
>x <- c(1:7,9)
>x
[1] 1 2 3 4 5 6 7 9
>y <-x^2
>print(y)
[1] 1 4 9 16 25 36 49 81
```

The above outputs are shown in Fig. B.6. The advantages of R are that it is not only free, but also has more than 100 add-on packages such as Bayesian statistics, deSolve (for solving differential equations), plot3D and many others.

B.4.2 Python

Python is a high-level, cross-platform, open-source programming language (https://www.python.org/). It is very versatile with over 100 add-on modules or extension packages such as Sympy (a library for symbolic mathematics), Matplotlib (for plotting 2D graphs), SciPy (for mathematics and scientific computing) and many others.

The core engine is an interpreter. Here is an example.

```
>>> 2**10
1024
>>> length=10
>>> width=2.5
```

```
>>> height=5
>>> area=length*width
>>> volume=area*height
>>> print(volume)
125.0
```

Python can enable us to do many things for both scientific computation, network communications and data analysis. Interested readers can see tutorials or documents at https://docs.python.org/3/tutorial/.

The software tools and packages for symbolic and numeric computation are vast, and it is not our intention here to provide a comprehensive list. As we have pointed out, the main purpose here is to highlight a few most commonly used packages that are relevant to engineering mathematics and students at most universities may have access to such packages. Various repositories or lists about other software packages and a wide range of tutorials can be also be found on the Internet. Interested readers can use such resources and also refer to more specialized textbooks on symbolic and numeric computation.

Appendix C

Answers to Exercises

This appendix provides the answers to all the exercises in the book.

CHAPTER 1

1.1 $\sqrt{3}$ and 2π are irrational numbers. The other numbers are rational.
1.2 $x^2 - 101x + 100 = (x-1)(x-100) = 0$. So $x = 1$ and $x = 100$.
1.3 In order for the equation to have no solution, $\Delta = b^2 - 4ac = 1^2 - 4k < 0$. This means that $4k > 1$ or $k > 1/4$.
1.4 By solving the equations, it is straightforward to check $x = 1$ and $y = 2$ are the solution.
1.5 To solve $|x-2| - 3 = 0$, we have to discuss two cases: $x > 2$ and $x < 2$ (x cannot be 2 because it does not satisfy the equation). For $x > 2$, we have

$$x - 2 - 3 = 0,$$

which gives $x = 5$. For $x < 2$, we have

$$-(x-2) - 3 = 0,$$

which becomes

$$-x + 2 - 3 = 0,$$

or

$$x = -1.$$

The equation has two solutions $x = 5$ and $x = -1$.

CHAPTER 2

2.1 The answers are 2, 4, 7, and 2, respectively.
2.2 You can expand the right-hand side to get $x^5 - 1$.
2.3 Using $a^2 - b^2 = (a+b)(a-b)$, we have

$$x^4 - 16 = (x^2)^2 - 4^2 = (x^2+4)(x^2-4) = (x^2+4)(x+2)(x-2),$$

which has two real roots ($x = -2$ and $x = 2$).
2.4 To factorize, we have

$$\begin{aligned} x^3 - 2x^2 - 2x + 1 &= (x^3 - 3x^2 + x) + (x^2 - 3x + 1) \\ &= (x^2 - 3x + 1)x + (x^2 - 3x + 1) = (x+1)(x^2 - 3x + 1) = 0. \end{aligned}$$

So either $x + 1 = 0$ or $x^2 - 3x + 1 = 0$, which leads to

$$x = -1, \quad \text{or} \quad x = \frac{3 \pm \sqrt{5}}{2}.$$

Engineering Mathematics with Examples and Applications

2.5 Since $x^3 - kx = x(x^2 - k) = 0$, we have either $x = 0$ or $x^2 - k = 0$. In order to have three different roots, k cannot be zero or negative. That is, $k > 0$ and the roots are

$$0, \ \sqrt{k}, \ -\sqrt{k}.$$

In order to ensure that all roots are integer, it requires that $k = n^2$ where $n = 1, 2, ...$ must be integers. So we have

$$k = 1, 4, 9, 16, \cdots.$$

CHAPTER 3

3.1 The answers are 7, 87, 1, and 2, respectively.
3.2 You can use the general formula to prove it. Alternatively, since

$$\binom{6}{4} = 15, \ \binom{5}{3} = 10, \ \binom{5}{4} = 5,$$

it is obvious that equality holds.
3.3 Expand the left-hand side and simply, we have

$$(x+1)^3 - (x-1)^3 = 6x^2 + 2.$$

3.4 Since

$$\binom{n}{k} = \frac{n!}{k!(n-k)!} = 6,$$

is one of the coefficients in the binomial expansion or a number in Pascal's triangle. Thus, the only possibility $n = 4$ and $k = 2$ to satisfying this equality.
3.5 Using the binomial expansion of $(a+b)^n$ when $a = b = 1$, we have

$$(1+1) = 2^n = \binom{n}{0} 1^n + \binom{n}{1} 1^{n-1} 1^1 + ... + \binom{n}{n-1} 1^1 1^{n-1} + \binom{n}{n} 1^n,$$

which gives

$$\binom{n}{0} + \binom{n}{1} + ... + \binom{n}{n-1} + \binom{n}{n} = 2^n.$$

CHAPTER 4

4.1 The formulas are $7(n-1)+1$, $(n-1)^2+5$, n^2+n, and $1/2^n$, respectively, for natural numbers ($n = 1, 2, 3, ...$).
4.2 They all prime numbers for $n = 0, 1, ..., 40$.
4.3 All the series are geometric, and the ratios are 1/10, 1/4, 1/100 (except the first term), 2/3 and $-1/2$, respectively. The sums of these series are 7/9, 4/3, 12/11, 36 and 128/3, respectively.
4.4 The sum of the series

$$1 + (\frac{1}{2} + \frac{1}{3}) + (\frac{1}{4} + \frac{1}{9}) + (\frac{1}{8} + \frac{1}{27}) + \cdots,$$

is equivalent to $A + B$ where

$$A = (1 + \frac{1}{2} + \frac{1}{2^2} + \frac{1}{2^3} + \cdots) = 2,$$

and

$$B = (\frac{1}{3} + \frac{1}{3^2} + \frac{1}{3^3} + \cdots) = \frac{1}{2}.$$

The sum of the original series is $2\frac{1}{2}$. Similarly, the sum of the series

$$1+(\frac{1}{2}-\frac{1}{3})+(\frac{1}{4}-\frac{1}{9})+(\frac{1}{8}-\frac{1}{27})+\cdots = A-B = 1\frac{1}{2}.$$

CHAPTER 5

5.1 The power functions are: $x^{5.5}$ and x^{π}. The exponential functions are 7^x, 2^{x^2}, π^x, and e^{-x}. Here, x^x is a mixed type, it can be considered as either.

5.2 The domains of 3^{-x}, e^{-2x} and $2x^2-3$ are the real numbers, while x^{-2} are all non-zero real numbers. For $x^{3/2} = (\sqrt{x})^3$, its domain is $x \geq 0$.

5.3 The domain of $\ln(x+5)$ is $x > -5$, and its range is the whole real number $\Re$. Both the domain and range of $\log_2(2^x) = x$ are $\Re$. The domain of $y = \ln^2(x^2+1)$ is $\Re$ and its corresponding range is $y \geq 0$. The domain of $\log_{10}(1/x)$ is $x > 0$, and its range is $\Re$. For $\log(1/(x-1))$, its domain is $x > 1$, and its range is $\Re$.

5.4 The final answers of the expressions are: 0, 1, $6+4/9$, $3+\ln 8$.

5.5 For such a nonlinear equation, it is better to use graphs and trial solutions. Indeed, $x = 1$ and $x = 2$ are two solutions.

CHAPTER 6

6.1 The final speed is $9000 \times \pi/30 = 300\pi$ rad/s. For 10 seconds, it turns $10 * 300\pi = 3000\pi$ radians. As $180° = \pi$ radians, we have $3000\pi \times 180/\pi = 540000°$ or 1500 rounds.

6.2 The length of the third side is 2, and the two other angles are 90° and 45°, respectively.

6.3 To prove these identities, we can use different identities as a combination. For example, by using $\sin 2A = 2\sin A\cos A$ and $\cos 2A = \cos^2 A - \sin^2 A = 1 - 2\sin^2 A$, we have

$$\begin{aligned}\sin(3A) &= \sin(2A + A) = \sin(2A)\cos A + \cos(2A)\sin A\\ &= 2\sin A\cos A\cos A + (1-2\sin^2 A)\sin A = 2\sin A\cos^2 A + \sin A - 2\sin^3 A.\end{aligned}$$

Since $\cos^2 A = 1 - \sin^2 A$, we have

$$\begin{aligned}\sin(3A) &= 2\sin A(1-\sin^2 A) + \sin A - 2\sin^3 A\\ &= 2\sin A - 2\sin^3 A + \sin A - 2\sin^3 A = 3\sin A - 4\sin^3 A.\end{aligned}$$

Other identities can be proved in a similar manner.

6.4 At the totality of a solar eclipse, the Earth, the Moon and the Sun for a straight line. Trigonometry suggests that the ratio of the distance to the Sun from the Earth to the distance of the Moon (from the Earth) is the same ratio of the Sun's diameter to the Moon's diameter. Similar arguments lead to the pinhole setting and ways to measure the Sun's diameter on the Earth.

6.5 An ellipse can be described in different ways such as parametric equations. For this question, t can vary from 0 to 2π.

CHAPTER 7

7.1 Use Euler's formula $e^{\theta i} = \cos\theta + i\sin\theta$ when necessary. Thus, $\exp[-\pi i/2] = -i$. The answers of these expressions are: $i+1$, $-i$, 1.718, i, $-1/2-i$, respectively.

7.2 From the definition of $\sinh(x) = (e^x - e^{-x})/2$ and $\cosh(x) = (e^x + e^{-x})/2$, we have

$$\sinh^2(x) = \frac{(e^x - e^{-x})^2}{2^2} = \frac{e^{2x} + e^{-2x} - 2e^x e^{-x}}{4} = \frac{e^{2x} + e^{-2x} - 2}{4},$$

and

$$\cosh^2(x) = \frac{(e^x + e^{-x})^2}{2^2} = \frac{e^{2x} + e^{-2x} + 2e^x e^{-x}}{4} = \frac{e^{2x} + e^{-2x} + 2}{4}.$$

So their sum is

$$\sinh^2(x) + \cosh^2(x) = \frac{e^{2x} + e^{-2x} - 2}{4} + \frac{e^{2x} + e^{-2x} + 2}{4} = \frac{2e^{2x} + 2e^{-2x}}{4} = \frac{e^{2x} + e^{-2x}}{2},$$

which is exactly $\cosh(2x)$ from its definition

$$\cosh(2x) = \frac{e^{2x} + e^{-2x}}{2},$$

which proves that $\cosh(2x) = \sinh^2(x) + \cosh^2(x)$.

7.3 You can use Matlab, Mathematica, MuPAD, Maxima, etc.

7.4 Hyperbolic functions can be used to describe the shape of a hanging rope, Mercator projection on a map, and even special relativity.

7.5 The parametric equations can be used to plot out a hyperbola. In theory, t can vary in the whole range of real numbers. To get a good plot, you can try t varying -2 to $+2$.

CHAPTER 8

8.1 For the given vectors, we have $\mathbf{u}+\mathbf{v}+\mathbf{w} = \begin{pmatrix} 1+2+2 \\ 2-1+3 \end{pmatrix} = \begin{pmatrix} 5 \\ 4 \end{pmatrix}$ and its modulus is $\sqrt{5^2+4^2} = \sqrt{41}$. Since $|\mathbf{u}| = \sqrt{1^2+2^2} = \sqrt{5}$, $|\mathbf{v}| = \sqrt{5}$, and $|\mathbf{w}| = \sqrt{2^2+3^2} = \sqrt{13}$. Thus, $|\mathbf{u}+\mathbf{v}+\mathbf{w}| = \sqrt{41} < |\mathbf{u}|+|\mathbf{v}|+|\mathbf{w}| = \sqrt{5}+\sqrt{4}+\sqrt{13}$.

8.2 The angle between $\mathbf{u}$ and $\mathbf{v}$ is 90° because their dot product is zero and thus they are orthogonal to each other. The angle between $\mathbf{v}$ and $\mathbf{w}$ is about 82.9°, while the angle between $\mathbf{u}$ and $\mathbf{w}$ is about 7.1°.

8.3 Let the starting point be the origin, the north be y-axis, and the east be x-axis. The movements of the balloon can be described by the sum of three vectors

$$\sqrt{2}\begin{pmatrix} 1 \\ -1 \end{pmatrix} + \begin{pmatrix} 0 \\ -0.5 \end{pmatrix} + \begin{pmatrix} 1.5 \\ 0 \end{pmatrix},$$

which means that the balloon is about 3.05 km from its starting point.

8.4 The angle θ against the stream is $\tan(\theta) = 3/4$ (about 37°).

8.5 The friction coefficient is $\mu = \tan(30°) = 0.5$.

CHAPTER 9

9.1 For the three matrices, we have

$$\mathbf{A}^T = \begin{pmatrix} 2 & 3 \\ 3 & 2 \end{pmatrix}, \quad \mathbf{B} = \begin{pmatrix} 1 & -3 \\ -2 & 5 \end{pmatrix}, \quad \bar{\mathbf{C}} = \begin{pmatrix} 1 & -i \\ 2 & 1 \end{pmatrix}, \quad \mathbf{C}^H = \begin{pmatrix} 1 & 2 \\ -i & 1 \end{pmatrix}.$$

9.2 The answers are

$$\mathbf{AB} = \begin{pmatrix} -7 & 11 \\ -3 & 4 \end{pmatrix}, \quad \mathbf{BA} = \begin{pmatrix} -4 & -1 \\ 9 & 1 \end{pmatrix}, \quad \mathbf{BC} = \begin{pmatrix} -3 & -2+i \\ 7 & 5-3i \end{pmatrix},$$

so

$$\mathbf{AB} - \mathbf{BA} = \begin{pmatrix} -3 & 12 \\ -12 & 3 \end{pmatrix}.$$

9.3 $\det(\mathbf{A}) = 2 \times 2 - 3 \times 3 = -5$, and $\det(\mathbf{C}) = 1 \times 1 - 2 \times i = 1 - 2i$.

9.4 The inverses are

$$\mathbf{A}^{-1} = \begin{pmatrix} -2/5 & 3/5 \\ 3/5 & -2/5 \end{pmatrix}, \quad \mathbf{B}^{-1} = \begin{pmatrix} -5 & -2 \\ -3 & -1 \end{pmatrix}.$$

9.5 The eigenvalues of $\mathbf{A}$ are $-1, 5$, and their corresponding eigenvectors are $(-\sqrt{2}/2, \sqrt{2}/2)^T$ and $(\sqrt{2}/2, \sqrt{2}/2)^T$, respectively.

9.6 Since $\mathbf{AB} = \begin{pmatrix} -7 & 11 \\ -3 & 4 \end{pmatrix}$, we have

$$(\mathbf{AB})^T = \begin{pmatrix} -7 & -3 \\ 11 & 4 \end{pmatrix}.$$

In addition, we have $\mathbf{A}^T = \begin{pmatrix} 2 & 3 \\ 3 & 2 \end{pmatrix}$ and $\mathbf{B}^T = \begin{pmatrix} 1 & -3 \\ -2 & 5 \end{pmatrix}$, thus we have

$$\mathbf{B}^T\mathbf{A}^T = \begin{pmatrix} 1 & -3 \\ -2 & 5 \end{pmatrix}\begin{pmatrix} 2 & 3 \\ 3 & 2 \end{pmatrix} = \begin{pmatrix} -7 & -3 \\ 11 & 4 \end{pmatrix} = (\mathbf{AB})^T.$$

CHAPTER 10

10.1 The derivatives of the expressions can be found as follows:

$$\begin{aligned} f'(x) &= [x^3+2x^2+3x]' = (x^3)' + (2x^2)' + (3x)' = 3x^2+4x+3, \\ g'(x) &= (xe^{-x})' = x'e^{-x} + x(e^{-x})' = e^{-x} + xe^{-x}(-1) = e^{-x} - xe^{-x}, \\ h'(x) &= [x^3 + \cos(x^2) + x\ln x]' = (x^3)' + [\cos(x^2)]' + [x\ln x]' \\ &= 3x^2 - \sin(x^2)\cdot(2x) + [x'\ln(x) + x(\ln x)'] \\ &= 3x^2 - 2x\sin(x^2) + \ln x + x\cdot\frac{1}{x} = 3x^2 - 2x\sin(x^2) + \ln x + 1. \end{aligned}$$

10.2 In order to find the first derivative of $f(x) = x^x$ for $x > 0$, we cannot use direct differentiation because there is no basic rule to deal with x^x. Since $x > 0$, we should take the logarithm of the function and we have

$$\ln f(x) = \ln x^x = x\ln x.$$

Taking the derivative of both sides with respect to x and using the chain rule and product rule, we get

$$[\ln f(x)]' = \frac{1}{f(x)}\cdot f'(x) = [x\ln x]' = \ln x + x\cdot\frac{1}{x},$$

which means that

$$\frac{f'(x)}{f(x)} = \ln x + 1,$$

or

$$f'(x) = f(x)[\ln x + 1].$$

Since $f(x) = x^x$, we finally have

$$f'(x) = [x^x]' = x^x[\ln x + 1].$$

10.3 You can either solve $y(x)$ first, or use the implicit differentiation by differentiating each term with respect to x, we have

$$y'(x) - [\sin(x)e^{-x}]' + [x^2y(x)]' = 0,$$

which leads to

$$y'(x) - (\cos(x)e^{-x} - \sin(x)e^{-x}) + 2xy(x) + x^2y'(x) = 0.$$

Solving $y'(x)$, we get

$$y'(x) = \frac{[\cos(x) - \sin(x)]e^{-x} - 2xy(x)}{1+x^2},$$

which is an implicit function of x and $y(x)$. If we really want to eliminate $y(x)$, we can re-arrange the original equation as

$$y(x) = \frac{\sin(x)e^{-x}}{1+x^2},$$

and substitute it into the expression of $y'(x)$, we finally obtain

$$y'(x) = \frac{[\cos(x) - \sin(x)]e^{-x} - 2x\sin(x)e^{-x}}{1+x^2}.$$

10.4 For the expression $f(x) = xe^{-x} + \sin(x^2)$, its first derivative is

$$f'(x) = (xe^{-x})' + [\sin(x^2)]' = e^{-x} - xe^{-x} + 2x\cos(x^2),$$

and the second derivative is

$$f''(x) = -2e^{-x} + xe^{-x} + 2\cos(x^2) - 4x^2\sin(x^2).$$

Similarly, we have

$$f'''(x) = 3e^{-x} - xe^{-x} - 12x\sin(x^2) - 8x^3\cos(x^2),$$
$$f''''(x) = -4e^{-x} + xe^{-x} - 12\sin(x^2) - 48x^2\cos(x^2) + 16x^4\sin(x^2),$$

and

$$f'''''(x) = 5e^{-x} - xe^{-x} - 120x\cos(x^2) + 32x^5\cos(x^2) + 160x^3\sin(x^2).$$

10.5 The expansion of $\exp(-x)$ can be obtained from

$$e^x = 1 + x + \frac{x^2}{2!} + \frac{x^3}{3!} + ... + \frac{x^n}{n!} + ...,$$

by replacing x with $-x$. We have

$$e^{-x} = 1 - x + \frac{x^2}{2!} - \frac{x^3}{3!} + ... + \frac{(-1)^n x^n}{n!} + ..., \quad \forall x \in \Re.$$

10.6 The expansion of $x^2e^{-x^2}$ can be obtained by using $-x^2$ in the series of e^x, we have

$$x^2e^{-x^2} = x^2\Big[1 - x^2 + \frac{x^4}{2!} - \frac{x^6}{3!} + ... + \frac{(-1)^n x^{2n}}{n!} + ...\Big]$$
$$= x^2 - x^4 + \frac{x^6}{2!} - \frac{x^8}{3!} + ... + \frac{(-1)^n x^{2n+2}}{n!} + ..., \quad \forall x \in \Re.$$

CHAPTER 11

11.1 The integrals can be obtained by direct integration, integration by parts or substitution. We have

$$\int (x - \frac{1}{x})dx = \int xdx - \int \frac{1}{x}dx = \frac{x^2}{2} - \ln x + C,$$

and

$$\int (x^2 + 2x + 1)dx = \int x^2 dx + 2\int xdx + \int 1dx = \frac{x^3}{3} + x^2 + x + C,$$

where C is an integration constant. Similarly, we have

$$\int (x + \cos(x) + e^x)dx = \int xdx + \int \cos(x)dx + \int e^x dx = \frac{1}{2}x^2 + \sin(x) + e^x + C.$$

For $\int \ln x dx$, we can use integration by parts $\int vdu = uv - \int udv$ or

$$\int vu'dx = uv - \int uv'dx.$$

Let $u' = 1$ and $v = \ln x$, we have $u = x$, $v' = 1/x$, $du = dx$ and $dv = dx/x$. Thus, we have

$$\int 1 \ln x = x \ln x - \int x \cdot \frac{1}{x} dx = x \ln x - x + C.$$

For $\int 2x \cos(x^2) dx$, we can use $u = x^2$ (thus $du = 2xdx$) and we have

$$\int 2x \cos(x^2) dx = \int \cos(u) du = \sin(u) + C = \sin(x^2) + C.$$

11.2 The definite integrals can be calculated in the same ways as before, but we have to take the integration limits into consideration. So we have

$$\int_0^1 (x^5 - x^4) dx = \int_0^1 x^5 dx - \int_0^1 x^4 dx = \Big[\frac{x^6}{6}\Big]_0^1 - \Big[\frac{x^5}{5}\Big]_0^1$$
$$= [\frac{1^6}{6} - \frac{0^6}{6}] - [\frac{1^5}{5} - \frac{0^5}{5}] = \frac{1}{6} - \frac{1}{5} = \frac{5}{30} - \frac{6}{30} = -\frac{1}{30}.$$

In the same manner, we have

$$\int_0^{\pi/2} = \Big[-\frac{1}{2} \cos(2x) \big]_0^{\pi/2} = -\frac{1}{2}\Big[\cos(2x)\Big]_0^{\pi/2}$$
$$= -\frac{1}{2}[\cos(2\pi/2) - \cos(0)] = -\frac{1}{2}[-1 - 1] = 1.$$

Since $\cos^2(x) = (1 + \cos(2x))/2$, we have

$$\int_0^{\pi} \cos^2(x) dx = \int_0^{\pi} [\frac{1 + \cos(2x)}{2} dx = \frac{1}{2} \int_0^{\pi} dx + \frac{1}{2} \int_0^{\pi} \cos(2x) dx$$
$$= \frac{1}{2}\Big[x\Big]_0^{\pi} + \frac{1}{2}\Big[\frac{\sin(2x)}{2}\Big]_0^{\pi} = \frac{\pi}{2} + \frac{1}{4}[\sin(2\pi) - \sin(0)] = \frac{\pi}{2}.$$

For $\int_0^{\infty} xe^{-x}$, we can use the integration by parts

$$\int uv' dx = uv - \int vu' dx.$$

Let $u = x$ and $v' = e^{-x}$, we have $u' = 1$, $du = dx$, and $v = -e^{-x}$. Thus, we get

$$\int xe^{-x} dx = -xe^{-x} - \int (-e^{-x}) dx = -xe^{-x} - e^{-x} = -(1 + x)e^{-x}.$$

Now the definite integral becomes

$$\int_0^{\infty} xe^{-x} dx = -\Big[(1 + x)e^{-x}\Big]_0^{\infty} = -[0 - (1 + 0)e^{-\infty}] = 1,$$

where we have used $\infty \exp[-\infty] = 0$.

11.3 In order to calculate the total area A under the curve $|x|e^{-|x|}$, we have to calculate the integral

$$A = \int_{-\infty}^{+\infty} |x|e^{-|x|} dx.$$

However, this integral involve the absolute function $|x|$, so we have to be careful. By looking the integrand carefully, we realize that

$$f(x) = |x|e^{-|x|},$$

is an even function since $f(-x) = f(x)$. This means that we can do only half of the integration limits, focusing on the branch with $x > 0$. Thus, we get

$$A = \int_{-\infty}^{+\infty} |x|e^{-|x|}dx = 2\int_{0}^{+\infty} xe^{-x}dx.$$

From the result of the previous exercise, we know that $\int_0^{+\infty} xe^{-x}dx = 1$, so we finally have

$$A = 2\int_{0}^{+\infty} xe^{-x}dx = 2.$$

11.4 For the bungee jumping, we can assume the initial velocity of the jumper is zero (thus the initial kinetic energy is also zero). After the jump, let us assume that the person will fall to the maximum at the height d above the river. At this point, the rope is at the maximum extension and the person just stops momentarily. The change (loss) of the potential energy is

$$E_p = (mgh - mgd) = (h - d)mg,$$

where g is the acceleration due to gravity. This energy will become the elastic potential energy stored in the rope. Since the force in the rope is kx when the rope has an extension x, and its elastic energy E_r is

$$E_r = \int_0^{h-d-L} kxdx = \frac{1}{2}k(h - d - L)^2.$$

From $E_r = E_p$, we have

$$\frac{1}{2}k(h - d - L)^2 = mg(h - d).$$

To be safe for jumping, it is required that $d > 0$ (otherwise, the person's head will dip into the river). From the given values of $m = 75$ kg, $L = 50$ m, $h = 100$ m, $k = 750$ N/m and $g = 10$ m/s^2, we have

$$\frac{1}{2} \times 750 \times (100 - d - 50)^2 = 75 \times 10(100 - d),$$

or

$$(50 - d)^2 = 20(100 - d),$$

which has two solutions $d = 6.8$ m and 73.2 m. As $d = 6.8 > 0$, so the jumper is safe. For further discussion, please discuss the meaning of these two solutions and why the solution 73.2 is not relevant in this case.

CHAPTER 12

12.1 The differential equation $y'(x) = \sin(x) - x^2$ can be solved by direct integration, and we have

$$y(x) = \int [\sin(x) - x^2]]dx = -\cos(x) - \frac{x^3}{3} + C,$$

where C is the integration constant to be determined by the initial condition. From the initial condition $y(0) = 1$, it means that

$$1 = -\cos(0) - \frac{0^3}{3} + C,$$

or $C = 2$. Thus, the final solution is

$$y(x) = 2 - \cos(x) - \frac{x^3}{3}.$$

12.2 This exercise has three differential equations.

- The first equation $y''(x)+3y'(x)+2y(x)=0$ can be solved by the method of characteristics by setting $y(x)=e^{\lambda x}$ where λ is an unknown. Thus, we have
$$\lambda^2 e^{\lambda x}+3\lambda e^{\lambda x}+2e^{\lambda x}=0,$$
or
$$(\lambda^2+3\lambda+2)e^{\lambda x}=0.$$
Since $e^{\lambda x}\neq 0$ (otherwise, the solution will be a trial zero solution), it requires that
$$\lambda^2+3\lambda+2=(\lambda+1)(\lambda+2)=0.$$
It has two solutions $\lambda_1=-1$ and $\lambda_2=-2$. Thus, the general solution of the original differential equation is
$$y(x)=Ae^{-x}+Be^{-2x},$$
where A and B are two unknown coefficients.
- The equation is linear, but its coefficient is not constant, so it cannot be solved using the characteristic method. However, it can be solved by either integrating factors or integration. By assuming $y(x)\neq 0$ and re-arranging, we have
$$\frac{y'(x)}{y(x)}=-x,$$
and its integration with respect to x becomes
$$\ln y(x)=-\frac{x^2}{2}+C,$$
where C is the integration constant. Taking exponential of both sides, we get
$$y(x)=e^{\ln y(x)}=e^{-\frac{x^2}{2}+C}=Be^{-x^2/2},$$
where $B=e^C$.
- Substituting $y(x)=e^{\lambda x}$ into $y''(x)+4y(x)=0$, we have
$$\lambda^2+4=0,$$
which gives $\lambda=\pm 2i$. This means that the solution can be written as
$$y(x)=Ae^{+2xi}+Be^{-2xi}.$$
Using the Euler's formula $e^{i\theta}=\cos\theta+i\sin\theta$, we have
$$y(x)=A[\cos(2x)+i\sin(2x)]+B[\cos(-2x)+i\sin(-2x)].$$
Since $\cos(-2x)=\cos(2x)$ and $\sin(-2x)=-\sin(2x)$, we have
$$y(x)=[A+B]\cos(2x)+[Ai-iB]\sin(2x)=C\cos(2x)+D\sin(2x),$$
where $C=A+B$ and $D=[A-B]i$ are two unknown coefficients.

12.3 This is a third-order differential equation with constant coefficients and proper initial conditions. As it is the third order, three initial conditions are needed to determine three unknown integration constants. Setting $y(x)=e^{\lambda x}$, we have
$$\lambda^3+\lambda^2+4\lambda+4=0,$$

which can be factorized as

$$\lambda^3 + \lambda^2 + 4\lambda + 4 = (\lambda + 1)(\lambda^2 + 4) = 0,$$

which means that it has three solutions $\lambda = -1$, $\lambda = \pm 2i$. Using the results from the previous exercise, the general solution becomes

$$y(x) = Ae^{-x} + C\cos(2x) + D\sin(2x).$$

To satisfy $y(0) = 1$, we have

$$1 = y(0) = Ae^{-0} + C\cos(2 \times 0) + D\sin(2 \times 0),$$

or

$$A + C = 1.$$

From the solution $y(x)$, we have its first and second derivatives as follows:

$$y'(x) = -Ae^{-x} - 2C\sin(2x) + 2D\cos(2x),$$

and

$$y''(x) = Ae^{-x} - 4C\cos(2x) - 4D\sin(2x).$$

From $y'(0) = 0$, we have

$$0 = y'(0) = -Ae^{-0} - 2C\sin(2 \times 0) + 2D\cos(2 \times 0),$$

which gives

$$2D - A = 0,$$

or simply $A = 2D$.
Similarly, from $y''(0) = 1$, we get

$$1 = y''(0) = Ae^{-0} - 4C\cos(2 \times 0) - 4D\sin(2 \times 0) = 0,$$

or simply

$$A - 4C = 1.$$

Now we have

$$A + C = 1, \quad A = 2D, \quad A - 4C = 1.$$

It is straightforward to obtain their solution as

$$A = 1, \quad C = 0, \quad D = \frac{1}{2}.$$

Therefore, the final solution is

$$y(x) = e^{-x} + \frac{\sin(2x)}{2}.$$

It leaves as a further exercise to check that it indeed satisfies the equation and the initial conditions.

12.4 Since $y'(x)+2y(x)=1-e^{-x}$ is a first-order differential equation, we can solve it either by the general formula introduced in this chapter or by an approach similar to the method of solving high-order linear ODEs. The homogeneous equation is

$$y'(x)+2y(x)=0,$$

thus its complementary function can be obtained by using $y(x)=e^{\lambda x}$. We have $\lambda+2=0$, and thus we have the complementary function as

$$y_c(x)=Ce^{-2x},$$

where C is a constant. As the right-hand side is $1-e^{-x}$, we should try a particular integral of the form

$$p(x)=A+Be^{-x}.$$

Thus, we have $p'(x)=-Be^{-x}$. Substituting these results into the original equation, we get

$$(-Be^{-x})+2[A+Be^{-x}]=1-e^{-x},$$

or

$$2A+Be^{-x}=1-e^{-x}.$$

This equality must hold for all x, thus it requires that their corresponding coefficients must be equal. This means that

$$2A=1, \quad B=-1.$$

The particular integral becomes

$$p(x)=\frac{1}{2}-e^{-x}.$$

Now the general solution for the differential equation becomes

$$y(x)=y_c(x)+p(x)=Ce^{-2x}+\frac{1}{2}-e^{-x}.$$

Using the initial condition $y(0)=1$, we have

$$1=y(0)=Ce^{-2\times 0}+\frac{1}{2}-e^{-0},$$

which gives $C=\frac{3}{2}$. Therefore, the final solution is

$$y(x)=\frac{3}{2}e^{-2x}+\frac{1}{2}-e^{-x}.$$

CHAPTER 13

13.1 The first derivatives of $f(x,y)=x^2+y^2+\sin(x^2y^2)$ are

$$\frac{\partial f}{\partial x}=2x+0+\cos(x^2y^2)y^2(2x)=2x+2xy^2\cos(x^2y^2),$$

and

$$\frac{\partial f}{\partial y}=0+2y+\cos(x^2y^2)x^2(2y)=2y+2yx^2\cos(x^2y^2).$$

13.2 For $g(x, y) = e^{-xy^2} + xy$, we have

$$\frac{\partial g}{\partial x} = e^{-xy^2}(-y^2) + y = y - y^2 e^{-xy^2},$$
$$\frac{\partial g}{\partial y} = e^{-xy^2}(-x)(2y) + x = x - 2xye^{-xy^2}.$$

Thus, we finally obtain

$$\frac{\partial^2 g}{\partial x^2} = y^4 e^{-xy^2}, \quad \frac{\partial^2 g}{\partial y^2} = (4x^2y^2 - 2x)e^{-xy^2}.$$

13.3 The divergence of $\mathbf{u}$ is

$$\nabla \cdot \mathbf{u} = \frac{\partial x^2}{\partial x} + \frac{\partial (y^2 - x^2)}{\partial y} + \frac{\partial z}{\partial z} = 2x + 2y + 1.$$

13.4 The curl of $\mathbf{v} = yz^n\mathbf{i} + xy\mathbf{j} + yz^2\mathbf{k}$ is

$$\text{curl } \mathbf{v} = \begin{vmatrix} \mathbf{i} & \mathbf{j} & \mathbf{k} \\ \frac{\partial}{\partial x} & \frac{\partial}{\partial y} & \frac{\partial}{\partial z} \\ yz^n & xy & yz^2 \end{vmatrix}$$
$$= (z^2 - 0)\mathbf{i} + (nyz^{n-1} - 0)\mathbf{j} + (y - z^n)\mathbf{k} = z^2\mathbf{i} + nyz^{n-1}\mathbf{j} + (y - z^n)\mathbf{k}.$$

Thus, the divergence of this curl is

$$\text{div}(\text{curl}(\mathbf{v})) = \nabla \cdot (\nabla \times \mathbf{v}) = \frac{\partial z^2}{\partial x} + \frac{\partial (nyz^{n-1})}{\partial y} + \frac{\partial (y - z^n)}{\partial z}$$
$$= 0 + nz^{n-1} - nz^{n-1} = 0.$$

13.5 For $\Psi = 3x^2y - y^3$, we have

$$\frac{\partial \Psi}{\partial x} = 6xy, \quad \frac{\partial \Psi}{\partial y} = 3x^2 - 3y^2, \quad \frac{\partial \Psi}{\partial z} = 0,$$

and

$$\frac{\partial^2 \Psi}{\partial x^2} = 6y, \quad \frac{\partial^2 \Psi}{\partial y^2} = -6y,$$
$$\nabla^2 \Psi = \frac{\partial^2 \Psi}{\partial x^2} + \frac{\partial^2 \Psi}{\partial y^2} + \frac{\partial^2 \Psi}{\partial z^2} = 6y - 6y + 0 = 0.$$

Similarly, the grad of Ψ is

$$\nabla \cdot \Psi = \frac{\partial \Psi}{\partial x}\mathbf{i} + \frac{\partial \Psi}{\partial y}\mathbf{j} + \frac{\partial \Psi}{\partial z}\mathbf{k} = 6xy\mathbf{i} + 3(x^2 - y^2)\mathbf{j} + 0\mathbf{k}.$$

Thus, the curl of this divergence is

$$\nabla \times (\nabla \Psi) = [\frac{\partial 0}{\partial y} - 3\frac{\partial (x^2 - y^2)}{\partial z}]\mathbf{i} - [\frac{\partial 0}{\partial x} - \frac{\partial (6xy)}{\partial z}]\mathbf{j} + [3\frac{\partial (x^2 - y^2)}{\partial x} - \frac{\partial (6xy)}{\partial y}]\mathbf{k}$$
$$= (6x - 6x)\mathbf{k} = 0.$$

CHAPTER 14

14.1 The arc length of $y = 2x^{3/2}/3$ from $x = 0$ to $x = 2$ is

$$L = \int_0^2 \sqrt{1 + y'(x)}dx = \int_0^2 \sqrt{1 + x}\, dx,$$

where we have used $y'(x) = \sqrt{x}$. Using $u = 1 + x$ and $du = dx$, we have the integration limits will be from $u = 1$ and $u = 3$. Thus, we have

$$L = \int_1^3 \sqrt{u}\, du = \frac{2u^{3/2}}{3}\Big|_1^3 = \frac{2}{3}[2^{3/2} - 1^{3/2}] = 2\sqrt{3} - \frac{2}{3} \approx 3.46.$$

14.2 The volume V is the double integral

$$V = \iint_D [5 + x^2 + y] dxdy, \quad 0 \le x \le 2, \quad -x \le y \le x.$$

We have

$$V = \int_0^2 \int_{-x}^x (5 + x^2 + y) dxdy = \int_0^2 \Big[(5 + x^2)y + \frac{y^2}{2}\Big]\Big|_{-x}^x dx = \int_0^2 [2x(5 + x^2)] dx$$
$$= 10\int_0^2 x dx + 2\int_0^2 x^3 dx = 10\frac{x^2}{2}\Big|_0^2 + 2\frac{x^4}{4}\Big|_0^2 = 28.$$

14.3 The integral in a circular domain

$$I = \iint_D e^{-(x^2+y^2)} dxdy, \quad x^2 + y^2 \le 1,$$

can be computed by changing into the polar coordinate (r, θ) using $r^2 = x^2 + y^2$ and $v = r^2$ so that $dv = 2rdr$, and we have

$$I = \int_0^{2\pi} \int_0^1 e^{-r^2} r dr d\theta = \frac{1}{2}\int_0^{2\pi} d\theta \int_0^1 e^{-v} dv$$
$$= \frac{1}{2} \cdot 2\pi \int_0^1 e^{-v} dv = \pi[-e^{-v}]\Big|_0^1 = \pi[1 - e^{-1}].$$

14.4 There are quite a few ways for showing that the area of an ellipse is $A = \pi ab$. Here we use double integrals. From the equation

$$\frac{x^2}{a^2} + \frac{y^2}{b^2} = 1,$$

we change it to the so-called distorted polar coordinates using $x = ar\cos(\phi)$ and $y = br\sin(\phi)$, which gives

$$\frac{x^2}{a^2} + \frac{y^2}{b^2} = r^2\cos^2(\phi) + r^2\sin^2(\phi) = r^2.$$

The Jacobian is

$$J = \frac{\partial(x, y)}{\partial(r, \phi)} = [(a\cos\phi)(br\cos\phi) - (b\sin\phi)(-ar\sin\phi)]$$
$$= abr[\cos^2\phi + \sin^2\phi] = abr,$$

where we have used

$$\frac{\partial x}{\partial r} = a\cos\phi, \quad \frac{\partial x}{\partial \phi} = -ar\sin\phi, \quad \frac{\partial y}{\partial r} = b\sin\phi, \quad \frac{\partial y}{\partial \phi} = br\cos\phi.$$

Thus, the area becomes

$$A = \iint_{\text{ellipse}} dxdy = \int_0^{2\pi}\Big(\int_0^1 J dr\Big) d\phi = \int_0^{2\pi} d\phi \int_0^1 (abr) dr$$
$$= 2\pi\Big(\frac{abr}{2}\Big)\Big|_0^1 = \pi ab.$$

CHAPTER 15

15.1 From $z = re^{i\theta}$ and the chain rule, we have

$$dz = dre^{i\theta} + rie^{i\theta}d\theta = e^{i\theta}dr + izd\theta,$$

which gives

$$\frac{dz}{z} = \frac{e^{i\theta}dr}{z} + \frac{izd\theta}{z} = \frac{dre^{i\theta}}{re^{i\theta}} + i\frac{zd\theta}{z} = \frac{dr}{r} + id\theta.$$

15.2 To compute $\ln(1+i)$, we first use $z = re^{i\theta}$ so that

$$\ln z = \ln(re^{i\theta}) = \ln r + \ln e^{i\theta} = \ln r + i\theta.$$

Since $1 + i = \sqrt{2}e^{i\pi/4}$, we have

$$\ln(1+i) = \ln\sqrt{2} + i\frac{\pi}{4} = \frac{1}{2}\ln 2 + i\frac{\pi}{4}.$$

Similarly, we have $i = e^{i\pi/2}$ and

$$\ln i = \ln e^{i\pi/2} = i\frac{\pi}{2}.$$

15.3 Using the result from the above first exercise, we get integral

$$I = \oint_\Gamma \frac{dz}{z} = \oint_\Gamma (\frac{dr}{r} + id\theta),$$

along half a unit circle $z = e^{i\theta}$. Let Γ_1 be the half of the unit circle above the real axis from -1 to $+1$ (with $\theta = \pi$ to 0) and Γ_2 be the lower half unit circle (with $\theta = \pi$ to 2π). Since on either path of the unit circle, $r = 1$ and $dr = 0$, we thus have

$$I_1 = \oint_{\Gamma_1} \frac{dz}{z} = \oint_\pi^0 id\theta = i(0-\pi) = -i\pi,$$

$$I_2 = \oint_{\Gamma_2} \frac{dz}{z} = \oint_\pi^{2\pi} = +i\pi.$$

If we reverse the path direction, the integral limits will be swapped and thus the integral value will have an extra negative sign. For example, if we integrate along the reverse direction of $\Gamma_3 = -\Gamma_1$, we have

$$I_3 = \oint_{\Gamma_3=-\Gamma_1} \frac{dz}{z} = \oint_0^\pi id\theta = +i\pi.$$

In addition, the integral along the whole unit circle with a positive oriented loop (counterclockwise) becomes

$$I = I_2 + I_3 = +i\pi + i\pi = 2\pi i.$$

15.4 For the contour integral

$$I = \oint_D \frac{e^z}{(z-\pi)^3}dz,$$

we have $f(z) = e^z$, $z_0 = \pi$ and $n = 2$ for the Cauchy integral formula:

$$\oint_\Gamma \frac{f(z)}{(z-z_0)^{n+1}}dz = \frac{2\pi i}{n!}f^{(n)}(z_0).$$

Since the pole $z = z_0 = \pi$ is within the square path, we can use the above formula and we get

$$I = \frac{2\pi i}{2!}f''(\pi) = \pi i \frac{d^2e^z}{dz^2}\Big|_{z=\pi} = i\pi e^\pi.$$

CHAPTER 16

16.1 This square wave is the mirror image (about a horizontal line at $f(t) = 1/2$) of the square wave discussed in the first example of Chapter 16, we have $g(t) = 1 - f(t)$. We can either follow the same procedure as that in the first example, or we can use the result

$$f(t) = \frac{1}{2} - \frac{2}{\pi}[\sin t + \frac{\sin 3t}{3} + \frac{\sin 5t}{5} + ...],$$

and we then get

$$g(t) = 1 - f(t) = 1 - \left\{\frac{1}{2} - \frac{2}{\pi}[\sin t + \frac{\sin 3t}{3} + \frac{\sin 5t}{5} + ...]\right\}$$
$$= \frac{1}{2} + \frac{2}{\pi}\Big[\sin t + \frac{\sin 3t}{3} + \frac{\sin 5t}{5} + \frac{\sin 7t}{t} + ...\Big].$$

16.2 From the example in Chapter 16 about the triangular waves, we know that

$$K = 1 + \frac{1}{3^2} + \frac{1}{5^2} + \frac{1}{7^2} + \frac{1}{9^2} + ... = \frac{\pi^2}{8}.$$

Let S be the sum of series

$$S = 1 + \frac{1}{2^2} + \frac{1}{3^2} + \frac{1}{4^2} + ... + \frac{1}{n^2} + ...$$
$$= \Big(1 + \frac{1}{3^2} + \frac{1}{5^2} + \frac{1}{7^2} + ...\Big) + \Big(\frac{1}{2^2} + \frac{1}{4^2} + \frac{1}{6^2} + \frac{1}{8^2} + ...\Big)$$
$$= \frac{\pi^2}{8} + \frac{1}{2^2}\Big(1 + \frac{1}{2^2} + \frac{1}{3^2} + \frac{1}{4^2} + ...\Big) = \frac{\pi^2}{8} + \frac{1}{4}S,$$

where we have used that the sum in the second part of the above equation is the original sum of series. So we have

$$S - \frac{1}{4}S = \frac{3S}{4} = \frac{\pi^2}{8}.$$

Thus, we finally have

$$S = \frac{4}{3} \cdot \frac{\pi^2}{8} = \frac{\pi^2}{6}.$$

16.3 Since $f(t) = |\sin t|$ is an even function because $f(-t) = |\sin(-t)| = |-\sin t| = f(t)$ with $T = \pi$, its Fourier series only contains cosine terms with $a_n \neq 0$ (and $b_n = 0$). First, let us calculate a_0 and we have

$$a_0 = \frac{1}{\pi}\int_{-\pi}^{\pi} |\sin t| dt = \frac{2}{\pi}\int_0^{\pi} \sin t dt$$
$$= \frac{2}{\pi}\Big[-\cos t\Big]\Big|_0^{\pi} = -\frac{2}{\pi}[\cos(\pi) - \cos 0)] = -\frac{2}{\pi}[-1 - 1] = \frac{4}{\pi}.$$

In addition, we have

$$a_n = \frac{1}{\pi}\int_{-\pi}^{\pi} |\sin t| \cos(nt) dt.$$

Since both $|\sin(t)|$ and $\cos(nt)$ are even functions, the integrand is also even. Thus, we get

$$a_n = \frac{2}{\pi}\int_0^{\pi} \sin t \cos(nt) dt.$$

It leaves as a further exercise to prove that

$$\int_0^{\pi} \sin(t)\cos(nt)dt = \begin{cases} 0 & \text{for odd } n \\ -\frac{2}{n^2-1} & \text{for even } n. \end{cases}$$

As n is even, we can set $n = 2k$ $(k = 1, 2, 3, ...)$ and we have

$$a_n = \frac{2}{\pi} \cdot \frac{(-2)}{n^2 - 1} = -\frac{4}{\pi} \cdot \frac{1}{4k^2 - 1}.$$

So the Fourier series becomes

$$f(t) = |\sin t| = \frac{4}{\pi} \cdot \frac{1}{2} - \frac{4}{\pi} \sum_{k=1}^{\infty} \frac{\cos(2kt)}{4k^2 - 1} = \frac{2}{\pi} \Big[1 - 2 \sum_{k=1}^{\infty} \frac{\cos(2kt)}{4k^2 - 1} \Big].$$

The Fourier series for $g(t) = Af(t)$ is simply the Fourier series multiplied each term by A.

CHAPTER 17

17.1 For $\sinh(at)$, its Laplace transform can be obtained by using either direct integration or the Laplace transform table. From $\sinh(at) = (e^{at} - e^{-at})/2$ and

$$\mathcal{L}[e^{at}] = \frac{1}{s+a},$$

we have

$$\mathcal{L}[\sinh(at)] = \frac{1}{2}\{\mathcal{L}[e^{at}] - \mathcal{L}[e^{-at}]\} = \frac{1}{2}[\frac{1}{s-a} - \frac{1}{s+a}]$$
$$= \frac{1}{2}\Big[\frac{(s+a)}{(s-a)(s+a)} - \frac{(s-a)}{(s+a)(s-a)}\Big] = \frac{1}{2} \cdot \frac{2a}{s^2 - a^2} = \frac{a}{s^2 - a^2}.$$

17.2 For the Laplace transform of $\sin(at + b)$, we first use

$$\sin(at + b) = \sin(at)\cos b + \cos(at)\sin b.$$

Then, we have

$$\mathcal{L}[\sin(at + b)] = \mathcal{L}[\sin(at)\cos b] + \mathcal{L}[\cos(at)\sin b]$$
$$= \cos b \mathcal{L}[\sin(at)] + \sin b \mathcal{L}[\cos(at)].$$

Using

$$\mathcal{L}[\sin(at)] = \frac{a}{s^2 + a^2}, \quad \mathcal{L}[\cos(at)] = \frac{s}{s^2 + a^2},$$

we have

$$\mathcal{L}[\sin(at + b)]] = \cos b \cdot \frac{a}{s^2 + a^2} + \sin b \cdot \frac{s}{s^2 + a^2} = \frac{a \cos b + s \sin b}{s^2 + a^2}.$$

17.3 The transfer function

$$H(s) = \frac{s^2 + 1}{s^2 + 3s + 2},$$

has two zeros $+i$ and $-i$ [from $s^2 + 1 = 0$] and two poles -1 and -2 [from $s^2 + 3s + 2 = (s + 1)(s + 2) = 0$].

17.4 To obtain the solution of $y'(t) - y(t) = e^{-t}$ with the initial condition $y(0) = 0$, we first apply Laplace transforms on both sides of the equation and we have

$$\mathcal{L}[y'(t)] - \mathcal{L}[y(t)] = \mathcal{L}[e^{-t}],$$

which gives

$$sY(s) - y(0) - Y(s) = \frac{1}{s+1},$$

or

$$sY(s) - Y(s) = \frac{1}{s+1},$$

where we have used $y(0) = 0$. Solving for $Y(s)$, we get

$$Y(s) = \frac{1}{(s-1)(s+1)},$$

which can be rewritten as partial fractions in the form of

$$Y(s) = \frac{1}{2}[\frac{1}{s-1} - \frac{1}{s+1}].$$

Now we use the Laplace transform pairs to get the inverse and the above equation becomes

$$y(t) = \mathcal{L}^{-1}[Y(s)] = \frac{1}{2}\{\mathcal{L}^{-1}[\frac{1}{s-1}] - \mathcal{L}^{-1}[\frac{1}{s+1}]\} = \frac{1}{2}[e^t - e^{-t}].$$

So the final solution is

$$y(t) = \frac{1}{2}e^t - \frac{1}{2}e^{-t} = \sinh(t).$$

CHAPTER 18

18.1 Data analysis suggests that the number of goals scored by a team in the World Cup obeys a Poisson distribution with a mean $\lambda = 2.5$ (Ugarte MD, Militino AF, Arnholt AT, *Probability and Statistics with R*, 2nd Edition, CRC Press, 2016; also, see https://en.wikipedia.org/wiki/Poisson_distribution).
The probability of n goals can be calculated by

$$P(n) = \frac{\lambda^n e^{-\lambda}}{n!} = \frac{2.5^n e^{-2.5}}{n!}.$$

Thus, the probability of no goal at all (i.e., $n = 0$) is

$$P(0) = \frac{2.5^0 e^{-2.5}}{0!} = e^{-2.5} \approx 0.0821.$$

Similarly, an exactly one goal match has a probability of

$$P(1) = \frac{2.5^1 e^{-2.5}}{1!} = 2.5e^{-2.5} \approx 0.2052.$$

Thus, the probability of scoring 2 or more goals $P(n \geq 2)$ satisfies

$$P(0) + P(1) + P(n \geq 2) = 1.$$

That is, the three events complete the sample space and thus their total probability is one. Therefore, we finally get

$$P(n \geq 2) = 1 - P(0) - P(1) = 1 - 0.0821 - 0.2052 = 0.7127.$$

18.2 The clean tile problem was first studied by French naturalist Buffon (see the Wolfram article at http://mathworld.wolfram.com/CleanTileProblem.html). For a coin of diameter d randomly dropped on a regular square grid (covered by square tiles) with tile edge length L, the probability of the coin lies inside a tile is related to area ratio

$$P = \frac{(L-d)^2}{L^2}.$$

Let Q be probability of the coin landing on a line between two adjacent tiles or at a corner (involving three or four tiles). We have

$$P + Q = 1.$$

Thus, we get

$$Q = 1 - P = 1 - \frac{(L-d)^2}{L^2}.$$

For example, if $L = 2d$, we have

$$P = \frac{(L-d)^2}{L^2} = \frac{(2d-d)}{(2d)^2} = \frac{1}{4}, \quad Q = 1 - \frac{1}{4} = \frac{3}{4}.$$

Interestingly, the probability of the coin landing on the corner of exactly four adjacent tiles is

$$P_{\text{corner}} = \frac{\text{area of the coin}}{\text{area of a tile}} = \frac{\pi(d/2)^2}{L^2} = \frac{\pi}{4}\frac{d^2}{L^2}.$$

In the case of $L = d$, we have

$$P_{\text{corner}} = \frac{\pi}{4} \cdot \frac{d^2}{d^2} = \frac{\pi}{4},$$

which can be used to estimate π by dropping multiple coins.

18.3 For Laplace's distribution

$$p(x) = \frac{1}{2a} \exp\Big(-\frac{|x-\mu|}{a}\Big), \quad a > 0,$$

we have its cumulative probability distribution

$$F(x) = \int_{-\infty}^{x} p(x)dx = \frac{1}{2a}\int_{-\infty}^{x} e^{-|x-\mu|/a}dx.$$

If $x < \mu$, we have

$$F(x) = \frac{1}{2a}\int_{-\infty}^{x} e^{-(\mu-x)/a}dx = \frac{1}{2a}\Big[ae^{-(\mu-x)/a}\Big|_{-\infty}^{x}\Big] = \frac{1}{2a}[ae^{(x-\mu)/a} - 0] = \frac{1}{2}e^{(x-\mu)/a}.$$

If $x \geq \mu$, we have

$$\begin{aligned} F(x) &= \frac{1}{2a}\Big[\int_{-\infty}^{\mu} e^{-(\mu-x)/a}dx + \int_{\mu}^{x} e^{-(x-\mu)/a}dx\Big] \\ &= \frac{1}{2a}\Big[e^{-\mu/a}\int_{-\infty}^{\mu} e^{x/a}dx + e^{\mu/a}\int_{\mu}^{x} e^{-x/a}dx\Big]. \end{aligned}$$

Thus we have

$$\begin{aligned} F(x) &= \frac{1}{2a} \cdot \Big[e^{-\mu/a}ae^{x/a}\Big|_{-\infty}^{\mu} + e^{\mu/a}(-a)e^{-x/a}\Big|_{\mu}^{x}\Big] \\ &= \frac{1}{2a}\Big[ae^{-\mu/a}(e^{\mu/a} - 0) - e^{\mu/a}a(e^{-x/a} - e^{-\mu/a})\Big] \\ &= \frac{1}{2a}[2a - ae^{-(x-\mu)/a}] = 1 - \frac{1}{2}e^{-(x-\mu)/a}. \end{aligned}$$

18.4 For the Rayleigh distribution, we have its probability density

$$r(x) = \frac{x}{\sigma^2}e^{-x^2/(2\sigma^2)},$$

we have its cumulative probability distribution

$$F(x) = \int_0^x \frac{x}{\sigma^2} e^{-x^2/(2\sigma^2)} dx = \frac{1}{2\sigma^2} \int_0^x e^{-x^2/(2\sigma^2)} d(x^2)$$
$$= \frac{1}{2\sigma^2}\Big[-2\sigma^2 e^{-x^2/(2\sigma^2)} \Big|_0^x \Big] = -[e^{-x^2/(2\sigma^2)} - e^{0^2/(2\sigma^2)}] = 1 - e^{-x^2/(2\sigma^2)}.$$

The mean of this distribution is

$$\mu = \int_0^\infty x \cdot \frac{x}{\sigma^2} e^{-x^2/(2\sigma^2)} dx = \frac{1}{\sigma^2} \int_0^\infty x^2 e^{-x^2/(2\sigma^2)} dx.$$

It leaves as a further exercise (using integration by parts and the definition of the error function) to show that

$$\int_0^\infty x^2 e^{-x^2/a} dx = \frac{\sqrt{\pi}}{4} a^{3/2},$$

and thus we have

$$\mu = \frac{1}{\sigma^2} \cdot \frac{\sqrt{\pi}}{4} (2\sigma^2)^{3/2} = \sigma \sqrt{\frac{\pi}{2}}.$$

So we have

$$F(x) = \begin{cases} \frac{1}{2} e^{(x-\mu)/a} & \text{if } x < \mu, \\ 1 - \frac{1}{2} e^{-(x-\mu)/a} & \text{if } x \geq \mu. \end{cases}$$

18.5 From the moment-generating function for the normal distribution

$$G(t) = e^{\mu t + \frac{1}{2}\sigma^2 t^2},$$

we have

$$\mu = m_1 = \frac{dG(t)}{dt}\Big|_{t=0} = e^{\mu t + \frac{1}{2}\sigma^2 t^2} \cdot [\mu + \frac{1}{2}\sigma^2 (2t)]\Big|_{t=0} = e^{0 + \frac{1}{2}\sigma^2 0^2} (\mu + \sigma^2 \times 0) = \mu.$$

The second moment m_2 is

$$m_2 = \frac{d^2 G(0)}{dt^2} = \Big[e^{\mu t + \frac{1}{2}\sigma^2 t^2} \sigma^2 + (\mu + \sigma t)^2 e^{\mu t + \frac{1}{2}\sigma^2 t^2} \Big]_{t=0}$$
$$= e^{0 + \frac{1}{2}\sigma^2 \times 0^2} \sigma^2 + (\mu + \sigma^2 \times 0)^2 e^{0 + \frac{1}{2}\sigma^2 \times 0^2} = \sigma^2 + \mu^2.$$

Thus, the variance is

$$\mathrm{Var}(X) = m_2 - \mu^2 = (\sigma^2 + \mu^2) - \mu^2 = \sigma^2.$$

CHAPTER 19

19.1 The sample mean of $x_i = 1, 2, 3, 4, 5, 6, 7, 8$ is $\bar{x} = 4.5$ and the sample variance is $S^2 = 6$.

19.2 In order to best fit a function $y = \exp[ax^2 + b\sin(x)]$, we first take the logarithm of both sides and we have

$$\ln y(x) = \ln e^{-ax^2 + b \sin x} = ax^2 + b\sin(x),$$

which can be fitted using the generalized linear regression method.

19.3 It seems that Robot B is always quicker than Robot A from a quick look at the data:
Robot A: $A_i = 23,\ 21,\ 19,\ 40,\ 20,\ 27,\ 25.$
Robot B: $B_i = 22,\ 18,\ 18,\ 25,\ 17,\ 21,\ 23.$

However, this impression does not guarantee that their differences are truly significant in the statistical sense. In fact, let us assume that these two robots have the same mean time with the null hypothesis:

$$\mu_A = \mu_B.$$

It is straightforward to calculate that

$$\bar{A} = 25, \quad \bar{B} = 20.57.$$

Thus combined sample variance with $n = 7$ can be computed by

$$S^2 = \frac{1}{n-1} \sum_{i=1}^{n} [(A_i - \bar{A}) - (B_i - \bar{B})]^2$$
$$= \frac{1}{6}\{[(23-25)-(22-20.57)]^2 + \dots + [(25-25)-(23-20.57)]^2\} = 24.62,$$

which means that $S = \sqrt{24.62} = 4.96$. So we have

$$t = \frac{\bar{A} - \bar{B}}{S/\sqrt{n}} = \frac{(25 - 20.57)}{4.96/\sqrt{7}} = 2.36.$$

For the degree of freedom is $n - 1 = 6$, the critical value to reject the null hypothesis is $t* = 2.45$ at the 95% interval. Since $t < t_*$, the null hypothesis holds. That is to say, the claim by the manufacturer (B is quicker than A) is essentially false because it has no statistical significance.

CHAPTER 20

20.1 Either the bisection method or Newton's method can be used to obtain $\sqrt{\pi} = 1.77245385$.

20.2 The solution to $e^{-x} - x^2 = 0$ is approximately 0.703467422498.

20.3 The integral $I = \int_0^1 \sqrt{1 - x^{3/2}} dx$ does not have any simple formula because it involves some hyper-geometrical function, and it can be evaluated by numerical integration to give $I = 0.73916994929$.

20.4 The equation of $y'(x) - xy(x) = x$ with $y(0) = 0$ has an analytical solution $y(x) = e^{x^2/2} - 1$. It can be solved by (for example) Euler's method.

CHAPTER 21

21.1 The determinant of $\mathbf{L} = \begin{pmatrix} 1 & 0 & 0 \\ 1 & 1 & 0 \\ 3 & 2 & 1 \end{pmatrix}$ is 1 and the inverse is

$$\mathbf{L}^{-1} = \begin{pmatrix} 1 & 0 & 0 \\ -1 & 1 & 0 \\ -1 & -2 & 1 \end{pmatrix}.$$

21.2 The LU decomposition of **A** is

$$\mathbf{L} = \begin{pmatrix} 1 & 0 & 0 \\ 1 & 1 & 0 \\ 3 & 2 & 1 \end{pmatrix}, \quad \mathbf{U} = \begin{pmatrix} 2 & 3 & 2 \\ 0 & -1 & 2 \\ 0 & 0 & 7 \end{pmatrix}, \quad \mathbf{LU} = \begin{pmatrix} 2 & 3 & 2 \\ 2 & 2 & 4 \\ 6 & 7 & 17 \end{pmatrix}.$$

The determinant of det(**A**) is -14, and the inverse of **A** is

$$\mathbf{A}^{-1} = \frac{1}{14} \begin{pmatrix} -6 & 37 & -8 \\ 10 & -22 & 4 \\ -2 & -4 & 2 \end{pmatrix}.$$

21.3 The linear system can be written as

$$\begin{pmatrix} 2 & 3 & 2 \\ 2 & 2 & 4 \\ 6 & 7 & 17 \end{pmatrix}\begin{pmatrix} x \\ y \\ z \end{pmatrix} = \begin{pmatrix} 10 \\ 10 \\ 37 \end{pmatrix}.$$

You can first use the LU-decomposition and the solve the resulting equations sequentially. The final solution is

$$x = 1, \quad y = 2, \quad z = 1.$$

21.4 Use either Jacobi or Gauss-Seidel iteration method to solve the linear system and re-arrange the order of the equations when necessary so as to make the matrix diagonally dominant. Then, it can be implemented in any programming language.

CHAPTER 22

22.1 For the worked example, the extra constraint $2x_1 + 3x_2 \le 70$ does not affect the feasible domain because $2x_1 + 3x_2 = 70$ gives a line outside the region imposed by other constraint. This extra constraint does not have any effect, and thus the solution will remain the same $x_1 = x_2 = 10$.

22.2 The constraint $x_1 \le x_2/2$ will make the feasible region smaller, and thus may affect the locality of the optimal solution. In fact, if we follow the similar procedure, we will obtain

$$x_1 = 5, \quad x_2 = 10, \quad f_{\text{opt}} = 2 \times 5 + 3 \times 10 = 40.$$

22.3 The simplex method can be used to solve the linear program and the optimal solution is

$$x = 4, \quad y = 14, \quad z = 0,$$

so that

$$f_{\text{opt}} = 2 \times 4 + 3 \times 14 + 4 \times 0 = 50.$$

22.4 This is a very small project management problem and can be modeled by linear programming. For two products (A with x units and B with y), we have the budget limit

$$5x + 10y \le 100.$$

The time constraint can be translated into

$$2x + 4y \le 50.$$

The packaging constraint is

$$2x + y \le 25.$$

The overall profit is the objective

$$P = 3x + 5y.$$

Obviously, we cannot produce negative numbers of products, which means $x, y \ge 0$. Thus, the linear programming problem becomes

$$\text{maximize } P = 3x + 5y,$$

subject to

$$\begin{cases} 2x + 4y \le 50, \\ 5x + 10y \le 100, \\ 2x + y \le 25, \\ x \ge 0, \\ y \ge 0. \end{cases}$$

Following the same procedure outlined in the chapter, we can obtain

$$x = 10, \quad y = 5, \quad f_{\rm opt} = 3 \times 10 + 5 \times 5 = 55.$$

CHAPTER 23

23.1 This function $f = (1-x)^2 + 100(y-x^2)^2$ is called Rosenbrock's function or the banana function because the global minimum $x = y = 1$ lies in a narrow, flat, banana-shaped valley. The stationary conditions are

$$\frac{\partial f}{\partial x} = 2(1-x)(-1) + 200(y-x^2)(-2x) = 0, \quad \frac{\partial f}{\partial y} = 200(y-x^2) = 0,$$

or

$$2(1-x) + 400x(y-x^2) = 0, \quad 200(y-x^2) = 0.$$

The solution is $x = 1$ and $y = x^2 = 1$ with the minimum of $f_* = 0$.

23.2 The constrained problem

$$f(x, y) = x^2 + y^2 + 2xy = (x+y)^2,$$

subject to

$$y - x^2 + 2 = 0,$$

has three optimal solutions $(x, y) = (1, -1)$, $(-2, 2)$ and $(-\frac{1}{2}, -\frac{7}{4})$. The solution $(-1/2, -7/4)$ corresponds to the global maximum $f_{\max} = 81/16$, while the two solutions $(1, -1)$ and $(-2, 2)$ correspond to the two minima with $f_{\min} = 0$. You can obtain the solution by the method of Lagrange multipliers and/or spotting the solution by using the fact that $x + y = 0$ for the minima.

23.3 For $f = x^2 + 5y^2 + xy$, we have its gradient

$$\nabla f = (\frac{\partial f}{\partial x}, \frac{\partial f}{\partial y})^T = (2x + y, \ 10y + x)^T,$$

and the Hessian

$$\mathbf{H} = \begin{pmatrix} \frac{\partial^2 f}{\partial x^2} & \frac{\partial^2 f}{\partial x \partial y} \\ \frac{\partial^2 f}{\partial y \partial x} & \frac{\partial^2 f}{\partial y^2} \end{pmatrix} = \begin{pmatrix} 2 & 1 \\ 1 & 10 \end{pmatrix}.$$

Its eigenvalues are $6 + \sqrt{17}$ and $6 - \sqrt{17}$. Since both eigenvalues are positive, the Hessian is positive definite and thus the function has a minimum at $x = y = 0$.

23.4 To find the minimum of $f = \ln[x^2 + (y-x)^2 + 10(z-y)^2 + 1]$, we can take its exponential and get

$$g = e^{\ln f} = x^2 + (y-x)^2 + 10(z-y)^2 + 1.$$

It is easy to solve this unconstrained optimization problem because it is a quadratic function (thus convex). We obtain its global minimum

$$x = y = z = 0, \quad g_{\min} = 1, \quad f_{\min} = 0.$$

23.5 This is number 4 of Xin-She Yang's functions, and it takes the following form

$$f(x_1, ..., x_n) = \Big[\sum_{i=1}^{n} \sin^2(x_i) - e^{-\sum_{i=1}^n x_i^2}\Big] e^{-\sum_{i=1}^n \sin^2 \sqrt{|x_i|}},$$

where $-10 \le x_i \le 10$ for $i = 1, 2, ..., n$, and $n \ge 2$ is an integer. It has a global minimum at $(0, 0, ..., 0)$ with $f_{\min} = -1$. However, it is not straightforward to find this optimal solution. The gradient is not easy to calculate due to $\sin^2(x_i)$ terms and $\sqrt{|x_i|}$ terms. You can try to solve it using Newton's method for the case of $n = 2$ to see if it works.

CHAPTER 24

24.1 The type of each PDE is given as follows:

- $xu_x + yu_y = xyu$ is a linear first-order PDE.
- $(u_t)^2 = u_x + xt$ is a nonlinear PDE.
- $u_t + u^2 u_x = 0$ is a nonlinear PDE.
- $u_{tt} = tu_{xx}$ is a mixed type. It is parabolic if $t > 0$ and hyperbolic for $t < 0$. This is the well-known Euler-Tricomi equation.
- $u_t = D\nabla^2 u + \gamma u(1-u)$ is a nonlinear parabolic equation, which is also known as a nonlinear reaction-diffusion equation.

24.2 The so-called traffic flow equation takes the form

$$\rho_t + \frac{\partial Q(\rho)}{\partial x} = 0, \quad Q(\rho) = u_0 \rho (1 - \rho/\rho_{\max}),$$

which is a nonlinear first-order PDE. Here, ρ represents the traffic density on a busy road. The characteristic flow velocity is $v = Q'(\rho)$, that is

$$v = Q'(\rho) = u_0[\rho - \rho^2/\rho_{\max}]' = u_0(1 - 2\rho/\rho_{\max}),$$

which gives a critical value defined by

$$1 - 2\rho/\rho_{\max} = 0,$$

or

$$\rho_c = \rho_{\max}/2.$$

For more details, please refer to the book by May A, *Traffic Flow Fundamentals*, Prentice-Hall, NJ (1990).

24.3 To get a general solution of

$$u_x + u^n u_y = 0, \quad (n \geq 1),$$

we can use the method of characteristics. We have

$$\frac{dx}{dt} = 1, \quad \frac{dy}{dt} = u^n, \quad \frac{du}{dt} = 0,$$

for a parameter t on the characteristic curves. The integral of $\frac{du}{dt} = 0$ gives

$$u = h,$$

where h is a function depending on the initial condition. The integral of $\frac{dx}{dt} = 1$ gives $x = t$ if we start with $t = 0$. Thus, we have

$$\frac{dy}{dt} = u^n = h^n,$$

whose integral is

$$y = h^n t + p,$$

where p is an unknown constant or parameter. By re-arranging the above, we get

$$p = y - h^n x = y - u^n x,$$

which is a constant along a characteristic curve. Thus, the general solution can be written as

$$u = h = h(y - u^n x).$$

24.4 For Laplace's equation in 3D

$$\nabla^2 u = \frac{\partial^2 u}{\partial x^2} + \frac{\partial^2 u}{\partial y^2} + \frac{\partial^2 u}{\partial z^2} = 0,$$

initial conditions and boundary conditions are needed. Thus, its solution can be very lengthy and interested readers can refer to more advanced literature listed at the end of the book. However, for a given solution, it is relatively straightforward to verify the solution by direct differentiation. From $u = 1/\sqrt{x^2+y^2+z^2}$, we have

$$\frac{\partial^2 u}{\partial x^2} = \frac{3x^2}{(x^2+y^2+z^2)^{5/2}} - \frac{1}{(x^2+y^2+z^2)^{3/2}},$$
$$\frac{\partial^2 u}{\partial y^2} = \frac{3y^2}{(x^2+y^2+z^2)^{5/2}} - \frac{1}{(x^2+y^2+z^2)^{3/2}},$$
$$\frac{\partial^2 u}{\partial z^2} = \frac{3z^2}{(x^2+y^2+z^2)^{5/2}} - \frac{1}{(x^2+y^2+z^2)^{3/2}},$$

and thus we have

$$\nabla^2 u = \frac{\partial^2 u}{\partial x^2} + \frac{\partial^2 u}{\partial y^2} + \frac{\partial^2 u}{\partial z^2} = \frac{3(x^2+y^2+z^2)}{(x^2+y^2+z^2)^{5/2}} - \frac{3}{(x^2+y^2+z^2)^{3/2}}$$
$$= \frac{3}{(x^2+y^2+z^2)^{3/2}} - \frac{3}{(x^2+y^2+z^2)^{3/2}} = 0.$$

CHAPTER 25

25.1 From $\mathbf{J} = \sigma \mathbf{E}$, we can simply write

$$\mathbf{J}_i = \sigma_{ij} E_j.$$

25.2 The rotation energy is simply

$$E_K = \frac{1}{2} I_{ij} \omega_i \omega_j.$$

25.3 To prove the identity, first use the definition of the cross product to show that

$$\mathbf{v} \times \mathbf{w} = (\mathbf{v} \times \mathbf{w})_i = \begin{vmatrix} \mathbf{e}_1 & \mathbf{e}_2 & \mathbf{e}_3 \\ v_1 & v_2 & v_3 \\ w_1 & w_2 & w_3 \end{vmatrix} = \epsilon_{ijk} v_j w_k \mathbf{e}_i,$$

then use the definition of the triple vector product and the properties of ϵ_{ijk} to show

$$\mathbf{u} \cdot (\mathbf{v} \times \mathbf{w}) = \epsilon_{ijk} u_i v_j w_k.$$

CHAPTER 26

26.1 Since $y = \sin(x)$, we have $y' = \cos(x)$ and $y'' = -\sin(x)$. Thus, the curvature can be calculated by

$$\kappa(x) = \left| \frac{-\sin(x)}{[1 + (\cos x)^2]^{3/2}} \right| = \frac{\sin x}{(1+\cos^2 x)^{3/2}}.$$

At $x = \pi/2$, we have $\cos(\pi/2) = 0$ and $\sin(\pi/2) = 1$, so we have

$$\kappa = \frac{1}{(1+0)^{3/2}} = 1.$$

26.2 From $I = \int_a^b [\frac{1}{2}y'^2(x) + x^2 y(x)]dx$, we have its integrand $f = \frac{1}{2}y'^2 + x^2 y$. Since

$$\frac{\partial f}{\partial y} = x^2, \quad \frac{\partial f}{\partial y'} = y',$$

we have the Euler-Lagrange equation

$$\frac{\partial f}{\partial y} - \frac{d}{dx}\left(\frac{\partial f}{\partial y'}\right) = x^2 - \frac{dy'}{dx} = x^2 - y'' = 0,$$

whose solution can be obtained by integrating twice with respect to x. Thus, we have

$$y'(x) = \frac{x^3}{3} + A,$$

and

$$y(x) = \frac{x^4}{12} + Ax + B,$$

where A and B are two integration constants.

26.3 From $I[u(x)] = \int_0^1 \frac{1}{2}[u'(x)]^2 dx$ and $J = [u(x)] = \int_0^1 u(x)dx = 1$, we have

$$f(x) = \frac{1}{2}u'^2, \quad g(x) = u(x),$$

and we can then define

$$\psi(x) = f(x) + \lambda g(x) = \frac{u'^2(x)}{2} + \lambda u(x).$$

Thus, we have

$$\frac{\partial \psi}{\partial u} = \lambda, \quad \frac{\partial \psi}{\partial u'} = u',$$

and the Euler-Lagrange equation becomes

$$\frac{\partial \psi}{\partial u} - \frac{d}{dx}\left(\frac{\partial \psi}{\partial u'}\right) = \lambda - \frac{du'}{dx} = \lambda - u'' = 0.$$

Integrating twice, we get

$$u(x) = \frac{\lambda x^2}{2} + Ax + B.$$

From the boundary condition $u(0) = 0$, we have

$$u(0) = 0 = \frac{\lambda 0^2}{2} + A \times 0 + B,$$

which gives $B = 0$. From $u(1) = 0$, we have

$$u(1) = 0 = \frac{\lambda}{2}1^2 + A \times 1,$$

which gives $A = -\lambda/2$. Now the solution becomes

$$u(x) = \frac{\lambda}{2}(x^2 - x).$$

From the constraint $\int_0^1 u(x)dx = 1$, we get

$$\int_0^1 \frac{\lambda}{2}(x^2 - x)dx = \frac{\lambda}{2}\Big[\frac{x^3}{3} - \frac{x^2}{2}\Big]_0^1 = 1,$$

which gives

$$\frac{\lambda}{2}(-\frac{1}{6}) = 1,$$

or

$$\lambda = -12.$$

So the final solution is

$$u(x) = -6(x^2 - x).$$

It is straightforward to verify that it indeed satisfies all the conditions.

CHAPTER 27

27.1 The types of equations are given as follows:

- $y(x) = \cos(x) + \frac{1}{3}\int_0^{\pi} xsy(s)ds$ is a linear integral equation of Fredholm type.
- $y(x) = x^2 e^{-x} + \int_0^x (x-s)y(s)ds$ is a linear Volterra integral equation.
- $y(x) = f(x) + \lambda \int_0^x (x-s)\sin[y(s)]ds$ is a nonlinear Volterra integral equation.
- $y'(x) = x + \frac{1}{2}\int_0^1 xsy^2(s)ds$ is an integro-differential equation of Fredholm type.

27.2 There are a few ways to solve this equation. Here, we use the successive approximation method in terms of a series or simply the series method From $y(x) = a + b\int_0^1 xy(s)ds$ with $a > 0$ and $b = 1$, we have

$$y_0(x) = a, \quad y_1(x) = 1 + \int_0^1 xy_0 ds = a(1+x),$$

$$y_2(x) = a + \int_0^1 xy_1(s)ds = a + \int_0^1 xa[1+s]ds = a + ax[1+\frac{1}{2}],$$

$$y_3(x) = a + \int_0^1 xy_2(s)ds = a + \int_0^1 xa[1+s(1+\frac{1}{2})]ds = a + ax[1+\frac{1}{2}+\frac{1}{2^2}].$$

Following the same procedure, we can get

$$y_{n+1}(x) = a + ax[1 + \frac{1}{2} + \frac{1}{2^2} + \dots + \frac{1}{2^n}].$$

As $n \to \infty$, we finally have

$$y(x) = a + ax\sum_{i=0}^{\infty}\frac{1}{2^i} = a(1+2x).$$

27.3 We will use the same method as used in the previous exercise to solve

$$y(x) = a + \int_0^x y(s)ds.$$

Using successive approximations, we have

$$y_0 = a, \quad y_1 = a + \int_0^x y_0(s)ds = a + \int_0^x a ds = a(1+x),$$

$$y_2 = a + \int_0^x a(1+s)ds = a(1 + x + \frac{x^2}{2}).$$

Following the same procedure, we have

$$y_n(x) = a\Big(1 + x + \frac{x^2}{2} + \ldots + \frac{x}{n!}\Big).$$

As $n \to \infty$, we have

$$y(x) = a\Big(1 + x + \frac{x^2}{2!} + \frac{x^3}{3!} + \ldots\Big) = ae^x.$$

CHAPTER 28

28.1 The Lorenz equations are a set of three ordinary differential equations

$$\frac{du}{dt} = \sigma(v - u), \quad \frac{dv}{dt} = u(\rho - w) - v, \quad \frac{dw}{dt} = uv - \beta w,$$

where $\sigma, \rho, \beta > 0$ are constants. Use any programming language to solve the system and see if you can reproduce the so-called Lorenz attractor (Lorenz EN, Deterministic nonperiodic flow, *J. Atmospheric Sciences*, vol. 20, no. 2, 130–141, 1963).

28.2 For bungee jumping, safety is crucially important and thus it is necessary to correctly calculate the key parameters such as the length of the rope given the weight $W = mg$ and height h of a jumper. Let L be the length of the rope, and k be the spring constant of the rope. If the jumper jumps at a height H on a bridge above the target level (e.g., river surface), the maximum extension can be estimated by using the energy conservation; that is, the potential energy is converted into the elastic energy in the rope

$$mg(L + D) = \frac{1}{2}kD^2, \tag{C.1}$$

where D is the extension of the rope. Obviously, if $L + D \le H - h$, then it is safe to do the bungee jump. However, the detailed motion can be described by a differential equation. If we take $H - L$ as $x = 0$ with the x-axis pointing downwards, then the governing equation becomes

$$m\frac{d^2x}{dt^2} = mg - f(x), \tag{C.2}$$

where

$$f(x) = \begin{cases} kx & x > 0, \\ 0 & x \le 0. \end{cases} \tag{C.3}$$

Since it is basically a free fall when $x \le 0$, the time taken for this free fall is $t_0 = \sqrt{2L/g}$. The rope starts to extend when $t > t_0$. The main characteristics of motion will be dominated by $x > 0$. The equilibrium state is at $x_* = mg/k$. With the proper initial condition such as the initial zero velocity, you should obtain a solution as follows:

$$x(t) = x_* - x_* \cos[\omega(t - t_0)] + \frac{\sqrt{2gL}}{\omega}\sin[\omega(t - t_0)].$$

28.3 The well-known SIR model has three equations:

$$\frac{dS}{dt} = -\beta IS, \quad \frac{dI}{dt} = \beta IS - \gamma I, \quad \frac{dR}{dt} = \gamma I,$$

where β and γ are constants. Among the total population N, S is the number of susceptible population to the flu, I is the number of the infectious population and R is the number of recovered in the population. Obviously, $S + N + R = N$. For more details, please see more advanced literature (Hethcote HW, The mathematics of Infectious Diseases, *SIAM Review*, vol. 42, No. 4, 599–653, 2000).

28.4 This is so-called Kolmogorov-Petrovskii-Piskunov (KPP) equation for reaction-diffusion and pattern formation. To solve this KPP equation

$$\frac{\partial u}{\partial t} = D(\frac{\partial^2 u}{\partial x^2} + \frac{\partial^2 u}{\partial y^2}) + \gamma u(1-u),$$

on a 2D grid with $D = 0.2$ and $\gamma = 0.5$, you can use finite difference methods with periodic boundary conditions for all four sides and random initial conditions. Then, beautiful patterns with rings and stripes will emerge as the system evolves.

28.5 To do the hand-written digit recognition, you need some machine learning techniques such as neural networks, support vector machines as well as a training data set. There are a vast set of literature that you can find and also some software packages on the Internet.

Bibliography

1. Abramowitz M, Stegun IA. *Handbook of Mathematical Functions*. 10th ed. Washington, DC: US Government Printing Office; 1972. National Bureau of Standards, Applied Mathematics Series; vol. 55.
2. Arfken GB, Weber HJ, Harris FE. *Mathematical Methods for Physicists: A Comprehensive Guide*. 7th ed. Waltham: Academic Press; 2012.
3. Armstrong M. *Basic Linear Geostatistics*. Berlin: Springer; 1998.
4. Atluri SN. *Methods of Computer Modeling in Engineering and the Sciences. Vol. I*. Forsyth: Tech Science Press; 2005.
5. Bird J. *Engineering Mathematics*. 7th ed. New York: Routeledge; 2014.
6. Bird J. *Higher Engineering Mathematics*. 7th ed. New York: Routeledge; 2014.
7. Carrrier GF, Pearson CE. *Partial Differential Equations: Theory and Technique*. 2nd ed. Waltham: Academic Press; 1988.
8. Carslaw HS, Jaeger JC. *Conduction of Heat in Solids*. 2nd ed. Oxford: Oxford University Press; 1986.
9. Chatterjee A, Siarry P. Nonlinear inertia variation for dynamic adaptation in particle swarm optimisation. *Comput Oper Res*. 2006;33:859–871.
10. Courant R, Hilbert D. *Methods of Mathematical Physics. 2 volumes*. New York: Wiley-Interscience; 1962.
11. Crank J. *Mathematics of Diffusion*. Oxford: Clarendon Press; 1970.
12. Croft A, Davison R, Hargreaves M, Flint J. *Engineering Mathematics: A Foundation for Electronic, Electrical, Communications and Systems Engineers*. 4th ed. Essex: Pearson Education; 2012.
13. Deb K. *Optimisation for Engineering Design: Algorithms and Examples*. New Delhi: Prentice-Hall; 1995.
14. Dorf RC, Bishop RH. *Modern Control Systems*. 12th ed. Essex: Pearson; 2011.
15. Flake GW. *The Computational Beauty of Nature: Computer Explorations of Fractals, Chaos, Complex Systems, and Adaptation*. Cambridge, Mass: MIT Press; 1998.
16. Farlow SJ. *Partial Differential Equations for Scientists and Engineers*. New York: John Wiley and Sons; 1982.
17. Fowler AC. *Mathematical Models in the Applied Sciences*. Cambridge: Cambridge University Press; 1997.
18. Gershenfeld N. *The Nature of Mathematical Modeling*. Cambridge: Cambridge University Press; 1998.
19. Gill PE, Murray W, Wright MH. *Practical Optimization*. Bingley: Emerald Group Publishing; 1982.
20. Goodman R. *Teach Yourself Statistics*. London: Teach Yourself Books; 1957.
21. Jeffrey A. *Advanced Engineering Mathematics*. San Diego: Academic Press; 2002.
22. James G. *Modern Engineering Mathematics*. 5th ed. Essex: Pearson Education; 2015.
23. Korn GA, Korn TM. *Mathematical Handbook for Scientists and Engineers*. New York: McGraw-Hill; 1968.
24. Koziel S, Yang XS. *Computational Optimization, Methods and Algorithms*. Heidelberg: Springer; 2011. Studies in Computational Intelligence; vol. 356.
25. Kreyszig E. *Advanced Engineering Mathematics*. 6th ed. New York: John Wiley & Sons; 1988.
26. Kreyszig E. *Advanced Engineering Mathematics: International Student Version*. 10th ed. New York: John Wiley and Sons; 2011.
27. Miersemann E. *Partial Differential Equations*. Amazon CreateSpace Independent Publishing Platform; 2014.
28. Moler CB. *Numerical Computing with MATLAB*. Philadelphia: SIAM; 2004.
29. Ockendon J, Howison S, Lacey A, Movchan A. *Applied Partial Differential Equations*. Oxford: Oxford University Press; 2003.
30. O'Neil PV. *Advanced Engineering Mathematics*. 7th ed. Stamford: Cengage Learning; 2012.
31. Pallour JD, Meadows DS. *Complex Variables for Scientists and Engineers*. London: Macmillan Publishing Co.; 1990.
32. Papoulis A. *Probability and statistics*. Englewood Cliffs: Prentice Hall International; 1990.
33. Pearson CE. *Handbook of Applied Mathematics*. 2nd ed. New York: Van Nostrand Reinhold; 1983.
34. Peterson JC. *Technical Mathematics with Calculus*. 3rd ed. New York: Thomson Learning; 2004.
35. Press WH, Teukolsky SA, Vetterling WT, Flannery BP. *Numerical Recipes in C++: The Art of Scientific Computing*. 2nd ed. Cambridge: Cambridge University Press; 2002.
36. Puckett EG, Colella P. *Finite Difference Methods for Computational Fluid Dynamics*. Cambridge: Cambridge University Press; 2005.
37. Rahman M. *Integral Equations and Their Applications*. Southampton: WIT Press; 2007.
38. Riley KF, Hobson MP, Bence SJ. *Mathematical Methods for Physics and Engineering*. 3rd ed. Cambridge: Cambridge University Press; 2006.
39. Singh K. *Engineering Mathematics Through Applications*. Basingstoke: Palgrave Macmillan; 2011.
40. Stroud KA, Booth DJ. *Engineering Mathematics*. 6th ed. New York: Palgrave Macmillan; 2007.
41. Weisstein EW. http://mathworld.wolfram.com.
42. Wikipedia. http://en.wikipedia.com.
43. Yang XS. *Nature-Inspired Metaheuristic Algorithms*. Frome: Luniver Press; 2008.
44. Yang XS. *Engineering Optimization: An Introduction with Metaheuristic Applications*. New Jersey: John Wiley and Sons; 2010.
45. Yang XS. *Mathematical Modelling with Multidisciplinary Applications*. Hoboken, NJ: John Wiley and Sons; 2013.
46. Yang XS. *Nature-Inspired Optimization Algorithms*. Waltham: Elsevier; 2014.
47. Yang XS. *Introduction to Computational Mathematics*. 2nd ed. Singapore: World Scientific Publishing; 2015.
48. Zill DG, Wright WS. *Advanced Engineering Mathematics*. 4th ed. Sudbury, MA: Jones & Bartlett Learning; 2009.
49. Zwillinger D. *CRC Standard Mathematical Tables and Formulae*. 32nd ed. Boca Raton, FL: CRC Press; 2012.

Index